T0188234

FUNDAMENTALS OF
QUANTUM
MECHANICS

FUNDAMENTALS OF
QUANTUM
MECHANICS

Şakir Erkoç

*Middle East Technical University,
Ankara, Turkey*

CRC Press
Taylor & Francis Group
Boca Raton London New York

CRC Press is an imprint of the
Taylor & Francis Group, an **informa** business

A TAYLOR & FRANCIS BOOK

CRC Press
Taylor & Francis Group
6000 Broken Sound Parkway NW, Suite 300
Boca Raton, FL 33487-2742

© 2007 by Taylor & Francis Group, LLC
CRC Press is an imprint of Taylor & Francis Group, an Informa business

First issued in paperback 2019

No claim to original U.S. Government works

ISBN 13: 978-0-367-45352-7 (pbk)
ISBN 13: 978-1-58488-732-4 (hbk)

Visit the Taylor & Francis Web site at
http://www.taylorandfrancis.com

and the CRC Press Web site at
http://www.crcpress.com

Library of Congress Cataloging-in-Publication Data

Erkoç, Sakir.
   Fundamentals of quantum mechanics / Sakir Erkoc.
      p. cm.
   Includes bibliographical references and index.
   ISBN 1-58488-733-8 (alk paper)
   1. Quantum theory. I. Title.

QC174.12.E75 2006
530.12--dc22   2006045544

Contents

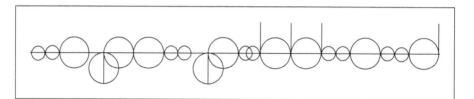

To my family, my teachers, and my students.

The author donated the royalty of this book to UNICEF for

The children of the World.

Preface

A unified account of the nonrelativistic quantum mechanics is presented in this book. The present volume is concerned mainly with the principles and formalism of quantum mechanics, and the development and application of general techniques for the solution of quantum mechanical problems. It is based on courses given by the author at Middle East Technical University, Ankara. It is suitable for study at third and/or fourth year level of an undergraduate course.

It is my intent that students should be able to read this volume and understand its contents without the need to supplement it by referring to more detailed discussions. One may expect that an introductory course on quantum mechanics cannot cover the whole field of quantum theory. This text is designed primarily for an introductory course in quantum mechanics at the undergraduate level. It provides sufficient breadth and depth both to familiarize the reader with the basic ideas and mathematical expressions of quantum mechanics and to form the basis for deeper understanding later.

Quantum mechanics provides not only the physical principles and mathematical methods used in all theoretical work on the microscopic physical world; it provides even the concepts and vocabulary of nearly all branches of theoretical physics.

Topics covered have been chosen as much for their pedagogical importance. The whole subjects covered in this volume are divided into three parts. The first part, in five chapters, contains the historical basis and the mathematical foundations of nonrelativistic quantum theory. Physical systems considered in this part are mainly in one–dimension. The second part, in seven chapters, contains basicaly the fundamentals of quantum theory in three dimensions. Many–particle systems, the motion of a particle in three–dimensions, angular and spin momenta, interaction of a charged particle with external fields, and matrix mechanical formulation of quantum mechanics are discussed in this part. The third part, in seven chapters, contains the approximation methods used in quantum mechanics, and scattering theory. Each part is expected to be finished in one semester.

I hope that the worked examples at the end of each chapter will clarify some details in the subjects covered in that chapter. The solved problems were selected according to their physical importance, and their pedagogical value. A total of 94 solved problems are included in this volume. They vary in difficulty; some are trivial while a few others may be considered as difficult. There are a few problems at the end of each chapter in addition to the worked examples. A total of 101 problems are included for practice for the students. The level of the problems are the same as the worked examples.

A final word to the students: Quantum mechanics is a physical theory, constructed out of concrete physical ideas. It is expressed mathematically because that is the only way to express this kind of thinking. Everything possible has been done to make this book easy mathematically. If you have trouble at any point, try to see whether the difficulty is in physics or in mathematics.

I have taken information from all the books that I found worthwhile and listed them in the bibliography and references. I have benefited from many books during the preparation of this lecture notes; however I would like to give special mention to the books of Dicke and Wittke, of Liboff, of Gasiorowicz, of Griffiths, of Shankar, and finally of Das and Melissions.

I would like to thank John Navas from Taylor and Francis (UK) and Amber Donley from Taylor and Francis (USA) for their kind instructive guidance. I would like to thank Professor Stephen Gasiorowicz for kindly giving permission to use the materials from his book. I would like to thank Prentice Hall (Pearson Education) for kindly giving permission to use the materials from Griffiths' book. I would like to thank Gordon and Breach (Taylor and Francis) for kindly giving permission to use the materials from Das and Melissions' book. I would like to thank Addison–Wesley for kindly giving permission to use the materials from Liboff's book. I would like to thank Plenum (Springer Science and Business Media) for kindly giving permission to use the materials from Shankar's book. I would like to thank Professor Ramamurti Shankar for kindly giving permission to use the materials from his book.

Şakir Erkoç
March 2006
METU, Ankara

PART – I
Fundamentals and

One–Dimensional Systems

Chapter 1

Historical Experiments and Theories

1.1 Dates of important discoveries and events

The important discoveries and events that occurred in the development of quantum mechanics may be ordered as follows:

- Blackbody radiation formulated by Max Planck in 1901.

- Photoelectric effect interpreted by Albert Einstein in 1905.

- Model of the atom proposed by Ernest Rutherford in 1911.

- Quantum theory of atomic spectra developed by Niels Bohr in 1913.

- Photon–electron scattering experiments done by Arthor Holly Compton in 1922.

- Exclusion principle proposed by Wolfgang Pauli in 1925.

- Matter waves proposed by Prince Louis Victor de Broglie in 1925.

- Wave equation proposed by Erwin Shrödinger in 1926.

- Electron diffraction experiments done by Clinton Joseph Davisson and L.H. Germer in 1927.

- Uncertainty principle proposed by Werner Karl Heisenberg in 1927.

- Interpretation of the wave function done by Max Born in 1927.

- Relativistic wave equation proposed by Paul A. M. Dirac in 1928.

In the following sections we will explain the details of some of these discoveries and events in chronological order.

1.2 Blackbody radiation

A blackbody is defined as one that absorbs all electromagnetic radiation, of whatever frequency, that is incident upon it. By thermodynamic arguments, it can be shown that such a body is also a better radiator of energy at every frequency than any other body at the same temperature.

When a body is heated, it is seen to radiate. In equilibrium the light ($\equiv$ electromagnetic radiation) emitted ranges over the whole spectrum of frequency ν, with a spectral distribution that depends both on the frequency or, equivalently, on the wavelength of the light λ, and on the temperature.

Blackbody radiation refers to the equilibrium radiant energy to be found inside a cavity whose walls are completely opaque and held at a fixed temperature T. Such a cavity is called a blackbody cavity. The interest in such radiation is because this radiation is independent of the nature of the walls of the cavity; the spectral properties of the radiation depend only on the temperature of the walls. Bodies such as a hot piece of iron or the sun are good approximations of a blackbody. By measuring the spectrum of their radient energy one can determine their temperature.

Since blackbody radiation was so simple, a theory for its spectrum was soon derived from classical mechanics and electromagnetism. The resultant spectral formula, called the Rayleigh–Jeans Law, proved to fail completely at high frequencies.

Rayleigh–Jeans Law:

Rayleigh and Jeans in 1900 tried to formulate the blackbody radiation by assuming totally classical descriptions. Completely opaque walls for a cavity can be described mathematically by assuming that all the radiation is reflected at the walls. The waves are standing inside the cavity. If the cavity is a cube with sides of length L, then the components of the wavelength of the radiation in each direction must exactly fit into L. Thus, the component of the wavelength of the frequency modes, say in x direction,

$$\lambda_{x,1} = 2L \quad , \quad \lambda_{x,2} = \frac{2L}{2} \quad , \quad \cdots \quad , \quad \lambda_{x,n} = \frac{2L}{n_x} \tag{1.1}$$

A similar discussion holds for the y and z directions. The wavenumbers (k_x, k_y, k_z) are related to the corresponding wavelengths by

$$k_x = \frac{2\pi}{\lambda_x} \quad , \quad k_y = \frac{2\pi}{\lambda_y} \quad , \quad k_z = \frac{2\pi}{\lambda_z} \tag{1.2}$$

The different vibration modes are therefore characterized by these integers (n_x, n_y, n_z) giving the total wavenumber

$$k^2 = k_x^2 + k_y^2 + k_z^2 = \left(\frac{2\pi}{2L}\right)^2 (n_x^2 + n_y^2 + n_z^2) \tag{1.3}$$

The wavenumber k is also related to the frequency ν by

$$2\pi\nu = \omega = ck \tag{1.4}$$

where c is the speed of light.

Now each of the vibrations of the electromagnetic field can be considered as a degree of freedom of the field; the different vibration modes are independent of each other. However, according to the equipartition principle of statistical mechanics we have for a temperature T an amount of energy $k_B T$ for each degree of freedom of the field. Here T is the temperature of the cavity wall and k_B is Boltzmann's constant, $k_B = 1.381 \times 10^{-16} \ erg/K$.

From the equipartition principle we can therefore write the formula for the energy dU in a frequency interval between ν and $\nu + d\nu$ as

$$dU = k_B T dN \tag{1.5}$$

where dN represents the number of modes of oscillations in the same frequency interval.

So to obtain the blackbody spectrum requires that we count the number of modes of oscillation corresponding to a frequency interval between ν and $\nu + d\nu$. It is easier to first obtain all modes up to a frequency ν. This is simply the number of points (the volume) inside one quadrant ($\frac{1}{8}$ since n_x, n_y, n_z are all positive) of a sphere whose radius r_k is

$$r_k^2 = n_x^2 + n_y^2 + n_z^2 = \left(\frac{2L}{2\pi}\right)^2 k^2 = \left(\frac{2L}{c}\right)^2 \nu^2 \tag{1.6}$$

The total number of modes inside the one quadrant of the sphere with radius r_k is therefore obtained as

$$N = \frac{1}{8} \cdot \frac{4}{3}\pi r_k^3 \cdot 2 = \frac{\pi}{3}\left(\frac{2L}{c}\nu\right)^3 \tag{1.7}$$

The factor of 2 is due to the fact that for light two independent polarizations for each vibration mode are possible. The number of modes dN in the frequency interval between ν and $\nu + d\nu$ is now given by:

$$dN = \pi \left(\frac{2L}{c}\right)^3 \nu^2 d\nu \tag{1.8}$$

Hence, the energy per unit volume in this frequency interval we get:

$$du = \frac{dU}{V} = \frac{1}{V}k_B T dN = \frac{8\pi k_B T}{c^3}\nu^2 d\nu \tag{1.9}$$

The volume of the cavity is simply $V = L^3$. So by applying the classical equipartition and classical electromagnetic theory we obtain for the energy density

$$\rho(\nu) = \frac{du}{d\nu} = \frac{8\pi k_B T}{c^3}\nu^2 = \frac{8\pi k_B T}{c^3 h^2}(h\nu)^2 = \frac{8\pi (k_B T)^3}{c^3 h^2}\left(\frac{h\nu}{k_B T}\right)^2 \tag{1.10}$$

In terms of wavelength, $\lambda = c/\nu$, one can express the energy density as

$$\rho(\lambda) = \rho(\nu)\frac{c}{\lambda^2} = \frac{8\pi k_B T \nu^2}{c^3}\frac{c}{\lambda^2} = \frac{8\pi k_B T \nu^2}{c^2 \lambda^2} = \frac{8\pi k_B T}{\lambda^4} \tag{1.11}$$

For low frequencies; this result agrees well with the experimental spectrum, but for high frequencies it fails. Furthermore, for a cavity of volume V at a temperature T the total radiant energy in this volume is given by

$$E = V \int_0^\infty \rho(\nu)d\nu = \frac{8\pi k_B T V}{c^3} \int_0^\infty \nu^2 d\nu = \infty \tag{1.12}$$

Clearly this shows that blackbody radiation due to the Rayleigh–Jeans Law fails at high frequencies, it gives the rate of radiation from a blackbody as infinite at all temperatures above absolute zero. On the other hand, this radiation law gives results in agreement with experiment in the limit of sufficiently small values of the frequency and sufficiently large values of the temperature. While this approximation is valid at low frequencies, it is seen to diverge at large frequencies, where the correct spectral distribution falls off to zero. Wien in 1893 expressed that ρ, as a function of wavelength λ, is of the form, **Wien's law**,

$$\rho(\lambda) = \frac{f(\lambda T)}{\lambda^5} \tag{1.13}$$

where f is an arbitrary continuous function of the product of wavelength λ and temperature T. Although this formula is valid over the whole spectrum of wavelength, it is incomplete in that $f(\lambda T)$ is undetermined. The complete explicit form for the spectral distribution ρ cannot be obtained from classical physics.

In 1901 Max Planck was able to derive a valid expression for the spectral distribution of blackbody radiation by making some assumptions.

Planck's Radiation Law:

To obtain the spectral distribution of blackbody radiation Planck applied the same tools as we did in the previous case, but in addition he made the radical assumption that radiation of frequency ν carries energy

$$\epsilon = h\nu \tag{1.14}$$

where, as determined by experiment, it was found that $h = 6.626 \times 10^{-27} \; erg \cdot s$ and is now known as Planck's constant and plays a fundamental role in all of modern physics. Planck's energy–frequency relation, $\epsilon = h\nu$, was a new equation in physics. Until Planck, the energy of a wave could be any number and was proportional to the square of the amplitude of the wave, $\epsilon \propto |\psi|^2$. After Planck, energy was quantized proportional to the frequency, $\epsilon \propto \nu$. A quantum of radiation of energy $h\nu$ is called a **photon**.

Planck combined classical statistical mechanics with his energy–frequency relation. Thus he made the following assumptions:

i– In accordance with classical statistical mechanics, the probability for an oscillator of energy ϵ, $P(\epsilon)$, to be excited is proportional to the Boltzmann factor

$$P(\epsilon) = e^{-\epsilon/k_B T} \tag{1.15}$$

ii– The energy of the oscillators is quantized and comes in quanta given by

$$\epsilon_n = nh\nu \;, \quad n = 1, 2, 3, \cdots \tag{1.16}$$

Combining these assumptions one can compute the average energy of an oscillator, $< \epsilon >$, from

$$< \epsilon > = \frac{\sum_i \epsilon_i P(\epsilon_i)}{\sum_i P(\epsilon_i)} \tag{1.17}$$

Substituting ϵ_i and $P(\epsilon_i)$ gives

$$< \epsilon > = \frac{\sum_n nh\nu e^{-nh\nu/k_B T}}{\sum_n e^{-nh\nu/k_B T}} = \frac{d}{d(1/k_B T)} \ln \left(\sum_n e^{-nh\nu/k_B T} \right) \tag{1.18}$$

$$= -\frac{d}{d(1/k_B T)} \ln \left(1 - e^{-h\nu/k_B T} \right)^{-1} = k_B T \frac{h\nu/k_B T}{e^{h\nu/k_B T} - 1} = \frac{h\nu}{e^{h\nu/k_B T} - 1}$$

The energy per unit volume du in a frequency interval between ν and $\nu + d\nu$ is equal to the number of modes of oscillations in this interval times the average energy of the oscillator.

$$du = < \epsilon > \frac{dN}{V} = \frac{h\nu}{e^{h\nu/k_B T} - 1} \frac{8\pi}{c^3} \nu^2 d\nu = \frac{8\pi}{c^3} \frac{h\nu^3}{e^{h\nu/k_B T} - 1} d\nu \tag{1.19}$$

The energy density then takes the form

$$\rho(\nu) = \frac{du}{d\nu} = \frac{8\pi}{c^3} \frac{h\nu^3}{e^{h\nu/k_BT} - 1} \tag{1.20}$$

This is the Planck's blackbody radiation formula. Note that for low frequencies namely $h\nu/k_BT \ll 1$, this law just goes into the Rayleigh–Jeans law as shown in Figure 1.1., since for these frequencies we have $e^{h\nu/k_BT} \approx 1 + h\nu/k_BT$.

The establishment of this formula and the introduction of the constant h was one of the most important developments in all of physics.

According to the Planck's law, for a cavity of volume V at a temperature T the total radiant energy in this volume is given by

$$E = V \int_0^\infty \rho(\nu)d\nu = V \int_0^\infty \frac{8\pi}{c^3} \frac{h\nu^3 d\nu}{e^{h\nu/k_BT} - 1} \tag{1.21}$$

$$= V \frac{8\pi k_B^3 T^3}{c^3 h^2} \int_0^\infty \frac{h^3\nu^3/k_B^3 T^3}{e^{h\nu/k_BT} - 1} d\nu = V \frac{8\pi k_B^3 T^3}{c^3 h^2} \frac{k_B T}{h} \int_0^\infty \frac{x^3 dx}{e^x - 1}$$

$$= V \frac{8\pi k_B^4 T^4}{c^3 h^3} \frac{\pi^4}{15} = V \left(\frac{8}{15} \frac{\pi^5 k_B^4}{h^3 c^3} \right) T^4 = VaT^4$$

The constant a has the value $a = 7.551 \times 10^{-15}\ erg/(cm^3 \cdot K^4)$. Thus the energy density, and hence the radiation rate from a blackbody, is proportional to the fourth power of the temperature, a fact which was long known and was first discovered by Stefan in 1879. The total energy density in a certain volume V depends on the temperature in accordance with **Stefan's law**

$$E = V \int_0^\infty \rho(\nu)d\nu = V \frac{4\sigma}{c} T^4 \tag{1.22}$$

Comparing this equality with the one obtained from the Planck's law, one gets for the value of Stefan's constant

$$\sigma = \frac{ac}{4} = \frac{2\pi^5 k_B^4}{15h^3 c^2} = 5.663 \times 10^{-5}\ erg/(cm^2 \cdot s \cdot K^4) \tag{1.23}$$

The Stefan's constant, σ, relates the radiation rate to the blackbody temperature, which previously had to be obtained from radiation rate measurements, but now could be derived from the constants of Planck's distribution law.

Planck's success in obtaining the correct distribution law of blackbody radiation on the basis of the assumption that radiation oscillators can have only certain discrete energies suggested that the same approach be tried to see if a theoretical explanation of the experimentally observed temperature dependence of the specific heats of solids could be obtained.

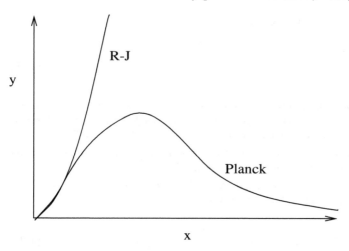

Fig. 1.1. Variation of energy density with respect to frequency in blackbody radiation.

$y = x^2$ (Rayleigh–Jeans law) , $y = x^3/(e^x - 1)$ (Planck's law).

$$x = \frac{h\nu}{k_B T} \quad , \quad y = \frac{c^3 h^2 \rho(\nu)}{8\pi k_B^3 T^3}$$

Summary of the laws of blackbody radiation:

Stefan's law (1879):

$$E = \left(V \int_0^\infty \rho(\nu) d\nu \right) = V \frac{4\sigma}{c} T^4$$

Wien's law (1893):

$$\rho(\lambda) = \frac{f(\lambda T)}{\lambda^5} \quad , \quad f(\lambda T) =? \quad \text{(unknown function)}$$

Rayleigh–Jean's law (1900):

$$\rho(\nu) = (k_B T) \frac{8\pi}{c^3} \nu^2$$

$$\rho(\lambda) = \rho(\nu) \frac{c}{\lambda^2} = \frac{8\pi k_B T}{\lambda^4} = \frac{8\pi k_B \lambda T}{\lambda^5}$$

$$E = V \int_0^\infty \rho(\nu) d\nu \to \infty$$

Planck's law (1901):

$$\epsilon = h\nu \; ; \; \epsilon_n = nh\nu \; , \; n = 1, 2, 3, \cdots$$

$$P(\epsilon) = e^{-\epsilon/k_B T}$$

$$<\epsilon> = \frac{\sum_n \epsilon_n P(\epsilon_n)}{\sum_n P(\epsilon_n)} = \frac{h\nu}{e^{h\nu/k_B T} - 1}$$

$$\rho(\nu) = <\epsilon> \frac{8\pi}{c^3}\nu^2 = \frac{8\pi}{c^3}\frac{h\nu^3}{e^{h\nu/k_B T} - 1}$$

$$E = V\int_0^\infty \rho(\nu)d\nu = V\frac{8\pi^5 k_B^4}{15h^3 c^3}T^4 = VaT^4$$

$$\sigma = \frac{ac}{4} = \frac{2\pi^5 k_B^4}{15h^3 c^2}$$

1.3   Photoelectric effect

In 1887 Hertz discovered that electrons could be ejected from solids by letting radiation fall onto the solid. Lenard and others found that the maximum energy of these photo–ejected electrons dependent only upon the frequency of the light falling on the surface, and not upon its intensity. Furthermore, it was found that for shorter wavelengths the maximum energy of the electrons was greater than for longer wavelengths.

In 1905 Einstein explained the photoelectric effect by making use of the ideas of Planck. He assumed that, as Planck proposed in the blackbody radiation law, radiation exists in the form of quanta of definite size, that is, that light consists of packets of energy of size $h\nu$. He also assumed that when light falls on a surface, individual electrons in the solid can absorb these energy quanta. Thus, the energy received by an electron depends only on the frequency of the light and is independent of its intensity. However, the number of ejected electrons are proportional to the intensity of the light source. The experimental setup is shown schematically in Figure 1.2.

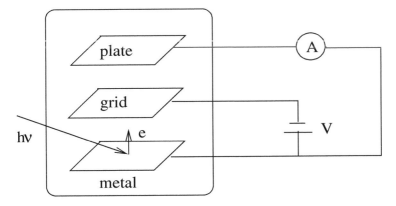

Fig. 1.2. The experimental setup showing the photoelectric effect.

The photoelectric current, which is the amount of charge arriving at the plate per unit time, is proportional to the rate of emission of electrons from the metal surface,

$$i = \frac{\Delta n e}{\Delta t} \tag{1.24}$$

To determine the velocity of the ejected electrons, a potential is applied to a grid mounted between the metal surface and the plate. The potential creates an electric field which decelerates the photoelectrons. Let the stopping voltage be V_s. At the stopping voltage, the initial kinetic energy of photoelectrons has been converted to potential energy.

$$\frac{1}{2}mv^2 = eV_s \tag{1.25}$$

Thus, by measuring i and V_s we know the number of electrons produced per second and their maximum kinetic energy. The experimental result shows that V_s is proportional to the frequency of the light and independent of the intensity, as shown in Figure 1.3. If the frequency ν is below a certain threshold value ν_0, no photoelectron current is produced. At frequencies greater than ν_0, the empirical equation for the stopping voltage is

$$V_s = k(\nu - \nu_0) \tag{1.26}$$

where k is a constant independent of the metal used, but ν_0 varies from one metal to another. Although there is no relation between V_s and the light intensity (I), it is found that the photoelectric current, and therefore the number of electrons liberated per second, is proportional to I. These results can not be explained by the wave theory. However, in the particle model, a photon of energy $h\nu$ strikes a bound electron, which may be absorbed by the photon energy. If $h\nu$ is greater than the binding energy (or work function) eV_0, the electron is liberated. Thus, the **threshold frequency** ν_0 is given by

$$\nu_0 = \frac{eV_0}{h} \tag{1.27}$$

Since V_0 is a characteristic of the particular metal, ν_0 depends upon the metal. For a photon of energy $h\nu$, the total energy of the struck electron is $h\nu$, with the excess over the potential energy eV_0 required to escape from the metal appearing as kinetic energy (see Figure 1.4),

$$\frac{1}{2}mv^2 = h\nu - eV_0 = eV_s \tag{1.28}$$

which is identical to the empirical relationship $V_s = k(\nu - \nu_0)$, with $k = h/e$.

Because the amount of energy absorbed by an electron is $(h\nu - eV_0)$ regardless of the rate at which photons impinge on the surface, the kinetic energy of the ejected electrons should be independent of the intensity of the light. So this experiment demonstrates the particle character of radiation.

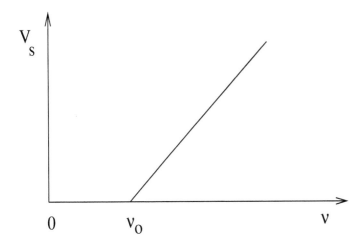

Fig. 1.3. Stopping potential versus frequency in photoelectric effect.

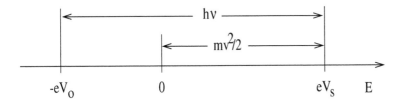

Fig. 1.4. Energy diagram in photoelectric effect.

According to Sommerfeld model a conductor is composed of fixed positive sites and free electrons. The positive ions generate a potential well with depth W in which the electrons are trapped. The electrons have energy from 0 to E_F, the Fermi energy; see Figure 1.5. The minimum work required to remove an electron from the metal is $W - E_F$, which is called the work function, namely

$$W - E_F = eV_0 = \Phi \tag{1.29}$$

Suppose that a photon of energy $h\nu$ hits an electron and ejects it with kinetic energy $\frac{1}{2}mv^2$. The most energetic electrons come from the top of the Fermi sea. The energy $\frac{1}{2}mv^2$ of such an electron ejected by a photon of energy $h\nu$ is given by

$$\frac{1}{2}mv^2 = h\nu - (W - E_F) = h\nu - \Phi \tag{1.30}$$

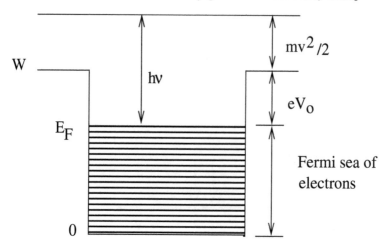

Fig. 1.5. Electronic structure of metals in Sommerfeld model.

Sommerfeld model of metals is used to exlain the phenomenon of **contact potential**, the finite potential that develops between two dissimilar metals, say A and B, which are brought into contact with each other. Let us suppose that $\Phi_A < \Phi_B$. When the metals are isolated and displaced far from each other, the common zero in potential of both metals corresponds to zero free–particle kinetic energy; see Figure 1.6(a).

When the metals are brought into contact with each other, electrons then fall from the Fermi level of metal A, which has a smaller work function, to the deeper lying Fermi level of metal B, until the tops of the two electron energy distributions are equalized. Having lost electrons, metal A is left electropositive with respect to metal B and a potential difference exists between the plates. This potential difference is called as the contact potential difference V_C betveen two contacting metals, which is approximated by the difference in work functions of the metals; see Figure 1.6(b).

$$eV_C = \Phi_B - \Phi_A \tag{1.31}$$

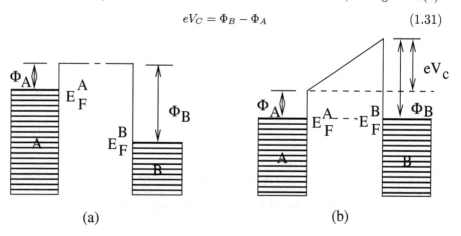

Fig. 1.6. Two metals in Sommerfeld model, (a) far removed from each other, (b) in contact with each other.

1.4 Quantum theory of spectra

In the early part of the 20th century, experiments to explore the structure of atoms were begun. The model due to J.J. Thomson, of an atom as a cloud of positive charge with bits of negative charge mixed together, was proven wrong by experiment. In scattering experiments Rutherford showed conclusively in 1911 that the atom consisted of a tiny positive core, called the nucleus, with electrons whirling about this nucleus. According to this planetary model atoms are unstable on the basis of classical physics. In fact, this planetary model gave a qualitative picture of the classical radiation spectrum, which was continuous. This also turns out to be wrong when compared with experiment.

From classical electromagnetic theory we find that a charge of magnitude e undergoing an acceleration a radiates energy at the rate

$$W = \frac{2}{3}\frac{e^2 a^2}{c^3} \tag{1.32}$$

Now consider an hydrogen atom consisting of an electron in a spherical orbit about a proton. To a good approximation the center of mass is located at the center of the proton. The acceleration is given by

$$a = \frac{v^2}{r} \tag{1.33}$$

where v is the speed of the electron and r the radius of its orbit. Equating the centrifugal force and force of electrostatic attraction we get

$$\frac{mv^2}{r} = \frac{e^2}{r^2} \tag{1.34}$$

Combining these two equations yields

$$a = \frac{e^2}{mr^2} \tag{1.35}$$

so that

$$W = \frac{2}{3}\frac{e^6}{m^2 c^3 r^4} \tag{1.36}$$

From this one can estimate the time t that it would take for an electron to lose all its kinetic energy (E_k) and spiral into the proton according to

$$t = \frac{E_k}{W} \tag{1.37}$$

Thus,

$$t = \frac{e^2}{2r} \cdot \frac{3}{2}\frac{m^2 c^3 r^4}{e^6} = \frac{3}{4}\frac{m^2 c^3 r^3}{e^4} \tag{1.38}$$

Using the numerical values $e \cong 1.6 \times 10^{-19}$ C, $m \cong 9.1 \times 10^{-31}$ kg, $r \cong 10^{-10}$ m, one can calculate $t \cong 4 \times 10^{-10}$ s. Clearly classical physics contradicts the stability of atoms. On the other hand classical result gives continuous spectrum.

The radiation frequency ν is determined by the angular frequency ω of rotation of the electron in its orbit according to

$$\omega = 2\pi\nu \tag{1.39}$$

Now using that the acceleration of the electron in its orbit is given by

$$a = \omega^2 r \tag{1.40}$$

and using $E = h\nu$ we get

$$W = \frac{2}{3}\frac{e^2 r^2}{c^3}(2\pi)^4 \nu^4 \tag{1.41}$$

So we conclude that the spectrum of the radiated energy is continuous. This also contradicts the experimental fact, namely that atomic spectra consist of discrete series of very sharp lines.

After studying the spectra of many atoms, Rydberg and Ritz independently discovered a very important result. They found that the discrete frequencies observed could be expressed more simply. All frequencies could be described by

$$\nu_{nm} = A_n - A_m \quad ; \quad n, m = 1, 2, 3, \cdots \tag{1.42}$$

that is, a difference of two terms. This so-called **Rydberg–Ritz combination principle** provided an important clue in the development of quantum mechanics. Bohr extended Planck's hypothesis and made some additional assumptions:

i– To get the observed stability of atoms, Bohr assumed that atoms exist only in certain definite states in which they do not radiate. These are the **stationary states**. These states are characterized by discrete values of the angular momentum as given by the relation

$$\oint p_\theta d\theta = nh \tag{1.43}$$

with n an integer greater than zero, $n = 1, 2, 3, \cdots$ and p_θ is the classical orbital angular momentum given by

$$p_\theta = mvr \tag{1.44}$$

The line integral follows the electron in one complete orbit about the nucleus, namely

$$2\pi p_\theta = nh \tag{1.45}$$

or

$$p_\theta = n\hbar \tag{1.46}$$

where

$$\hbar = \frac{h}{2\pi} \tag{1.47}$$

The energy is therefore automatically quantized since only these stationary states occur and not all possible states.

ii– To get discrete spectra Bohr assumed Planck's law in the form

$$E_m - E_n = h\nu \tag{1.48}$$

Here E_m and E_n are the energies associated with two stationary states. Thus, he made the implicit further assumption that the energy changes discontinuously from one state to another. This explains discrete spectra both for emission and absorption. This also gives the Rydberg–Ritz combination principle immediately since from the combination principle

$$\nu = A_m - A_n \tag{1.49}$$

Thus the terms A_m can be identified with E_m/h.

iii– The correspondence principle was one of Bohr's most useful assumptions. It states that in the limit of large quantum numbers the classical predictions must be recovered at least asymptotically. One can show this by taking

$$A_n = \frac{a}{(n+b)^2} \tag{1.50}$$

for almost all atomic spectra. Thus

$$E_n = \frac{ah}{(n+b)^2} \tag{1.51}$$

Then

$$\Delta E = E_n - E_m = ah \left[\frac{1}{(n+b)^2} - \frac{1}{(m+b)^2} \right] \tag{1.52}$$

$$\xrightarrow{n,m \to \infty} ah \left[\frac{m^2 - n^2}{m^2 n^2} \right] \to 0$$

Hence, $\Delta E_n \to 0$ and we get a continuum of energies for large quantum numbers.

The reformulation of hydrogen atom after Bohr's assumptions is as follows: We assume the electron in a hydrogen atom is in a circular orbit about a force center given by the Coulomb attraction of the proton. Also we choose the zero of energy to correspond to an unbound electron with zero kinetic energy. Thus in the orbit E is negative. Classically the frequency ω at which the electron radiates is given by the angular frequency of rotation of the electron in its orbit. Thus

$$\omega = \frac{v}{r} \tag{1.53}$$

The total energy of the electron in its orbit is

$$E = \frac{1}{2}mv^2 - \frac{e^2}{r} = \frac{p_\theta^2}{2mr^2} - \frac{e^2}{r} \tag{1.54}$$

Equating the Coulomb force of attraction and the centrifugal force gives

$$\frac{mv^2}{r} = \frac{e^2}{r^2} = \frac{p_\theta^2}{mr^3} \tag{1.55}$$

Thus

$$\frac{e^2}{r} = \frac{p_\theta^2}{mr^2} = \frac{n^2\hbar^2}{mr^2} \tag{1.56}$$

$$r_n = \frac{n^2\hbar^2}{me^2} \tag{1.57}$$

These are the quantized values of r at which the electron persists without radiating. The values of the energy at these radii are

$$E_n = \frac{p_\theta^2}{2mr^2} - \frac{e^2}{r} = -\frac{p_\theta^2}{2mr^2} = -\frac{n^2\hbar^2}{2m}\left(\frac{me^2}{n^2\hbar^2}\right)^2 = -\frac{R}{n^2} \qquad (1.58)$$

where R is the Rydberg constant

$$R = \frac{me^4}{2\hbar^2} \cong 2.18 \times 10^{-11} \; erg = 13.6 \; eV \qquad (1.59)$$

The negative quality of the energy reflects the fact that we are dealing with **bound states**. When $n = 1$, the atom is in the ground state and has energy, $-R$. The value of r when the atom is in the ground state is

$$r_1 = a_0 = \frac{\hbar^2}{me^2} = 5.29 \times 10^{-9} \; cm = 0.529 \; \text{Å} \qquad (1.60)$$

This is the fundamental length in physics. It is called the **Bohr radius**.

The frequencies so generated from the relation

$$\nu = \frac{E_m - E_n}{h} = \nu_{mn} \qquad (1.61)$$

agree with experiment. Experimentally we can not measure or observe the absolute energies of atoms; we can measure only the differences. The spectral lines seen in the spectrum correspond to the energy differences. Characteristically, the spectrum divide into various series of lines. In the hydrogen atom case the Lyman series is comprised of frequencies generated by transitions to the ground state

$$\nu_L = \frac{E_n - E_1}{h} \quad , \quad n > 1 \qquad (1.62)$$

The Balmer series is generated by transitions to the second state, namely to the first excited state:

$$\nu_B = \frac{E_n - E_2}{h} \quad , \quad n > 2 \qquad (1.63)$$

and so forth. Remaining series are Paschen, Brackett, and Pfund. The possible transitions in Hydrogen atoms are shown schematically in Figure 1.7; the corresponding spectral lines are shown in Figure 1.8.

Summary of the Bohr theory:

Atomic observables are quantized:

Quantum number : $n = 1, 2, 3, \cdots$

Angular momentum: $p_\theta = n\hbar$

Radius of electron orbit: $r_n = n^2 a_0$

Energy of bound electron : $E_n = -R/n^2$

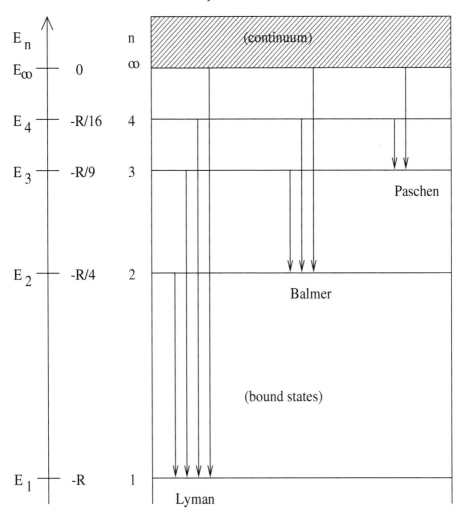

Fig. 1.7. Hydrogen atom energy levels and series of transitions in Bohr model.

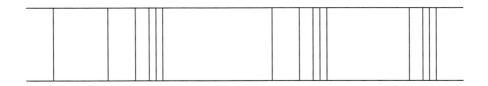

Fig. 1.8. Series of lines in the Hydrogen spectrum.

1.5   The Compton effect

The photoelectric effect indicated that photons were somehow particle–like carrying a definite amount of energy given by $E = h\nu$. By scattering X–rays (photons) off free electrons, in 1922 A.H. Compton showed that photons are definitely particle–like, carrying a definite momentum and scattering like point particles. The situation is as shown in Figure 1.9.

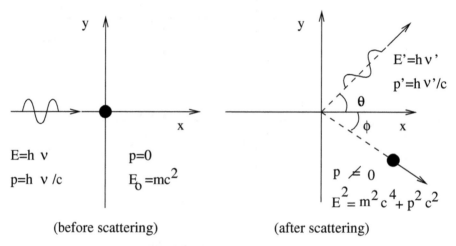

<table>
<tr><td>(before scattering)</td><td>(after scattering)</td></tr>
</table>

Fig. 1.9. Compton scattering.

The incoming photon carries energy $E = h\nu$ and momentum $p = h\nu/c$ in the x–direction (the whole collision occurs in a plane), while the electron is initially at rest (rest mass energy $E = mc^2$).

The scattered photon carries energy $E' = h\nu'$ and total momentum $p' = h\nu'/c$ in the direction given by the angle θ. The electron, which was originally at rest, recoils with momentum p in the direction given by the angle ϕ. Applying conservation of energy and momentum as for point particles we get

Conservation of energy:
$$h\nu + mc^2 = h\nu' + E \tag{1.64}$$

Conservation of momentum:
$$\frac{h\nu}{c} = \frac{h\nu'}{c}\cos\theta + p\cos\phi \tag{1.65}$$

$$0 = \frac{h\nu'}{c}\sin\theta - p\sin\phi \tag{1.66}$$

In this experiment Compton measured the change in wavelength of the scattered X–rays as a function of the scattering angle θ. According to classical electromagnetic theory no change in wavelength should occur. The above equation can be written only if X–rays were point particles.

By squaring and adding the momentum equations we get

$$h^2\nu^2 + h^2\nu'^2 - 2h^2\nu\nu'\cos\theta = c^2p^2 \tag{1.67}$$

Rearranging the energy equation and squaring yields

$$h^2\nu^2 + h^2\nu'^2 - 2h^2\nu\nu' = E^2 + m^2c^4 - 2Emc^2 \tag{1.68}$$

Subtracting the last two equations and using the energy–momentum relation

$$E^2 = c^2p^2 + m^2c^4 \tag{1.69}$$

we get

$$-2h^2\nu\nu'(1 - \cos\theta) = -2mc^2(E - mc^2) \tag{1.70}$$

From the energy equation we have $E - mc^2 = h(\nu - \nu')$. Thus, we finally get

$$\frac{h}{mc}(1 - \cos\theta) = c\left(\frac{\nu - \nu'}{\nu\nu'}\right) = \frac{c}{\nu'} - \frac{c}{\nu} = \lambda' - \lambda \tag{1.71}$$

So the increase in wavelength is given by

$$\Delta\lambda = \lambda' - \lambda = \frac{h}{mc}(1 - \cos\theta) = \lambda_c(1 - \cos\theta) \tag{1.72}$$

The quantity $\lambda_c = h/(mc)$ is known as the **Compton wavelength** of the electron. Compton's measurements showed that the change in wavelength of the photon in scattering given by the equation $\Delta\lambda = \lambda_c(1 - \cos\theta)$ agres with the experimental results. Thus, a photon has particle properties although it is originally a wave (electromagnetic wave).

1.6   Matter waves, the de Broglie hypothesis

For a consistent explanation of certain experiments (photoelectric and Compton experiments, for instance) it is necessary to ascribe particle (photon) behavior to light. The energy of such a photon of frequency ν is $E = h\nu$. Its momentum is

$$p = \frac{E}{c} = \frac{h\nu}{c} \tag{1.73}$$

This formula can also be written in terms of wavelength λ. The relation between λ and ν for light is particularly simple. It is

$$\lambda\nu = c \tag{1.74}$$

In terms of wavenumber k and angular frequency ω,

$$k = \frac{2\pi}{\lambda} \quad , \quad \omega = 2\pi\nu \tag{1.75}$$

Thus, one can write

$$E = \hbar\omega \quad , \quad p = \hbar k \quad , \quad \omega = ck \quad ; \quad \hbar = \frac{h}{2\pi} \tag{1.76}$$

The equation giving the relation between ω and k is known as the **dispersion relation**. It relates a linear dependence between ω and k. The significance of this is that the phase velocity, $c = \omega/k$, of a monochromatic wave of frequency ω is independent of ω or k. It is the constant c (speed of light). If a wave packet composed of a collection of waves of different wavelengths or wavenumbers is constructed, it propagates with no distortion (dispersion). All component waves have the same speed, c.

The equation $E = \hbar\omega$ and $p = \hbar k$ reveals that photons, which are in essence particles, are identified by two wave parameters; wavenumber k and frequency ω. The difference between a photon and other particles such as electron, proton, etc. is that the photon is special in that it has zero rest mass and travels only at the speed of light. For a nonrelativistic particle of kinetic energy

$$E = \frac{p^2}{2m} \tag{1.77}$$

the wavelength for the corresponding matter wave is

$$\lambda = \frac{h}{p} \quad \text{or} \quad p = \hbar k \tag{1.78}$$

which are equally relevant to photons. The equation $\lambda = h/p$ is, in essence, the **de Broglie hypothesis**. It ascribes a wave property to particles. While the Planck hypothesis, which assigned a particle quality to electromagnetic waves, had strong experimental motivation, the de Broglie hypothesis, when first introduced in 1925, attracted much attention. According to the Bohr hypothesis for stationary orbits of the electron in the hydrogen atom the quantization of circular orbits of radius r and momentum p was expressed as

$$2\pi p_\theta = 2\pi r p = nh \tag{1.79}$$

In terms of the de Broglie wavelength λ, the last equation reads

$$2\pi r = n\lambda \tag{1.80}$$

The stationary orbits in the Bohr model have an integer number of wavelengths precisely fitting the circumference. This is the classical criterion for the existence of standing waves on a circle (see Figure 1.10). Thus, the de Broglie hypothesis returns the stationary orbit radii of the Bohr theory. This result lends support to the idea that the electron has something wavy associated with it, this property being characterized by the de Broglie wavelength. The experimental verification of the wave property of electrons was also obtained by observing electron diffraction.

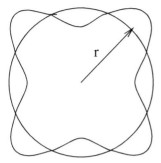

Fig. 1.10. Stationary wave on a circular orbit $2\pi r = n\lambda$, here $n = 4$.

1.7   The Davisson–Germer experiment

In 1927 C.J. Davisson and L.H. Germer showed that electron beams could be diffracted when scattered from crystals, and displayed the predicted wave properties by de Broglie. In the Davisson–Germer experiment, a beam of monoenergetic electrons is directed to strike the surface of a crystal of nickel normally, and the number of electrons $N(\theta)$ scattered at an angle θ, namely the scattered intensity $I(\theta)$, to the incident direction are measured; see Figure 1.11.

The electron beam energy employed was 54 eV. The scattered intensity is shown in Figure 1.12. It falls from a maximum at $\theta = 0°$ to a minimum near $35°$, then rises to a peak near $50°$.

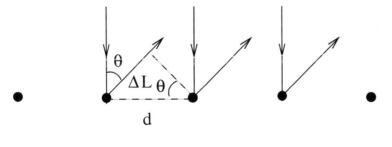

Fig. 1.11. Diffraction of electrons from a crystal surface.

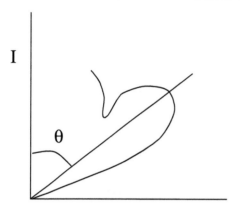

Fig. 1.12. Intensity distribution of scattered electrons in diffraction experiments.

The strong scattering at $\theta = 0°$ is expected from either a particle or a wave theory, but the peak at $50°$ can only be explained by constructive interference of the waves scattered by the regular crystal lattice. If the atoms in the crystal are spaced at a distance d, then the **Bragg condition** for constructive interference at a scattering angle θ is

$$\Delta L = d \sin \theta = n\lambda \quad , \quad n = 1, 2, 3, \cdots \qquad (1.81)$$

For nickel crystal $d = 2.15$ Å. Assuming that the peak at $50°$ corresponds to first–order diffraction $(n = 1)$, the corresponding electron wavelength must be $\lambda = 2.15 \cdot \sin 50° = 1.65$ Å. From the de Broglie hypothesis, $\lambda = h/p$, the wavelength of a 54 eV electron is 1.67 Å, which agrees with the value of 1.65 Å within the experimental error. Subsequent experiments have confirmed the variation of λ with momentum predicted by the de Broglie formula $\lambda = h/p$. Diffraction has also been demonstrated when atoms and neutrons are scattered by crystals. In all cases agreement has been found with the de Broglie hypothesis.

1.8   Heisenberg's uncertainty principle

In classical mechanics given the initial coordinates and velocity of a particle, $\mathbf{r}(0)$ and $\dot{\mathbf{r}}(0)$, respectively, and knowing all the forces on the particle, the orbit $\mathbf{r}(t)$ is exactly determined. The same holds true for a system of particles. This is the principle of **determinism**. However in quantum mechanics there is not any exact event, all events are **probabilistic** and there is an uncertainty in each event. Uncertainty principle was proposed by Heisenberg for the first time in 1927.

According to Heisenberg if the momentum of a particle is known precisely, it follows that the position of the same particle is completely unknown. Quantitatively, if an identical experiment involving an electron is performed many times, and in each run of the experiment the position (x) of the electron is measured, then although the experimental setup is identical (same electron momentum) in each run, measurement of the position of the electron does not give the same result. Let the average of these measurements be $< x >$. The mean–square deviation in the position measurements may be taken as the square of the standard deviation Δx,

$$(\Delta x)^2 \equiv < (x - < x >)^2 > = < x^2 - 2x < x > + < x >^2 > \qquad (1.82)$$

$$= < x^2 > - < x >^2$$

If Δx is small compared to some typical length in the experiment, one is more certain to find the value $x = < x >$ in any given run. If Δx is large, it is not certain what the measurement of x will yield (see Figure 1.13). For this reason Δx is also called the uncertainty in x.

Similarly, one may speak of an uncertainty in any physically observable quantity; magnetic field **B**, energy E, momentum **p**, and so forth.

$$\Delta B_\alpha = [< B_\alpha^2 > - < B_\alpha >^2]^{1/2} \quad , \quad \alpha = x, y, z \qquad (1.83)$$

$$\Delta E = [< E^2 > - < E >^2]^{1/2} \qquad (1.84)$$

$$\Delta p_\alpha = [< p_\alpha^2 > - < p_\alpha >^2]^{1/2} \quad , \quad \alpha = x, y, z \qquad (1.85)$$

Heisenberg's uncertainty relations for position and momentum (parallel components) are expressed as

$$\Delta x \Delta p_x \geq \hbar \quad , \quad \Delta y \Delta p_y \geq \hbar \quad , \quad \Delta z \Delta p_z \geq \hbar \qquad (1.86)$$

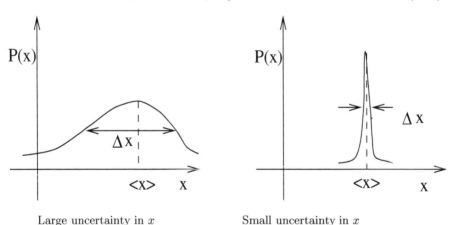

Large uncertainty in x Small uncertainty in x

Fig. 1.13. Uncertainty in position.

$P(x)dx$ = probability of finding an electron in the interval x, $x + dx$.

If it can be said with certainty what the position of a particle is ($\Delta x = 0$), then there is total uncertainty regarding the momentum of the particle ($\Delta p_x = \infty$). Observables obeying a relation such as Eq. (1.86) are called **complementary variables** or conjugate variables. Coordinates and momenta (x, p_x), energy and time (E, t), a component L_z of angular momentum of a particle and its angular position ϕ in the perpendicular (xy) plane are the examples for complementary variables. The energy–time uncertainty relation is expressed as

$$\Delta E \Delta t \geq \hbar \qquad (1.87)$$

The uncertainty relation between ϕ and L_z is expressed as

$$\Delta \phi \Delta L_z \geq \hbar \qquad (1.88)$$

The uncertainty relations between position and wave number (parallel components) appear as

$$\Delta x \Delta k_x \geq 1 \quad , \quad \Delta y \Delta k_y \geq 1 \quad , \quad \Delta z \Delta k_z \geq 1 \qquad (1.89)$$

The uncertainty relation between frequency and time appears as

$$\Delta \nu \Delta t \geq 1 \qquad (1.90)$$

The uncertainty principle states that the order of magnitude of the product of the uncertainties in the knowledge of the two variables must be at least $\hbar$. The uncertainty principle shows us why classical theory is not applicable to phenomena on a microscopic scale.

The mathematical derivations of these uncertainty relations appear explicitly as

$$\Delta x \Delta p_x \geq \frac{\hbar}{2} \quad , \quad \Delta y \Delta p_y \geq \frac{\hbar}{2} \quad , \quad \Delta z \Delta p_z \geq \frac{\hbar}{2} \qquad (1.91)$$

$$\Delta x \Delta k_x \geq \frac{1}{2} \quad , \quad \Delta y \Delta k_y \geq \frac{1}{2} \quad , \quad \Delta z \Delta k_z \geq \frac{1}{2}$$

$$\Delta E \Delta t \geq \frac{\hbar}{2} \quad , \quad \Delta \phi \Delta L_z \geq \frac{\hbar}{2} \quad , \quad \Delta \nu \Delta t \geq \frac{1}{2\pi}$$

1.9   Difference between particles and waves

Let us consider the double–slit experiment. A continuous spray of particles is fired from the source S. They strike the wall or pass through the two slits A and B. An intensity I_A [number / (unit area · second)] emerges from A and an intensity I_B emerges from B. When striking the screen, the two streams of particles superimpose and the net intensity is

$$I = I_A + I_B \qquad (1.92)$$

This is nothing more than the statement that numbers of particles add; see Figure 1.14.

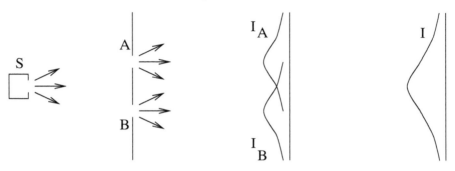

Fig. 1.14. Particle double–slit experiment. $I = I_A + I_B$.

Now let us consider the same experimental setup, but instead of a source of particles, let S represent a source of waves, say water waves. Waves are characterized by an amplitude function ψ such that the absolute square of this function gives the intensity I,

$$I = |\psi|^2 \tag{1.93}$$

Let the two propagating waves have amplitudes ψ_A and ψ_B, respectively. These functions have the representations

$$\psi_A = |\psi_A| e^{i\alpha_A} \quad , \quad \psi_B = |\psi_B| e^{i\alpha_B} \tag{1.94}$$

where α is the phase of the wave. The intensities of the waves are

$$I_A = |\psi_A|^2 \quad , \quad I_B = |\psi_B|^2 \tag{1.95}$$

At a common point in space, the two wave amplitudes superimpose to give the resultant amplitude.

$$\psi = \psi_A + \psi_B \tag{1.96}$$

The corresponding resultant intensity is

$$I = |\psi|^2 = |\psi_A + \psi_B|^2 = (\psi_A + \psi_B)^*(\psi_A + \psi_B) \tag{1.97}$$

$$= |\psi_A|^2 + |\psi_B|^2 + |\psi_A \psi_B| \left[e^{i(\alpha_A - \alpha_B)} + e^{-i(\alpha_A - \alpha_B)} \right]$$

$$= I_A + I_B + 2\sqrt{I_A I_B} \cos(\alpha_A - \alpha_B)$$

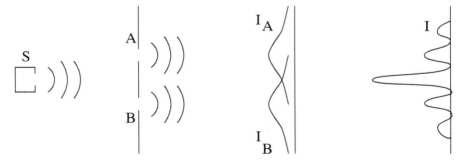

Fig. 1.15. Wave double–slit experiment. $I = I_A + I_B + \Delta$.

Comparing the intensity expressions for particles and waves, we note that the wave intensity carries the additional term Δ,

$$\Delta = 2\sqrt{I_A I_B} \cos(\alpha_A - \alpha_B) \tag{1.98}$$

This is an interference term. Δ oscillates and gives a pattern of the form as shown in Figure 1.15. Hence, the difference between particles and waves is that waves exhibit interference, particles do not. So we have the following important rule: *When two noninteracting beams of particles combine in the same region of space, intensities add; when waves interact, amplitudes add.* The intensity is then proportional to the time average of the absolute square of the resultant amplitude.

1.10 Interpretation of the wavefunction

When discussing the double–slit wave experiment, we introduced an amplitude function ψ, the square of whose modulus, $|\psi|^2$, was set equal to the intensity, $I = |\psi|^2$, of the wave.

Born suggested in 1927 that, when referred to the propagation of particles, $|\psi|^2$ is more appropriately termed a **probability density**. The function ψ is called the **wave function** or the **state function** of the particle. Quantitatively, the Born postulate states the following: The wave function for a particle (in Cartesian space) $\psi(x, y, z, t)$ is such that

$$|\psi|^2 dxdydz = Pdxdydz \tag{1.99}$$

where $Pdxdydz$ is the probability that measurement of the particle's position at the time t finds it in the volume element $dxdydz$ about point (x, y, z).

The wave function ψ generates the interference pattern. Where $|\psi|^2$ is large, the probability that a particle is found there is large. When enough particles are present, they distribute themselves in the probability pattern outlined by the density function $|\psi|^2$.

For problems where it can be said with certainty that the particle is somewhere in a given volume V,

$$\int_V |\psi|^2 dxdydz = 1 \tag{1.100}$$

This is a standard property that probability density satisfy. It is a mathematical expression of the certainty that the particle is in the volume V.

1.11 Worked examples

Example - 1.1 : (a) Write the total energy for the hydrogen atom in terms of r and p^2. (b) Use the uncertainty principle to write an estimate of the energy of the hydrogen atom in terms of r (treat this as a one–dimensional problem, write the uncertainty principle for p and r). (c) Estimate the ground state energy and the radius of the atom in the ground state.

Solution : (a)

$$E = \frac{p^2}{2m} - \frac{e^2}{r}$$

(b) $< p > \sim \Delta p \; , < r > \sim \Delta r \; ; \; \Delta p \Delta r \sim \hbar$

$$E \cong \frac{\hbar^2}{2m < r >^2} - \frac{e^2}{< r >}$$

(c)

$$\frac{dE}{d < r >} = -\frac{\hbar^2}{m < r >^3} + \frac{e^2}{< r >^2} = 0 \quad \rightarrow \quad < r >= \frac{\hbar^2}{me^2} = a_0$$

$$E_0 = \frac{1}{2} < V >= -\frac{1}{2}\frac{e^2}{< r >^2} = -\frac{e^2}{2a_0} = -\frac{e^4 m}{2\hbar^2}$$

Example - 1.2 : A particle is described by the wave function in one dimension $\psi = Ae^{-x/2}$. (a) Normalize the function in the interval $0 < x < \infty$. (b) Calculate the probability of finding the particle in the interval $0 < x < 1$.

Solution : (a)

$$\int_0^\infty |\psi|^2 dx = 1 \quad , \quad \int_0^\infty A^2 e^{-x} dx = A^2 \int_0^\infty e^{-x} dx = A^2 = 1 \quad , \quad A = 1$$

(b)

$$\int_0^1 |\psi|^2 dx = \int_0^1 e^{-x} dx = -e^{-x}|_0^1 = 1 - e^{-1}$$

Example - 1.3 : (a) Use the Bohr quantization rules to calculate the energy states for a potential given by $V(r) = V_0 (r/a)^k$, where V_0, a, and k are constants. (b) Show that the energy values in part (a) approach $E_n \cong cn^2$, where c is a constant.

Solution : (a)

$$F = -\frac{d}{dr} V(r) = -m\frac{v^2}{r} \quad \rightarrow \quad -\frac{d}{dr}\left(V_0 \frac{r^k}{a^k}\right) = -m\frac{v^2}{r}$$

$$k\frac{V_0}{a}\left(\frac{r}{a}\right)^{k-1} = m\frac{v^2}{r}$$

From Bohr quantization rule: $mvr = n\hbar$, $v = n\hbar/mr$, therefore,

$$k\frac{V_0}{a}\left(\frac{r}{a}\right)^{k-1} = \frac{m}{r}\frac{n^2\hbar^2}{m^2r^2} = \frac{n^2\hbar^2}{mr^3}$$

Rearranging the last equality, we may write

$$kV_0a^2\left(\frac{r}{a}\right)^{k+2} = \frac{n^2\hbar^2}{m}$$

or

$$\left(\frac{r}{a}\right)^{k+2} = \frac{n^2\hbar^2}{m}\frac{1}{kV_0a^2}$$

$$r = a\left(\frac{n^2\hbar^2}{mkV_0a^2}\right)^{1/(k+2)}$$

Therefore, energy may be expressed as the following,

$$E = \frac{1}{2}mv^2 + V_0\left(\frac{r}{a}\right)^k = \frac{1}{2}kV_0\left(\frac{r}{a}\right)^k + V_0\left(\frac{r}{a}\right)^k$$

$$= V_0(\frac{1}{2}k + 1)\left(\frac{r}{a}\right)^k = V_0(\frac{1}{2}k + 1)\left(\frac{n^2\hbar^2}{mkV_0a^2}\right)^{k/(k+2)}$$

(b) For $k \ll 1$, $k/(k+2) \rightarrow 1$, therefore energy becomes

$$E_n \approx V_0k\frac{n^2\hbar^2}{mkV_0a^2} = \left(\frac{\hbar^2}{ma^2}\right)n^2 = cn^2$$

Example - 1.4 : Neutron diffraction can be used to determine crystal structures. (a) Estimate a suitable value for the velocity of the neutrons. (b) Calculate the kinetic energy of the neutron in eV for this velocity. (c) It is common practice in this type of experiment to select a beam of monoenergetic neutrons from a gas of neutrons at temperature T. Estimate a suitable value for T.

Solution : (a) To obtain information on the crystal structure, neutrons are diffracted by the crystal in accordance with Bragg's law $n\lambda = 2d\sin\theta$. Here λ is the wavelength of the neutrons, d is the distance between the planes of diffracting atoms, θ is the glancing angle between the direction of the incident neutrons and the planes of atoms, and n is an integer. The equation cannot be satisfied unless $\lambda < 2d$. On the other hand, if $\lambda \ll 2d$, θ is very small. So it is necessary for λ to be of the same order as d, which is of the order of the interatomic spacing in the crystal. We may take $\lambda = d = 2$ Å (a typical value). The de Broglie relation between λ and the velocity v of the neutron is $\lambda = h/(m_nv)$, where the mass of the neutron is $m_n = 1.675 \times 10^{-27}$ kg. Thus $v = h/(m_n\lambda) = 2000$ m/s.

(b) The kinetic energy of the neutrons is $E = m_nv^2/2 = 3.3 \times 10^{-21}$ $J = 0.02$ eV for the above velocity.

(c) Using $E = k_BT$, the above value of E corresponds to $T = 240$ K, which is of the order of room temperature. Such neutrons are called thermal neutrons.

Example - 1.5 : The Stefan–Boltzmann law states that the total energy density in blackbody radiation at all frequencies is proportional to the fourth power of the temperature,

$$W = \int_0^\infty \rho(\nu)d\nu = \sigma T^4$$

Use the Planck radiation equation to obtain this result and determine the constant σ.

Solution : Inserting Planck's equation for $\rho(\nu)$ we obtain

$$W = \int_0^\infty \frac{8\pi\nu^3}{c^3} \left(\frac{h}{e^{h\nu/k_B T} - 1}\right) d\nu$$

Let $x = h\nu/k_B T$, then the integral becomes

$$W = \frac{8\pi k_B^4 T^4}{c^3 h^3} \int_0^\infty \frac{x^3 dx}{(e^x - 1)} = \frac{8\pi k_B^4 T^4}{c^3 h^3} \frac{\pi^4}{15}$$

$$= \left(\frac{8\pi^5 k_B^4}{15 c^3 h^3}\right) T^4 = \sigma T^4$$

and

$$\sigma = 7.565 \times 10^{-16} \ Jm^{-3}K^{-4}$$

The rate of emission from a surface at the temperature T is given by

$$\mathcal{E}(T) = \frac{c}{4}W = \frac{c}{4}\sigma T^4 = \sigma' T^4$$

Here $\sigma' = 5.669 \times 10^{-8} \ Wm^{-2}K^{-4}$. This is the Stefan–Boltzmann constant that is often quoted in the literature.

Example - 1.6 : A photon of energy $h\nu$ strikes the surface of a metal. If the energy of the photon is greater than the work function $e\phi$, the electron may be ejected with a maximum kinetic energy of $\frac{1}{2}mv^2$. A repeller potential can be used to measure $\frac{1}{2}mv^2$ by finding the minimum voltage V_0 necessary to stop the electrons. Show how Planck's constant, h, may be measured from a plot of stopping voltage versus ν.

Solution : The energy balance equation gives

$$h\nu = \frac{1}{2}mv^2 + e\phi$$

Hence

$$\frac{1}{2}mv^2 = h\nu - e\phi$$

The repeller potential energy necessary to just stop all of the electrons (stopping voltage) must be equal in magnitude to the kinetic energy of the ejected electrons. Thus

$$eV_0 = \frac{1}{2}mv^2 = h\nu - e\phi$$

This equation is a line equation; a plot of eV_0 versus ν gives a straight line, the slope of which is h and the intercept on the ν axis gives the work function $e\phi = h\nu_0$, where ν_0 is the threshold frequency of the metal.

1.12   Problems

Problem - 1.1 : Consider a particle with energy $E = p^2/2m$ moving in one dimension. The uncertainty in its location is Δx. Show that if $\Delta x \Delta p > \hbar$, then $\Delta E \Delta t > \hbar$.

Problem - 1.2 : The workfunction of zinc is 3.6 eV. What is the energy of the most energetic photoelectron emitted by ultraviolet light of wavelength 2500 $\mathring{A}$?

Problem - 1.3 : In a photon–electron collision experiment laser light has been backscattered from energetic electrons to obtain very high–energy photons. Find the energy of the backscattered photons.

Problem - 1.4 : The most accurate values of the sizes of atomic nuclei come from measurements of electron scattering. Estimate roughly the energies of electrons that provide useful information.

Problem - 1.5 : A plot of the energy density $\rho(\lambda)$ of the radiation in equilibrium with a blackbody versus λ shows a maximum at $\lambda = \lambda_{max}$. The Wien displacement law states that $\lambda_{max} = A/T$ where A is a constant and T is the absolute temperature. Show that Planck's radiation equation is consistent with this law and determine the constant A.

Problem - 1.6 : Quantum phenomena are often negligible in the macroscopic world. Show this numerically for the following cases: (a) The amplitude of the zero-point oscillation for a pendulum of length $l = 1$ m and mass $m = 1$ kg. (b) The tunneling probability for a marble of mass $m = 5$ g moving at a speed of 10 cm/s against a rigid obstacle of height $H = 5$ cm and width $w = 1$ cm. (c) The diffraction of a tennis ball of mass $m = 0.1$ kg moving at a speed $v = 0.5$ m/s by a window of size 1×1.5 m^2.

Chapter 2

Axiomatic Structure of Quantum Mechanics

2.1 The necessity of quantum theory

Classical mechanics (CM) is applied to a wide range of dynamical systems, including the EM field of interaction with matter. On the other hand, quantum mechanics (QM) is more suitable for the description of phenomena on the atomic scale.

The necessity of QM is shown by experimental results: The phenomena of interference and diffraction (explained by wave theory), the phenomena of photoelectric emission and scattering by free electrons (show that light is composed of small particles, called photons, have each a definite energy and momentum, a fraction of a photon is never observed). Experiments show that this behaviour is not peculiar to light, but is general.

All material particles have wave properties, which can be exhibited under suitable conditions. From a classical point of view matter is made up of a large number of small parts and one would postulate laws for the behaviour of these parts, from which the laws of the matter in bulk could be deduced. However, one can never explain the structure of matter by this way. So, big and small systems are relative concepts; it is not possible to explain the big in terms of the small. Therefore the size of a system becomes important.

Science is concerned only with observable things and that we can observe an object only by letting it interact with some outside influence. An act of observation is accompanied by some disturbance of the object observed. An object is **big** if disturbance is negligible, then we apply CM to it, otherwise it is **small** if disturbance is not negligible then we apply QM to it.

There is an unavoidable indeterminacy in the calculation of observational results, the theory enabling us to calculate in general only the probability of our obtaining a particular result when we make an observation. One may conclude that all calculations in CM are exact, but in QM they are probabilistic.

2.2 Function spaces

The concept of a space of functions is one of the important and basic subjects in quantum mechanics. Specifically we will discuss briefly the Hilbert space. This serves the purpose of giving a geometrical quality to some of the abstract concepts of quantum mechanics.

We recall that in Cartesian 3–space a vector $\mathbf{V}$ is a set of three numbers, called components (V_x, V_y, V_z). Any vector in this space can be expanded in terms of the three unit vectors $\hat{e}_x$, $\hat{e}_y$, $\hat{e}_z$. Under such conditions one terms the triad $\hat{e}_x$, $\hat{e}_y$, $\hat{e}_z$, a **basis**.

$$\mathbf{V} = V_x\hat{e}_x + V_y\hat{e}_y + V_z\hat{e}_z \tag{2.1}$$

The vectors $\hat{e}_x, \hat{e}_y, \hat{e}_z$ are said to **span** the vector space.

The inner (or dot) product of two vectors ($\mathbf{U}$ and $\mathbf{V}$) in the space is defined as

$$\mathbf{V} \cdot \mathbf{U} = V_x U_x + V_y U_y + V_z U_z \tag{2.2}$$

The length of the vector is

$$V = \sqrt{\mathbf{V} \cdot \mathbf{V}} = \sqrt{V_x^2 + V_y^2 + V_z^2} \tag{2.3}$$

A **Hilbert space** is much the same type of object. Its elements are functions instead of three–dimensional vectors. The similarity is so close that the functions are sometimes called vectors. A Hilbert space $\mathcal{H}$ has the following properties:

1. The space is linear. A function space is linear under the following conditions:

   (a) If a is a constant and ϕ is any element of the space, then $a\phi$ is also an element of the space.

   (b) If ϕ and ψ are any two elements of the space, then $\phi + \psi$ is also an element of the space.

2. There is an inner product, $< \psi|\phi >$, for any two elements in the space. For functions defined in the interval $a \leq x \leq b$ (in one–dimension), we may take

$$< \phi|\psi >= \int_a^b \phi^*\psi dx \tag{2.4}$$

3. Any element of $\mathcal{H}$ has a norm (length) that is related to the inner product as follows:

$$(\text{norm of } \phi)^2 =\| \phi \|^2 =< \phi|\phi > \tag{2.5}$$

4. $\mathcal{H}$ is complete. Every Cauchy sequence of functions in $\mathcal{H}$ converges to an element of $\mathcal{H}$. A Cauchy sequence $\{\phi_n\}$ is such that $\| \phi_n - \phi_l \| \to 0$ as n and l approach infinity. Loosely speaking, a Hilbert space contains all its limit points.

An example of a Hilbert space is given by the set of functions defined on the interval $(0 \leq x \leq L)$ with finite norm

$$\| \phi \|^2 = \int_0^L \phi^*\phi dx \ < \ \infty \ (\mathcal{H}_1) \tag{2.6}$$

Another example is the space of functions commonly referred to by mathematicians as L^2–**space**. This is the set of square–integrable functions defined on the whole x interval.

$$\| \phi \|^2 = \int_{-\infty}^{\infty} \phi^*\phi dx \ < \ \infty \ (\mathcal{H}_2) \tag{2.7}$$

We interpret the function $\phi(x)$ as a vector with infinitely many components. These components are the values that ϕ assumes at each distinct value of its independent variable x.

Just as the inner product between $\mathbf{U}$ and $\mathbf{V}$ is a sum over the products of parallel components, so is the inner product between ϕ and ψ a sum over parallel components. This sum is nothing but the integral of the product of ϕ and ψ. The reason we complex–conjugate the first vector is to ensure that the length of a vector ϕ is real.

Thus we see that Hilbert space is closely related to a vector space. Mathematicians call it **an infinite–dimensional vector space** (also, a complete, normed, linear vector space). Elements of this space have length and one can form an inner product between any two elements. We recall that if two vectors $\mathbf{U}$ and $\mathbf{V}$ in three–dimensional vector space are orthogonal to each other, their inner product vanishes. Similarly, two vectors in Hilbert space, ϕ and ψ, are said to be orthogonal if

$$< \phi|\psi >= 0 \tag{2.8}$$

Further, we recall that the three unit vectors $\hat{e}_x$, $\hat{e}_y$, and $\hat{e}_z$ span 3–space. Similarly, there is a set of vectors that span Hilbert space. For instance, the Hilbert space whose elements all have the property given by

$$\| \phi \|^2 = \int_0^L \phi^* \phi \, dx \quad < \quad \infty \quad (\mathcal{H}_1) \tag{2.9}$$

is spanned by the sequence of functions $\{\phi_n\}$, which are the eigenfunctions of the Hamiltonian relevant to the one–dimensional box problem. This means that any function ϕ in this Hilbert space may be expanded in a series of the sequence $\{\phi_n\}$

$$\phi(x) = \sum_{n=1}^{\infty} a_n \phi_n(x) \tag{2.10}$$

The geometrical interpretation of this relation may be pictured as shown in Figure 2.1.

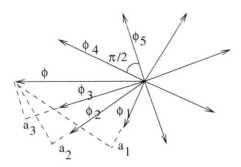

Fig. 2.1. Geometrical representation of the projection of ϕ onto $\phi_i =< \phi_i|\phi >= a_i$.

The coefficient a_n is the projection of ϕ onto the vector ϕ_n. If the basis vectors

$\{\phi_n\}$ form an orthogonal set, then we write

$$< \phi_n | \phi_{n'} >= 0 \quad (n \neq n') \tag{2.11}$$

Furthermore, if ϕ_n is a unit vector; that is, it has unit length, then we write

$$< \phi_n | \phi_n >=\| \phi_n \|^2= 1 \tag{2.12}$$

These latter two statements may be combined into the single equation

$$< \phi_n | \phi_{n'} >= \delta_{nn'} \tag{2.13}$$

where $\delta_{nn'}$ is the Kronecker delta ($\delta_{nn'} = 0$ for $n \neq n'$, $\delta_{nn'} = 1$ for $n = n'$). Any sequence of functions that obey $< \phi_n | \phi_{n'} >= \delta_{nn'}$ is called an orthonormal set. We can show that a_n is the projection of ϕ onto ϕ_n:

$$\phi = \sum_n a_n \phi_n \quad \rightarrow \quad |\phi >= \sum_n a_n |\phi_n > \tag{2.14}$$

$$< \phi_{n'} | \phi >= \sum_n a_n < \phi_{n'} | \phi_n >= \sum_n a_n \delta_{n'n} = a_{n'} \tag{2.15}$$

or

$$a_n =< \phi_n | \phi > \tag{2.16}$$

The coefficient a_n is the inner product between the basis vector ϕ_n and the vector ϕ. Since ϕ_n is a unit vector, a_n is the projection of ϕ onto ϕ_n.

2.3   Postulates of quantum mechanics

Quantum mechanics may be considered as a method of obtaining a (theoretical) description of a physical system. This could be possible in two steps: First, one isolates certain relevant characteristics of the system which are susceptible to measurement. Next, one forms an abstraction and refers to a collection of such measurements, say carried out at a particular time, as a **state** of the system. So there are measurable quantities, which are called **observables**, and states which summarize a set of measurements carried out at a fixed time.

The object of quantum mechanics is to identify these two general attributes of a system (observables and states) with the constituents of some appropriate mathematical construct. There are several postulates which describe how this is to be done in the case of nonrelativistic quantum mechanics.

Therefore quantum mechanics is based on certain postulates. Postulates are not proven, they are just accepted.

Postulate − 1 : It is assumed for a system consisting of a particle moving in a conservative field of force (produced by an external potential) that there is an associated wave function (w.f.), ψ, that this w.f. determines everything that can be known about the system, and that it is a single valued function of the coordinates of the particle and of the time, $\psi(\mathbf{r}, t)$. In general it is a complex function, and may be multiplied by an arbitrary complex number without changing its physical significance. $|\psi|^2$, and not ψ itself, is the quantity of measurable physical significance.

Postulate − 2 : With every physical observable (the energy of the system, the position coordinates of the particle, linear and angular momenta of the system, mass, the number of particles, etc.) there is associated an operator. Denote by Q the operator associated with the observable q. Then a measurement of q gives a result which is one of the eigenvalues of the eigenvalue equation

$$Q\psi_n = q_n\psi_n \qquad (2.17)$$

This measurement constitutes an interaction between the system and the measuring apparatus. If the state function was ψ_n prior to the measurement, the result q_n is certain to be obtained from an exact measurement of the observable associated with the operator Q. After a measurement yielding the value q_n, the state function is ψ_n. This is equivalent to the condition that a measurement be repeatable; a measurement giving a result q_n will, if repeated immediately, give with certainty the same result.

Examples of operators:

Operator	Its function
$\hat{D} = \frac{\partial}{\partial x}$	$\hat{D}\phi(x) = \frac{\partial\phi(x)}{\partial x}$
$\hat{\Delta} = -\frac{\partial^2}{\partial x^2} = -\hat{D}^2$	$\hat{\Delta}\phi(x) = -\frac{\partial^2\phi(x)}{\partial x^2}$
$\hat{M} = \frac{\partial^2}{\partial x \partial y}$	$\hat{M}\phi(x,y) = \frac{\partial^2\phi(x,y)}{\partial x \partial y}$
$\hat{I} =$ operation that leaves ϕ unchanged (identity operator)	$\hat{I}\phi = \phi$
$\hat{Q} = \int_a^b dx'$	$\hat{Q}\phi(x) = \int_a^b \phi(x')dx'$
$\hat{F} =$ multiplication by $F(x)$	$\hat{F}\phi(x) = F(x)\phi(x)$
$\hat{B} =$ division by the number 7	$\hat{B}\phi(x) = \frac{1}{7}\phi(x)$
$\hat{\Theta} =$ operator that annihilates ϕ	$\hat{\Theta}\phi = 0$
$\hat{P} =$ operator that changes ϕ to a specific polynomial of ϕ	$\hat{P}\phi = \phi^2 - 2\phi + 3$
$\hat{G} =$ operator that changes ϕ to the number 9	$\hat{G}\phi = 9$

Definition − 1 : An operator is **Hermitian** if

$$\int \psi_a^* Q\psi_b dv = \int (Q\psi_a)^* \psi_b dv \qquad (2.18)$$

where ψ_a and ψ_b are arbitrary normalizable functions. The integration is over the entire space. The operator x associated with the measurement of the x–component of the position of a particle is Hermitian. The operator $p_x = -i\hbar\frac{\partial}{\partial x}$ associated with the x–component of the momentum of the particle can also be seen to be Hermitian. Since p_x is a Hermitian operator, the square of p_x is also, and so is any power of p_x, as shown by

$$\int \psi_a^* p_x^2 \psi_b dv = \int (p_x \psi_a)^* p_x \psi_b dv = \int (p_x^2 \psi_a)^* \psi_b dv$$

Also, a linear combination of Hermitian operators is a Hermitian operator.

A number of elementary results, which follow directly from the postulates, can be expressed in the form of simple theorems. These theorems are fundamental to the whole structure of the quantum–mechanical formalism.

Theorem $-$ 1 : The eigenvalues of a Hermitian operator are all real.

Proof: $Q\psi_n = q_n\psi_n$

$$\int \psi_n^* Q\psi_n dv = \int \psi_n^* q_n \psi_n dv = q_n \int \psi_n^* \psi_n dv$$

$$\int (Q\psi_n)^* \psi_n dv = \int (q_n\psi_n)^* \psi_n dv = q_n^* \int \psi_n^* \psi_n dv$$

Therefore $q_n^* = q_n$, and q_n is real. The eigenvalues have been interpreted as the result of physical measurements, and such results are real numbers.

Postulate $-$ 3 : Any operator associated with a physically measurable quantity is Hermitian.

Definition $-$ 2 : Two wave functions are said to be **orthogonal** when

$$\int \psi_a^* \psi_b dv = 0 \tag{2.19}$$

the integration being over all space.

Definition $-$ 3 : A set of functions is linearly independent if the linear equation

$$\sum_j c_j \psi_j = 0 \tag{2.20}$$

implies that all $c_j = 0$. If the functions are not linearly independent, they are said to be linearly dependent.

Definition $-$ 4 : An eigenvalue q of an eigenvalue equation is m th–order degenerate if there are m linearly independent eigenfunctions corresponding to this eigenvalue.

Theorem $-$ 2 : Two eigenfunctions of an operator are orthogonal to each other if the corresponding eigenvalues are unequal.

Proof:

$$\int \psi_n^* Q\psi_m dv = \int (Q\psi_n)^* \psi_m dv$$

$$= q_n^* \int \psi_n^* \psi_m dv = q_n \int \psi_n^* \psi_m dv = q_m \int \psi_n^* \psi_m dv$$

$$(q_n - q_m) \int \psi_n^* \psi_m dv = 0$$

Therefore $\int \psi_n^* \psi_m dv = 0$ if $q_n \neq q_m$

Theorem − 3 : If an eigenvalue q of the operator Q is degenerate, any linear combination of the linearly independent eigenfunctions is also an eigenfunction;

$$Q\left(\sum_n c_n \psi_n\right) = q\left(\sum_n c_n \psi_n\right) \tag{2.21}$$

Definition − 5 : A set of functions constitutes a **complete set** of linearly independent eigenfunctions corresponding to the eigenvalue q if with any other eigenfunction of the eigenvalue q the set is linearly dependent. In other words, the set of functions is complete (or closed) if there is no other function which falls in the set of linearly independent functions. One may also define the complete set of functions as the following: A set $\{\psi_i\}$ of functions is complete if and only if any function in Hilbert space can be written as a linear combination of functions from the set $\{\psi_i\}$. A complete set of orthonormal functions forms a **basis**.

Theorem − 4 : If ψ_j $(j = 1, ..., m)$ constitute a complete set of eigenfunctions with the eigenvalue q of m th−order degeneracy for some operator, then any other eigenfunction of this eigenvalue may be expanded in terms of this complete set.

Proof: Let $a\psi - \sum_{j=1}^m c_j \psi_j = 0$. If $a = 0$, then $c_j = 0$ for all j. Since $\{\psi_j\}$; $j = 1, ..., m$ form a linearly independent and complete set, the solution for $a \neq 0$ is

$$\psi = \frac{1}{a} \sum_{j=1}^m c_j \psi_j$$

Theorem − 5 : Linear combinations of the ψ_j may be taken to form a set of m mutually orthogonal functions. These m mutually orthogonal functions are also linearly independent and can be used to expand any other eigenfunction corresponding to a particular eigenvalue. This theorem can be verified by making use of the Schmidt orthogonalization procedure.

Schmidt orthogonalization procedure:

Designate the set of independent functions corresponding to the eigenvalue q as ψ_j $(j = 1, ..., m)$. Choose any one of these functions, say ψ_1, as the first member of a new set ϕ_j $(j = 1, ..., m)$: $\phi_1 \equiv \psi_1$.

Designate $\int |\phi_1|^2 dv \equiv c_{11}$, $\int \phi_1^* \psi_2 dv \equiv c_{12}$.

Take $\phi_2 \equiv \frac{c_{12}}{c_{11}}\phi_1 - \psi_2$. Clearly, $\int \phi_1^* \phi_2 dv = 0$.

Designate $\int |\phi_2|^2 dv \equiv c_{22}$, $\int \phi_1^* \psi_3 dv \equiv c_{13}$, $\int \phi_2^* \psi_3 dv \equiv c_{23}$.

Take $\phi_3 \equiv \frac{c_{13}}{c_{11}}\phi_1 + \frac{c_{23}}{c_{22}}\phi_2 - \psi_3$, clearly, $\int \phi_1^* \phi_3 dv = \int \phi_2^* \phi_3 dv = 0$.

This procedure can be extended to obtain $\phi_4, ..., \phi_m$. This procedure can be used to obtain a complete orthogonal set of eigenfunctions for any Hermitian operator.

The general form:

$$\phi_m = \sum_{i=1}^{m-1} \frac{c_{im}}{c_{ii}} \phi_i - \psi_m \tag{2.22}$$

$$c_{ii} = \int |\phi_i|^2 dv \quad , \quad c_{im} = \int \phi_i^* \psi_m dv \quad , \quad \int \phi_i^* \phi_j dv = 0 \quad (i \neq j) \qquad (2.23)$$

Linearly independent functions:

A set of functions $\psi_n(x)$; $(n = 1, 2, \cdots, N)$ is said to be linearly independent if the linear combination

$$\sum_{n=1}^{N} \lambda_n \psi_n(x) = 0 \qquad (2.24)$$

for all x is only satisfied when

$$\lambda_1 = \lambda_2 = \cdots = \lambda_N = 0 \qquad (2.25)$$

For example, the two functions e^x and $\sin x$ are linearly independent since $\lambda_1 e^x + \lambda_2 \sin x = 0$ for all x is only satisfied by $\lambda_1 = \lambda_2 = 0$. The two functions e^x and $3e^x$ are not linearly independent (or, are linearly dependent) since $\lambda_1 e^x + \lambda_2 3e^x = 0$ is true for all x if $\lambda_1 = -3\lambda_2 \neq 0$.

From a geometrical point of view in Hilbert space, if two functions (or vectors) ψ_1 and ψ_2 in a Hilbert space $\mathcal{H}$ are linearly independent, they do not lie along the same axis (or line) in $\mathcal{H}$. We can generalize this for N vectors; if the set of N vectors ψ_n is such that all members are linearly independent, no two elements of this set lie on the same axis.

Postulate − 4 : The set of functions ψ_j which are eigenfunctions of the eigenvalue equation, $Q\psi_j = q_j \psi_j$, form, in general, an infinite set of linearly independent functions. A linear combination of these functions of the form

$$\psi = \sum_j c_j \psi_j \qquad (2.26)$$

can be used to express an infinite number of possible functions. The infinite set of functions formed by the eigenfunctions of any operator playing a role in quantum mechanics can be used to expand a wave function which is a suitable physical wave function.

If ψ is a physically acceptable wave function, it can be expanded in eigenfunctions of any observable of the system. If the complete set of linearly independent eigenfunctions for an operator has been chosen to be orthogonal and the eigenfunctions are all square–integrable and have been normalized to unity, then we have

$$\int \psi_j^* \psi_k dv = \delta_{jk} \qquad (2.27)$$

Such a set of functions is said to constitute a complete orthonormal set. The expansion coefficients $\{c_j\}$ are evaluated for such a set from

$$c_j = \int \psi_j^* \psi dv \qquad (2.28)$$

Definition − 6 : If there exists a complete set of linearly independent state functions ψ_j such that ψ_j is an eigenfunction of both of the operators R and S corresponding to physical observables, the corresponding observables are said to be **compatible** (or simultaneously measurable).

By **compatible observables** is meant that both R and S are completely pre-dictable for the complete set of states ψ_j. For example, position and momentum measurements are not compatible. However, the three components of the momentum are simultaneously measurable and hence compatible.

Definition – 7 : If $Q\psi = R\psi$ for any arbitrary function in the set of physically permissible wave functions, the operators are equivalent; $Q \equiv R$.

Theorem – 6 : If two observables are compatible, their operators commute.

Proof: $S\psi_j = s_j\psi_j$, $R\psi_j = r_j\psi_j$. Therefore
$(RS - SR)\psi_j = 0$ or $(RS - SR)\sum_j c_j\psi_j \equiv (RS - SR)\psi = 0$.
ψ may be any arbitrary function of the class of all wave functions of importance in quantum mechanics. Therefore, the commutation of the operators R and S :

$$[R, S] \equiv RS - SR = 0 \tag{2.29}$$

The expression $RS - SR$ is known as the **commutator** of the operators R and S.

Theorem – 7 : If two operators Q and R commute and either Q or R has nondegenerate eigenvalues, its eigenfunctions are also eigenfunctions of the other operator.

Proof: $Q\psi_j = q_j\psi_j$, where q_j is assumed to be nondegenerate; then $Q(R\psi_j) = q_j(R\psi_j)$. Here the function $R\psi_j$ is an eigenfunction of the operator Q. The function $R\psi_j$ can be expressed in terms of ψ_j, such that $R\psi_j = r_j\psi_j$. This shows that the wave function ψ_j is simultaneously an eigenfunction of both Q and R.

Theorem – 8 : If Q and R are operators which commute with each other, there exists a complete set of eigenstates which are simultaneously eigenstates of both Q and R.

The case of degenerate eigenvalues (or The case of discrete eigenvalues):

Assume that $Q\psi_j = q\psi_j$, where q is an m th–order degenerate eigenvalue of Q. Operating on $Q\psi_j = q\psi_j$ with R and making use of the commutation relation leads to

$$Q(R\psi_j) = q(R\psi_j)$$

The function $R\psi_j$ is an eigenfunction of Q. Let

$$R\psi_j = \sum_{k=1}^{m} q_{jk}\psi_k$$

multiplied by a constant c_j and summed over j give

$$R\sum_{j=1}^{m} c_j\psi_j = \sum_{j,k} c_j q_{jk}\psi_k$$

Assume that

$$\sum_j c_j q_{jk} = rc_k$$

r is a constant, such that

$$|q_{jk} - r\delta_{jk}| = 0$$

Expansion of this determinant leads to an m th–order equation for r which has m roots. With each root r_k there is associated a solution c_j^k for the c's.

Defining

$$\phi_k \equiv \sum_j c_j^k \psi_j \quad ; \quad R\phi_k = r_k\phi_k \quad , \quad Q\phi_k = q_k\phi_k$$

Therefore, the functions ϕ_k constitute a complete set of simultaneous eigenfunctions of R and Q. A complete orthonormal set of functions can be obtained from the ϕ_k by using the Schmidt orthogonalization procedure. Consequently, for this case the normalization of the eigenfunctions and the discreteness of the eigenvalues are related, one following from the other.

The case of continuous distribution of eigenvalues:

The eigenvalue equation can be written as $Q\psi_q = q\psi_q$. Here the eigenvalue q, which takes on a continuous distribution of values, is also used as a subscript to designate the eigenfunction with which it is associated. The theorem of the orthogonality of eigenfunctions corresponding to different eigenvalues takes the form

$$\int \psi_{q'}^*\psi_q dv = 0 \quad , \quad q \neq q' \tag{2.30}$$

For $q' = q$, the integral is divergent, in as much as the wave function is known to be unnormalizable. We can then define the orthogonality integral as delta function

$$\int \psi_{q'}^*\psi_q dv = \delta(q - q') \tag{2.31}$$

In a similar manner, the expansion hypothesis for the case of a continuous distribution of eigenvalues can be written as

$$\psi = \int \phi(q)\psi_q dq \tag{2.32}$$

If there is both a continuous and a discrete range of eigenvalues for the operator Q, the expansion hypothesis is written as

$$\psi = \sum_q \phi_q\psi_q + \int \phi(q)\psi_q dq \tag{2.33}$$

where the summation is over the discrete range of the eigenvalues and the integration is over the continuous range of the eigenvalues. If it is assumed that the wave function ψ can be and is normalized to unity,

$$\int |\psi|^2 dv = \sum_q |\phi_q|^2 + \int |\phi(q)|^2 dq = 1 \tag{2.34}$$

If

$$\psi(x) = \int_{-\infty}^{\infty} \phi(p)\psi_p(x)dp \tag{2.35}$$

the inverse transformation can then be written as

$$\phi(p) = \int_{-\infty}^{\infty} \psi_p^*\psi(x)dx \tag{2.36}$$

and

$$\int_{-\infty}^{\infty} |\psi|^2 dx = \int_{-\infty}^{\infty} |\phi(p)|^2 dp = 1 \tag{2.37}$$

Consider the expansion of an arbitrary wave function in terms of eigenfunctions of some particular operation,

$$\psi(r) = \sum_q \phi_q \psi_q + \int \phi(q)\psi_q dq \tag{2.38}$$

Use of the orthonormal property of these eigenfunctions gives

$$\int \psi_q^* \psi dv = \phi_q \quad or \quad \phi(q) \tag{2.39}$$

One can write

$$\psi(r) = \sum_q \left[\int \psi_q^* \psi dr' \right] \psi_q(r) + \int \left[\int \psi_q^* \psi dr' \right] \psi_q dq \tag{2.40}$$

$$= \int \left[\sum_q \psi_q^*(r')\psi_q(r) + \int \psi_q^*(r')\psi_q(r)dq \right] \psi(r')dr'$$

$$= \int \delta(r - r')\psi(r')dr'$$

This is the **closure relation**.

Some properties of commutators:

An important operation in quantum mechanics is the commutator between two operators, A and B. It is written $[a, b]$ and is defined as

$$[A, B] = AB - BA \tag{2.41}$$

An immediate property of the commutator is that

$$[A, B] = -[B, A] \tag{2.42}$$

If $[A, B] = 0$ the two operators are said to commute (A and B are compatible) with each other. That is, $AB = BA$. Any operator commutes with any constant a.

$$[A, a] = 0 \tag{2.43}$$

$$[A, aB] = [aA, B] = a[A, B] \tag{2.44}$$

Any operator A commutes with its own square, A^2.

$$[A, A^2] = (AA^2 - A^2A) = (AAA - AAA) = 0 \tag{2.45}$$

The meaning of this relation is that, no matter what A is, when $[A, A^2]$ operates on any function $g(x)$, one gets zero,

$$[A, A^2]g(x) = 0 \tag{2.46}$$

More generally, A commutes with any function of A, $f(A)$.

$$[f(A), A] = 0 \tag{2.47}$$

As an example of this rule, consider the following commutator involving the momentum operator, p.

$$[e^p, p] = \left[\sum_{n=0}^{\infty} \frac{p^n}{n!}, p \right] = \sum \frac{1}{n!} [p^n, p] \tag{2.48}$$

$$= [1, p] + [p, p] + \frac{1}{2!} [p^2, p] + \cdots = 0$$

It follows that

$$[e^p, p]g(x) = 0 \tag{2.49}$$

where $g(x)$ represents any function of x.

One of the most important commutators in physics is that between the coordinate, x, and the momentum, p.

$$[x, p]g(x) = i\hbar \left(-x \frac{\partial}{\partial x} + \frac{\partial}{\partial x} x \right) g(x) \tag{2.50}$$

$$= i\hbar \left(-x \frac{\partial g}{\partial x} + x \frac{\partial g}{\partial x} + g \right) = i\hbar g(x)$$

It follows that

$$[x, p] = i\hbar \tag{2.51}$$

In other words, the operator $[x, p]$ has the sole effect of a simple multiplication by the constant $i\hbar$.

If A, B, and C are three distinct operators, we may write the following equalities

$$[AB, C] = ABC - CAB = ABC - ACB + ACB - CAB \tag{2.52}$$

$$= A[B, C] + [A, C]B$$

$$[A, [B, C]] + [B, C, A]] + [C, [A, B]] = 0 \tag{2.53}$$

As an immediate consequence

$$[x, p^2] = [x, p]p + p[x, p] = 2i\hbar p \tag{2.54}$$

so that

$$[x, p^2]g(x) = 2\hbar^2 \frac{\partial g}{\partial x} \tag{2.55}$$

In a similar vein,

$$[x^2, p] = x[x, p] + [x, p]x = 2i\hbar x \tag{2.56}$$

The operator $[x^2, p]$ multiplies by $2i\hbar x$.

If A and B are Hermitian, so is $i[A, B]$. If A and B each commutes with $[A, B]$, then one can write the equality

$$e^A e^B = e^{A+B} e^{\frac{1}{2}[A,B]} \tag{2.57}$$

One can write the general relation

$$e^A B e^{-A} = B + [A, B] + \frac{1}{2!}[A, [A, B]] + \frac{1}{3!}[A, [A, [A, B]]] + \cdots \tag{2.58}$$

Another important property of commutators is that

$$[A, B]^+ = (AB)^+ - (BA)^+ = B^+ A^+ - A^+ B^+ = [B^+, A^+] \qquad (2.59)$$

Postulate − 5 : If a system is described by a wave function ψ, the expectation value of any observable q with corresponding operator Q is given by

$$< q >= \int \psi^* Q \psi dv \qquad (2.60)$$

If

$$\psi = \sum_j c_j \psi_j \;\;, \;\; Q\psi_j = q_j \psi_j \;\;, \;\; \psi_j^* \psi_k = \delta_{jk}$$

then one can write

$$\int \psi^* \psi dv = \sum_{j,k} c_j^* c_k \int \psi_j^* \psi_k dv = \sum_{j,k} c_j^* c_k \delta_{jk} = 1$$

Therefore

$$\sum_j |c_j|^2 = 1$$

Thus,

$$< q >= \sum_j q_j |c_j|^2$$

From these relations, $|c_j|^2$ can be interpreted as the probability of finding the system in the state designated by the subscript j.

In a measurement in which q is determined, the probability that the result q_j will be obtained is given by $P_j = |c_j|^2$. If the result q_j is a degenerate eigenvalue, the probability of obtaining this result is found by summing P_j over all subscripts j corresponding to this particular eigenvalue. One can obtain an explicit expression for the probability in the form

$$c_j = \int \psi_j^* \psi dv \;\;, \;\; P_j = |c_j|^2 = |\int \psi_j^* \psi dv|^2$$

Postulate − 6 : The development in time of the wave function ψ, given its form at an initial time and assuming the system is left undisturbed, is determined by the Schrödinger equation

$$H\psi = i\hbar \frac{\partial}{\partial t}\psi \qquad (2.61)$$

where the Hamiltonian operator H is formed from the corresponding classical Hamiltonian function by substituting for the classical observables their corresponding operators. This form of the Schrödinger equation is known as the time–dependent Schrödinger equation.

The behavior of an isolated system is **causal**. This means that the state of a system at time t is completely determined by the state of the same system at time t_0 ($t > t_0$) provided the system is undisturbed by measurement at intermediate times. The behavior of an isolated system is independent of the absolute scale of time, that is, the dynamics is invariant under time translations. Hence, the change of a state of an isolated system is **smooth**.

If $\psi(x, t_0)$ and $\psi(x, t)$ represent the states of an isolated system at times t_0 and t ($t > t_0$), respectively, then this postulate implies that

$$\psi(x, t) = U(t, t_0)\psi(x, t_0) \tag{2.62}$$

Clearly we must have $U(t, t) = I$ and $U(t, t_0) = U(t - t_0)$. $U(t, t_0)$ is known as the **time evolution operator**. Each $U = U(t - t_0)$ for a definite time interval is a symmetry operator, therefore each U is unitary.

$$U(t, t_0)^+ = [U(t, t_0)]^{-1} = U(t_0, t) \tag{2.63}$$

Let us (formally) expand U in a power series:

$$U(t + \delta t, t) = U(t, t) + \left[\frac{\partial U(t', t)}{\partial t'}\right]_{t'=t} \delta t + \mathcal{O}[(\delta t)^2] \tag{2.64}$$

Let us define

$$\left[\frac{\partial U(t', t)}{\partial t'}\right]_{t'=t} \equiv -\frac{i}{\hbar} H(t) \tag{2.65}$$

which requires that $H(t)$ be an operator with the dimensions of an energy (it is actually the Hamiltonian oparetor). However,

$$U(t + \delta t, t) = U(\delta t, 0) = U(\delta t) \tag{2.66}$$

which implies that $H(t)$ is independent of t. We have then

$$U(\delta t) = I - \frac{i}{\hbar} H\delta t + \mathcal{O}[(\delta t)^2] \tag{2.67}$$

Let δt be infinitesimal. Then the unitary of U implies

$$U(\delta t)U^+(\delta t) = I \tag{2.68}$$

or

$$\left[I - \frac{i}{\hbar} H\delta t + \mathcal{O}[(\delta t)^2]\right] \left[I + \frac{i}{\hbar} H^+\delta t + \mathcal{O}[(\delta t)^2]\right] = I$$

Because δt is infinitesimal the terms linear in δt must vanish separately, therefore one gets

$$\frac{i}{\hbar}(H^+ - H)\delta t = 0 \tag{2.69}$$

which implies that H is Hermitian, $H^+ = H$.

Consider $\psi(x, t) = U(t, t_0)\psi(x, t_0)$. Let us (formally) differentiate this w.r. to t:

$$\frac{\partial}{\partial t}\psi(x, t) = \frac{\partial}{\partial t}U(t, t_0)\psi(x, t_0) \tag{2.70}$$

Setting $t = t_0$ we have

$$i\hbar\frac{\partial}{\partial t}\psi(x, t) = H\psi(x, t) \tag{2.71}$$

which is the time–independent Schrödinger equation. This also implies an equation of motion for the operator U. It is obvious from $\psi(x, t) = U(t, t_0)\psi(x, t_0)$ that

$$i\hbar\frac{\partial}{\partial t}\psi(x, t) = H\psi(x, t) = HU(t, t_0)\psi(x, t_0) \tag{2.72}$$

But

$$ih\frac{\partial}{\partial t}\psi(x,t) = ih(\frac{\partial}{\partial t}U(t,t_0))\psi(x,t_0) \tag{2.73}$$

so that

$$ih\frac{\partial}{\partial t}U(t,t_0) = HU(t,t_0) \tag{2.74}$$

This equation may be solved, subject to the boundary condition $U(t,t) = I$, to yield

$$U(t,t_0) = e^{-\frac{i}{\hbar}H(t-t_0)} \tag{2.75}$$

Suppose that

$$H\psi(x,t_0) = E\psi(x,t_0) \tag{2.76}$$

Then

$$\psi(x,t) = e^{-\frac{i}{\hbar}E(t-t_0)}\psi(x,t_0) \tag{2.77}$$

is also an eigenstate of H with the same eigenvalue. It is evident that we can regard H as a generator of displacements in time. The generator of displacements in time, H, represents the total energy of an isolated system and moreover it is an observable.

Postulate – 7 : The operators of quantum theory are such that their commutators are proportional to the corresponding classical Poisson brackets according to the prescription

$$[Q,R] \equiv (QR - RQ) \rightleftharpoons i\hbar\{q,r\} \tag{2.78}$$

where $\{q,r\}$ is the classical Poisson bracket for the observables q and r. Poisson bracket of two functions F and G of the canonical variables q and p is defined as

$$\{F,G\} \equiv \sum_i \left(\frac{\partial F}{\partial q_i}\frac{\partial G}{\partial p_i} - \frac{\partial F}{\partial p_i}\frac{\partial G}{\partial q_i}\right) \tag{2.79}$$

The variables in the Poisson bracket are to be replaced by operators. This postulate represents an important bridge between classical and quantum mechanics.

Summary of Postulates:

1. The state of a quantum mechanical system is characterized by a square integrable wave function, ψ, which is a member of the Hilbert space.

2. The only results of measurements of a physical quantity are the eigenvalues, q, of the operator, Q, associated with it; $Q\psi = q\psi$.

3. Associated with a dynamical variable $Q(r,p,t)$ there is a Hermitian operator $Q(r,\hbar\nabla/i,t)$.

4. If a normalized state of a quantum system can be written as $\psi = \sum_j c_j\psi_j$ where the ψ_j's are the eigenfunctions of a Hermitian operator Q associated with a physical observable, i.e., $Q\psi_j = q_j\psi_j$, then $|c_j|^2$ is the probability of finding the value q_j when measurements of Q are made.

5. The mean (average) value of Q when the system is in a state ψ is $< Q >= \int \psi^* Q\psi dv / \int \psi^* \psi dv$.

6. $i\hbar\partial\psi/\partial t = H\psi$.

7. $QR - RQ = i\hbar\{q,r\}$.

2.4   The Kronecker delta and the Dirac delta function

In quantum mechanics one frequently encounters mathematical expressions containing summations over one or more indices. In many cases, the expressions for these summations can be greatly simplified by the use of a symbol known as the **Kronecker delta**, δ_{nm}. This symbol has two indeces and is defined by the properties

$$\delta_{nm} = 1 \ \text{ if } \ n = m \ ; \ \ \delta_{nm} = 0 \ \text{ if } \ n \neq m \tag{2.80}$$

Another mathematical concept that will also prove to be of great usefulness is the **Dirac delta function**. It is an improper function; it can be given a satisfactory meaning by the use of suitable limiting procedures. Considering the equation

$$F(\mathbf{k}) = \left(\frac{1}{2\pi}\right)^{3/2} \int_{-\infty}^{\infty} f(\mathbf{r}) e^{-i\mathbf{k}\cdot\mathbf{r}} d\mathbf{r} \tag{2.81}$$

This can be thought of as an expansion of an arbitrary function $f(\mathbf{r})$ in terms of exponential periodic functions (plane waves) of $\mathbf{k} \cdot \mathbf{r}$. It is not possible to obtain a Fourier transform of the exponential function itself, since for such a function the condition of (square) integrability is not satisfied. Although a plane wave thus does not have a true Fourier transform, one can define the improper Dirac delta function so as to perform the role of such a transform. To do this, we write the exponential periodic function as

$$f(x) = e^{ik_0 x} = \lim_{\alpha \to 0} e^{(-\alpha x^2 + ik_0 x)} \tag{2.82}$$

For any finite, real, positive value of α, the Fourier integral of $f(x)$ exists. This allows the calculation of the **Fourier transform** of $f(x)$:

$$F(k) = \lim_{\alpha \to 0} \frac{1}{\sqrt{2\pi}} \int_{-\infty}^{\infty} e^{(-\alpha x^2 + ik_0 x)} e^{-ikx} dx \tag{2.83}$$

The resulting limit vanishes for $k \neq k_0$ and diverges for $k = k_0$, giving as the form of the improper function $F(k)$:

$$F(k) = 0 \ \text{ if } \ k \neq k_0 \ ; \ \ F(k) = \infty \ \text{ if } \ k = k_0$$

Singular as this function is, it is still possible to define its integral over all k by performing the integration before taking the limit:

$$\int_{-\infty}^{\infty} F(k)dk = \lim_{\alpha \to 0} \frac{1}{\sqrt{2\pi}} \int_{-\infty}^{\infty} dk \int_{-\infty}^{\infty} e^{[-\alpha x^2 + i(k_0 - k)x]} dx \tag{2.84}$$

This suggests the definition of a new singular function which is called the Dirac delta function:

$$\delta(k) = \frac{1}{2\pi} \int_{-\infty}^{\infty} e^{-ikx} dx \tag{2.85}$$

The function $\delta(k)$ has the properties

$$\delta(k) = 0 \ \text{ if } \ k \neq 0 \ ; \ \ \delta(k) = \infty \ \text{ if } \ k = 0 \ ; \ \ \int_{-\infty}^{\infty} \delta(k)dk = 1$$

In any computations involving the delta function, it is assumed that the computations are made prior to taking the limit. When dealing with regular well–behaved functions, the limiting process must be taken after the computations are made. **The Dirac delta function is meaningful only under integral sign**, where this limiting technique can be used.

Some properties of the delta function are:

$$\delta(x) = \delta(-x) \tag{2.86}$$

$$\int f(x)\delta(x-a)dx = f(a) \tag{2.87}$$

$$\delta(ax) = \frac{1}{a}\delta(x) \ , \ \ a > 0 \tag{2.88}$$

$$\int \delta(x-x_1)\delta(x_1-x_2)dx_1 = \delta(x-x_2) \tag{2.89}$$

$$f(x)\delta(x-a) = f(a)\delta(x-a) \tag{2.90}$$

It is also possible to define the derivative of the delta function:

$$\delta'(k) = \frac{1}{2\pi} \int_{-\infty}^{\infty} ike^{-ikx}dx \tag{2.91}$$

Some of the formal properties of the derivative of the delta function are:

$$-\delta'(k) = \delta'(-k) \tag{2.92}$$

$$\int f(x)\delta'(x-a)dx = -f'(a) \tag{2.93}$$

The definition of the delta function can be easily extended to three dimensions to give the delta function of the vector variable $\mathbf{k}$:

$$\delta(\mathbf{k}) \equiv \delta(k_x)\delta(k_y)\delta(k_z) \tag{2.94}$$

$$= \frac{1}{(2\pi)^3} \int_{-\infty}^{\infty}\int_{-\infty}^{\infty}\int_{-\infty}^{\infty} e^{[i(k_x x + k_y y + k_z z)]}dxdydz = \frac{1}{(2\pi)^3} \int_{-\infty}^{\infty} e^{i\mathbf{k}\cdot\mathbf{r}}d\mathbf{r}$$

A step–function may be expressed in terms of a delta function

$$\theta(x-a) \equiv \int_{-\infty}^{x} \delta(y-a)dy \tag{2.95}$$

$$\theta(x-a) = 0 \ \ if \ \ x < a \ ; \ \ \theta(x-a) = 1 \ \ if \ \ x > a$$

Therefore the derivative of a step–function is a delta function

$$\frac{d}{dx}\theta(x-a) = \delta(x-a) \tag{2.96}$$

Some other definitions of delta function:

$$\delta(x) = \frac{1}{2\pi} \lim_{L\to\infty} \int_{-L}^{L} e^{ikx}dx \tag{2.97}$$

$$= \lim_{L \to \infty} \frac{1}{2\pi} \frac{e^{iLx} - e^{-iLx}}{ix} = \lim_{L \to \infty} \frac{\sin Lx}{\pi x}$$

$$\delta(x) = \lim_{a \to 0} \frac{1}{\pi} \frac{a}{x^2 + a^2} \tag{2.98}$$

$$\delta(x) = \lim_{a \to 0} \frac{\alpha}{\sqrt{\pi}} e^{-\alpha^2 x^2} \tag{2.99}$$

Consider the function $\Delta(x, a)$ defined by

$$\begin{aligned} \Delta(x, a) &= 0 & x < -a \\ &= \frac{1}{2a} & -a < x < a \\ &= 0 & x > a \end{aligned}$$

then

$$\delta(x) = \lim_{a \to 0} \Delta(x, a) \tag{2.100}$$

These functions are defined as the limit of a sequence of properly behaved functions. However, in practice it is usually possible to calculate in a perfectly straightforward manner with these functions, as though they were well–behaved functions. For example, consider the equation

$$F(k) = \frac{1}{\sqrt{2\pi}} \int_{-\infty}^{\infty} f(x) e^{-ikx} dx \tag{2.101}$$

Multiply both sides by $e^{-ikx'}/\sqrt{2\pi}$ and integrate over all k. The result is

$$\frac{1}{\sqrt{2\pi}} \int_{-\infty}^{\infty} F(k) e^{-ikx'} dk = \frac{1}{\sqrt{2\pi}} \int_{-\infty}^{\infty} \int_{-\infty}^{\infty} f(x) e^{ik(x'-x)} dk dx$$

$$= \int_{-\infty}^{\infty} f(x) \delta(x - x') dx = f(x')$$

or

$$f(x) = \frac{1}{\sqrt{2\pi}} \int_{-\infty}^{\infty} F(k) e^{-ikx} dk \tag{2.102}$$

This is the Fourier inversion theorem; if one of the relations holds so does the other. We say that the functions $F(k)$ and $f(x)$ are Fourier transforms of each other. In a similar fashion, the relation

$$\int |f(x)|^2 dx = \int |F(k)|^2 dk \tag{2.103}$$

is obtained.

The eigenfunctions of position are δ functions,

$$x\delta(x - a) = a\delta(x - a) \tag{2.104}$$

These functions form a complete orthonormal set and any function $\psi(x)$ can be considered as an expansion in terms of this set,

$$\psi(x) = \int_{-\infty}^{\infty} \delta(x - a)\psi(a) da \tag{2.105}$$

2.5 Dirac notation

To begin to explain the Dirac notation we consider an abstract vector space of states. The linear superposition of states is defined but not an inner product. The elements of the space are called **kets** and are denoted by $|\ >$. If we want to specify a specific ket we insert a label $|a>$. This specification of kets is completely basis independent. This means that we do not explicitly write any wave function $\psi_a(x)$ but only write the symbol $|a>$. The label a usually refers to the eigenvalue of some operator A,

$$A|a>= a|a> \tag{2.106}$$

Corresponding to the space of kets we introduce the dual space of continuous linear functions defined on the kets. This is called the space of **bras** and they are written $<\ |$. Specific bras are labelled in the same manner as kets. Furthermore if A is an operator on the space of kets then the corresponding operator on the space of bras is A^+. Thus the eigenvalue equation corresponding to kets is given on the space of bras by

$$<a|A^+ = a^* <a| \tag{2.107}$$

Since the bras are linear functions over the kets they give a mapping from the kets into the complex numbers. We write this as $<\ |\ >$ or for two specific ones as $<a|b>$. In terms of functions the corresponding expression is

$$<a|b>= (\psi_a, \psi_b) = \int_{-\infty}^{\infty} \psi_a^*(x)\psi_b(x)dx \tag{2.108}$$

The completeness relation for these states is given by
$1 = \sum_n |n><n|$ if the label n is discrete or
$1 = \int |k> dk <k|$ if the index k is continuous.
These equations correspond to the equations
$\delta(x-y) = \sum_n \psi_n(x)\psi_n^*(y)$ if the index n is discrete or
$\delta(x-y) = \int dk\psi_k(x)\psi_k^*(y)$ if the index n is continuous.
The matrix element (or the expectation value) of an operator is now written as $<a|A|b>$. A acts to right, A^+ acts to left. In terms of wave functions these are written as $(\psi_a, A\psi_b)$ or $(A^+\psi_a, \psi_b)$.

To establish precisely the connection between Dirac's bra, ket notation and the usual wave–function formalism consider a specific eigenket $|n>$ of the Hamiltonian H. Thus,

$$H|n>= E_n|n> \tag{2.109}$$

Now let the ket $|x>$ be an eigenket of the position operator x_{op} so that

$$x_{op}|x>= x|x> \tag{2.110}$$

The eigenfunctions $\phi_n(x)$ of the Hamiltonian H in configuration space are given by

$$\phi_n(x) =<x|n> \quad , \quad \phi_n^*(x) =<n|x> \tag{2.111}$$

The orthogonality relation of two eigenfunctions ϕ_n and ϕ_m now follows from the orthogonality of the ket $|n>$ and the bra $<m|$, namely

$$<m|n>= \delta_{nm} \tag{2.112}$$

and the closure condition

$$\int |x> dx <x| = 1 \tag{2.113}$$

Consider

$$(\phi_m, \phi_n) = \int \phi_m^*(x)\phi_n(x)dx = \int <m|x> dx <x|n> = <m|n> = \delta_{mn}$$

The connection between time–independent Schrödinger equation and the eigenket equation may be written as follows. If p is the momentum operator

$$p|x> = -i\hbar \frac{d}{dx}|x> \tag{2.114}$$

If $H = \frac{p^2}{2m} + V(x)$, we take the inner product of eigenket equation with the bra $<x|$ to get

$$<x|H|n> = E_n <x|n> = E_n\phi_n(x) \tag{2.115}$$

The left hand side of this equation can be rewritten as

$$<x|H|n> = \int <x|H|y> dy <y|n> = \int <x|H|y> \phi_n(y)dy \tag{2.116}$$

But using the explicit form for the Hamiltonian we have

$$<x|V(x)|y> = V(y) <x|y> = V(y)\delta(x-y) \tag{2.117}$$

$$<x|\frac{p^2}{2m}|y> = -\frac{\hbar^2}{2m} <x|\frac{d^2}{dy^2}|y> = -\frac{\hbar^2}{2m}\frac{d^2}{dy^2} <x|y> \tag{2.118}$$

$$= -\frac{\hbar^2}{2m}\frac{d^2}{dy^2}\delta(x-y)$$

Substituting we get

$$<x|H|n> = \int \left[\left(-\frac{\hbar^2}{2m}\frac{d^2}{dy^2} + V(y) \right) \delta(x-y) \right] \phi_n(y)dy \tag{2.119}$$

$$= -\frac{\hbar^2}{2m}\frac{d^2}{dy^2}\phi_n(x) + V(x)\phi_n(x)$$

so that the eigenket equation reads explicitly

$$-\frac{\hbar^2}{2m}\frac{d^2}{dy^2}\phi_n(x) + V(x)\phi_n(x) = E_n\phi_n(x) \tag{2.120}$$

which is the usual form of the Schrödinger equation.

If we have the expression $AB|\eta >$ and we wish to write it in terms of wave functions then we can use the basis set $\{|x >\}$ consisting of the eigenkets of the position operator. The expression is then rewritten as follows

$$<x|AB|\eta> = \int <x|A|y> dy <y|B|z> dz <z|\eta> \tag{2.121}$$

$$= \int dydz <x|A|y> <y|B|z> \phi_\eta(z)$$

Let $\{u_n\}$ be a complete basis and let $\{|n>\}$ and $\{<n|\}$ denote the corresponding complete sets of kets and bras respectively so that

$$u_n(x) =< x|n > \quad , \quad u_n^*(x) =< n|x > \tag{2.122}$$

Now any wave function $\psi(x)$ and corresponding ket $|\psi>$ can be expanded as

$$\psi = \sum_n a_n u_n \quad or \quad |\psi >= \sum_n a_n|n > \tag{2.123}$$

In both cases we have

$$a_n = (u_n, \psi) =< n|\psi > \tag{2.124}$$

Thus

$$|\psi >= \sum_n |n >< n|\psi > \tag{2.125}$$

which implies

$$\sum_n |n >< n| = 1 \tag{2.126}$$

This is the **completeness relation** expressed in bra, ket notation.

If we now consider matrix elements of any operator A between wave functions ϕ and ψ we can write the whole expression in bra and ket notation. The matrix element $(\phi, A\psi)$ can be written as

$$< \phi|A\psi >= \sum_{m,n} < \phi|m >< m|A|n >< n|\psi > \tag{2.127}$$

Then $< m|A|n >$ is a matrix for the operator A in the standard basis we have used for labelling our bras and kets.

Formulated in this way, quantum mechanics is historically referred to as **matrix mechanics** in contrast to the Schrödinger formulation which is called **wave mechanics**. Clearly they are just two different versions of the same thing.

2.6   Worked examples

Example - 2.1 : Show that the differential operator $p = \frac{\hbar}{i}\frac{d}{dx}$ is linear and Hermitian in the space of all differentiable wave functions $< x|\phi >= \phi(x)$, say, which vanish at both ends of an interval (a, b).

Solution : We have that

$$< \psi|p|\phi >= \frac{\hbar}{i} \int_a^b \psi^* \frac{d\phi}{dx} dx$$

Integrating by parts we obtain

$$< \psi|p|\phi >= \frac{\hbar}{i}(\phi\psi)|_a^b - \frac{\hbar}{i} \int_a^b \frac{d\psi^*}{dx} \phi \, dx$$

$$= -\frac{\hbar}{i} \int_a^b \frac{d\psi^*}{dx} \phi dx = < \phi|p|\psi >^*$$

Example - 2.2 : Suppose that $\{u_i\}$ constitutes a complete set of eigenfunctions for the two linear operators R and P. Show that $[R, P] = 0$.

Solution : We are given $Ru_i = r_i u_i$, $Pu_i = p_i u_i$; $i = 1, 2, \cdots, n$ where r_i and p_i are the eigenvalues of R and P, respectively. We must prove $[R, P] = 0$ or $(RP - PR)f = 0)$. Expanding f in the complete set of functions we then have

$$(RP - PR) \sum_i c_i u_i = \sum_i c_i (RP - PR) u_i$$

$$= \sum_i c_i (Rp_i - Pr_i) u_i = \sum_i c_i (r_i p_i - p_i r_i) u_i = 0$$

Since $f = \sum_i c_i u_i \neq 0$, we must have $(RP - PR) = [R, P] = 0$. Note that the existence of a common eigenfunction of P and R is not sufficient to conclude that $[R, P] = 0$.

Example - 2.3 : A Hermitian operator A is said to be positive–definite if, for any vector $|u>$, $< u|A|u >\geq 0$. Show that the operator $A = |a >< a|$ is Hermitian and positive–definite.

Solution : We have that

$$< u|A|v >=< u|a >< a|v >=< v|a >^* < a|u >^* =< v|A|u >^*$$

and also that

$$< u|A|u >=< u|a >< a|u >= |< u|a >|^2 \geq 0$$

Example - 2.4 : Show that a necessary and sufficient condition for two linear operators A and B to be equal (to within a phase factor), that is, for $A = Be^{i\alpha}$, is that $|< u|A|v >| = |< u|B|v >|$ should hold for any pair of linearly independent kets $|u>$ and $|v>$.

Solution : The necessity of the condition is evident. To prove its sufficiency, let us consider a representation in which $A_{ij} =< i|A|j >$ and $B_{ij} =< i|B|j >$ are the matrix elements of the operators A and B. By

$$|< u|A|v >| = |< u|B|v >| \quad (1)$$

with $|u >= |i >$ and $|v >= |j >$, we have that

$$|A_{ij}| = |B_{ij}| \quad (2)$$

for any $|i >$ and $|j >$. On the other hand, with $|u >= |i >$ and $|v >= x_j |j > +x_l |l >$, where x_j and x_l are arbitrary complex numbers. From (1) we find that

$$|A_{ij} x_j + A_{il} x_l| = |B_{ij} x_j + B_{il} x_l| \quad (3)$$

Taking into account (2), (3) can be written as

$$\text{Re}[x_j x_l^*(A_{ij}A_{il}^* - B_{ij}B_{il}^*)] = 0 \quad (4)$$

Since the complex number $x_j x_l^*$ is arbitrary, it follows from (4) that

$$A_{ij}A_{il}^* - B_{ij}B_{il}^* = 0 \quad (5)$$

From (2) and (5) we then have that

$$\frac{A_{ij}}{B_{ij}} = \frac{A_{il}}{B_{il}}$$

which means that the ratio A_{ij}/B_{ij} does not depend on j. On repeating the same argument after interchanging rows and columns, we find that the ratio A_{ij}/B_{ij} does not depend on i either. Taking into account (2), we conclude that

$$\frac{A_{ij}}{B_{ij}} = e^{i\alpha}$$

where α is a real number independent of i and of j; that is, the two operators A and B are equal to within a constant phase factor.

Example - 2.5 : Show that

$$e^A e^B = e^{A+B} e^{\frac{1}{2}[A,B]} \quad (1)$$

if $[[A, B], A] = [[A, B], B] = 0$.

Solution : Consider the operator $T(s) = e^{As}e^{Bs}$ and differentiate it with respect to s:

$$\frac{dT}{ds} = Ae^{As}e^{Bs} + e^{As}Be^{Bs} = (A + e^{As}Be^{-As})T(s)$$

Since the operators $[B, A]$ and A commute, we find, by using

$$[B, A^n] = nA^{n-1}[B, A]$$

$$[B, e^{-As}] = \sum_n (-1)^n \frac{s^n}{n!}[B, A^n]$$

$$= \sum_n (-1)^n \frac{s^n}{(n-1)!}A^{n-1}[B, A] = -e^{-As}[B, A]s$$

and hence that

$$e^{As}Be^{-As} = B - [B, A]s$$

Therefore

$$\frac{dT(s)}{ds} = (A + B + [A, B]s)T(s) \quad (2)$$

and $T(s)$ is thus the solution of this differential equation with the initial condition $T(0) = 1$. Since the operator $A + B$ and $[A, B]$ commute, equation (2) can be integrated as if they were merely numbers, to give the solution

$$T(s) = e^{(A+B)s}e^{\frac{1}{2}[A,B]s^2}$$

The identity (1) follows by putting $s = 1$.

Example - 2.6 : Show that if a unitary operator U can be written in the form $U = 1 + i\varepsilon F$, where ε is a real infinitesimally small number, then the operator F is Hermitian.

Solution : Since U is unitary, we have that

$$(1 - i\varepsilon F^+)(1 + i\varepsilon F) = (1 + i\varepsilon F)(1 - i\varepsilon F^+)$$

and, retaining only the terms to first order in ε, it follows that $F = F^+$.

Example - 2.7 : If H is an operator, u is a function of x, and E is a constant, the equation $Hu = Eu$ indicates that u is an eigenfunction of H, and E is the corresponding eigenvalue. In addition to satisfying this eigenvalue equation the function u must also be well–behaved in order to be an acceptable eigenfunction. By well–behaved we mean that u is single valued, finite, and continuous and that the first derivative of u is continuous. Which of the following functions are well–behaved? For those functions that are not well behaved, indicate the reason. (a) $u = x$, $x \geq 0$, and $u = 0$ otherwise; (b) $u = x^2$; (c) $u = e^{-|x|}$; (d) $u = e^{-x}$; (e) $u = \cos(x)$; (f) $u = \sin|x|$; (g) $u = e^{-x^2}$; (h) $u = 1 - x^2$, $-1 \leq x \leq +1$, $u = 0$ otherwise.

Solution : (a) $u = x$, $x \geq 0$, is not well behaved, u does not remain finite as $x \rightarrow \infty$ and the first derivative at $x = 0$ is not continuous.

(b) $u = x^2$ is not well behaved since u does not remane finite as $|x| \rightarrow \infty$.

(c) $u = e^{-|x|}$ is not well behaved since the first derivative is not continuous at $x = 0$.

(d) $u = e^{-x}$ is not well behaved since u does not remane finite as $x \rightarrow -\infty$.

(e) $u = \cos(x)$ ia a well–behaved function.

(f) $u = \sin|x|$ is not well behaved since the first derivative is not continuous at $x = 0$.

(g) $u = e^{-x^2}$ ia a well–behaved function.

(h) $u = 1 - x^2$, $-1 \leq x \leq +1$, is not well behaved since the first derivative is not continuous at $x = \mp$. However, u may be a good approximation to a well–behaved function which is continuous at $|x| = 1$ and very small for $|x| > 1$.

Example - 2.8 : A linear operator R has the following properties:

$$R(u + v) = Ru + Rv \quad , \quad R(cu) = cRu$$

where c is a complex number. Which of the following operators are linear? (a) $Au = \lambda u$, $\lambda = $ constant; (b) $Bu = u^*$; (c) $Cu = u^2$; (d) $Du = \frac{du}{dx}$; (e) $Eu = \frac{1}{u}$; (f) $LP = [H, P] = HP - PH$.

Solution : (a) $A(u + v) = \lambda(u + v) = Au + Av$, $A(cu) = \lambda cu = cAu$. Therefore A is a linear operator.

(b) $B(cu) = c^* Bu \neq cBu$. B is not a linear operator.

(c) $C(u + v) = (u + v)^2 = u^2 + 2uv + v^2$ but $Cu + Cv = u^2 + v^2 \neq C(u + v)$. Therefore, C is not a linear operator.

(d) $D(u + v) = \frac{du}{dx} + \frac{dv}{dx} = Du + Dv$, $D(cu) = c\frac{du}{dx} = cDu$. D is a linear operator.

(e) $E(cu) = \frac{1}{cu} = \frac{1}{c}Eu$, E is not a linear operator.

(f) $L(P+Q) = H(P+Q)-(P+Q)H = (HP-PH)+(HQ-QH) = [H,P]+[H,Q] = LP + LQ$,
$L(cP) = HcP - cPH = c(HP - PH) = cL$. Here L is an operator which acts on operators. It is a linear operator.

Example - 2.9 : It can be shown that the eigenfunctions of an Hermitian operator which belong to different eigenvalues are orthogonal. When k eigenfunctions correspond to the same eigenvalue they are said to belong to a k–fold degenerate set. Suppose that the functions ϕ_1 and ϕ_2 are degenerate eigenfunctions of H which are linearly independent and normalized but not orthogonal. (a) Show that the function $u = c_1\phi_1 + c_2\phi_2$ is an eigenfunction of H which has the same eigenvalue as ϕ_1 and ϕ_2. (b) Construct two linear combinations of ϕ_1 and ϕ_2 which are orthogonal to each other, and hence demonstrate that there is always sufficient freedom to choose degenerate functions which are orthogonal.

Solution : (a) We are given that $H\phi_1 = \lambda\phi_1$, $H\phi_2 = \lambda\phi_2$. Therefore

$$H(c_1\phi_1 + c_2\phi_2) = \lambda(c_1\phi_1 + c_2\phi_2) \quad \text{or} \quad Hu = \lambda u$$

Since linear combinations of degenerate functions give additional degenerate eigenfunctions, we see that the initial degenerate set is not unique.

(b) We are given the linearly independent set of functions $\{\phi_1, \phi_2\}$, and we wish to obtain the orthogonal set $\{u_1, u_2\}$. Let us set $u_1 = \phi_1$ and $u_2 = c_1\phi_1 + c_2\phi_2$. Now we find the constants c_1 and c_2 so that: $< u_2|u_1 >= 0$ and $< u_2|u_2 >= 1$. Expanding these integrals we find:

$$< u_1|u_2 >=< \phi_1|c_1\phi_1 + c_2\phi_2 >= c_1 < \phi_1|\phi_1 > +c_2 < \phi_1|\phi_2 >= 0$$

Since $< \phi_1|\phi_1 >= 1$, $c_1 = -c_2 < \phi_1|\phi_2 >$. Therefore,

$$< u_2|u_2 >=< c_1\phi_1 + c_2\phi_2|c_1\phi_1 + c_2\phi_2 >= 1$$

$$= c_1^2 + c_2^2 + c_1c_2 < \phi_1|\phi_2 > +c_2c_1 < \phi_2|\phi_1 >$$

Substituting c_1 in $< u_2|u_2 >$ we get

$$1 = c_2^2 < \phi_1|\phi_2 >^2 +c_2^2 - c_2^2 < \phi_1|\phi_2 >^2 -c_2^2 < \phi_2|\phi_1 >< \phi_1|\phi_2 >$$

$$1 = c_2^2(1 - | < \phi_1|\phi_2 > |^2) \quad \text{or} \quad c_2 = (1 - | < \phi_1|\phi_2 > |^2)^{-1/2}$$

Here we used $< \phi_1|\phi_2 >^* = < \phi_2|\phi_1 >$ and $< \phi_1|\phi_2 >< \phi_1|\phi_2 >^* = | < \phi_1|\phi_2 > |^2$. Therefore, the orthogonal and normalized set (orthonormal set) is:

$$u_1 = \phi_1$$

$$u_2 = (\phi_2 - < u_1|\phi_2 > u_1)/(1 - | < u_1|\phi_2 > |^2)^{1/2}$$

The quantity $< \phi_2|u_1 >$ may be interpreted as the projection of the function ϕ_2 on the function u_1. This procedure (Schmidt orthogonalization) can be continued for a set of k functions to give the general term:

$$u_n = \left(\phi_n - \sum_{j=1}^{n-1} < u_j|\phi_n > u_j \right) / \left(1 - \sum_{j=1}^{n-1} | < u_j|\phi_n > |^2 \right)^{1/2}$$

2.7   Problems

Problem - 2.1 : Given three operators A, B, and C, express the commutator $[AB, C]$ in terms of the commutators $[A, C]$ and $[B, C]$.

Problem - 2.2 : A and B are operators which commute, and $u_{A'}$ is an eigenfunction of A which has the eigenvalue A'. (a) Show that $u_{A'}$ is also an eigenfunction of B in the case that $u_{A'}$ is nondegenerate. (b) If $u_{A'_i}$ $(i = 1, \cdots, n)$ are linearly independent functions having the same eigenvalue A', show that linear combinations of these functions can be chosen which are simultaneously eigenfunctions of B.

Problem - 2.3 : Show that if A and B are two operators satisfying the relation $[[A, B], A] = 0$, then the relation $[A^m, B] = mA^{m-1}[A, B]$ holds for all positive integers m.

Problem - 2.4 : Show that, for any two operators A and L,

$$e^L A e^{-L} = A + [L, A] + \frac{1}{2!}[L, [L, A]] + \frac{1}{3!}[L, [L, [L, A]]] + \cdots$$

Problem - 2.5 : Verify Kubo's identity

$$[A, e^{-\beta H}] = e^{-\beta H} \int_0^\beta e^{\lambda H}[A, H]e^{-\lambda H} d\lambda$$

where A and H are any two operators.

Problem - 2.6 : The translation operator $\Omega(a)$ is defined to be such that $\Omega(a)\phi(x) = \phi(x + a)$. Show that (a) $\Omega(a)$ may be expressed in terms of the operator $p = -i\hbar\frac{d}{dx}$, (b) $\Omega(a)$ is unitary.

Problem - 2.7 : Consider the operator $R = -\frac{d^2}{dx^2}$ and the eigenvalue equation $Ru = \lambda u$. Write out the possible eigenfunctions and discuss the conditions under which they are well behaved.

Problem - 2.8 : Which of the following pairs of operators can have simultaneous eigenfunctions? (a) p and T, (b) p and $V(x)$, (c) H and p, where p is the momentum, T and $V(x)$ are the kinetic and potential energies, respectively, and $H = T + V(x)$.

Chapter 3

Observables and Superposition

3.1 Free particle

The operator that corresponds to the observable linear momentum is

$$\hat{p} = -i\hbar\nabla \tag{3.1}$$

To find the eigenfunctions and eigenvalues of the momentum operator let us consider a free particle which is constrained to move in one dimension (x). Then the momentum has only one nonvanishing component, p_x. The corresponding operator is

$$\hat{p}_x = -i\hbar\frac{\partial}{\partial x} \tag{3.2}$$

The eigenvalue equation for this operator is

$$-i\hbar\frac{\partial}{\partial x}\psi = p_x\psi \tag{3.3}$$

The values p_x represent the possible values that measurement of the x component of momentum will yield. The eigenfunction $\psi(x)$ corresponding to a specific value of momentum (p_x) is such that $|\psi|^2 dx$ is the probability of finding the particle (with momentum p_x) in the interval x, $x + dx$. Since the particle we consider is a free particle, it is unconfined (along the x–axis). For this case there is no boundary condition on ψ and the solution to the equation

$$\frac{\partial\psi}{\partial x} = \frac{ip_x}{\hbar}\psi = ik\psi \tag{3.4}$$

where

$$k = \frac{p}{\hbar} \tag{3.5}$$

k is known as the wave number, it is a continuous quantity.

$$\psi = Ae^{ikx} \tag{3.6}$$

The eigenfunction e^{ikx} is a periodic function (in x). To find its wavelength λ, we set

$$e^{ikx} = e^{ik(x+\lambda)} \tag{3.7}$$

$$1 = e^{ik\lambda} = \cos k\lambda + i\sin k\lambda \tag{3.8}$$

which is satisfied if

$$\cos k\lambda = 1 \quad , \quad \sin k\lambda = 0 \tag{3.9}$$

The first nonvanishing solution to these equations is

$$k\lambda = 2\pi \quad , \quad (\lambda = \frac{2\pi}{k}) \tag{3.10}$$

which is equivalent to the de Broglie relation

$$k = \frac{p}{\hbar} = \frac{p}{h}2\pi \quad \rightarrow \quad p = \frac{kh}{2\pi} = \frac{h}{\lambda} \tag{3.11}$$

We conclude that the eigenfunction of the momentum operator corresponding to the eigenvalue p has a wavelength that is the de Broglie wavelength $\frac{h}{p}$. In quantum mechanics it is convenient to speak in terms of wavenumber k instead of momentum p. In this notation we say that the eigenfunctions and eigenvalues of the momentum operator are

$$\psi_k = Ae^{ikx} \quad , \quad p = \hbar k \tag{3.12}$$

The subscript k on ψ_k denotes that there is a continuum of eigenfunction and eigenvalues, $\hbar k$, which yield nontrivial solutions to the eigenvalue equation

$$-i\hbar\frac{\partial\psi}{\partial x} = p\psi \tag{3.13}$$

The Hamiltonian operator for a free particle with mass m, whose energy is purely kinetic, is

$$H = \frac{p^2}{2m} = -\frac{\hbar^2}{2m}\nabla^2 \tag{3.14}$$

Constraining the particle to move in one dimension, the time–independent Schrödinger equation becomes

$$-\frac{\hbar^2}{2m}\frac{\partial^2}{\partial x^2}\psi = E\psi \tag{3.15}$$

It yields the possible energies E which the particle may have. In terms of the wave vector, k

$$k^2 = \frac{2mE}{\hbar^2} \tag{3.16}$$

The Schrödinger equation takes the form

$$\frac{\partial^2\psi}{\partial x^2} + k^2\psi = 0 \tag{3.17}$$

For a free particle there are no boundary conditions and we may write the solution to this wave equation as

$$\psi = Ae^{ikx} + Be^{-ikx} \tag{3.18}$$

This is the eigenfunction of H which corresponds to the energy eigenvalue

$$E = \frac{\hbar^2 k^2}{2m} \tag{3.19}$$

We have found above that $p = \hbar k$. This is the same $\hbar k$ that appears in $E = \frac{\hbar^2 k^2}{2m}$, since for a free particle

$$E = \frac{p^2}{2m} = \frac{\hbar^2 k^2}{2m} \tag{3.20}$$

Note that the eigenfunction of H ($\psi = Ae^{ikx} + Be^{-ikx}$), with $B = 0$, is also an eigenfunction of p ($\psi = Ae^{ikx}$). That H and p for a free particle have common eigenfunctions is a special case of a more general theorem discussed before (in Chapter 2). In other words, H and p commute, $[H, p] = 0$. The following argument demonstrates this fact. Let

$$p\psi = \hbar k\psi \tag{3.21}$$

Let us see if ψ is also an eigenfunction of H (for a free particle).

$$H\psi = \frac{p}{2m}(p\psi) = \frac{p(\hbar k\psi)}{2m} = \frac{\hbar k}{2m}p\psi = \frac{(\hbar k)^2}{2m}\psi \tag{3.22}$$

It follows that ψ is also an eigenfunction of H. Both the energy and momentum eigenvalues for the free particle comprise a continuum of values:

$$E = \frac{\hbar^2 k^2}{2m} \quad , \quad p = \hbar k \tag{3.23}$$

That is, these are valid eigenvalues for any wavenumber k. The eigenfunction (of both H and p) corresponding to these eigenvalues is

$$\psi_k = Ae^{ikx} \tag{3.24}$$

If the free particle is in this state, measurement of its momentum will definitely yield $\hbar k$, and measurement of its energy will definitely yield $\frac{\hbar^2 k^2}{2m}$. If the particle is in the state ψ_k, the probability density relating to the probability of finding the particle in the interval x, $x + dx$, is

$$|\psi_k|^2 = |A|^2 = \text{constant} \tag{3.25}$$

The probability density is the same constant value for all x. That means we would be equally likely to find the particle at any point from $x = -\infty$ to $x = +\infty$. This is a statement of maximum uncertainty which is in agreement with the Heisenberg uncertainty principle. In the state ψ_k, it is known with absolute certainty that measurement of momentum yields $\hbar k$. Therefore, for the state ψ_k, $\Delta p = 0$, whence $\Delta x = \infty$.

3.2   Particle in a box

Let us now consider a point mass m, constrained to move on an infinitely thin, frictionless wire which is strung tightly between two impenetrable walls a distance a apart. This model system can also be described as the motion of a particle in a one–dimensional infinitely deep well, in which the particle does not interact with the walls of the well. The corresponding potential has the form

$$V(x) = \infty \quad , \quad x \leq 0 \ , \ x \geq a \quad \text{(region 1)} \tag{3.26}$$

$$V(x) = 0 \quad , \quad 0 < x < a \quad \text{(region 2)}$$

and is shown in Figure 3.1. This configuration is known as the **one–dimensional box.**

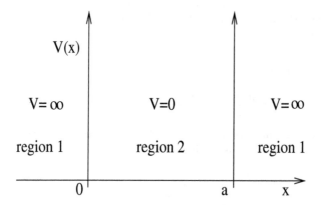

Fig. 3.1. Potential corresponding to the one–dimensional box.

The Hamiltonian for this problem is the following operator:

$$H = \frac{p^2}{2m} + \infty = \infty \quad \text{(region 1)} \tag{3.27}$$

$$H = \frac{p^2}{2m} \quad \text{(region 2)}$$

In region 1 the time–independent Schrödinger equation gives $\psi = 0$. The fact that $\psi = 0$ in region 1 implies that there is zero probability that the particle is found there ($|\psi|^2 = 0$). In region 2 the time–independent Schrödinger equation is

$$-\frac{\hbar^2}{2m}\frac{\partial^2}{\partial x^2}\psi = E\psi \tag{3.28}$$

Since ψ is a continuous function, it must have the values (boundary conditions)

$$\psi(0) = \psi(a) = 0 \tag{3.29}$$

Let us rewrite the Schrödinger equation in the following form

$$\frac{\partial^2 \psi}{\partial x^2} + k^2 \psi = 0 \quad , \quad k^2 = \frac{2mE}{\hbar^2} \tag{3.30}$$

The solution to this wave equation appears as

$$\psi = A\sin kx + B\cos kx \tag{3.31}$$

The boundary conditions, $\psi(0) = \psi(a) = 0$, give

$$B = 0 \quad , \quad A\sin ka = 0 \tag{3.32}$$

From the second of these equations we may write

$$ka = n\pi \quad , \quad n = 0, 1, 2, \cdots \tag{3.33}$$

This is seen to be equivalent to the requirement that an integral number of half–wavelength, $n(\lambda/2)$, fits into the width a. The spectrum of eigenvalues and eigenfunctions is discrete. Therefore we label the eigenvalues as E_n and the eigenfunctions as ψ_n. The eigenvalues for the one–dimensional box problem are

$$E_n = \frac{\hbar^2 k^2}{2m} = \frac{\hbar^2}{2m}\left(\frac{n\pi}{a}\right)^2 = \frac{n^2\hbar^2\pi^2}{2ma^2} = n^2\left(\frac{\hbar^2\pi^2}{2ma^2}\right) = n^2 E_1 \qquad (3.34)$$

The energy eigenvalues are positive. The lowest energy, the ground state energy, is

$$E_1 = \frac{\hbar^2\pi^2}{2ma^2} \qquad (3.35)$$

The eigenfunctions are simply expressed as

$$\psi_n = A\sin k_n x = A\sin\left(\frac{n\pi x}{a}\right) \qquad (3.36)$$

To find the constant A in the wave function, we normalize ψ_n.

$$\int_0^a |\psi_n|^2\,dx = A^2\int_0^a \sin^2\left(\frac{n\pi x}{a}\right)\,dx \qquad (3.37)$$

$$= A^2\left(\frac{a}{n\pi}\right)\int_0^{n\pi} \sin^2\theta\,d\theta = A^2\frac{a}{2}$$

One can find that $A = \sqrt{\frac{2}{a}}$. It follows that the normalized eigenfunctions for the one–dimensional box problem are

$$\psi_n = \sqrt{\frac{2}{a}}\sin\left(\frac{n\pi x}{a}\right) \qquad (3.38)$$

These eigenfunctions form an orthonormal basis with an infinite number of elements.

$$\int_0^a \psi_m(x)^*\psi_n(x)\,dx = \delta_{mn} \qquad (3.39)$$

The relative positions of the eigenenergies, and the corresponding eigenfunctions and probability densities for a particle in a one–dimensional box, are shown in Figure 3.2.

The eigenstate corresponding to $n = 0$ is $\psi = 0$. This, together with the solution in region 1, gives $\psi = 0$ over the whole x–axis. There is zero probability of finding the particle anywhere. This is equivalent to the statement that the particle does not exist in the $n = 0$ state.

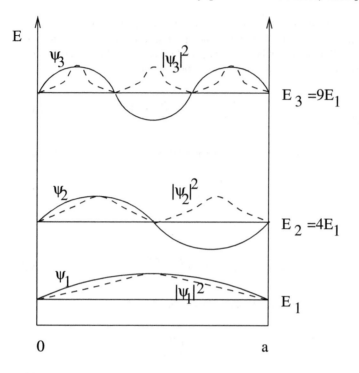

Fig. 3.2. Eigenenergies, E_n, eigenfunctions, $\psi_n(x)$, and probability densities, $|\psi_n(x)|^2$, for the one–dimensional box problem.

Another argument that disallows the $n = 0$ state follows from the uncertainty principle. The energy corresponding to $n = 0$ is $E = 0$. Since the energy in region 2 is entirely kinetic, this, in turn, implies that the particle is in a state of absolute rest ($\Delta p = 0$), a prohibited state of affairs for a particle constrained to move in a finite region.

If we choose the boundries of the well as follows (see Figure 3.3.),

$$V(x) = 0 \quad , \quad |x| < \frac{a}{2} \tag{3.40}$$

$$V(x) = \infty \quad , \quad |x| \geq \frac{a}{2}$$

then we get two different sets of eigen solutions. The boundary conditions now take the form

$$\psi(-\frac{a}{2}) = \psi(+\frac{a}{2}) = 0 \tag{3.41}$$

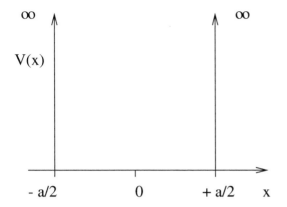

Fig. 3.3. One–dimensional infinite potential well.

There is no solution outside the box as before. Inside the box we have the same wave equation

$$-\frac{\hbar^2}{2m}\frac{\partial^2\psi}{\partial x^2} = E\psi \quad \rightarrow \quad \frac{\partial^2\psi}{\partial x^2} + \frac{2mE}{\hbar^2}\psi = 0 \tag{3.42}$$

Taking again

$$k^2 = \frac{2mE}{\hbar^2} \tag{3.43}$$

the wave equation takes the form

$$\frac{\partial^2\psi}{\partial x^2} + k^2\psi = 0 \tag{3.44}$$

whose solutions are again $\sin kx$ and $\cos kx$,

$$\psi(x) = A\sin kx + B\cos kx \tag{3.45}$$

However, the boundry conditions imply that for the sine solution, denoted by $\psi^{(-)}(x)$, which is an odd function,

$$\psi^{(-)}(x) = A\sin kx \quad , \quad B = 0 \tag{3.46}$$

$$ka = n\pi \quad , \quad n = 1,2,3,\cdots \tag{3.47}$$

so that the eigenvalues take the form

$$E_n^{(-)} = \frac{n^2\pi^2\hbar^2}{2m(a/2)^2} = \frac{(2n)^2\pi^2\hbar^2}{2ma^2} \tag{3.48}$$

The corresponding normalized eigenfunctions are

$$\psi_n^{(-)} = \sqrt{\frac{2}{a}}\sin\left(\frac{2n\pi x}{a}\right) \tag{3.49}$$

The cosine solution, denoted by $\psi^{(+)}(x)$, which is an even function,

$$\psi^{(+)}(x) = B\cos kx \quad , \quad A = 0 \tag{3.50}$$

$$ka = (n + \frac{1}{2})\pi \quad , \quad n = 0, 1, 2, 3, \cdots \tag{3.51}$$

so that the eigenvalues take the form

$$E_n^{(+)} = \frac{(n + \frac{1}{2})^2 \pi^2 \hbar^2}{2m(a/2)^2} = \frac{(2n + 1)^2 \pi^2 \hbar^2}{2ma^2} \tag{3.52}$$

The corresponding normalized eigenfunctions are

$$\psi_n^{(+)}(x) = \sqrt{\frac{2}{a}} \cos\left(\frac{(2n + 1)\pi x}{a}\right) \tag{3.53}$$

The ($\pm$) signs in the eigen solutions refer to the even/odd property under the reflection $x \rightarrow -x$.

The eigen solutions have the property that

$$\int_{-\frac{a}{2}}^{+\frac{a}{2}} \psi_m^{(+)}(x)\psi_n^{(+)}(x)dx = \int_{-\frac{a}{2}}^{+\frac{a}{2}} \psi_m^{(-)}(x)\psi_n^{(-)}(x)dx = \delta_{mn} \tag{3.54}$$

$$\int_{-\frac{a}{2}}^{+\frac{a}{2}} \psi_m^{(+)}(x)\psi_n^{(-)}(x)dx = 0 \tag{3.55}$$

that is, they satisfy the orthonormality conditions. The discrete energy spectrum is the same in both cases.

3.3   Ensemble average

Consider a particle in a one–dimensional box. Imagine a large number of identical replicates of the system, which form an ensemble. If each such box is in the same initial state $\psi(x, 0)$, after an interval of time t, each box will again be in a common state $\psi(x, t)$; see Figure 3.4.

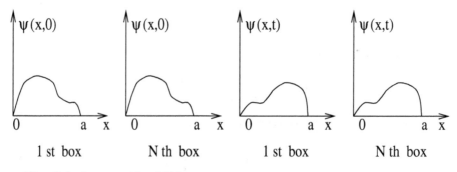

Fig. 3.4. An ensemble of N boxes.

The measured energy of the particle in each identical box at time t may not be the same.

If the probability of finding the value E_n in a given measurement of energy is $P(E_n)$, then the average over measurements of all members of the ensemble in the limit as this number becomes large is given by the expression

$$< E >= \sum_{all\ E_n} P(E_n)E_n \tag{3.56}$$

We can generalize this expression for all physical observables.

$$< A >= \sum_n P(a_n)a_n \tag{3.57}$$

For example, the average particle position in a one–dimensional box of length a is given by

$$< x >= \int_0^a P(x)x dx \tag{3.58}$$

In this case the integral is a sum over the continuum of values x may assume.

The average of a dynamical observable $\hat{A}$ in a certain state ψ is given by the expectation value

$$< \hat{A} >= \int_\tau \psi^* \hat{A}\psi d\tau =< \psi|\hat{A}|\psi > \tag{3.59}$$

Specifically, for the energy we have

$$< E >=< \psi|\hat{H}|\psi > \tag{3.60}$$

The state ψ may be expanded in terms of the eigenstates of $\hat{H}$, which obey the eigenvalue equation

$$H\phi_n = E_n\phi_n \tag{3.61}$$

For a particle in a box

$$\phi_n = \sqrt{\frac{2}{a}} \sin\left(\frac{n\pi x}{a}\right) \tag{3.62}$$

The expansion of ψ in these eigenstates appears as

$$\psi(x,t) = \sum_{n=1}^\infty b_n(t)\phi_n(x) \quad \text{or} \quad |\psi >= \sum_{n=1}^\infty b_n|\phi_n > \tag{3.63}$$

The state ψ is that of the system at the time t, so that it is, in general, a function of x and t. Since ϕ_n is a function of x only, the coefficients of expansion b_n may, in general, be functions of time.

Expectation value of energy is calculated as

$$< E >=< \psi|\hat{H}|\psi >=< \sum_m b_m\phi_m|\hat{H}| \sum_n b_n\phi_n > \tag{3.64}$$

$$= \sum_m \sum_n b_m^* b_n < \phi_m|\hat{H}|\phi_n >= \sum_m \sum_n b_m^* b_n E_n < \phi_m|\phi_n >$$

$$= \sum_m \sum_n b_m^* b_n E_n \delta_{mn} = \sum_{n=1}^\infty |b_n|^2 E_n$$

Comparing this equation with

$$< E >= \sum_n P(E_n) E_n \qquad (3.65)$$

we can write

$$\sum_n |b_n|^2 E_n = \sum_n P(E_n) E_n \qquad (3.66)$$

and

$$P(E_n) = |b_n|^2 \qquad (3.67)$$

The square of the modulus of b_n, $|b_n|^2$, is the probability that, at the time t, measurement of the energy of the particle which is in the state $\psi(x,t)$ yields the value E_n. If the states ψ and ϕ_n are normalized, we have

$$< \psi|\psi >=< \sum_m b_m \phi_m | \sum_n b_n \phi_n >= \sum_m \sum_n b_m^* b_n < \phi_m|\phi_n > \qquad (3.68)$$

$$= \sum_m \sum_n b_m^* b_n \delta_{mn} = \sum_n |b_n|^2 = 1$$

In this case the coefficient $|b_n|^2$ is an absolute probability. If the states ψ and ϕ_n are not normalized, the correct expression for the probability that measurement finds E_n is

$$P(E_n) = \frac{|b_n|^2| < \phi_n|\phi_n > |^2}{\sum_n |b_n|^2| < \phi_n|\phi_n > |^2} \qquad (3.69)$$

The coefficients b_n are calculated as follows:

$$|\psi >= \sum_n b_n|\phi_n > \qquad (3.70)$$

$$< \phi_m||\psi >= \sum_n b_n < \phi_m|\phi_n >= \sum_n b_n \delta_{mn} = b_m \qquad (3.71)$$

The coefficient b_m is the projection of ψ onto ϕ_m. The physical interpretation of b_n is that $|b_n|^2$ is the probability that measuring E finds the value E_n when the system is in the state ψ. This prescription is true for any dynamical observable.

3.4   Hilbert–space interpretation

When we look in Hilbert space, $\{\phi_n\}$ is one set of states and ψ is another state. The system is in the state ψ. Measurement of an observable A causes the state ψ to fall to one of the ϕ_n states.

Consider a particle of mass m is in a one–dimensional box of width a. At $t = 0$ the particle is in the state

$$\psi(x,0) = \frac{3\phi_2 + 4\phi_9}{\sqrt{25}} \quad or \quad |\psi >= \frac{3|\phi_2 > +4|\phi_9 >}{\sqrt{25}} \qquad (3.72)$$

The ϕ_n functions are the orthonormal eigenstates of $\hat{H}$:

$$\phi_n = \sqrt{\frac{2}{a}} \sin\left(\frac{2\pi x}{a}\right) \tag{3.73}$$

Let us check the normalization of ψ:

$$< \psi|\psi >= \frac{1}{25}(3 < \phi_2| + 4 < \phi_9|)(3|\phi_2 > +4|\phi_9 >) \tag{3.74}$$

$$= \frac{1}{25}(9 < \phi_2|\phi_2 > +12 < \phi_2|\phi_9 > +12 < \phi_9|\phi_2 > +16 < \phi_9|\phi_9 >)$$

$$= \frac{1}{25}(9 \cdot 1 + 12 \cdot 0 + 12 \cdot 0 + 16 \cdot 1) = \frac{1}{25}(9 + 16) = \frac{25}{25} = 1$$

and ψ is normalized. According to the superposition principle we can write

$$\psi = \sum_n b_n\phi_n = \frac{3\phi_2 + 4\phi_9}{\sqrt{25}} \tag{3.75}$$

From this equality we find that

$$b_2 = \frac{3}{\sqrt{25}} \quad , \quad b_9 = \frac{4}{\sqrt{25}} \tag{3.76}$$

and $b_n = 0$ for $n \neq 2$ or 9. Therefore, the probability $P(E_n)$ that measurement of E at $t = 0$ finds the value E_n is

$$P(E_2) = |b_2|^2 = \frac{9}{25} \quad , \quad P(E_9) = |b_9|^2 = \frac{16}{25} \tag{3.77}$$

and $P(E_n) = 0$ for $n \neq 2$ or 9. Although the state $\psi(x,0)$ is a precise superposition of well–defined eigenstates of the observables being measured, one is not certain what measurement will yield.

There is nothing in classical mechanics that is similar to this concept. Any uncertainty in classical mechanics arises from uncertain initial data. On the other hand, in quantum mechanics, although the initial state $\psi(x,0)$ is prescribed with perfect accuracy, one is never certain in which eigenstate, ϕ_n, measurement will leave the system. However, once E is measured and, say, the value E_9 is found, then one knows with absolute certainty that the state of the system immediately after this measurement is ϕ_9.

3.5   The initial square wave

Let us consider a free particle in one–dimension. Suppose that at $t = 0$ the system is in the state (see Figure 3.5),

$$\psi(x,0) = \frac{1}{\sqrt{a}} \quad \text{for} \quad |x| < \frac{a}{2} \tag{3.78}$$

$$= 0 \quad \text{elsewhere}$$

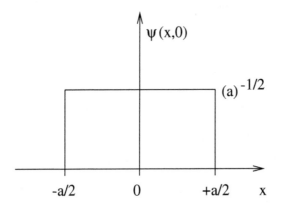

Fig. 3.5. A square wave in one–dimension.

If the momentum of the particle is measured, what are the possible values that will be found, and with what probability will these values occur? To answer these questions we must first expand $\psi(x,0)$ in a superposition of the eigenstates of $\hat{p}$:

$$\phi_k = \frac{1}{\sqrt{2\pi}} e^{ikx} \tag{3.79}$$

$$\psi(x,0) = \int_{-\infty}^{\infty} b(k)\phi_k dk \tag{3.80}$$

The coefficients can be calculated as

$$b(k) = \int_{-\infty}^{\infty} \phi_k^* \psi(x,0) dx = \frac{1}{\sqrt{2\pi}} \int_{-\infty}^{\infty} e^{-ikx} \psi(x,0) dx \tag{3.81}$$

$$= \frac{1}{\sqrt{2\pi}} \int_{-\infty}^{\infty} e^{-ikx} \frac{dx}{\sqrt{a}} = \frac{1}{\sqrt{2\pi a}} \int_{-\frac{a}{2}}^{+\frac{a}{2}} e^{-ikx} dx$$

$$= \frac{1}{\sqrt{2\pi a}} \frac{2}{k} \left(\frac{e^{ika/2} - e^{-ika/2}}{2i} \right) = \sqrt{\frac{2}{\pi a}} \frac{\sin(ka/2)}{k}$$

This coefficient is the projection of the state $\psi(x,0)$ onto the eigenstate ϕ_k.

$|b(k)|^2 dk$ is the probability that measurement of momentum yields $p = \hbar k$ in the interval $\hbar k$, $\hbar(k+dk)$. The corresponding probability density, in momentum space, is (see Figure 3.6.)

$$|b|^2 = \frac{2}{\pi a} \frac{\sin^2(ka/2)}{k^2} \tag{3.82}$$

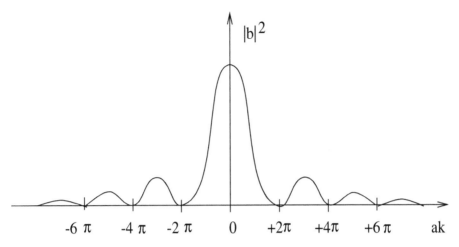

Fig. 3.6. Probability density of a square wave in momentum space.

This function has its maximum at $k = 0$; it drops to zero at $\frac{ka}{2} = \pi$ or equivalently at $p = \hbar k = \frac{2\pi\hbar}{a}$. The momentum values $\mp n\frac{2\pi\hbar}{a}$ with n an integer and $n > 1$ are never found, for at these values $b(k) = 0$.

The interval of momentum values that measurements are most likely to uncover has the approximate width $\Delta k = \frac{4\pi}{a}$ or $\Delta p = \hbar\delta k = \frac{4\pi\hbar}{a}$. On the other hand, it is uniformly probable that measurement of x finds the particle anywhere in the interval $(-\frac{a}{2}, +\frac{a}{2})$, of width $\Delta x = a$. Combining these uncertainties one gets

$$\Delta x \Delta p = a\frac{4\pi\hbar}{a} = 4\pi\hbar \approx \hbar \tag{3.83}$$

This shows that the Heisenberg uncertainty principle works.

3.6   Particle beam

Let us suppose that the free–particle system in the previous example is composed of N noninteracting particles. Every particle is in the state $\psi(x,0)$ given by

$$\psi(x,0) = \frac{1}{\sqrt{a}} \quad \text{for} \quad |x| < \frac{a}{2} \tag{3.84}$$

$$= 0 \quad \text{elsewhere}$$

The density ρ (number of particles per unit length) is related to ψ as

$$\text{number of particles in } dx = \rho dx = N|\psi|^2 dx \tag{3.85}$$

The total number in the whole system is

$$N = \int_{-\infty}^{\infty} \rho(x) dx = N \int_{-\frac{a}{2}}^{+\frac{a}{2}} |\psi|^2 dx = N \tag{3.86}$$

Suppose that we now ask how many particles have momentum in the interval $(-\frac{2\pi\hbar}{a}, +\frac{2\pi\hbar}{a})$, or, equivalently, how many have wavenumber in the interval $(-\frac{2\pi}{a}, +\frac{2\pi}{a})$. For a single particle, the probability of finding a particle with momentum in the interval $\hbar k$ to $\hbar(k+dk)$ is

$$P(k)dk = |b(k)|^2 dk \tag{3.87}$$

provided that

$$\int_{-\infty}^{\infty} |b(k)|^2 dk = 1 \tag{3.88}$$

The number of particles that have momentum in the interval $\hbar k$, $\hbar(k+dk)$ is

$$\rho(k)dk = N|b(k)|^2 dk \tag{3.89}$$

The total number of particles in the whole system is

$$N = \int_{-\infty}^{\infty} \rho(k)dk = N \int_{-\infty}^{\infty} |b(k)|^2 dk = N \tag{3.90}$$

For the example at hand

$$\int_{-\infty}^{\infty} |b(k)|^2 dk = \frac{2}{\pi a} \int_{-\infty}^{\infty} \frac{\sin^2(ka/2)}{k^2} dk \tag{3.91}$$

$$= \frac{1}{\pi} \int_{-\infty}^{\infty} \frac{\sin^2 \eta}{\eta^2} d\eta = 1$$

The number of particles ΔN in the system with momentum in the interval $(-\frac{2\pi\hbar}{a}, +\frac{2\pi\hbar}{a})$ is given by the integral

$$\Delta N = N \int_{-\frac{2\pi}{a}}^{+\frac{2\pi}{a}} \frac{2}{\pi a} \frac{\sin^2(ka/2)}{k^2} dk = \frac{N}{\pi} \int_{-\pi}^{+\pi} \frac{\sin^2 \eta}{\eta^2} d\eta = 0.903N \tag{3.92}$$

Almost 90% of the particles are in this momentum interval.

3.7   Superposition and uncertainty

Let us consider a single free particle in the state $\psi(x,0)$ given by

$$\psi(x,0) = \frac{1}{\sqrt{a}} \quad \text{for} \quad |x| < \frac{a}{2} \tag{3.93}$$

$$= 0 \quad \text{elsewhere}$$

Suppose at $t = 0$ we measure the momentum of the particle. We know from previous examples that the values $p = \mp n\frac{2\pi\hbar}{a}$ with $n > 1$ are never found; any other value may occur with corresponding probability density $|b(k)|^2$. Let us assume that the measurement finds the particle to have the momentum

$$p = \frac{\pi\hbar}{a} \tag{3.94}$$

The state of the particle after the measurement will be

$$\psi = \frac{1}{\sqrt{2\pi}} e^{\frac{i\pi x}{a}} \tag{3.95}$$

The corresponding energy is measured as

$$E = \frac{(\pi\hbar/a)^2}{2m} \tag{3.96}$$

Suppose that we now measure the position of the particle. The probability density is

$$P = |\psi|^2 = \frac{1}{2\pi} \tag{3.97}$$

which is a constant. It is uniformly probable to find the particle anywhere along the whole x axis. The uncertainty in x is $\Delta x = \infty$. For this same state it is certain that measurement of momentum finds the value $\frac{\pi\hbar}{a}$, so that $\Delta p = 0$. This example also agrees with the Heisenberg uncertainty principle; see Figure 3.7. In the state

$$\psi = \phi_{k_0} = \frac{1}{\sqrt{2\pi}} e^{ik_0 x} \tag{3.98}$$

$$\Delta p = 0 \quad , \quad \Delta x = \infty \tag{3.99}$$

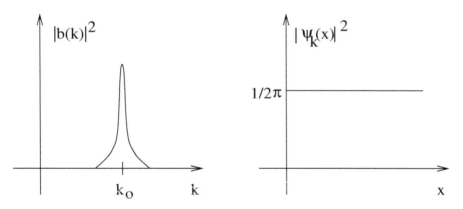

Fig. 3.7. Uncertainty in momentum and position of a particle in the state $\psi_k(x) = \frac{1}{\sqrt{2\pi}} e^{ik_0 x}$.

Let us assume that we place a uniform array of detectors along the x axis. One of them is placed at $x = x'$. The state of the particle after measuring its position will be the eigenstate of the position operator corresponding to the eigenvalue x'.

$$\psi = \delta(x - x') \tag{3.100}$$

Now we measure momentum again. To be able to calculate this we expand ψ in the eigenstates of $\hat{p}$,

$$\psi = \int b(K)\phi_k dk \quad , \quad \phi_k = \frac{1}{\sqrt{2\pi}} e^{ikx} \tag{3.101}$$

$$b(k) = \int \phi_k^* \psi dx = \frac{1}{\sqrt{2\pi}} \int e^{-ikx} \delta(x - x') dx = \frac{1}{\sqrt{2\pi}} e^{-ikx'} \tag{3.102}$$

The corresponding momentum probability density is

$$P(k) = |b(k)|^2 = \frac{1}{2\pi} \tag{3.103}$$

It is uniformly probable to find the particle with any momentum along the whole k axis. The uncertainty in momentum is $\Delta p = \infty$ for the state $\psi = \delta(x - x')$, for which $\Delta x = 0$; this also agrees with the uncertainty principle (see Figure 3.8). In the state

$$\psi = \delta(x - x') \tag{3.104}$$

$$\Delta x = 0 \quad , \quad \Delta p = \infty \tag{3.105}$$

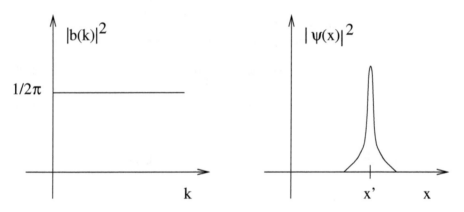

Fig. 3.8. Uncertainty in momentum and position of a particle in the state $\psi(x) = \delta(x - x')$.

We may state the superposition principle as the following: The superposition principle requires us to assume that between states there exist peculiar relationships such that wherever the system is definitely in one state we can consider it as being partly in each of two or more other states. The original state must be regarded as the result of a kind of superposition of the two or more new states.

3.8   Degeneracy of states

Suppose there are two linearly independent eigenfunctions (ψ_1 and ψ_2) of the operator A which both correspond to the eigenvalue a.

$$A\psi_1 = a\psi_1 \quad , \quad A\psi_2 = a\psi_2 \tag{3.106}$$

Under such circumstances we say that the eigenvalue a is doubly degenerate. The eigenfunctions ψ_1 and ψ_2 are degenerate. The most general eigenfunction of A

corresponding to the eigenvalue a may be expressed as

$$\psi_a = \alpha\psi_1 + \beta\psi_2 \tag{3.107}$$

with α and β arbitrary constants.

$$A\psi_a = A(\alpha\psi_1 + \beta\psi_2) = \alpha a\psi_1 + \beta a\psi_2 = a(\alpha\psi_1 + \beta\psi_2) \tag{3.108}$$

In Hilbert space the two functions ψ_1 and ψ_2 span a plane (two–dimensional subspace). Equation $\psi_a = \alpha\psi_1 + \beta\psi_2$ indicates that any vector ψ_a in this plane is an eigenfunction of A corresponding to the eigenvalue a.

If two operators A and B commute and $A\psi_1 = a\psi_1$,

$$BA\psi_1 = B(a\psi_1) = aB\psi_1 \quad\Longrightarrow\quad A(B\psi_1) = a(B\psi_1) \tag{3.109}$$

$B\psi_1$ is an eigenstate of A with eigenvalue a.

There are some α and β such that

$$B\psi_1 = b(\alpha\psi_1 + \beta\psi_2) \tag{3.110}$$

here ψ_1 need not be an eigenfunction of B. So we have the following rule: If $[A, B] = 0$, and a is a degenerate eigenvalue of A, the corresponding eigenfunctions of A are not necessarily eigenfunctions of B.

Degenerate operators have more eigenstates than nondegenerate operators. Consider a free particle in one dimension again. The eigenvalue $E_k = \frac{\hbar^2 k^2}{2m}$ of the Hamiltonian $H = \frac{p^2}{2m}$ is doubly degenerate. All of the following functions are eigenfunctions of H corresponding to this eigenvalue

$$(\psi_1, \psi_2, \psi_3) = \{\cos kx, \sin kx, e^{ikx}\} \tag{3.111}$$

This is not a linearly independent set. However, any two are, so that for the free particle the eigenvalue $E_k = \frac{\hbar^2 k^2}{2m}$ is doubly degenerate.

The two linearly independent functions, say,

$$(\psi_1, \psi_2) = \{\cos kx, \sin kx\} \tag{3.112}$$

both have the eigenvalue $E_k = \frac{\hbar^2 k^2}{2m}$. Although $[p, H] = 0$, for the free particle, the set of functions $\{\cos kx, \sin kx\}$, being degenerate eigenfunctions of energy, need not be eigenfunctions of p; in fact, they are not.

Another linearly independent set of degenerate eigenstates corresponding to the eigenenergies $\frac{\hbar^2 k^2}{2m}$ is (ψ_2, ψ_3). Of these, ψ_3 is an eigenstate of p and ψ_2 is not. Of the set (ψ_1, ψ_3), ψ_1 is not an eigenstate of p, and again ψ_3 is.

When there are n linearly independent eigenstates of an operator A that all correspond to the same eigenvalue, the eigenvalue is n–fold degenerate. If $[A, B] = 0$, and A has n degenerate eigstates, one can form n linear combinations which are n linearly independent eigenstates of both A and B.

For instance, from the two degenerate eigenstates $\{\cos kx, \sin kx\}$ in the free particle problem, we can form

$$\psi_+ = \psi_1 + i\psi_2 = \cos kx + i\sin kx = e^{ikx} \tag{3.113}$$

$$\psi_- = \psi_1 - i\psi_2 = \cos kx - i\sin kx = e^{-ikx} \tag{3.114}$$

These two functions are common eigenstates of H and p. They remain degenerate eigenstates of H but are nondegenerate eigenstates of p.

3.9 Commutators and uncertainty

Any function of the form $\psi = Ae^{ikx}$ is a common eigenstate of both p and H for a free particle. If the particle is in this state, it is certain that measurement of p gives $\hbar k$ and measurement of energy gives $\frac{\hbar^2 k^2}{2m}$. Since $\psi = Ae^{ikx}$ is a common eigenstate of p and H, measurement of p gives $\hbar k$ and leaves the particle in the state $\psi = Ae^{ikx}$. Subsequent measurement of E gives $\frac{\hbar^2 k^2}{2m}$ and also leaves the particle in the state $\psi = Ae^{ikx}$.

The operators H and p are compatible, that is, they commute, $[p, H] = 0$. Quantum mechanics allows p and E to be simultaneously specified. Although there exists a state in which both energy and momentum may be specified simultaneously, the same is not true for the observable x and p. There is no state in which measurement is certain to yield definite values of x and p. Measurement of p leaves the system in an eigenstate of p, $\psi = Ae^{ikx}$. Subsequent measurement of x is infinitely uncertain. The state $\psi = Ae^{ikx}$ is not an eigenstate of x. Conversely, measurement of x that finds x' leaves the system in the eigenstate of x, $\psi = \delta(x - x')$. When the particle is in this state, measurement of momentum is infinitely uncertain.

For the free particle, there are states in which the uncertainty in E and p obeys the relation

$$\Delta E \Delta p = 0 \tag{3.115}$$

On the other hand, the uncertainty in p and x obeys the relation

$$\Delta x \Delta p \geq \frac{\hbar}{2} \tag{3.116}$$

These uncertainty relations have their origin in the compatibility properties, $[A, B] = 0$, of the operators that correspond to the observables being measured. Suppose that two observables A and B are not compatible,

$$[A, B] = C \neq 0 \tag{3.117}$$

for example, displacement and kinetic energy. Then one can show the following: If measurement of A, in the state ψ, is uncertain by the amount ΔA, then measurement of B is uncertain by the amount ΔB, such that

$$\Delta A \Delta B \geq \frac{1}{2} | < C > | \tag{3.118}$$

We know that

$$(\Delta A)^2 \equiv < (A - < A >)^2 > = < A^2 > - < A >^2 \tag{3.119}$$

Expectation values in

$$\Delta A \Delta B \geq \frac{1}{2} | < C > | \tag{3.120}$$

are calculated in the state ψ, namely,

$$< C > = < \psi | C | \psi > \tag{3.121}$$

$$(\Delta A)^2 = < \psi | A^2 | \psi > - < \psi | A | \psi >^2 \tag{3.122}$$

$$(\Delta B)^2 = <\psi|B^2|\psi> - <\psi|B|\psi>^2 \tag{3.123}$$

If A and B do not commute, then the eigenstates ψ_a of A which the system goes into on measurement of A is not necessarily an eigenstate of B. Subsequent measurement of B will give any of the spectrum of eigenvalues of B with a corresponding probability distribution

$$P(b) = |<\psi_b|\psi_a>|^2 \tag{3.124}$$

which is obtained from the coefficients in the expansion of ψ_a in the eigenstates ψ_b of B. Remeasurement of A is then in no way certain of finding the system in the state ψ_a.

3.10   Worked examples

Example - 3.1 : Show that whenever a quantum mechanical state is formed by superposition of energy eigenstates, the mean energy of the system remains constant but the configuration of the system changes in time.

Solution : Let

$$\psi = \sum_n c_n \psi_n e^{-\frac{i}{\hbar}E_n t} \quad , \quad H\psi_n = E_n \psi_n \quad , \quad <\psi_n|\psi_m> = \delta_{nm}$$

Mean energy of the system:

$$<H> = \frac{<\psi|H|\psi>}{<\psi|\psi>}$$

$$= \frac{<\sum_n c_n \psi_n e^{-\frac{i}{\hbar}E_n t}|H|\sum_m c_m \psi_m e^{-\frac{i}{\hbar}E_m t}>}{<\sum_n c_n \psi_n e^{-\frac{i}{\hbar}E_n t}|\sum_m c_m \psi_m e^{-\frac{i}{\hbar}E_m t}>}$$

$$= \frac{\sum_{n,m} c_n^* c_m e^{\frac{i}{\hbar}(E_n - E_m)t} E_m \delta_{nm}}{\sum_{n,m} c_n^* c_m e^{\frac{i}{\hbar}(E_n - E_m)t} \delta_{nm}} = \frac{\sum_n |c_n|^2 E_n}{\sum_n |c_n|^2}$$

$<H>$ is independent of time. Configuration of the system:

$$\psi^* \psi = \sum_n c_n^* \psi_n^* e^{\frac{i}{\hbar}E_n t} \sum_m c_m \psi_m e^{-\frac{i}{\hbar}E_m t}$$

$$= \sum_{n,m} c_n^* c_m e^{\frac{i}{\hbar}(E_n - E_m)t}$$

$\psi^* \psi$ is varying with time.

Example - 3.2 : Show that if $|nr>$ denotes the eigenvectors of an observable, then the closure relation $\sum_{n,r} |nr><nr| = 1$ holds.

Solution : Let P_A be the projection operator on the subspace spanned by the vector $|nr>$. Since A is an observable, we have $P_A = 1$. On the other hand, the projection on a subspace is equal to the sum of all projections onto vectors forming a base in this subspace. In particular, $P_A = \sum_{n,r} P_{nr}$ where P_{nr} is the projection operator on the vector $|nr>$. Thus we have that $\sum_{n,r} P_{nr} = 1$. But, $P_{nr} = |nr><nr|$, and hence $\sum_{n,r} |nr><nr| = 1$. For any orthonormal set of vectors, the closure relation and the completeness condition are equivalent.

Example - 3.3 : Show that the momentum eigenfunctions for a free particle moving in one dimension is

$$u_{p_x} = \frac{1}{\sqrt{h}} e^{\frac{i}{\hbar} p_x x}$$

Solution : The eigenvalue equation for momentum in one–dimension is

$$-i\hbar \frac{\partial}{\partial x} u = p_x u$$

which gives on indefinite integration

$$u_{p_x} = C e^{\frac{i}{\hbar} p_x x}$$

Then by use of

$$\delta(\gamma) = \frac{1}{2\pi} \int_{-\infty}^{\infty} e^{i\gamma x} dx \quad \text{and} \quad \int u_f^* u_{f'} d\tau = \delta(f - f')$$

we may determine the normalization C

$$\int u_{p_x'}^* u_{p_x} dx = \delta(p_x - p_x') = \int_{-\infty}^{\infty} C^2 e^{\frac{i}{\hbar} x (p_x - p_x')} dx$$

$$= 2\pi \hbar C^2 \left(\frac{1}{2\pi} \int_{-\infty}^{\infty} e^{iq(p_x - p_x')} dq \right) \quad , \quad q = \frac{x}{\hbar}$$

$$= 2\pi \hbar C^2 \delta(p_x - p_x')$$

Thus

$$C = \frac{1}{\sqrt{2\pi\hbar}} = \frac{1}{\sqrt{h}} \quad \text{and} \quad u_{p_x} = \frac{1}{\sqrt{h}} e^{\frac{i}{\hbar} p_x x}$$

Example - 3.4 : Consider the function $f(x) = \sin(kx)$. (a) Is $f(x)$ an eigenstate of the momentum? If so, what is the eigenvalue? (b) Is $f(x)$ an eigenstate of the kinetic energy? If so, what is the eigenvalue?

Solution : (a)

$$f(x) = \sin(kx) \quad , \quad \hat{p} = -i\hbar \frac{\partial}{\partial x}$$

$$\hat{p} f(x) = -i\hbar \frac{\partial}{\partial x} \sin(kx) = -i\hbar k \cos(kx)$$

$$\hat{p} f(x) \neq p f(x)$$

therefore, $f(x)$ is not an eigenfunction of $\hat{p}$.

(b)

$$T = \frac{p^2}{2m} = -\frac{\hbar^2}{2m}\frac{\partial^2}{\partial x^2}$$

$$Tf(x) = -\frac{\hbar^2}{2m}\frac{\partial^2}{\partial x^2}\sin(kx) = \frac{\hbar^2 k^2}{2m}\sin(kx)$$

$$Tf(x) = \frac{\hbar^2 k^2}{2m}f(x)$$

therefore $f(x)$ is an eigenfunction of T with eigenvalue $\frac{\hbar^2 k^2}{2m}$.

Example - 3.5 : Suppose that an energy level E_a is three–fold degenerate. ψ_1, ψ_2, and ψ_3 are the orthonormal degenerate eigenfunctions corresponding to this level. A new function is formed from the linear combinations

$$\phi(x) = N(2\psi_1 + \psi_2 + 3\psi_3)$$

Find N and show that $\phi(x)$ is also eigenfunction of the energy operator H with the same eigenvalue E_a.

Solution :

$$H\psi_1 = E_a\psi_1 \quad , \quad H\psi_2 = E_a\psi_2 \quad , \quad H\psi_3 = E_a\psi_3$$

$$<\phi|\phi> = N^2 < 2\psi_1 + \psi_2 + 3\psi_3|2\psi_1 + \psi_2 + 3\psi_3 >$$

$$= N^2(4 + 1 + 9) = 14N^2 = 1 \quad \rightarrow \quad N = \frac{1}{\sqrt{14}}$$

$$H\phi = N(2\psi_1 + \psi_2 + 3\psi_3) = N(2H\psi_1 + H\psi_2 + 3H\psi_3)$$

$$= N(2E_a\psi_1 + E_a\psi_2 + 3E_a\psi_3) = E_aN(2\psi_1 + \psi_2 + 3\psi_3) = E_a\phi$$

3.11 Problems

Problem - 3.1 : If for a system there are two incompatible physical quantities which are conserved, show that the energy levels of the system are in general degenerate.

Problem - 3.2 : Show that the validity of the relation

$$<\alpha|\beta> = \sum_i <\alpha|u_i><u_i|\beta>$$

for any arbitrary vectors $|\alpha>$ and $|\beta>$ is a necessary and sufficient condition for the system of orthonormal vectors $\{|u_i>\}, i = 1, 2, \cdots$ to be complete.

Problem - 3.3 : The momentum eigenfunction for a particle moving in one dimension is

$$\phi_p(x) = \frac{1}{\sqrt{h}}e^{\frac{i}{\hbar}px}$$

The energy eigenfunction for a particle in a one–dimensional box of length a is

$$u(x) = \sqrt{\frac{2}{a}} \sin\left(\frac{n\pi x}{a}\right)$$

If $u(x)$ is expanded in terms of $\phi_p(x)$, the expansion coefficient may be interpreted as the momentum probability amplitude; its square gives the probability distribution function for momentum. Determine the momentum probability distribution for the given function $u(x)$.

Problem - 3.4 : The function $f(x) = \sqrt{30/a^5}\, x(a - x)$ is in the Hilbert space of the functions $\phi_n(x) = \sqrt{2/a}\sin(n\pi x/a)$ defined in the interval $0 \le x \le a$. Here a is a constant. (a) Express $f(x)$ as a superposition of $\phi_n(x)$. (b) Calculate the expansion coefficients in (a). (c) calculate the probability of finding the system described by $f(x)$ being in the ground state.

Problem - 3.5 : A particle in a one–dimensional box with walls at $x = (0, a)$ is in the superposition state

$$\psi(x) = a_1 e^{i\lambda_1} \psi_1(x) + a_3 e^{i\lambda_3} \psi_3(x)$$

where λ_1 and λ_3 are arbitrary real phase factors, a_1 and a_3 are real coefficients, and $a_1^2 + a_3^2 = 1$. (a) What is the functional dependence of the probability density, $\psi^*(x)\psi(x)$, on λ_1, λ_3? (b) Calculate the integral $\int_0^a \psi^*(x)\psi(x)dx$.

Problem - 3.6 : You are given a real operator $\hat{A}$ satisfying the quadratic equation

$$\hat{A}^2 - 3\hat{A} + 2 = 0$$

This is the lowest–order equation that $\hat{A}$ obeys. (a) What are the eigenvalues of $\hat{A}$? (b) What are the eigenstates of $\hat{A}$? (c) Prove that $\hat{A}$ is an observable.

Chapter 4

Time Development and Conservation Theorems

4.1 Time development of state functions, The discrete case

Let us consider the solution $\psi(x,0)$ to the initial–value problem at time $t = 0$. The state $\psi(x,t)$ at $t > 0$ is in the form

$$\psi(x,t) = e^{-\frac{i}{\hbar}Ht}\psi(x,0) \tag{4.1}$$

If ϕ_n is an eigenfunction of H with eigenvalue E_n, that is $H\phi_n = E_n\phi_n$, then we can write

$$e^{-\frac{i}{\hbar}Ht}\phi_n = e^{-\frac{i}{\hbar}E_n t}\phi_n \tag{4.2}$$

Here we expand the exponential operator as

$$e^{-\frac{i}{\hbar}Ht} = 1 + \left(-\frac{i}{\hbar}Ht\right) + \frac{1}{2!} + \left(-\frac{i}{\hbar}Ht\right)^2 + \cdots \tag{4.3}$$

Consider the solution of a particle in a 1D box with walls at $(0,a)$, which is initially in an eigenstate of the Hamiltonian of this system

$$\psi_n(x,0) = \phi_n(x) \tag{4.4}$$

Then the state at time t is

$$\psi_n(x,t) = e^{-\frac{i}{\hbar}Ht}\phi_n = e^{-i\omega_n t}\phi_n(x) \tag{4.5}$$

$$\hbar\omega_n = E_n = n^2 E_1 \tag{4.6}$$

The time–dependent eigenstates, $\psi_n(x,t)$ of H, are called stationary states. The expectation value of any operator (which does not contain the time explicitly) is constant in a stationary state. As an example of a stationary state, consider the $n = 3$ eigenstate of the problem at hand,

$$\psi_3(x,t) = e^{-\frac{i}{\hbar}E_3 t}\phi_3(x) = e^{-i\omega_3 t}\phi_3(x) \tag{4.7}$$

$$= e^{-\frac{i}{\hbar}9E_1 t}\sqrt{\frac{2}{a}}\sin\left(\frac{3\pi x}{a}\right)$$

The eigenstate ψ_3 oscillates with the frequency $\omega = \frac{9E_1}{\hbar}$. Both real and imaginary parts of $\psi_3(x,t)$ are standing waves. The expectation value of energy in this state is constant and equal to $9E_1$, namely $<E> = 9E_1$.

Suppose, on the other hand, that $\psi(x,0)$ is not an eigenstate of H. Under such circumstances, to determine the time development of $\psi(x,0)$, one calls on the superposition principle and writes $\psi(x,0)$ as a linear superposition of the eigenstates of H.

$$\psi(x,0) = \sum b_n \phi_n(x) \quad , \quad b_n = < \phi_n | \psi(x,0) > \tag{4.8}$$

The calculation of $\psi(x,t)$ becomes

$$\psi(x,t) = e^{-\frac{i}{\hbar}Ht}\psi(x,0) = e^{-\frac{i}{\hbar}Ht}\sum b_n \phi_n(x) \tag{4.9}$$

$$= \sum b_n e^{-\frac{i}{\hbar}Ht}\phi_n(x) = \sum b_n e^{-i\omega_n t}\phi_n(x) \quad , \quad \hbar\omega_n = E_n = n^2 E_1$$

This solution indicates that each component amplitude $b_n\phi_n$ oscillates with the corresponding angular frequency ω_n. Consider the specific example in which the initial state is

$$\psi(x,0) = \sqrt{\frac{2}{a}}\frac{1}{\sqrt{5}}\left[2\sin\left(\frac{\pi x}{a}\right) + \sin\left(\frac{2\pi x}{a}\right)\right] \tag{4.10}$$

This state is simply the superposition of the two eigenstates ϕ_1 and ϕ_2 (see Figure 4.1), that is

$$\psi(x,0) = \sum b_n \phi_n(x) \tag{4.11}$$

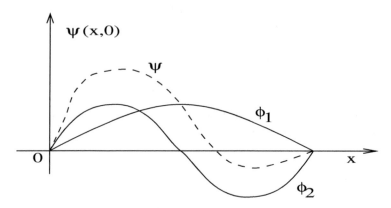

Fig. 4.1. Superposition of ϕ_1 and ϕ_2; $\psi = b_1\phi_1 + b_2\phi_2$.

In the expansion one obtains $b_1 = 2/\sqrt{5}$, $b_2 = 1/\sqrt{5}$, $b_n = 0$ (for all other n). The state of the system at $t > 0$ is given by

$$\psi(x,t) = \sqrt{\frac{2}{a}}\frac{1}{\sqrt{5}}\left[2e^{-i\omega_1 t}\sin\left(\frac{\pi x}{a}\right) + e^{-i\omega_2 t}\sin\left(\frac{2\pi x}{a}\right)\right] \tag{4.12}$$

We can write this expansion in the form

$$\psi(x,t) = \sum b'_n(t)\phi_n(x) \tag{4.13}$$

so that $b'_n(t)$ now includes the exponential time factor

$$b'_n(t) = e^{-i\omega_n t}b_n \tag{4.14}$$

Suppose that the energy is measured at $t > 0$. The expectation value of energy becomes

$$< E > = \sum |b'_n(t)|^2 E_n \tag{4.15}$$

The square of the coefficient of expansion $b_n(t)$ gives the probability that measurement of E at the time t finds the value E_n.

$$P(E_n) = |b'_n(t)|^2 = e^{+i\omega_n t} e^{-i\omega_n t} b_n^* b_n = |b_n|^2 \tag{4.16}$$

$$= \text{constant in time}$$

Therefore,

$$< E > = \sum |b'_n(t)|^2 E_n = \sum |b_n|^2 E_n \tag{4.17}$$

$$= \text{constant in time}$$

For any isolated system, in any initial state, the probability of finding a specific energy E_n is constant in time, and the expectation value of the energy, $< E >$, is constant in time. For the state given above,

$$\psi(x,t) = \sqrt{\frac{2}{a}} \frac{1}{\sqrt{5}} \left[2e^{-i\omega_1 t} \sin\left(\frac{\pi x}{a}\right) + e^{-i\omega_2 t} \sin\left(\frac{2\pi x}{a}\right) \right] \tag{4.18}$$

the probability distribution is

$$P(E_1) = \frac{4}{5} \quad , \quad P(E_2) = \frac{1}{5} \quad , \quad P(E_n) = 0 \text{ (for all other } n) \tag{4.19}$$

For the initial state $\psi(x,0)$ at any time $t > 0$, the probability that measurement of energy finds the value E_1 is 4/5. Similarly, the probability that measurement finds the value E_2 is 1/5. The expectation value of E at $t > 0$ for the initial state $\psi(x,0)$ is

$$< E >_{t>0} = \frac{1}{5} < 2e^{-i\omega_1 t}\phi_1 + e^{-i\omega_2 t}\phi_2|H|2e^{-i\omega_1 t}\phi_1 + e^{-i\omega_2 t}\phi_2 > \tag{4.20}$$

$$= \frac{1}{5}(4E_1 + E_2) = \frac{8}{5}E_1 = < E >_{t=0}$$

The cross terms vanish due to orthogonality of the eigenstates of H, and one finds that the expectation value of energy is constant in time.

4.2 The continuous case, Wave packets

Let us consider a free particle moving in 1D. Let the particle be initially in a localized state $\psi(x,0)$. Since the eigenstates of the Hamiltonian for a free particle comprise a continuum, the representation of $\psi(x,0)$ as a superposition of energy eigenstates is an integral.

$$\psi(x,0) = \frac{1}{\sqrt{2\pi}} \int_{-\infty}^{\infty} b(k)e^{ikx} dk \tag{4.21}$$

$$b(k) = \frac{1}{\sqrt{2\pi}} \int_{-\infty}^{\infty} \psi(x,0)e^{-ikx}dx \qquad (4.22)$$

The state of the particle at $t > 0$ takes the form

$$\psi(x,t) = e^{-\frac{i}{\hbar}Ht}\psi(x,0) = e^{-\frac{i}{\hbar}Ht}\frac{1}{\sqrt{2\pi}}\int_{-\infty}^{\infty}b(k)e^{ikx}dk \qquad (4.23)$$

$$= \frac{1}{\sqrt{2\pi}}\int_{-\infty}^{\infty}b(k)e^{-i\omega t}e^{ikx}dk = \frac{1}{\sqrt{2\pi}}\int_{-\infty}^{\infty}b(k)dk$$

$$\hbar\omega = \frac{\hbar^2 k^2}{2m} = E_k \qquad (4.24)$$

While the component amplitude of the wave function of a particle in a box oscillate as standing waves, $e^{-i\omega_n t}$, the k–component amplitude of the free particle state function propagate, $e^{i(kx-\omega t)}$. For each value of k, the integrand of

$$\psi(x,t) = \frac{1}{\sqrt{2\pi}}\int_{-\infty}^{\infty}b(k)e^{i(kx-\omega t)}dk \qquad (4.25)$$

appears as

$$b(k)e^{i(kx-\omega t)} = b(k)e^{i(kx-\frac{\omega}{k}t)} \qquad (4.26)$$

The phase of this component, $(x - \frac{\omega}{k}t)$, is constant on the propagating surface, $x = \frac{\omega}{k}t$. This is a surface of constant phase. It propagates with the phase velocity

$$v = \frac{\omega}{k} = \frac{\hbar\omega}{\hbar k} = \frac{p^2/2m}{p} = \frac{p}{2m} = \frac{\hbar k}{2m} \qquad (4.27)$$

The components with larger wavenumbers (shorter wavelengths), $k = 2\pi/\lambda$, propagate with larger speeds. The long–wavelength components propagate more slowly.

Suppose that at $t = 0$, the state $\psi(x,0)$ is an eigenstate of H. As time increases, each k component propagates with a distinct phase velocity. The initial state begins to distort. In this case one speaks of a propagating wave packet. To have a wave packet propagate, it is necessary that the average momentum of the particle in the initial state does not vanish,

$$<p>_{t=0} = <\psi(x,0)|p|\psi(x,0)> \neq 0 \qquad (4.28)$$

and the packet is localized in space, $|\psi(x,0)|^2 \neq 0$, only over a small domain. The velocity with which such a packet moves is called the group velocity,

$$v_g = \left.\frac{\partial\omega}{\partial k}\right|_{k_{max}} = \left.\frac{\partial\hbar\omega}{\hbar\partial k}\right|_{k_{max}} = \left.\frac{\partial(\hbar^2k^2/2m)}{\hbar\partial k}\right|_{k_{max}} \qquad (4.29)$$

$$= \frac{\hbar k_{max}}{m} = \frac{<p>}{m} = v_{CL}$$

The packet moves with the classical velocity $<p>/m$. The meaning of k_{max} is that the amplitude $|b(k)|^2$ is maximum $k = k_{max}$

$$\hbar k_{max} = <p> = \int_{-\infty}^{\infty}|b(k)|^2\hbar k dk \qquad (4.30)$$

4.3 Particle beam

Consider a beam of particles each of which has momentum $\hbar k_0$. The beam is chopped, producing a pulse with length a containing N particles; see Figure 4.2.

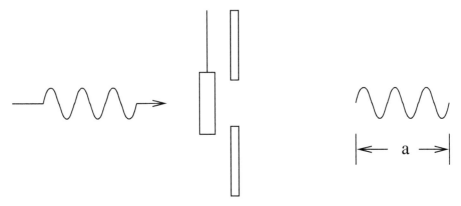

Fig. 4.2. A chopped beam of particles represented by the wave e^{ik_0x}.

The state function for each particle at the instant after the pulse is produced is

$$\psi(x,0) = \begin{cases} \frac{1}{\sqrt{a}}e^{ik_0x} & -\frac{a}{2} \le x \le +\frac{a}{2} \\ 0 & \text{elsewhere} \end{cases} \tag{4.31}$$

The momentum of any one of the particles measured at $t > 0$ may be predicted theoretically as follows:

The general form of the wave function at $t = 0$ is

$$\psi(x,0) = \frac{1}{\sqrt{2\pi}} \int_{-\infty}^{\infty} b(k)e^{ikx}dk \tag{4.32}$$

The expansion coefficient is

$$b(k) = \frac{1}{\sqrt{2\pi}} \int_{-\infty}^{\infty} \psi(x,0)e^{-ikx}dx \tag{4.33}$$

Substituting the state function we get

$$b(k) = \frac{1}{\sqrt{2\pi a}} \int_{-\frac{a}{2}}^{\frac{a}{2}} e^{ik_0x}e^{-ikx}dx \tag{4.34}$$

$$= \sqrt{\frac{2}{\pi a}} \frac{\sin[(k-k_0)a/2]}{(k-k_0)}$$

The state at time $t > 0$ is

$$\psi(x,t) = \frac{1}{\pi\sqrt{a}} \int_{-\infty}^{\infty} \frac{\sin[(k-k_0)a/2]}{(k-k_0)} e^{i(kx-\omega t)}dk \tag{4.35}$$

$\hbar\omega = \hbar^2 k^2/2m$. The amplitude square $|b(k)|^2$ is shown in Figure 4.3., which represents the momentum probability density $P(k) = |b(k)|^2$. This probability density is constant in time.

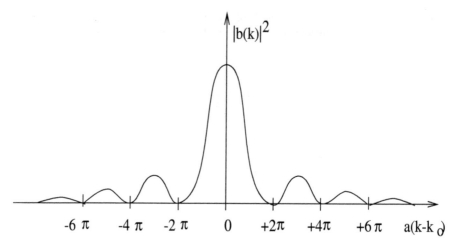

Fig. 4.3. Momentum probability density of a chopped beam of particles.

$P(k)$ gives the probability that measurement of momentum of any of the particles yields a value in the interval $\hbar k$ to $\hbar(k + dk)$.

At any time $t > 0$, it is most likely that measurement of momentum of any particle in the pulse finds the value $p = \hbar k_{max} = \hbar k_0$. This value was the only momentum the particles had before the beam was chopped.

At any time $t > 0$, the momentum values

$$\hbar k = hbar k_0 + n\frac{2\pi\hbar}{a} \quad , \quad n = 1, 2, 3, \cdots \tag{4.36}$$

have zero probability of being found. These momentum eigenstates do not enter into the superposition construction of $\psi(x, 0)$.

The number of particles found with momentum in the interval $\hbar(k - k_0) - \hbar k_0$ to $\hbar(k - k_0) + \hbar k_0$ is

$$\Delta N = N \int_{k-2k_0}^{k} |b(k)|^2 dk \tag{4.37}$$

This number is also constant in time.

4.4   Gaussian wave packet

Let us consider the propagation of a Gaussian wave packet. The initial state is

$$\psi(x, 0) = \frac{1}{\sqrt{a\sqrt{2\pi}}} e^{ik_0 x} e^{-\frac{x^2}{4a^2}} \tag{4.38}$$

The corresponding initial probability density is

$$P(x,0) = \psi^*\psi = \frac{1}{a\sqrt{2\pi}}e^{-\frac{x^2}{2a^2}} \qquad (4.39)$$

$$\int_{-\infty}^{\infty} P(x)dx = 1 \qquad (P \text{ is normalized}) \qquad (4.40)$$

The initial uncertainty in position of a particle in the state $\psi(x,0)$ is $\Delta x = a$. The average momentum is $< p >= \hbar k_0$. The initial Gaussian state function $\psi(x,0)$ represents a particle localized within a spread of a about the origin and moving with an average momentum $\hbar k_0$. The momentum amplitude corresponding to this initial state is

$$b(k) = \frac{1}{\sqrt{2\pi}}\int_{-\infty}^{\infty} \psi(x,0)e^{-ikx}dx \qquad (4.41)$$

$$= \frac{1}{\sqrt{a(2\pi)^{3/2}}}\int_{-\infty}^{\infty} e^{-\frac{x'^2}{4a^2}}e^{ix'(k_0-k)}dx' = \sqrt{\frac{2a}{\sqrt{2\pi}}}e^{-a^2(k_0-k)^2}$$

The initial momentum probability density is $P(k) = |b(k)|^2$ is normalized, centered at $k = k_0$, and has a spread $\Delta k = \frac{1}{2a}$. Therefore, in the initial Gaussian state,

$$\Delta x \Delta p|_{Gaus.} = \Delta x \hbar \Delta k = \frac{\hbar}{2} = \Delta x \Delta p|_{min}. \qquad (4.42)$$

The product of uncertainties has its minimum value in a Gaussian packet.

4.5 Free particle propagator

The construction of $\psi(x,t)$ from the initial state

$$\psi(x,0) = \frac{1}{\sqrt{a\sqrt{2\pi}}}e^{ik_0x}e^{-\frac{x^2}{4a^2}} \qquad (4.43)$$

may be obtained as follows:

$$\psi(x,t) = \frac{1}{\sqrt{2\pi}}\int_{-\infty}^{\infty} b(k)e^{i(kx-\omega t)}dk \qquad (4.44)$$

where

$$b(k) = \frac{1}{\sqrt{2\pi}}\int_{-\infty}^{\infty} \psi(x,0)e^{-ikx}dx \qquad (4.45)$$

Therefore,

$$\psi(x,t) = \frac{1}{2\pi}\int_{-\infty}^{\infty}\int_{-\infty}^{\infty} dx'dk\, e^{-ikx'}\psi(x',0)e^{i(kx-\omega t)} \qquad (4.46)$$

$$= \frac{1}{2\pi}\frac{1}{\sqrt{a\sqrt{2\pi}}}\int_{-\infty}^{\infty} dx' e^{ik_0x'}e^{-\frac{x'^2}{4a^2}}\int_{-\infty}^{\infty} dk\, e^{ik(x-x')}e^{-i\frac{k^2a^2}{\tau}t}$$

where

$$\frac{k^2 a^2}{\tau} = \omega = \frac{\hbar k^2}{2m} \tag{4.47}$$

We can also construct $\psi(x,t)$ using free particle propagator, $K(x',x;t)$,

$$\psi(x,t) = \int_{-\infty}^{\infty} dx' \psi(x,0) K(x',x;t) \tag{4.48}$$

Comparing the equations Eq. (4.46) and Eq. (4.48) we may write the explicit form of $K(x',x;t)$ for free particle as

$$K(x',x;t) = \frac{1}{2\pi} \int_{-\infty}^{\infty} dk e^{ik(x-x')} e^{-i\frac{k^2 a^2}{\tau}t} \tag{4.49}$$

Using the integral

$$\int_{-\infty}^{\infty} e^{-uy^2} e^{vy} dy = \sqrt{\frac{\pi}{u}} e^{v^2/4u} \quad (\text{Re}u > 0) \tag{4.50}$$

the free particle propagator $K(x',x;t)$ takes the final form

$$K(x',x;t) = \left(\frac{\tau}{i4\pi a^2 t}\right)^{1/2} e^{i\frac{(x-x')^2}{4a^2 t}\tau} = \left(\frac{m}{2\pi i\hbar t}\right)^{1/2} e^{i\frac{m(x-x')^2}{2\hbar t}} \tag{4.51}$$

The wave function $\psi(x,t)$ gives the probability amplitude related to finding the particle at x at the instant t. If the particle was at x' at $t = 0$, then the probability that it is found at x at $t > 0$ depends on the probability that the particle propagated from x' to x in the interval t. In the equation

$$\psi(x,t) = \int_{-\infty}^{\infty} dx' \psi(x',0) K(x',x;t) \tag{4.52}$$

we may interpret $K(x',x;t)$ as the probability amplitude that a particle initially at x' propagates to x in the interval t. The explicit form of $\psi(x,t)$ for a free particle is in the form:

$$\psi(x,t) = \int_{-\infty}^{\infty} dx' \psi(x',0) K(x',x;t) \tag{4.53}$$

$$= \int_{-\infty}^{\infty} dx' \frac{1}{\sqrt{a\sqrt{2\pi}}} e^{ik_0 x'} e^{-\frac{x'^2}{4a^2}} \sqrt{\frac{m}{2\pi i\hbar t}} e^{i\frac{m(x-x')^2}{2\hbar t}}$$

$$= \frac{1}{\sqrt{a\sqrt{2\pi}}} \frac{1}{\sqrt{1+i\frac{t}{\tau}}} e^{i\frac{\tau}{t}\left(\frac{x}{2a}\right)^2} e^{-\frac{\left(\frac{i\tau}{4a^2 t}\right)\left(x-\frac{\hbar k_0}{m}t\right)}{\left(1+i\frac{t}{\tau}\right)}}$$

The corresponding probability density is

$$P(x,t) = |\psi(x,t)|^2 = \frac{1}{a\sqrt{2\pi}\sqrt{1+\frac{t^2}{\tau^2}}} e^{-\frac{\left(x-\frac{\hbar k_0 t}{m}\right)^2}{2a^2\left(1+\frac{t^2}{\tau^2}\right)}} \tag{4.54}$$

If we compare this form with the initial probability density,

$$P(x,0) = |\psi(x,0)|^2 = \frac{1}{a\sqrt{2\pi}}e^{-\frac{x^2}{2a^2}} \tag{4.55}$$

three modifications appear:

i – It has become wider,

$$a \rightarrow a\sqrt{1+\frac{t^2}{\tau^2}} \tag{4.56}$$

ii – The center of symmetry of the packet is now at $x = v_0 t$, where $v_0 = \hbar^2 k/m$. This shows that the probability density of a Gaussian wave packet propagates with a velocity that is directly related to the expectation value of momentum of the particle in the Gaussian state.

iii – The height of the density function has diminished,

$$\frac{1}{a\sqrt{2\pi}} \rightarrow \frac{1}{a\sqrt{2\pi}\sqrt{1+\frac{t^2}{\tau^2}}} \tag{4.57}$$

The area under the curve P, at any time, remains unity.

$$\int_{-\infty}^{\infty} P(x,t)dx = 1 \tag{4.58}$$

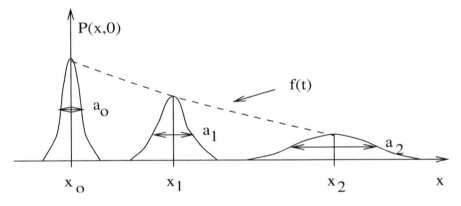

Fig. 4.4. Evolution of Gaussian wave in time. The spread (width), a, and the position, x, of the wave packet at different times: $t = 0$, $a_0 = a$, $x_0 = 0$; $t = 1$, $a_1 = \sqrt{2}a$, $x_1 = v_0\tau$; $t = 2$, $a_2 = \sqrt{5}a$, $x_2 = 2v_0\tau$. The envelope of the peaks, $f(t)$, varies with time, t, as $f(t) = 1/\sqrt{1+t^2/\tau^2}$.

4.6   The limiting cases of the Gaussian wave packets

There are two limits that can be taken on the probability density $P(x,t)$ related to the Gaussian wave packet. The first evolves from the initial state

$$P(x,0) = |\psi(x,0)|^2 = \delta(x) \tag{4.59}$$

A valid representation of the delta function is given by the limit

$$\delta(x) = \lim_{a \to 0} \frac{1}{a\sqrt{2\pi}} e^{-\frac{x^2}{2a^2}} \tag{4.60}$$

Measurement of the position of a particle which finds the value $x = 0$ leaves the particle in the state $\psi = \delta(x)$. This state is not normalizable. However, the state given by Eqs. 59 and 60 is a little less sharply peaked than $\psi = \delta(x)$ and is normalizable.

To obtain the probability density $P(x, t)$ which follows from the initial value, Eq. 59, we look at

$$P(x,t) = |\psi(x,t)|^2 = \frac{1}{a\sqrt{2\pi}\sqrt{1 + \frac{t^2}{\tau^2}}} e^{-\frac{(x - \frac{\hbar k_0}{m} t)^2}{2a^2 (1 + \frac{t^2}{\tau^2})}} \tag{4.61}$$

in the limit, $a \to 0$.

$$\lim_{a \to 0} P(x,t) = \lim_{a \to 0} \frac{2ma}{t\hbar\sqrt{2\pi}} e^{-\frac{2a^2 (x - \frac{\hbar k_0}{m} t)^2}{\frac{t^2 \hbar^2}{m^2}}} \tag{4.62}$$

$$= \lim_{a \to 0} \frac{2ma}{t\hbar\sqrt{2\pi}} [1 + \mathcal{O}(a^2)]$$

We see that for all $t > 0$, P vanishes uniformly for all x, in the limit $a \to 0$. This means that the momentum probability density $|b(k)|^2$ corresponding to such a state is flat.

The second limit changes $P(x, t)$ to the classical probability relating to a point particle of mass m moving with velocity $\hbar k_0/m$. This is obtained by setting $\hbar \to 0$ in $P(x, t)$, except where $\hbar$ appears in $p_0 = \hbar k_0$.

$$\lim_{\hbar \to 0} P(x,t) = \lim_{\hbar \to 0} \frac{1}{a\sqrt{2\pi}\sqrt{1 + \frac{t^2}{\tau^2}}} e^{-\frac{(x - \frac{\hbar k_0 t}{m})^2}{2a^2 (1 + \frac{t^2}{\tau^2})}} \tag{4.63}$$

$$= \frac{1}{a\sqrt{2\pi}} e^{-\frac{(x - \frac{p_0}{m} t)^2}{2a^2}}$$

here $\hbar k_0 = p_0 = $ constant. For this probability to relate to a point particle we impose the additional constraint, $a \to 0$. This gives

$$\lim_{a \to 0} \frac{1}{a\sqrt{2\pi}} e^{-\frac{(x - \frac{p_0}{m} t)^2}{2a^2}} = \delta(x - \frac{p_0}{m} t) = P_{CL}(x,t) \tag{4.64}$$

The probability of finding the particle at t is zero everywhere except on the classical trajectory $x = p_0 t/m$. This is another example of the correspondence principle at work.

4.7   Time development of expectation values

Time development of the expectation value of an observable, $< A >$, is related with the time–dependent Schrödinger equation. We wish to calculate $d < A > /dt$. Since $< A >$ has all its space dependence integrated out, it is at most a function of time. We may therefore write

$$\frac{d < A >}{dt} = \frac{\partial < A >}{\partial t} \tag{4.65}$$

In the state $\psi(x, t)$, this expression becomes

$$\frac{d}{dt} < \psi|A|\psi > = \int dx \frac{\partial}{\partial t}(\psi^* A\psi) \tag{4.66}$$

where

$$\frac{\partial}{\partial t}(\psi^* A\psi) = \left(\frac{\partial \psi^*}{\partial t}\right) A\psi + \psi^* \frac{\partial A}{\partial t}\psi + \psi^* A \frac{\partial \psi}{\partial t} \tag{4.67}$$

Considering the time–dependent Schrödinger equation

$$i\hbar \frac{\partial \psi}{\partial t} = H\psi \tag{4.68}$$

$$\frac{\partial \psi}{\partial t} = \frac{-i}{\hbar} H\psi \quad, \quad \frac{\partial \psi^*}{\partial t} = \frac{i}{\hbar} H\psi^*$$

$$\frac{\partial}{\partial t}(\psi^* A\psi) = \left(\frac{i}{\hbar} H\psi^*\right) A\psi + \psi^* \frac{\partial A}{\partial t}\psi + \psi^* A \left(-\frac{i}{\hbar} H\psi\right) \tag{4.69}$$

$$= \frac{i}{\hbar}\left(H\psi^* A\psi + \frac{\hbar}{i}\psi^* \frac{\partial A}{\partial t}\psi - \psi^* AH\psi\right)$$

Substituting this expansion in Eq. (4.66) gives

$$\frac{d}{dt} < A > = \frac{i}{\hbar}\left(< H\psi|A\psi > + \frac{\hbar}{i} < \psi|\frac{\partial A}{\partial t}\psi > - < \psi|AH\psi >\right) \tag{4.70}$$

Since H is Hermitian, we may rewrite the last equation as

$$\frac{d}{dt} < A > = \frac{i}{\hbar}(< \psi|HA\psi > - < \psi|AH\psi >) + < \psi|\frac{\partial A}{\partial t}\psi > \tag{4.71}$$

or

$$\frac{d < A >}{dt} = \left\langle \frac{i}{\hbar}[H, A] + \frac{\partial A}{\partial t}\right\rangle \tag{4.72}$$

If A does not contain the time explicitly, then the last term on the right hand side vanishes, $\partial A/\partial t = 0$,

$$\frac{d < A >}{dt} = \left\langle \frac{i}{\hbar}[H, A]\right\rangle \tag{4.73}$$

In the event that A commutes with H, the quantity $< A >$ is constant in time and A is called a constant of the motion. For a free particle, p commutes with H, $[H, p] = 0$, and $< p >$ is constant in time for any state (wave packet). Since H commutes with itself, $[H, H] = 0$, $< H >$, the expectation value of the energy is always constant in time.

Motion of wave packets:

Let a particle moving in one dimension be in the presence of the potential $V(x)$. The Hamiltonian of the particle is

$$H = \frac{p^2}{2m} + V(x) \tag{4.74}$$

How does $<x>$ vary in time? Eq. (4.73) gives

$$\frac{d<x>}{dt} = \frac{i}{\hbar} < [H,x] > = \frac{i}{\hbar} < [\frac{p^2}{2m}, x] > \tag{4.75}$$

$$= \frac{i}{2m\hbar} < p[p,x] + [p,x]p > = \frac{i}{2m\hbar} < -2i\hbar p > = \frac{<p>}{m}$$

or, equivalently

$$m\frac{d<x>}{dt} = <p> \tag{4.76}$$

This equation yields the same relation between expected values of displacement, $<x>$, and momentum, $<p>$, as in the classical case. Eq. (4.76) cannot hold for the eigenvalue of x and p, since such an equation implies that $x(t)$ and $p(t)$ are simultaneously known.

The reduction of quantum mechanical equations to classical forms when averages are taken, such as $md<x>/dt = <p>$, is known as **Ehrenfest's principle**. Newton's second law follows from the commutator $[H,p]$, which for the Hamiltonian $H = \frac{p^2}{2m} + V(x)$ is

$$[H,p] = i\hbar\frac{\partial V}{\partial x} \tag{4.77}$$

Using Eq. 4.72, one obtains

$$\frac{d<p>}{dt} = -\left\langle \frac{\partial V}{\partial x} \right\rangle \tag{4.78}$$

which is the x component of the vector relation

$$\frac{d<\mathbf{p}>}{dt} = - < \nabla V(x,y,z) > = < \mathbf{F}(x,y,z) > \tag{4.79}$$

where $\mathbf{F}$ is the force at (x,y,z). In any state $\psi(x,t)$, the time development of the averages of x and p follows the laws of classical dynamics, with the force at any given point replaced by its expectation in the state $\psi(x,t)$.

As a conclusion Ehrenfest's principle demonstrates that the Newtonian laws of motion, in the form

$$\frac{d\mathbf{r}}{dt} = \frac{\mathbf{p}}{m} \quad , \quad \frac{d\mathbf{p}}{dt} = -\nabla V = \mathbf{F} \tag{4.80}$$

are satisfied exactly by the average motion of a wave packet described by a wave function ψ which is a solution of the Schrödinger equation. The components of $\mathbf{r}$ and $\mathbf{p}$ with a normalized wave function ψ satisfy the equations

$$\frac{d<x>}{dt} = \frac{<p_x>}{m} \quad , \quad \frac{d<p_x>}{dt} = < -\frac{\partial V}{\partial x} > = < F_x > \tag{4.81}$$

In general,

$$\frac{d<\mathbf{r}>}{dt} = \frac{<\mathbf{p}>}{m} \quad , \quad \frac{d<\mathbf{p}>}{dt} = < -\nabla V > = < \mathbf{F} > \tag{4.82}$$

These equations are the quantum equivalent of the classical equations, Eq. (4.80). Thus, the expectation values of position, momentum, and force obey Newton's second law of motion exactly. This is an expression of the correspondence principle for wave packets.

4.8 Conservation of energy and momentum

The principle of conservation of energy in classical mechanics states that the energy of an isolated system or a conservative system is constant in time. A conservative system is one whose dynamics are describable in terms of a potential function. A particle in a one–dimensional box is a conservative system. Suppose that at $t = 0$, the state of the particle is

$$\psi(x,0) = \frac{1}{\sqrt{25}}(3\phi_1 + 4\phi_5) \tag{4.83}$$

Measurement of the energy of the particle at time $t = 0$ has a $\frac{9}{25}$ probability of finding the value E_1 and a $\frac{16}{25}$ probability of finding the value $E_5 = 5^2 E_1 = 25E_1$. At $t > 0$ the state, Eq. (4.83), becomes

$$\psi(x,t) = \frac{1}{\sqrt{25}}\left[3\phi_1(x)e^{-\frac{i}{\hbar}E_1 t} + 4\phi_5(x)e^{-\frac{i}{\hbar}E_5 t}\right] \tag{4.84}$$

The probability that measurement yields E_1 is

$$P(E_1) = \frac{1}{25}\left[\left(3e^{-\frac{i}{\hbar}E_1 t}\right)^* \left(3e^{-\frac{i}{\hbar}E_1 t}\right)\right] = \frac{9}{25} \tag{4.85}$$

A similar calculation of $P(E_5)$ yields the constant value $\frac{16}{25}$. In other words, in the state given, one cannot say with certainty what the energy is at $t > 0$. The energy is conserved in the average sense. It follows from Eq. (4.73) that $< H >=< E >=$ constant. For the above example, at any instant in time the expectation value of the energy is

$$< E >= \frac{9E_1 + 16E_5}{25} = \frac{9E_1 + 16 \times 25E_1}{25} = \frac{409E_1}{25} \tag{4.86}$$

$$= 16.36E_1 = \text{constant}$$

For a free particle, p also commutes with H; hence we can conclude from Eq. (4.73) that $< p >=$ constant. The energy and total momentum of an isolated system are constants of the motion.

Conservation theorems in physics are closely related to symmetry principles. The laws of physics do not depend on the time at which they are applied. Newton's second law, Maxwell's equations, and so on, do not change their structure with time. This symmetry of time, that is homogeneity, gives rise to the conservation of energy. Homogeneity of time implies that H is not an explicit function of time, which implies that $d < E > /dt = 0$. Same conclusion is reached for any isolated system.

Conservation of momentum for an isolated system depends on the homogeneity of space. The dynamical laws of an isolated system of particles can only depend on the relative orientation of particles, not on the distances from these particles to some arbitrary chosen origin. Equivalently, the Hamiltonian of the system can always be transformed so that it does not contain these variables, namely the coordinates of the center of mass, for example.

The relation between the homogeneity of space and conservation of linear momentum can be shown by applying the momentum operator p to any differentiable function, $f(x)$

$$\mathcal{D}(\xi)f(x) = e^{\frac{i}{\hbar}\xi p_x}f(x) = f(x+\xi) \tag{4.87}$$

For infinitesimal displacement ($\xi \to 0$), the displacement operator becomes

$$\mathcal{D}(\xi) = I + \frac{i}{\hbar}\xi p_x \tag{4.88}$$

or, equivalently

$$p_x = \frac{\hbar}{i\xi}\left(\mathcal{D}(\xi) - I\right) \tag{4.89}$$

here I is the identity operator. The Hamiltonian of an isolated system cannot depend on displacement of the system from an origin at an arbitrary point in space. Therefore, the displacement operator $\mathcal{D}$ commutes with H, whence p_x does also. Here again considering Eq. (4.73), we obtain the constancy of $< p_x >$.

Let us consider now a system rotating about a fixed axis. Suppose that there is a property of the system which is dependent on the system's rotational orientation ϕ about the rotation axis, say z–axis. Let the measure of this property be $f(\phi)$. After rotation of the system through the angle $\Delta\phi$, $\phi \to \phi + \Delta\phi$ and $f(\phi) = f(\phi + \Delta\phi)$. This transformation of function is effected by the rotation operator $R_{\Delta\phi}$:

$$R_{\Delta\phi}f(\phi) = f(\phi + \Delta\phi) \tag{4.90}$$

$$R_{\Delta\phi} = e^{\frac{i}{\hbar}\Delta\phi L_z} \tag{4.91}$$

Here L_z is the z–component of the total angular momentum of the system. Since the Hamiltonian of the isolated system cannot depend on ϕ, it is insensitive to the rotation operator, $R_{\Delta\phi}$, that is, $[H, R_{\Delta\phi}] = 0$, hence, $[H, L_z] = 0$, and we conclude that $< L_z >$ is constant. The argument demonstrating the constancy of L_z carries over to L, the total angular momentum of the system.

Let us now consider an infinitesimal rotation about the z–axis (see Figure 4.5)

$$\mathbf{r}' = \mathbf{r} + \delta\mathbf{r} = \mathbf{r} + \delta\hat{\phi} \times \mathbf{r} \tag{4.92}$$

$$R_{\delta\phi}f(\mathbf{r}) = f(\mathbf{r} + \delta\mathbf{r}) = f(\mathbf{r}) + \delta\mathbf{r} \cdot \nabla f(\mathbf{r}) \tag{4.93}$$

$$= f(\mathbf{r}) + \delta\hat{\phi} \times \mathbf{r} \cdot \nabla f(\mathbf{r}) = f(\mathbf{r}) + \delta\hat{\phi} \cdot \mathbf{r} \times \nabla f(\mathbf{r})$$

$$= f(\mathbf{r}) + \frac{i}{\hbar}\delta\hat{\phi} \cdot \mathbf{r} \times \mathbf{p}f(\mathbf{r}) = f(\mathbf{r}) + \frac{i}{\hbar}\delta\hat{\phi} \cdot \mathbf{L}f(\mathbf{r})$$

$$= (I + \frac{i}{\hbar}\delta\hat{\phi} \cdot \mathbf{L})f(\mathbf{r})$$

For a finite rotational displacement $\Delta\phi = n\delta\phi$,

$$R_{\Delta\phi} = (R_{\delta\phi})^n = (I + \frac{i}{\hbar}\delta\hat{\phi} \cdot \mathbf{L})^n \tag{4.94}$$

as $n \to \infty$, or $\Delta\phi / \delta\phi \to \infty$

$$R_{\Delta\phi} = \lim_{\frac{\Delta\phi}{\delta\phi} \to \infty} \left(I + \frac{i}{\hbar} \delta\hat{\phi} \cdot \mathbf{L} \right)^{\frac{\Delta\phi}{\delta\phi}} = e^{\frac{i}{\hbar} \Delta\hat{\phi} \cdot \mathbf{L}} \qquad (4.95)$$

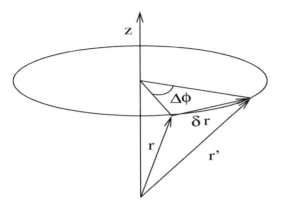

Fig. 4.5. Infinitesimal rotation about the z–axis.

In summary, with p and L denoting, respectively, the total linear and angular momentum of an isolated system whose Hamiltonian is H, the following symmetry–conservation principles hold:

Homogeneity of space (conservation of linear momentum):

$$[H, p] = 0 \quad \to \quad \frac{d}{dt} <p> = 0 \qquad (4.96)$$

Isotropy of space (conservation of angular momentum):

$$[H, L] = 0 \quad \to \quad \frac{d}{dt} <L> = 0 \qquad (4.97)$$

Homogeneity of time (conservation of energy):

$$\frac{\partial H}{\partial t} = 0 \quad \to \quad \frac{d}{dt} <E> = 0 \qquad (4.98)$$

4.9 Conservation of parity

Parity is related with symmetry. It is a mathematical concept. Consider a physical system and its mirror image. Physical laws apply to the system and to its mirror image as well. This is a symmetry principle. In quantum mechanics this principle is associated with a conservation law, conservation of parity. Parity is a property of a function. A function $f(x)$ has odd parity if

$$f(-x) = -f(x) \tag{4.99}$$

A function $f(x)$ has even parity if

$$f(-x) = f(x) \tag{4.100}$$

The parity operator $\wp$ is defined as

$$\wp f(x) = f(-x) \tag{4.101}$$

Let $g(x)$ be an eigenfunction of $\wp$ with eigenvalue α, then

$$\wp g(x) = g(-x) = \alpha g(x) \tag{4.102}$$

To find α we operate again with $\wp$

$$\wp \wp g(x) = \wp g(-x) = g(x) = \alpha^2 g(x) \tag{4.103}$$

Hence

$$\alpha^2 = 1 \quad , \quad \alpha = \mp 1 \tag{4.104}$$

For $\alpha = +1$, $g(-x) = g(x)$; for $\alpha = -1$, $g(-x) = -g(x)$. Any even function is an eigenfunction of $\wp$ with $+1$, and any odd function is an eigenfunction of $\wp$ with eigen value -1. The order of degeneracy of $\alpha = \mp 1$ is infinite. There are no other eigen values of $\wp$.

Consider a particle (m) moving in one–dimension interacts with another stationary particle (M) which is at the position $x = 0$. The potential of interaction between the particles is $V(x)$ (see Figure 4.6).

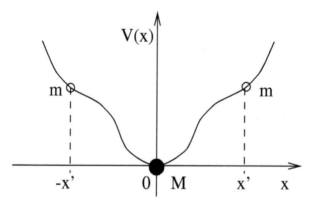

Fig. 4.6. Interaction of two particles in one–dimension.

Suppose that the (moving) particle is at a position $x' > 0$. The image of the particle seen in a mirror which intersects the x–axis normally at $x = 0$ is at $x = -x' < 0$. The temporal behavior of the image particle will be the same as that for the laboratory particle if $V(x) = V(-x)$. The potential seen by the image particle is $V(-x)$.

The Hamiltonian for the particle in the laboratory system is

$$H = \frac{p^2}{2m} + V(x) \tag{4.105}$$

For $V(x)$ an even function, $\wp$ commutes with $V(x)$,

$$\wp V(x)g(x) = V(-x)g(-x) = V(x)\wp g(x) \tag{4.106}$$

Therefore,

$$[\wp, V(x)] = 0 \tag{4.107}$$

$\wp$ commutes with the kinetic energy part of H.

$$\wp T g(x) = \wp \frac{p^2}{2m} g(x) = \frac{p^2}{2m} g(-x) = T \wp g(x) \tag{4.108}$$

Therefore,

$$[\wp, T] = 0 \tag{4.109}$$

Therefore, $\wp$ commutes with the Hamiltonian,

$$[H, \wp] = 0 \tag{4.110}$$

Together with Eq. (4.73), this gives the conservation principle, $< \wp >=$ constant. The parity of the state of a system is a constant of the motion.

As an example, consider a one–dimensional box centered at the origin, so that its walls are at $x = \mp\frac{a}{2}$. The eigensolutions of the Hamiltonian for this system are

$$\tilde{\phi}_n(x) = \sqrt{\frac{2}{a}} \sin\left(\frac{n\pi x}{a}\right) \quad , \quad n = 2, 4, 6, \cdots \quad ; \quad E_n = n^2 E_1 \tag{4.111}$$

$$\phi_n(x) = \sqrt{\frac{2}{a}} \cos\left(\frac{n\pi x}{a}\right) \quad , \quad n = 1, 3, 5, \cdots \quad ; \quad E_n = n^2 E_1 \tag{4.112}$$

The eigen states $\tilde{\phi}_n(x)$ are odd while $\phi_n(x)$ are even.

$$\wp \tilde{\phi}_n(x) = -\tilde{\phi}_n(x) \quad , \quad \wp \phi_n(x) = \phi_n(x) \tag{4.113}$$

Suppose that at $t = 0$ the particle is in the state

$$\psi(x, 0) = \frac{1}{\sqrt{45}} \left(6\tilde{\phi}_2(x) + 3\phi_1(x)\right) \tag{4.114}$$

$$= \frac{1}{\sqrt{45}} \sqrt{\frac{2}{a}} \left[6\sin\left(\frac{2\pi x}{a}\right) + 3\cos\left(\frac{\pi x}{a}\right)\right]$$

At $t > 0$

$$\psi(x, t) = \frac{1}{\sqrt{45}} \sqrt{\frac{2}{a}} \left[6\sin\left(\frac{2\pi x}{a}\right) e^{-\frac{i}{\hbar} E_2 t} + 3\cos\left(\frac{\pi x}{a}\right) e^{-\frac{i}{\hbar} E_1 t}\right] \tag{4.115}$$

The expectation value of $\wp$ at $t = 0$ is

$$< \wp >=< \psi(x, 0)|\wp|\psi(x, 0) > \tag{4.116}$$

$$\wp\psi(x,0) = \frac{1}{\sqrt{45}}\left(-6\tilde{\phi}_2(x) + 3\phi_1(x)\right) \tag{4.117}$$

$$< \wp > = \frac{1}{45} < 6\tilde{\phi}_2 + 3\phi_1 | -6\tilde{\phi}_2 + 3\phi_1 > = \frac{1}{45}(-36 + 9) = -\frac{27}{45} \tag{4.118}$$

Since $[\wp, H] = 0$, this is the value of $< \wp >$ for all time. The meaning of $(-)$ value in the expectation value of parity is that the probability of finding the system in the odd state is most likely.

4.10   Worked examples

Example - 4.1 : Obtain the Fourier transform $F(k)$ for the wave packet function $\psi(x)$ defined by

$$\psi(x) = \begin{cases} he^{ik_0x} & , -\frac{d}{2} \le x \le \frac{d}{2} \\ 0 & , \text{otherwise} \end{cases}$$

Solution : By definition

$$F(k) = \int_{-\infty}^{\infty} \psi(x)e^{-ikx}dx = \int_{-\frac{d}{2}}^{\frac{d}{2}} he^{i(k_0-k)x}dx$$

$$= h\frac{e^{i(k_0-k)x}}{i(k_0-k)}\bigg|_{\frac{d}{2}}^{\frac{d}{2}} = hd\frac{\sin\left[(k_0-k)\frac{d}{2}\right]}{\left[(k_0-k)\frac{d}{2}\right]}$$

Example - 4.2 : Find the Fourier transforms of (a) r^{-1}, (b) $r^{-1}e^{-\lambda r}$ where λ is a real positive constant.

Solution : (a)

$$\text{F.T.}\left(\frac{1}{r}\right) = \frac{1}{(2\pi)^{3/2}}\int e^{-i\mathbf{k}\cdot\mathbf{r}}\frac{1}{r}d\mathbf{r} = \frac{1}{(2\pi)^{1/2}}\int_0^\infty rdr\frac{2\sin(kr)}{kr}$$

$$= \frac{2}{(2\pi)^{1/2}}\lim_{\alpha\to 0}\int_0^\infty \frac{\sin(kr)}{k}e^{-\alpha r}dr = \sqrt{\frac{2}{\pi}}\frac{1}{k^2}$$

(b)

$$\text{F.T.}\left(\frac{e^{-\lambda r}}{r}\right) = \frac{1}{(2\pi)^{3/2}}\int e^{-i\mathbf{k}\cdot\mathbf{r}}\frac{e^{-\lambda r}}{r}d\mathbf{r}$$

$$= \frac{1}{(2\pi)^{1/2}}\int_0^\infty e^{-\alpha r}dr\frac{2\sin(kr)}{k} = \frac{1}{(2\pi)^{1/2}}\left(\frac{2}{k^2+\lambda^2}\right)$$

Example - 4.3 : Evaluate the commutator $[e^{-i\mathbf{k}\cdot\mathbf{P}}, \mathbf{p}]$.

Solution :

$$[e^{-i\mathbf{k}\cdot\mathbf{P}}, \mathbf{p}]\psi = [e^{-i\mathbf{k}\cdot\mathbf{P}}, -i\hbar\nabla]\psi$$

$$= e^{-i\mathbf{k}\cdot\mathbf{P}}(-i\hbar\nabla)\psi - (-i\hbar\nabla)e^{-i\mathbf{k}\cdot\mathbf{P}}\psi = \hbar\mathbf{k}e^{-i\mathbf{k}\cdot\mathbf{P}}\psi$$

Therefore

$$[e^{-i\mathbf{k}\cdot\mathbf{P}}, \mathbf{p}] = \hbar\mathbf{k}e^{-i\mathbf{k}\cdot\mathbf{P}}$$

Example - 4.4 : Given

$$\psi(q,0) = \frac{1}{(2\pi\sigma^2)^{1/4}}e^{-\frac{q^2}{4\sigma^2}}$$

obtain $\psi(q,t)$ by the time–evolution operator method.

Solution :

$$\psi(q,t) = e^{-\frac{i}{\hbar}Ht}\psi(q,0) \quad , \quad H = -\frac{\hbar^2}{2m}\frac{\partial^2}{\partial q^2}$$

Using the identity

$$\frac{\partial^2}{\partial q^2}\left(\frac{1}{\sqrt{\sigma^2}}e^{-\frac{q^2}{4\sigma^2}}\right) = \frac{\partial}{\partial(\sigma^2)}\left(\frac{1}{\sqrt{\sigma^2}}e^{-\frac{q^2}{4\sigma^2}}\right)$$

$$\psi(q,t) = e^{\frac{i\hbar}{2m}t\frac{\partial^2}{\partial q^2}}\left(\frac{\sigma^2}{2\pi}\right)^{1/4}\frac{1}{\sqrt{\sigma^2}}e^{-\frac{q^2}{4\sigma^2}}$$

$$= e^{\frac{i\hbar}{2m}t\frac{\partial}{\partial(\sigma^2)}}\left(\frac{\sigma^2}{2\pi}\right)^{1/4}\frac{1}{\sqrt{\sigma^2}}e^{-\frac{q^2}{4\sigma^2}}$$

$$= \left(\frac{\sigma^2}{2\pi}\right)^{1/4}\frac{1}{\sqrt{\sigma^2 + \frac{i\hbar}{2m}t}}e^{-\frac{q^2}{4(\sigma^2 + \frac{i\hbar}{2m}t)}}$$

here we used the relation

$$e^{\alpha\frac{\partial}{\partial z}}f(z) = f(z+\alpha)$$

The probability density at time t is

$$|\psi(q,t)|^2 = \frac{1}{\sqrt{2\pi}\alpha(t)}e^{-\frac{q^2}{2\alpha^2(t)}}$$

where

$$\alpha(t) = \sigma\left[1 + \frac{\hbar^2 t^2}{4m^2\sigma^4}\right]^{1/2}$$

Example - 4.5 : Prove the following: (a) $\frac{d}{dt} < x >= \frac{<p_x>}{m}$, (b) $\frac{d}{dt} < p_x >=< F_x >$. These are called Ehrenfest's theorem.

Solution : (a)

$$\frac{d}{dt} < x >= \frac{1}{i\hbar} < [x, H] >= \frac{1}{2mi\hbar} < [x, p_x^2] >= \frac{< p_x >}{m}$$

(b)

$$\frac{d}{dt} < p_x >= \frac{1}{i\hbar} < [p_x, H] >= \frac{1}{i\hbar} < [p_x, V(x)] >$$

$$= - < \frac{\partial V}{\partial x} >=< F_x >$$

Example - 4.6 : Show for a bound state if the normalization of the wave function is to be constant in time, it is necessary and sufficient that the Hamiltonian be Hermitian.

Solution : The normalization of a wave function would be constant if $N = \int |\psi(\mathbf{r}, t)|^2 d\mathbf{r}$ is a constant, or $\frac{dN}{dt} = 0$, i.e.,

$$\int \left(\psi^* \frac{\partial \psi}{\partial t} + \psi \frac{\partial \psi^*}{\partial t} \right) d\mathbf{r} = 0$$

(integrating over entire configuration space)

$$\int [\psi^* H\psi - (H\psi)^* \psi] d\mathbf{r} = 0$$

Since ψ is any arbitrary quantum mechanical state the above equation is true for any ψ. Hence H must be Hermitian.

Example - 4.7 : A pulse 1 meter long contains 1000 α–particles. At $t = 0$, each particle is in the state

$$\psi(x, 0) = \frac{1}{10} e^{ik_0 x} \quad , \quad -50 \leq x \leq 50 \quad , \quad k_0 = \frac{\pi}{50}$$

$\psi(x, 0) = 0$ otherwise. (a) At $t = 0$, how many particles have momentum in the interval $(0, \hbar k_0)$? (b) At which values of momentum will particles not be found at $t = 0$?

Solution : (a)

$$\Delta N = N \int_0^{k_0} |b(k)|^2 dx$$

$$b(k) = \frac{1}{\sqrt{2\pi}} \int_{-50}^{50} \psi(x, 0) e^{-ikx} dx = \frac{1}{\sqrt{2\pi}} \int_{-50}^{50} \frac{1}{10} e^{ik_0 x - ikx} dx$$

$$= \frac{1}{10\sqrt{2\pi}} \left(\frac{e^{i(k_0 - k)50} - e^{-i(k_0 - k)50}}{i(k_0 - k)} \right) = \frac{1}{5\sqrt{2\pi}} \frac{\sin[(k_0 - k)50]}{(k_0 - k)}$$

Therefore,

$$|b(k)|^2 = \frac{1}{50\pi} \frac{\sin^2[(k_0 - k)50]}{(k_0 - k)^2}$$

$$\Delta N = N \int_0^{k_0} \frac{1}{50\pi} \frac{\sin^2[(k_0 - k)50]}{(k_0 - k)^2} dk$$

$$= N \frac{50}{\pi} \int_0^{k_0} \frac{\sin^2[(k_0 - k)50]}{(k_0 - k)^2} dk$$

Let $(k_0 - k)50 = z \rightarrow -50dk = dz$

$$\Delta N = N \frac{50}{\pi} (-\frac{1}{50}) \int_{50k_0}^0 \frac{\sin^2 z}{z^2} dz$$

$$= \frac{N}{\pi} \int_0^\pi \frac{\sin^2 z}{z^2} dz \cong 0.45N = 450$$

(b) The sine function vanishes at the values $(k_0 - k)50 = n\pi$, $n = 1, 2, 3, \cdots$
Therefore, $k = k_0 - \frac{n\pi}{50}$,

$$p = \hbar k = \hbar k_0 - \frac{n\pi\hbar}{50} = \hbar\frac{\pi}{50} - n\hbar\frac{\pi}{50} = \frac{\hbar\pi}{50}(1 - n)$$

Example - 4.8 : Let a particle move in the potential $V(x) = Ax^n$, where A is constant and n is a finite integer. Determine the value of n for which the Ehrenfest's equation gives the classical relation.

Solution : Ehrenfest's equation:

$$\frac{d <p>}{dt} = - < \frac{\partial V}{\partial x} >= -An < x^{n-1} >$$

Classical relation:

$$\frac{dp}{dt} = -\frac{\partial V}{\partial x} = -Anx^{n-1}$$

To obtain the classical form we must equate $< x^{n-1} >=< x >^{n-1}$; this is valid only for $n = 2$.

4.11   Problems

Problem - 4.1 : Obtain the Fourier transform for the function $f(x) = (\delta/\pi)(\delta^2 + x^2)^{-1}$ where δ is a positive nonzero constant. This function is known as a Lorentzian function.

Problem - 4.2 : Evaluate the commutator $[H, x]$, H is a Hamiltonian operator in one dimension.

Problem - 4.3 : Show that if the Hamiltonian H of a system does not contain time explicitly,

$$|\psi(t) >= e^{-\frac{i}{\hbar}Ht}|\psi(0) >$$

Problem - 4.4 : Given

$$\psi(q, 0) = \frac{1}{(2\alpha)^{1/4}} e^{-\pi\alpha q^2 + ik_0 q}$$

where k_0 is a constant, find $\psi(q, t)$. Show that the above wave function at $t = 0$ is a minimum uncertainty product wave function.

Problem - 4.5 : If a quantum mechanical system is in an energy eigenstate, show that $< T >= \frac{1}{2} < \mathbf{r} \cdot \nabla V >$ (Virial theorem) where T and V are respectively the kinetic and potential energy terms.

Problem - 4.6 : The state function $\psi(x, t)$ must obey the time–dependent wave equation. In the event that $|\psi(x, t)|^2$ is independent of time we say that ψ describes a stationary state. (a) $\psi(x, t)$ can be expanded as

$$\psi(x, t) = \sum_n c_n \phi_n(t) u_n(x)$$

where the c_n's are constants and u_n's are eigenfunctions of H. Determine the form of $\phi_n(t)$ by direct substitution into the wave equation. (b) What condition on the c_n's must be satisfied in order for ψ to represent a stationary state?

Problem - 4.7 : A particle is in the state

$$|\phi> = \frac{1}{\sqrt{3}}(|\psi_1> + i|\psi_2> - |\psi_3>)$$

where $|\psi_i>$ are energy eigenstates; $H|\psi_i> = E_i|\psi_i>$. (a) Find $< \psi_2|\phi>$. (b) What is the probability that measurement of the energy in the state $|\phi>$ will give the value E_2? (c) If the energy is measured and the value E_2 is found, what is the probability of finding E_3 in a later measurement?

Problem - 4.8 : For a certain system, the operator corresponding to the physical quantity A does not commute with the Hamiltonian. It has eigenvalues a_1 and a_2, corresponding to eigenfunctions $\phi_1 = (u_1 + u_2)/\sqrt{2}$ and $\phi_2 = (u_1 - u_2)/\sqrt{2}$, where u_1 and u_2 are eigenfunctions of the Hamiltonian with eigenvalues E_1 and E_1. If the system is in the state $\psi = \phi_1$ at time $t = 0$, what is the expectation value of A at time t?

Problem - 4.9 : A free particle of mass m moves in one dimension. The initial wave function of the particle is $\psi(x, 0)$. After a sufficiently long time t the wave function of the particle spreads to reach a unique limiting form given by

$$\psi(x, t) = \sqrt{m/\hbar t}\exp\left(-i\pi/4\right)\exp\left(imx^2/2\hbar t\right)\phi(mx/\hbar t).$$

Give a reasonable physical interpretation of the limiting value of $|\psi(x, t)|^2$.
Hint: Note that when $\alpha \to \infty$, $\exp\left(-i\alpha u^2\right) \to \sqrt{\pi/\alpha}\exp\left(-i\pi/4\right)\delta(u)$.

Chapter 5

Bound and Unbound States in One–Dimension

5.1 One–dimensional Schrödinger equation

Before going further to discuss the one–dimensional problems in quantum mechanics, let us first describe the general properties of the one–dimensional Schrödinger equation.

The time–independent Schrödinger equation for a particle of mass m moving in one dimension in a potential field $V(x)$ appears as

$$\left[-\frac{\hbar^2}{2m}\frac{\partial^2}{\partial x^2} + V(x)\right]\psi(x) = E\psi(x) \tag{5.1}$$

We may rewrite this equation as

$$\frac{\partial^2\psi}{\partial x^2} = -\frac{2m}{\hbar^2}(E - V)\psi = -k^2\psi \tag{5.2}$$

One can write from the last equality that

$$\frac{\hbar^2 k^2}{2m} = E - V \tag{5.3}$$

Since total energy is separated into kinetic and potential parts (partitioning of energy)

$$E = T + V \tag{5.4}$$

We can identify $\hbar^2 k^2/2m$ as the kinetic energy of the particle

$$T = \frac{\hbar^2 k^2}{2m} \tag{5.5}$$

This identification is especially relevant if $E > V$. More generally, there are three distinct possibilities. These are $E > V$, $E = V$, and $E < V$ (see Figure 5.1).

In the region, where $E > V$, the kinetic energy is positive and the corresponding classical motion is permitted. In the region, where $E < V$, the kinetic energy is negative and the classical motion is forbidden. The points where $E = V$ are the classical turning point, at these points the kinetic energy is zero.

In the regions where the kinetic energy is negative, the Schrödinger equation becomes

$$\frac{d^2\psi}{dx^2} = \kappa^2\psi \tag{5.6}$$

where

$$\frac{\hbar^2 \kappa^2}{2m} = V - E > 0 \tag{5.7}$$

$$\text{Kinetic energy} = -\frac{\hbar^2 \kappa^2}{2m} = E - V < 0 \tag{5.8}$$

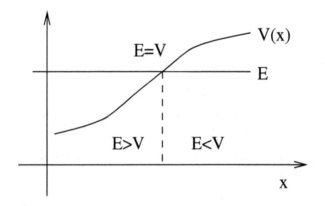

Fig. 5.1. Regions of space in 1D with respect to E and V.

Considering the analytic properties of the second derivative of a function we may analize the properties of the Schrödinger equation as follows:

When the kinetic energy is positive, the Schrödinger equation takes the form

$$\text{Kinetic energy} = \frac{\hbar^2 k^2}{2m} = E - V > 0 \tag{5.9}$$

and ψ has the following properties: $\partial^2 \psi / \partial x^2 < 0$ in the upper half–plane, so ψ is concave downward; $\partial^2 \psi / \partial x^2 > 0$ in the lower half–plane, so ψ is concave upward. These conditions permit oscillating solutions.

When the kinetic energy is negative, the Schrödinger equation takes the form

$$\text{Kinetic energy} = \frac{\hbar^2 \kappa^2}{2m} = V - E > 0 \tag{5.10}$$

and ψ has the following properties: $\partial^2 \psi / \partial x^2 > 0$ in the upper half–plane, so ψ is concave upward; $\partial^2 \psi / \partial x^2 < 0$ in the lower half–plane, so ψ is concave downward. These conditions permit decaying solutions. At a turning point $\partial^2 \psi / \partial x^2 = 0$ and ψ has a constant slope (see Figure 5.2).

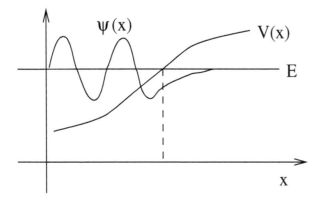

Fig. 5.2. General properties of E, $V(x)$, and $\psi(x)$ in 1D.

5.2 The simple harmonic oscillator

The configuration of a classical simple harmonic oscillator (SHO) can be described as shown in Figure 5.3.

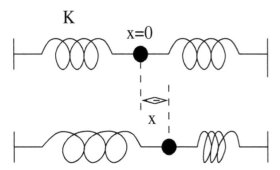

Fig. 5.3. Classical simple harmonic oscillator in 1D.

The classical equation of motion of a particle of mass m is given by Hooke's law, $F = -Kx$, or

$$m\frac{d^2x}{dt^2} = -Kx \tag{5.11}$$

The spring constant is K. In terms of the natural frequency ω_0

$$\omega_0^2 = \frac{K}{m} \tag{5.12}$$

The equation of motion takes the form

$$\frac{d^2x}{dt^2} + \omega_0^2 x = 0 \quad , \quad \ddot{x} + \omega_0^2 x = 0 \tag{5.13}$$

Multiplying this equation by $\dot{x}$ gives

$$\dot{x}\ddot{x} + \omega_0^2 \dot{x}x = 0 \quad \longrightarrow \quad \frac{d}{dt}\left[\frac{1}{2}(\dot{x}^2 + \omega_0^2 x^2)\right] = 0 \tag{5.14}$$

Integrating

$$\int \frac{d}{dt}\left[\frac{1}{2}(\dot{x}^2 + \omega_0^2 x^2)\right] dt = 0 \quad \longrightarrow \quad \int d\left[\frac{1}{2}(\dot{x}^2 + \omega_0^2 x^2)\right] = 0 \tag{5.15}$$

or

$$\frac{1}{2}(\dot{x}^2 + \omega_0^2 x^2) - C = 0 \tag{5.16}$$

The constant C may be taken as the constant of the motion, E/m

$$\frac{E}{m} = \frac{1}{2}(\dot{x}^2 + \omega_0^2 x^2) \tag{5.17}$$

Therefore, the total energy of the oscillator becomes

$$E = \frac{1}{2}m\dot{x}^2 + \frac{1}{2}m\omega_0^2 x^2 = \frac{1}{2}m\dot{x}^2 + \frac{1}{2}Kx^2 \tag{5.18}$$

The potential energy is (see Figure 5.4)

$$V = \frac{1}{2}Kx^2 \tag{5.19}$$

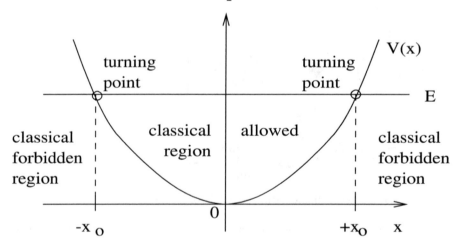

Fig. 5.4. Regions of space for SHO potential in 1D.

When the particle comes to rest ($T = 0$) the energy is entirely potential

$$E = V = \frac{1}{2}Kx_0^2 \tag{5.20}$$

Such points ($\mp x_0$) are turning points. For $x^2 > x_0^2$, the kinetic energy T is negative, so that classically this is a forbidden region.

$$T = E - V = \frac{1}{2}K(x_0^2 - x^2) \tag{5.21}$$

$$T < 0 \quad \text{for} \quad x^2 > x_0^2 \tag{5.22}$$

From the quantum mechanical point of view the harmonic oscillator problem is simply a particle in a potential problem. The Hamiltonian for a particle of mass m in the potential

$$V = \frac{1}{2}Kx^2 \tag{5.23}$$

is

$$H = \frac{p^2}{2m} + \frac{1}{2}Kx^2 \tag{5.24}$$

The corresponding Schrödinger equation appears as

$$-\frac{\hbar^2}{2m}\frac{\partial^2\psi}{\partial x^2} + \frac{1}{2}Kx^2\psi = E\psi \tag{5.25}$$

In the classical allowed region, $E > \frac{1}{2}Kx^2$, this equation may be written

$$\frac{\partial^2\psi}{\partial x^2} = -\frac{2m}{\hbar^2}(E - \frac{1}{2}Kx^2)\psi \tag{5.26}$$

or

$$\psi_{xx} = -k^2\psi \tag{5.27}$$

$$\frac{\hbar^2 k^2}{2m} = E - \frac{1}{2}Kx^2 > 0 \quad ; \quad k^2 = \frac{2m}{\hbar^2} \quad ; \quad \psi_{xx} = \frac{\partial^2\psi}{\partial x^2} \tag{5.28}$$

The wave function ψ is oscillatory in this region. In the classical forbidden region where $x^2 > x_0^2$, $E < \frac{1}{2}Kx^2$ and the Schrödinger equation becomes

$$\psi_{xx} = \kappa^2\psi \tag{5.29}$$

$$\frac{\hbar^2\kappa^2}{2m} = \frac{1}{2}Kx^2 - E > 0 \quad ; \quad \kappa^2 = \frac{2m}{\hbar^2}(V - E) \tag{5.30}$$

so the wave function ψ is nonoscillatory in this region. At the points far from the turning points, $V = \frac{1}{2}Kx^2 \gg E$, the Schrödinger equation becomes

$$\psi_{xx} = \kappa^2\psi \cong \frac{2m}{\hbar^2}V\psi = \frac{2m}{\hbar^2}\frac{1}{2}Kx^2\psi \tag{5.31}$$

$$= \frac{mK}{\hbar^2}x^2\psi = \beta^4 x^2\psi$$

where β is the characteristic wavenumber

$$\beta^4 = \frac{mK}{\hbar^2} = \frac{m^2 K}{\hbar^2 m} = \frac{m^2\omega_0^2}{\hbar^2} \quad ; \quad \beta^2 \equiv \frac{m\omega_0}{\hbar} \tag{5.32}$$

In terms of the nondimensional displacement

$$\xi = \beta x \tag{5.33}$$

the equation $\psi_{xx} = \beta^4 x^2\psi$ appears as

$$\psi_{\xi\xi} = \xi^2\psi \tag{5.34}$$

Using change of variable one can show this equality,

$$\frac{\partial \psi}{\partial \xi} = \frac{\partial \psi}{\partial x} \frac{\partial x}{\partial \xi} = \frac{\partial \psi}{\partial \xi} \frac{1}{\beta}$$

$$\frac{\partial^2 \psi}{\partial \xi^2} = \frac{\partial}{\partial \xi}\left(\frac{\partial \psi}{\partial x}\frac{1}{\beta}\right) = \frac{\partial^2 \psi}{\partial x^2}\frac{\partial x}{\partial \xi}\frac{1}{\beta} = \frac{\partial^2 \psi}{\partial x^2}\frac{1}{\beta^2}$$

Therefore,

$$\frac{\partial^2 \psi}{\partial x^2} = \beta^2 \frac{\partial^2 \psi}{\partial \xi^2} = \beta^4 x^2 \psi \qquad \longrightarrow \qquad \frac{\partial^2 \psi}{\partial \xi^2} = \beta^2 x^2 \psi$$

In the region under consideration, $\xi > 1$, the solution to Eq. (5.34) appears as

$$\psi_\mp \sim A e^{\mp \frac{1}{2}\xi^2} = A e^{\mp \frac{1}{2}\beta^2 x^2} \tag{5.35}$$

The $(+)$ solution (ψ_+) does not satisfy the normalization

$$\int_{-\infty}^{\infty} \psi_+^* \psi_+ dx \quad \longrightarrow \quad \infty \tag{5.36}$$

The $(-)$ solution (ψ_-) satisfies the required conditions

$$\psi_- = \psi \sim A e^{-\frac{1}{2}\xi^2} = A e^{-\frac{1}{2}\beta^2 x^2} \tag{5.37}$$

The character of the wave function changes from oscillatory for $x^2 < x_0^2$ to decaying for $x^2 > x_0^2$, so the turning points $x = \mp x_0$ are also physically relevant in quantum mechanics; see Figure 5.5.

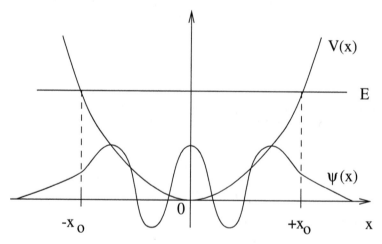

Fig. 5.5. Behaviour of wave function in various regions of space for SHO potential in 1D.

General solution of the SHO problem:

The general formulation of the solution to the Schrödinger equation for the harmonic oscillator

$$-\frac{\hbar^2}{2m}\frac{\partial^2 \psi}{\partial x^2} + \frac{1}{2}Kx^2\psi = E\psi \tag{5.38}$$

is known as the algebraic method. The following operators are defined in the algebraic method in terms of position and momentum operators,

$$\hat{a} = \frac{\beta}{\sqrt{2}}\left(\hat{x} + i\frac{\hat{p}}{m\omega_0}\right) \quad , \quad \hat{a}^+ = \frac{\beta}{\sqrt{2}}\left(\hat{x} - i\frac{\hat{p}}{m\omega_0}\right) \tag{5.39}$$

$\hat{a}$ is non–Hermitian,

$$\hat{a} \neq \hat{a}^+ \tag{5.40}$$

$$\beta = \sqrt{\frac{m\omega_0}{\hbar}} \quad , \quad [\beta] = \frac{1}{\text{Length}}$$

Using the fundamental commutator relation $[\hat{x}, \hat{p}] = i\hbar$, we write the properties of these operators as

$$[\hat{a}, \hat{a}^+] = 1 \quad , \quad \hat{a}\hat{a}^+ = 1 + \hat{a}^+\hat{a} \tag{5.41}$$

The position and momentum operators can be expressed in terms of these operators as

$$\hat{x} = \frac{1}{\sqrt{2}\beta}(\hat{a} + \hat{a}^+) \quad , \quad \hat{p} = \frac{m\omega_0}{i\sqrt{2}\beta}(\hat{a} - \hat{a}^+) \tag{5.42}$$

The Hamiltonian for the SHO becomes

$$H = \frac{p^2}{2m} + \frac{1}{2}Kx^2 = \hbar\omega_0\left(\hat{a}^+\hat{a} + \frac{1}{2}\right) = \hbar\omega_0\left(\hat{N} + \frac{1}{2}\right) \tag{5.43}$$

We can show this equality as the following:
The squares of $\hat{x}$ and $\hat{p}$ are

$$x^2 = \frac{1}{2\beta^2}(\hat{a} + \hat{a}^+)^2 \quad , \quad p^2 = -\frac{m^2\omega^2}{2\beta^2}(\hat{a} - \hat{a}^+)^2 \tag{5.44}$$

After substituting these into the Hamiltonian, we get

$$H = \frac{p^2}{2m} + \frac{1}{2}Kx^2 \tag{5.45}$$

$$= \frac{1}{2m}\left(-\frac{m^2\omega^2}{2\beta^2}\right)(\hat{a} - \hat{a}^+)^2 + \frac{1}{2}K\left(\frac{1}{2\beta^2}\right)(\hat{a} + \hat{a}^+)^2$$

$$= -\frac{m\omega_0^2}{4\beta^2}(\hat{a} - \hat{a}^+)^2 + \frac{K}{4\beta^2}(\hat{a} + \hat{a}^+)^2$$

$$= -\frac{m\omega_0^2}{4\beta^2}(\hat{a} - \hat{a}^+)^2 + \frac{m\omega_0^2}{4\beta^2}(\hat{a} + \hat{a}^+)^2$$

$$= \frac{m\omega_0^2}{4\beta^2}\left[(\hat{a} + \hat{a}^+)^2 - (\hat{a} - \hat{a}^+)^2\right] = \frac{m\omega_0^2}{4\beta^2}[2(\hat{a}\hat{a}^+ + \hat{a}^+\hat{a})]$$

Using the relations $\hat{a}\hat{a}^+ = 1 + \hat{a}^+\hat{a}$ and $\beta^2 = \frac{m\omega_0}{\hbar}$, we obtain

$$H = \frac{m\omega_0^2}{4\beta^2}[2(\hat{a}^+\hat{a} + 1 + \hat{a}^+\hat{a})] = \frac{\hbar\omega_0}{4}2(2\hat{a}^+\hat{a} + 1) \tag{5.46}$$

$$= \hbar\omega_0\left(\hat{a}^+\hat{a} + \frac{1}{2}\right) = \hbar\omega_0\left(\hat{N} + \frac{1}{2}\right)$$

The eigenvalues of H depend on the eigenvalues of the operator

$$\hat{N} = \hat{a}^+\hat{a} \tag{5.47}$$

Let ψ_n be the eigenfunction of $\hat{N}$ corresponding to the eigenvalue n, so that

$$\hat{N}\psi_n = n\psi_n \tag{5.48}$$

Here n is just a number, therefore $\hat{N}$ is usually called as the **number operator**. Consider the effect of the operator $\hat{N}$ on $\hat{a}\psi_n$,

$$\hat{N}\hat{a}\psi_n = \hat{a}^+\hat{a}\hat{a}\psi_n = (\hat{a}\hat{a}^+ - 1)\hat{a}\psi_n = \hat{a}(\hat{a}^+\hat{a} - 1)\psi_n \tag{5.49}$$

$$\hat{N}\hat{a}\psi_n = \hat{a}(\hat{N} - 1)\psi_n = \hat{a}(n - 1)\psi_n = (n - 1)\hat{a}\psi_n \tag{5.50}$$

It follows that $\hat{a}\psi_n$ is the eigenfunction of $\hat{N}$ which corresponds to the eigenvalue $(n - 1)$. That is

$$\hat{a}\psi_n = \psi_{n-1} \tag{5.51}$$

Similarly, $\hat{a}\psi_{n-1} = \psi_{n-2}$, and so on. Because of this property, $\hat{a}$ is called an annihilation or stepdown or demotion operator.

In a similar manner, if we consider the effect of the operator $\hat{N}$ on $\hat{a}^+\psi_n$,

$$\hat{N}\hat{a}^+\psi_n = \hat{a}^+\hat{a}\hat{a}^+\psi_n = \hat{a}^+(1 + \hat{a}^+\hat{a})\psi_n = \hat{a}^+(1 + \hat{N})\psi_n \tag{5.52}$$

$$= \hat{a}^+(1 + n)\psi_n = (n + 1)\hat{a}^+\psi_n$$

This equation implies that $\hat{a}^+\psi_n$ is the eigenfunction of $\hat{N}$ corresponding to the eigenvalue $(n + 1)$.

$$\hat{a}^+\psi_n = \psi_{n-1} \tag{5.53}$$

Similarly, $\hat{a}^+\psi_{n+1} = \psi_{n+2}$, and so on. The operator $\hat{a}^+$ is called a creation or stepup or promotion operator.

Since the Hamiltonian for the harmonic oscillator is the sum of the squares of two Hermitian operators, $(p^2$ and $x^2)$,

$$<H> \geq 0 \tag{5.54}$$

In the eigenstate ψ_n,

$$H\psi_n = \hbar\omega_0(\hat{N} + \frac{1}{2})\psi_n = \hbar\omega_0(n + \frac{1}{2})\psi_n \tag{5.55}$$

$$<\psi_n|H\psi_n> = \hbar\omega_0(n + \frac{1}{2}) \geq 0 \tag{5.56}$$

This implies that the eigenvalues n must obey the condition

$$n \geq -\frac{1}{2} \tag{5.57}$$

That is, all eigenstates of H, or equivalently $\hat{N}$, corresponding to eigenvalues $n < -1/2$ must vanish identically. For harmonic oscillator such states do not exist. This condition is guaranteed if we set

$$\hat{a}\psi_0 = 0 \tag{5.58}$$

With $\hat{a}\psi_n = \psi_{n-1}$ we obtain

$$\hat{a}\psi_0 = \psi_{-1} = 0 \quad , \quad \hat{a}(\psi_{-1}) = \psi_{-2} = 0 \tag{5.59}$$

Therefore,

$$\hat{N}\psi_0 = \hat{a}^+\hat{a}\psi_0 = 0\psi_0 = 0 \tag{5.60}$$

and we may conclude that $\hat{N}\psi_0 = 0$; it follows that

$$\hat{N}\hat{a}^+\psi_0 = \hat{a}^+\hat{a}\hat{a}^+\psi_0 = \hat{a}^+(\hat{a}^+\hat{a} + 1)\psi_0 = \hat{a}^+\psi_0 \qquad (5.61)$$

$$\hat{N}\hat{a}^+\psi_0 = 1\hat{a}^+\psi_0 = \psi_1 \qquad (5.62)$$

The eigenvalue of $\hat{N}$ corresponding to ψ_1 is the integer 1. This result allows one to conclude that the index n, which labels the eigenfunction ψ_n, and the eigenvalue of $\hat{N}$, is indeed an integer.

$$H\psi_n = \hbar\omega_0 \left(\hat{N} + \frac{1}{2}\right)\psi_n = \hbar\omega_0 \left(n + \frac{1}{2}\right)\psi_n \qquad (5.63)$$

From this equality one finds that the energy eigenvalues of the simple harmonic oscillator (one–dimensional SHO) are

$$E_n = \hbar\omega_0 \left(n + \frac{1}{2}\right) \quad ; \quad n = 0, 1, 2, \cdots \qquad (5.64)$$

The energy levels are equally spaced by the interval $\hbar\omega_0 = h\nu_0$; see Figure 5.6.

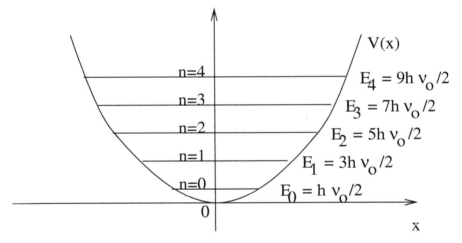

Fig. 5.6. Relative positions of energy levels in SHO in 1D.

Energy difference between any two states labeled by n' and n, say

$$\Delta E = E_{n'} - E_n = \hbar\omega_0 \left(n' + \frac{1}{2}\right) - \hbar\omega_0 \left(n + \frac{1}{2}\right) \qquad (5.65)$$

$$= \hbar\omega_0(n' - n) = \hbar\omega_0 s$$

Since n' and n are integers, so their difference, s, is also an integer. Therefore the energy difference is the integral multiples of the natural frequency of the oscillator, ω_0.

The lowest energy level of the harmonic oscillator, corresponding to the label $n = 0$, $E_0 = \frac{1}{2}\hbar\omega_0$, is called as the zero–point energy of the oscillator, which is the ground state energy of the oscillator.

Eigenfunctions of the SHO Hamiltonian:

The step operators may be expressed in terms of the nondimensional displacement ξ

$$\xi = \beta x \quad , \quad \beta = \sqrt{\frac{m\omega_0}{\hbar}}$$

$$\hat{a} = \frac{\beta}{\sqrt{2}}\left(x + \frac{i}{m\omega_0}p\right) = \frac{\beta}{\sqrt{2}}\left(x + \frac{\hbar}{m\omega_0}\frac{\partial}{\partial x}\right) = \frac{1}{\sqrt{2}}\left(\xi + \frac{\partial}{\partial \xi}\right) \tag{5.66}$$

$$\hat{a}^+ = \frac{\beta}{\sqrt{2}}\left(x - \frac{i}{m\omega_0}p\right) = \frac{\beta}{\sqrt{2}}\left(x - \frac{\hbar}{m\omega_0}\frac{\partial}{\partial x}\right) = \frac{1}{\sqrt{2}}\left(\xi - \frac{\partial}{\partial \xi}\right) \tag{5.67}$$

The time–independent Schrödinger equation, $H\psi = E\psi$, becomes

$$\hbar\omega_0\left(\hat{a}^+\hat{a} + \frac{1}{2}\right)\psi = E\psi \tag{5.68}$$

or

$$\hbar\omega_0(2\hat{a}^+\hat{a} + 1)\psi = 2E\psi \tag{5.69}$$

or

$$(2\hat{a}^+\hat{a} + 1)\psi = \frac{2E}{\hbar\omega_0}\psi \tag{5.70}$$

or

$$\left(2\hat{a}^+\hat{a} + 1 - \frac{2E}{\hbar\omega_0}\right)\psi = 0 \tag{5.71}$$

Substituting $\hat{a}$ and $\hat{a}^+$ expressed in terms of ξ gives

$$\left[2\frac{1}{\sqrt{2}}\left(\xi - \frac{\partial}{\partial \xi}\right)\frac{1}{\sqrt{2}}\left(\xi + \frac{\partial}{\partial \xi}\right) + 1 - \frac{2E}{\hbar\omega_0}\right] = 0 \tag{5.72}$$

$$\left(\xi - \frac{\partial}{\partial \xi}\right)\left(\xi + \frac{\partial}{\partial \xi}\right)\psi + \psi - \frac{2E}{\hbar\omega_0}\psi = 0 \tag{5.73}$$

$$\xi^2\psi + \xi\frac{\partial\psi}{\partial\xi} - \frac{\partial}{\partial\xi}(\xi\psi) - \frac{\partial^2\psi}{\partial\xi^2} + \psi - \frac{2E}{\hbar\omega_0}\psi = 0 \tag{5.74}$$

or

$$\xi^2\psi - \frac{\partial^2\psi}{\partial\xi^2} - \frac{2E}{\hbar\omega_0}\psi = 0 \tag{5.75}$$

or

$$\frac{\partial^2\psi}{\partial\xi^2} + \left(\frac{2E}{\hbar\omega_0} - \xi^2\right)\psi = 0 \quad \longrightarrow \quad \psi_{\xi\xi} + \left(\frac{2E}{\hbar\omega_0} - \xi^2\right)\psi = 0 \tag{5.76}$$

The ground state wave function of the SHO Hamiltonian obeys $\hat{a}\psi_0 = 0$, or equivalently,

$$\frac{1}{\sqrt{2}}\left(\xi + \frac{\partial}{\partial\xi}\right)\psi_0 = 0 \tag{5.77}$$

This differential equation has the solution

$$\psi_0 = A_0 e^{-\frac{1}{2}\xi^2} \tag{5.78}$$

The normalization constant A_0 may be calculated as

$$\int_{-\infty}^{\infty}|\psi_0|^2 d\xi = A_0^2 \int_{-\infty}^{\infty} e^{-\xi^2} d\xi = A_0^2\sqrt{\pi} = 1 \tag{5.79}$$

so

$$A_0 = \frac{1}{\pi^{1/4}} \tag{5.80}$$

In terms of the dimensional displacement x, the normalized ground state wave function is

$$\psi_0(x) = B_0 e^{-\frac{1}{2}\xi^2} = B_0 e^{-\frac{1}{2}\beta^2 x^2} \tag{5.81}$$

Normalization with respect to x gives

$$\int_{-\infty}^{\infty} |\psi_0(x)|^2 dx = \int_{-\infty}^{\infty} \frac{B_0^2}{\beta} e^{-\xi^2} d\xi = \frac{B_0^2}{\beta} = 1 \tag{5.82}$$

Therefore,

$$B_0 = \sqrt{\frac{\beta}{\sqrt{\pi}}} = \left(\frac{\beta^2}{\pi}\right)^{1/4} \tag{5.83}$$

$$\psi_0(x) = \left(\frac{\beta^2}{\pi}\right)^{1/4} e^{-\frac{1}{2}\beta^2 x^2} \tag{5.84}$$

The ground state wave function is a purely exponentially decaying wave function. It has no oscillatory component (no nodes).

Considering the ground state solution,

$$\psi_0(\xi) = \pi^{-1/4} e^{-\frac{1}{2}\xi^2}$$

the remaining normalized eigenstates of the SHO Hamiltonian are generated with the aid of the stepup operator $\hat{a}^+$, as follows:

$$\psi_1 = \hat{a}^+ \psi_0 \tag{5.85}$$

$$\psi_2 = \frac{1}{\sqrt{2}} \hat{a}^+ \psi_1 = \frac{1}{\sqrt{2}} (\hat{a}^+)^2 \psi_0 \tag{5.86}$$

$$\vdots$$

$$\psi_n = \frac{1}{\sqrt{n!}} (\hat{a}^+)^n \psi_0 \tag{5.87}$$

With $\hat{a}^+$ written in terms of ξ,

$$\hat{a}^+ = \frac{1}{\sqrt{2}} \left(\xi - \frac{\partial}{\partial \xi}\right)$$

the equation for ψ_1 above becomes

$$\psi_1 = A_1 \left(\xi - \frac{\partial}{\partial \xi}\right) e^{-\frac{1}{2}\xi^2} = A_1 2\xi e^{-\frac{1}{2}\xi^2} \tag{5.88}$$

The normalization constant A_1 is calculated to be

$$A_1 = (2\sqrt{\pi})^{-1/2} \tag{5.89}$$

The nth eigenstate is given by the formula

$$\psi_n = A_n \left(\xi - \frac{\partial}{\partial \xi}\right)^n e^{-\frac{1}{2}\xi^2} \tag{5.90}$$

The nth–order differential operator $(\hat{a}^+)^n$, when acting on the exponential form $e^{-\frac{1}{2}\xi^2}$, reproduces the same exponential factor, multiplied by an nth–order polynomial in ξ

$$\left(\xi - \frac{\partial}{\partial \xi}\right)^n e^{-\frac{1}{2}\xi^2} = H_n(\xi)e^{-\frac{1}{2}\xi^2} \tag{5.91}$$

Thus the nth eigenstate of the SHO Hamiltonian may be written together with its eigenvalue as

$$\psi_n(\xi) = A_n H_n(\xi)e^{-\frac{1}{2}\xi^2} \tag{5.92}$$

$$E_n = \left(n + \frac{1}{2}\right)\hbar\omega_0 \tag{5.93}$$

The nth–order polynomials $H_n(\xi)$ are well–known functions, which are called Hermite polynomials. From

$$\psi_1 = A_1 2\xi e^{-\frac{1}{2}\xi^2}$$

we see that

$$H_1(\xi) = 2\xi$$

The first six Hermite polynomials are the following:

n	H_n
0	1
1	2ξ
2	$4\xi^2 - 2$
3	$8\xi^3 - 12\xi$
4	$16\xi^4 - 48\xi^2 + 12$
5	$32\xi^5 - 160\xi^3 + 120\xi$

Hermite polynomials can be simplified to the form

$$H_n(\xi) = (-1)^n e^{\xi^2}\frac{d^n}{d\xi^n}\left(e^{-\xi^2}\right) \tag{5.94}$$

The normalization constant can be generalized to the form

$$A_n = \frac{1}{\sqrt{2^n n!\sqrt{\pi}}} \tag{5.95}$$

The nth–order Hermite polynomial H_n enters in the eigenfunctions ψ_n of the quantum mechanical harmonic oscillator as

$$\psi_n(\xi) = A_n H_n(\xi)e^{-\frac{1}{2}\xi^2} \tag{5.96}$$

H_n is a solution to Hermite's equation

$$H_n'' - 2\xi H_n' + 2nH_n = 0 \tag{5.97}$$

The step operators $\hat{a}$ and $\hat{a}^+$ connect the eigenfunctions of the SHO $\psi_{n-1}, \psi_n, \psi_{n+1}$ as the following:

$$\hat{a}\psi_n(\xi) = \sqrt{n}\psi_{n-1}(\xi) \quad \text{or} \quad \hat{a}|n> = \sqrt{n}|n-1> \tag{5.98}$$

$$\hat{a}^{+}\psi_n(\xi) = \sqrt{n+1}\psi_{n+1}(\xi) \quad \text{or} \quad \hat{a}^{+}|n> = \sqrt{n+1}|n+1> \tag{5.99}$$

Let us see how $\hat{N} = \hat{a}^{+}\hat{a}$ acts on ψ_n:

$$\hat{N}|n> = \hat{a}^{+}\hat{a}|n> = \hat{a}^{+}(\sqrt{n}|n-1>) = \sqrt{n}(\hat{a}^{+}|n-1>) \tag{5.100}$$

$$= \sqrt{n}(\sqrt{n}|n>) = n|n>$$

The eigenfunctions of the SHO form an orthonormal basis set,

$$\int_{-\infty}^{\infty} \psi_n^{*}(\xi)\psi_m(\xi)d\xi = \delta_{nm} \tag{5.101}$$

Let us now calculate the expectation value of x and p in the nth eigenstate $|n>$.

$$<x> = <n|x|n> = <n|\frac{\hat{a}+\hat{a}^{+}}{\sqrt{2}\beta}|n> = \frac{1}{\sqrt{2}\beta} <n|\hat{a}+\hat{a}^{+}|n> \tag{5.102}$$

$$= \frac{1}{\sqrt{2}\beta} <n|(\sqrt{n}|n-1> +\sqrt{n+1}|n+1>)$$

$$= \frac{1}{\sqrt{2}\beta}(\sqrt{n} <n|n-1> +\sqrt{n+1} <n|n+1>) = 0$$

$$<p> = <n|p|n> = <n|\frac{m\omega_0(\hat{a}-\hat{a}^{+})}{i\sqrt{2}\beta}|n> \tag{5.103}$$

$$= \frac{m\omega_0}{i\sqrt{2}\beta} <n|\hat{a}-\hat{a}^{+}|n> = \frac{m\omega_0}{i\sqrt{2}\beta} <n|(\sqrt{n}|n-1> -\sqrt{n+1}|n+1>)$$

$$= \frac{m\omega_0}{i\sqrt{2}\beta}(\sqrt{n} <n|n-1> -\sqrt{n+1} <n|n+1>) = 0$$

The average value of x in any eigenstate ψ_n vanishes. This is due to the symmetry of the probability density $P = |\psi_n|^2$ about the origin; see Figure 5.7.

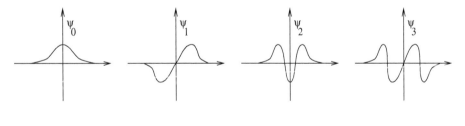

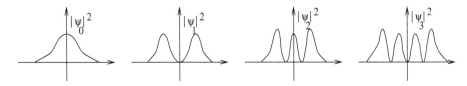

Fig. 5.7. The first four eigenfunctions (ψ_n) of the SHO and the corresponding probability densities ($|\psi_n|^2$).

The average value of p in any eigenstate ψ_n vanishes; this is due to the fact that the probability of finding the particle with momentum $\hbar k$ is equal to that of finding the particle with momentum $-\hbar k$. The probability amplitude of the momentum eigenstates $b(k)$ are a symmetric function of k, $b(k) = b(-k)$.

SHO and the correspondence principle:

Let us calculate the classical probability density P, corresponding to a one–dimensional spring with natural frequency w_0. Let the particle be at the origin at $t = 0$ with velocity $x_0 w_0$. The displacement at the time t is then given by

$$x = x_0 \sin(w_0 t) \tag{5.104}$$

$$\dot{x} = x_0 w_0 \cos(w_0 t) \tag{5.105}$$

This gives correct initial data: $x(0) = 0$, $\dot{x}(0) = x_0 w_0$. The product $P(x)dx$ is the probability of finding the particle in the interval dx about the point x at any time. If T_0 is the period of oscillation

$$T_0 = \frac{2\pi}{w_0} \tag{5.106}$$

then

$$P dx = \frac{dt}{T_0} = \frac{w_0 dt}{2\pi} \tag{5.107}$$

where

$$dt = \frac{dx}{\dot{x}} = \frac{dx}{x_0 w_0 \cos(w_0 t)} = \frac{dx}{x_0 w_0 \sqrt{1 - \sin^2(w_0 t)}} \tag{5.108}$$

$$= \frac{dx}{w_0 \sqrt{x_0^2 - x^2}}$$

so that

$$P dx = \frac{w_0}{2\pi} dt = \frac{dx}{2\pi \sqrt{x_0^2 - x^2}} \tag{5.109}$$

$$P^{CL} = \frac{1}{2\pi \sqrt{x_0^2 - x^2}} \tag{5.110}$$

The integral of $P dx$ over the interval $-x_0 < x < +x_0$ is

$$\int_{-x_0}^{+x_0} P(x) dx = \int_{-x_0}^{+x_0} \frac{dx}{2\pi \sqrt{x_0^2 - x^2}} = 1 \tag{5.111}$$

Plot of $P^{CL}(x)$ looks like as shown in Figure 5.8.

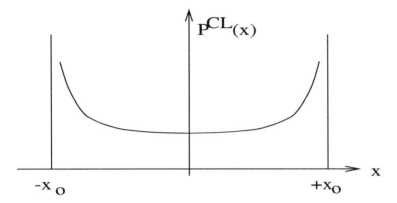

Fig. 5.8. Classical probability distribution in position for SHO in 1D.

The quantum mechanical probability density corresponding to a state with $n \gg 1$ looks like as shown in Figure 5.9. The quantum mechanical probability density

$$P_n^{QM} = |\psi_n|^2 \tag{5.112}$$

superimposes on the classical probability density P^{CL}. For the case $n \gg 1$,

$$\lim_{n \to \infty} < P_n^{QM} > = P^{CL} \tag{5.113}$$

where

$$< P_n^{QM} > = \frac{1}{2\epsilon} \int_{x-\epsilon}^{x+\epsilon} |\psi_n(x)|^2 dx \tag{5.114}$$

is the local average; it represents the average of P^{QM} in a small interval (2ϵ) centered at x.

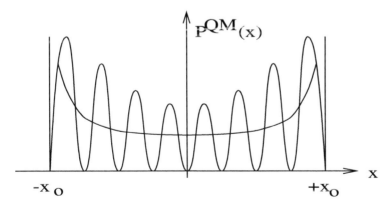

Fig. 5.9. Quantum probability distribution in position for SHO in 1D.

The harmonic oscillator in momentum space:

The eigenfunctions of the momentum operator p in one–dimensional space can be expressed as

$$\psi_k(x) = \frac{1}{\sqrt{2\pi}} e^{ikx} \tag{5.115}$$

The corresponding momentum value is $p = \hbar k$. Any function in momentum space may be expressed as

$$\psi(x) = \int_{-\infty}^{\infty} b(k)\psi_k \, dk \tag{5.116}$$

where

$$b(k) = \int_{-\infty}^{\infty} \psi_k^* \psi \, dx \tag{5.117}$$

The wave function $\psi(x)$ gives the probability density in coordinate space

$$P(x) = |\psi(x)|^2 \tag{5.118}$$

The momentum coefficient $b(k)$ gives the probability density in momentum space

$$P(k) = |b(k)|^2 \tag{5.119}$$

$P(k)dk$ is the probability of finding the particle to have momentum in the interval $\hbar k$ to $\hbar(k + dk)$ in momentum space. Any information contained in $\psi(x)$ can be obtained from the knowledge of $b(k)$ and vice versa. Given the Hamiltonian of a system, $\psi(x)$ is determined.

Let us consider the time–independent Schrödinger equation for the SHO,

$$\left(-\frac{\hbar^2}{2m}\frac{\partial^2}{\partial x^2} + \frac{1}{2}Kx^2\right)\psi(x) = E\psi(x) \tag{5.120}$$

Using

$$\psi(x) = \int_{-\infty}^{\infty} b(k)\psi_k \, dk \tag{5.121}$$

and

$$\frac{\partial \psi_k}{\partial k} = ix\psi_k \quad , \quad \frac{\partial^2 \psi_k}{\partial k^2} = -x^2\psi_k \tag{5.122}$$

we can write the Schrödinger equation

$$\int_{-\infty}^{\infty} dk\, b(k) \left(\frac{\hbar^2 k^2}{2m} - \frac{1}{2}K\frac{\partial^2}{\partial k^2}\right)\psi_k = E\int_{-\infty}^{\infty} dk\, b(k)\psi_k \tag{5.123}$$

integrating by parts twice and taking

$$(b(k))_{k=\mp\infty} = 0 \tag{5.124}$$

we can write

$$\int_{-\infty}^{\infty} dk\, b(k) \left(-\frac{1}{2}K\frac{\partial^2}{\partial k^2}\right)\psi_k \quad \rightarrow \quad \int_{-\infty}^{\infty} dk\, \psi_k \left(-\frac{1}{2}K\frac{\partial^2}{\partial k^2}\right)b(k) \tag{5.125}$$

The Schrödinger equation then takes the form

$$\int_{-\infty}^{\infty} dk\psi_k \left[\left(\frac{\hbar^2 k^2}{2m} - \frac{1}{2}K\frac{\partial^2}{\partial k^2} - E \right) b(k) \right] = 0 \tag{5.126}$$

From this equality we conclude that $b(k)$ satisfies the k–dependent Schrödinger equation

$$\left(\frac{\hbar^2 k^2}{2m} - \frac{1}{2}K\frac{\partial^2}{\partial k^2} \right) b(k) = Eb(k) \tag{5.127}$$

This equation is also called the Schrödinger equation in momentum representation.

The time–dependent Schrödinger equation in momentum representation may be expressed as

$$i\hbar\frac{\partial}{\partial t}b(k,t) = H(k)b(k,t) \tag{5.128}$$

The time dependent function $b(k,t)$ may be obtained from the initial value $b(k,0)$ as

$$b(k,t) = e^{-\frac{i}{\hbar}Ht}b(k,0) \tag{5.129}$$

5.3   Unbound states

If a wave function ψ represents a bound state (in 1D), then $|\psi|^2 \to 0$, $|x| \to \infty$ for all t. A wave function that does not obey this condition represents an unbound state. The square modulus of a bound state gives a finite integral over the infinite interval

$$\int_{-\infty}^{\infty} |\psi|^2 dx < \infty \tag{5.130}$$

The square modulus of an unbound state gives a finite integral over any finite interval

$$\int_a^b |\psi|^2 dx < \infty \quad , \quad |b - a| < \infty \tag{5.131}$$

The eigenstate of the momentum operator

$$\psi_n(x) = \frac{1}{\sqrt{2\pi}}e^{ikx} \tag{5.132}$$

represents an unbound state. The eigenfunction of the SHO Hamiltonian

$$\psi_n(\xi) = A_n H_n(\xi)e^{-\frac{1}{2}\xi^2} \tag{5.133}$$

represents a bound state. Unbound states are relevant to scattering problems. Since

$$\int_{-\infty}^{\infty} |\psi|^2 dx$$

diverges for unbound states, it is convenient to normalize the wave function for scattering problems in terms of the particle density ρ. For 1D scattering problem we take

$$|\psi|^2 dx = \rho dx = dN \tag{5.134}$$

Fundamentals of quantum mechanics; Erkoç

$$= \text{\# of particles in the interval } dx$$

$$\int_a^b |\psi|^2 dx = N \tag{5.135}$$

$$= \text{\# of particles in the interval } (b-a)$$

For a 1D beam of N particles per unit length, all moving with momentum $p = \hbar k_0$, the wave function is written

$$\psi = \sqrt{N}e^{i(k_0 x - \omega t)} \tag{5.136}$$

$$|\psi|^2 = N \tag{5.137}$$

$$\frac{\hbar^2 k_0^2}{2m} = \hbar \omega \tag{5.138}$$

When $|\psi|^2$ is referred to particle density, it is proportional to probability density also. For uniform beams, $|\psi|^2$ is constant, which implies that it is uniformly probable to find particles anywhere along the beam. This is consistent with the uncertainty principle. For instance, for the wave function given in Eq. (5.136), the momentum of any particle in the beam is $\hbar k_0$, whence its position is maximally uncertain.

Continuity equation:

One–dimensional barrier problems involve incident, reflected, and transmitted current densities, J_{inc}, J_{ref}, and J_{trans}, respectively (see Figure 5.10).

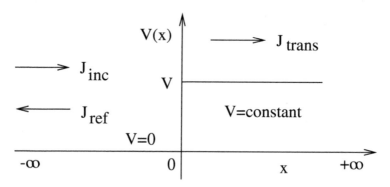

Fig. 5.10. Current densities for a step potential in 1D.

In three–dimensions the number density (ρ) and the current density ($\mathbf{J}$) are related through the continuity equation

$$\frac{\partial \rho}{\partial t} + \nabla \cdot \mathbf{J} = 0 \tag{5.139}$$

If we integrate this equation over a volume V,

$$\int_V \frac{\partial \rho}{\partial t} dV = -\int_V \nabla \cdot \mathbf{J} dV \tag{5.140}$$

From Gauss's theorem, considering the surface which encloses the volume, we write

$$\frac{\partial}{\partial t} \int_V \rho dV = - \int_S \mathbf{J} \cdot d\mathbf{S} \qquad (5.141)$$

Since

$$N = \int_V \rho dV \qquad (5.142)$$

we can rewrite Eq. (5.141) as

$$\frac{\partial N}{\partial t} = - \int_S \mathbf{J} \cdot d\mathbf{S} \qquad (5.143)$$

This equation says that the number of particles in the volume V changes by virtue of a net flux of particles out of (or into) the volume V. It is a statement of the conservation of matter.

If particles are moving only in the x–direction, $\mathbf{J} = (J_x, 0, 0)$, and the continuity equation becomes

$$\frac{\partial \rho}{\partial t} + \frac{\partial J_x}{\partial x} = 0 \qquad (5.144)$$

Since $\rho = |\psi|^2$, we relate J_x to ψ as follows:

The wave function for particles in the beam obeys the Schrödinger equation,

$$i\hbar \frac{\partial \psi}{\partial t} = H\psi \qquad (5.145)$$

We may rewrite this equation in the forms:

$$\frac{\partial \psi}{\partial t} = -\frac{i}{\hbar} H\psi \qquad (5.146)$$

and

$$\frac{\partial \psi^*}{\partial t} = \frac{i}{\hbar} H\psi^* \qquad (5.147)$$

The time derivative of the particle density $\psi^*\psi$ is

$$\frac{\partial}{\partial t}(\psi^*\psi) = \psi^* \frac{\partial \psi}{\partial t} + \psi \frac{\partial \psi^*}{\partial t} = \psi^* \left(-\frac{i}{\hbar} H\psi \right) + \psi \left(\frac{i}{\hbar} H\psi^* \right) \qquad (5.148)$$

For a 1D Hamiltonian

$$H = \frac{p^2}{2m} + V(x) \qquad (5.149)$$

$$H\psi = E\psi \quad \longrightarrow \quad -\frac{\hbar^2}{2m} \frac{\partial^2 \psi}{\partial x^2} + V\psi = E\psi \qquad (5.150)$$

$$H\psi^* = E\psi^* \quad \longrightarrow \quad -\frac{\hbar^2}{2m} \frac{\partial^2 \psi^*}{\partial x^2} + V\psi^* = E\psi^* \qquad (5.151)$$

$$\frac{\partial}{\partial t}(\psi^*\psi) = \psi^* \left[-\frac{i}{\hbar} \left(-\frac{\hbar^2}{2m} \frac{\partial^2 \psi}{\partial x^2} + V\psi \right) \right] \qquad (5.152)$$

$$+ \psi \left[\frac{i}{\hbar} \left(-\frac{\hbar^2}{2m} \frac{\partial^2 \psi^*}{\partial x^2} + V\psi^* \right) \right] = \frac{i\hbar}{2m} (\psi^* \psi_{xx} - \psi \psi_{xx}^*)$$

or

$$\frac{\partial}{\partial t}(\psi^*\psi) - \frac{i\hbar}{2m}(\psi^*\psi_{xx} - \psi\psi_{xx}^*) = 0 \tag{5.153}$$

or

$$\frac{\partial}{\partial t}(\psi^*\psi) + \frac{\hbar}{2mi}(\psi^*\psi_{xx} - \psi\psi_{xx}^*) = 0 \tag{5.154}$$

or

$$\frac{\partial}{\partial t}(\psi^*\psi) + \frac{\partial}{\partial x}\left[\frac{\hbar}{2mi}\left(\psi^*\frac{\partial\psi}{\partial x} - \psi\frac{\partial\psi^*}{\partial x}\right)\right] = 0 \tag{5.155}$$

Comparing this equation with

$$\frac{\partial}{\partial t} + \frac{\partial J_x}{\partial x} = 0 \tag{5.156}$$

we can write

$$J_x = \frac{\hbar}{2mi}\left(\psi^*\frac{\partial\psi}{\partial x} - \psi\frac{\partial\psi^*}{\partial x}\right) \tag{5.157}$$

or

$$J_x = \frac{i\hbar}{2m}\left(\psi\frac{\partial\psi^*}{\partial x} - \psi^*\frac{\partial\psi}{\partial x}\right) \tag{5.158}$$

The dimension of J_x is [number $\times$ unit distance / unit time]. In three dimensions the current density is written

$$\mathbf{J} = \frac{i\hbar}{2m}(\psi\nabla\psi^* - \psi^*\nabla\psi) \tag{5.159}$$

The dimension of $\mathbf{J}$ is [number / (unit area $\times$ unit time)].

Transmission and reflection coefficients:

For one–dimensional scattering problems, the particles in the beam are in plane–wave states with definite momentum. Given the wave functions relevant to incident, reflected, and transmitted beams, one may calculate the corresponding current densities according to

$$J_x = \frac{i\hbar}{2m}\left(\psi\frac{\partial\psi^*}{\partial x} - \psi^*\frac{\partial\psi}{\partial x}\right) \tag{5.160}$$

The transmission coefficient T and the reflection coefficient R are defined as

$$T \equiv \left|\frac{J_{trans}}{J_{inc}}\right| \quad , \quad R \equiv \left|\frac{J_{ref}}{J_{inc}}\right| \tag{5.161}$$

These 1D barrier problems are closely related to problems on the transmission and reflection of electromagnetic plane waves through media of varying index of refraction (see Figure 5.11).

$$\mathcal{E} = \mathcal{E}_0 e^{i(kx-\omega t)} \quad \longrightarrow \quad \frac{\partial^2\mathcal{E}}{\partial x^2} - \left(\frac{n}{c}\right)^2\frac{\partial^2\mathcal{E}}{\partial t^2} = 0 \tag{5.162}$$

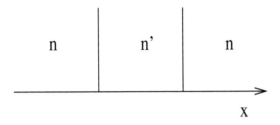

Fig. 5.11. Media of varying index of refraction in 1D.

In the quantum mechanical case, the scattering is also of waves from a potential barrier; see Figure 5.12.

$$\psi = Ae^{i(kx-\omega t)} \qquad \longrightarrow \qquad i\hbar\frac{\partial\psi}{\partial t} = H\psi \tag{5.163}$$

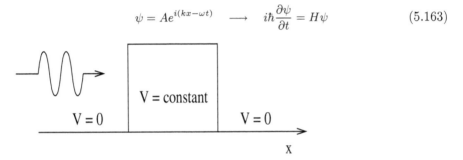

Fig. 5.12. Quantum mechanical scattering of a wave from a potential barrier.

For 1D barrier problems there are three relevant beams. Particles in the incident beam have momentum

$$p_{inc} = \hbar k_1 \tag{5.164}$$

Particles in the reflected beam have the opposite momentum

$$p_{ref} = \hbar k_1 \tag{5.165}$$

If the environment in the region of the transmitted beam is different from that of the incident beam, the momenta in these two regions will differ. Particles in the transmitted beam will have momentum

$$p_{trans} = \hbar k_2 \neq \hbar k_1 \tag{5.166}$$

If the potential is constant in the regions of the incident and transmitted beams, (see Figure 5.13), the wave functions in these regions describe free particles, and we may write

$$\psi_{inc} = Ae^{i(k_1 x - \omega_1 t)} \tag{5.167}$$

$$\hbar\omega_1 = E_{inc} = \frac{\hbar^2 k_1^2}{2m} \tag{5.168}$$

$$\psi_{ref} = Be^{i(k_1 x + \omega_1 t)} \tag{5.169}$$

$$\hbar\omega_1 = E_{ref} = E_{inc} \tag{5.170}$$

$$\psi_{trans} = Ce^{i(k_2 x - \omega_2 t)} \tag{5.171}$$

$$\hbar\omega_2 = E_{trans} = \frac{\hbar^2 k_2^2}{2m} + V_0 = E_{inc} = \hbar\omega_1 \qquad (5.172)$$

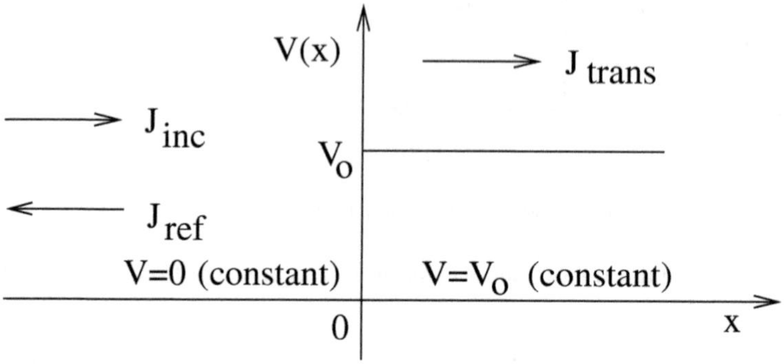

V(x)   ⟶ J trans

⟶ J inc V_0

⟵ J ref

V=0 (constant)   V=V$_0$ (constant)

0   X

Fig. 5.13. Scattering from a potential step.

Energy is conserved across the potential barrier so that frequency remains constant ($\omega_1 = \omega_2$). The change in wavenumber k corresponds to changes in momentum and kinetic energy. Using Eq. (5.157) permits calculation of the currents.

$$J_{inc} = \frac{\hbar}{2mi} 2ik_1 |A|^2 = \frac{\hbar k_1}{m} |A|^2 \qquad (5.173)$$

$$J_{trans} = \frac{\hbar}{2mi} 2ik_2 |C|^2 = \frac{\hbar k_2}{m} |C|^2 \qquad (5.174)$$

$$J_{ref} = \frac{\hbar}{2mi} (-2ik_1) |B|^2 = -\frac{\hbar k_1}{m} |B|^2 \qquad (5.175)$$

These relations are equivalent to the classical prescription for particle current,

$$J = \rho v \qquad (5.176)$$

with

$$\rho = |\psi|^2 \quad , \quad v = \frac{\hbar k}{m} \qquad (5.177)$$

These formulas give the T and R coefficients

$$T = \left| \frac{J_{trans}}{J_{inc}} \right| = \left| \frac{C}{A} \right|^2 \frac{k_2}{k_1} \qquad (5.178)$$

$$R = \left| \frac{J_{ref}}{J_{inc}} \right| = \left| \frac{B}{A} \right|^2 \qquad (5.179)$$

To calculate C/A and B/A explicitly, one must solve the Schrödinger equation across the region of the potential barrier.

5.4   One–dimensional barrier problems

In a one–dimensional scattering experiment, the intensity and energy of the particles in the incident beam are known in addition to the structure of the potential barrier $V(x)$. Three fundamental scattering configurations are shown below. The energy of the particles in the beam is denoted by E.

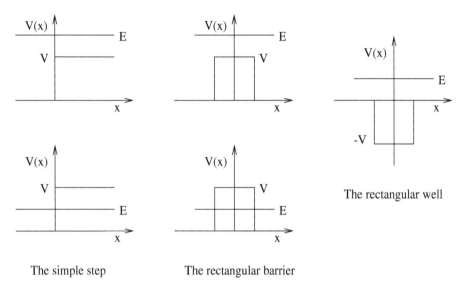

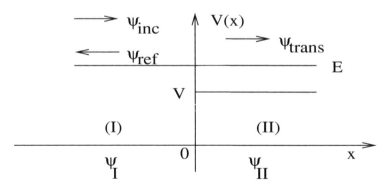

The simple step   The rectangular barrier

Fig. 5.14. One–dimensional potential models for unbound states.

The simple step:

Let us consider the simple step for the case $E > V$ (see Figure 5.15).

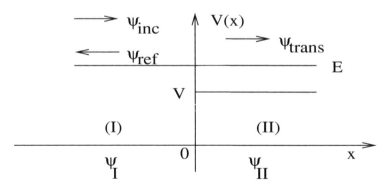

Fig. 5.15. Simple step potential in 1D; $E > V$.

We wish to obtain the space–dependent wave function ψ for all x. The potential function $V(x)$ is

$$V(x) = 0 \qquad \text{for } x < 0$$
$$= V = \text{const. for } x \geq 0 \qquad (5.180)$$

The incident beam comes from $x = -\infty$. In region I, where $V = 0$, the time–

independent Schrödinger equation appears as

$$-\frac{\hbar^2}{2m}\frac{\partial^2\psi}{\partial x^2} = E\psi \tag{5.181}$$

In this region the energy is entirely kinetic. If we set

$$\frac{\hbar^2 k_1^2}{2m} = E \tag{5.182}$$

then the Schrödinger equation becomes

$$\frac{\partial^2\psi}{\partial x^2} = -k_1^2\psi \tag{5.183}$$

In region II, where $V = $ constant $\neq 0$, the time–independent Schrödinger equation appears as

$$-\frac{\hbar^2}{2m}\frac{\partial^2\psi}{\partial x^2} + V\psi = E\psi \tag{5.184}$$

or

$$-\frac{\hbar^2}{2m}\frac{\partial^2\psi}{\partial x^2} = (E - V)\psi \tag{5.185}$$

The kinetic energy decreases by V and is given by

$$\frac{\hbar^2 k_2^2}{2m} = E - V \tag{5.186}$$

In terms of k_2, the Schrödinger equation appears as

$$\frac{\partial^2\psi}{\partial x^2} = -k_2^2\psi \tag{5.187}$$

Writing ψ_I for the solution to $\psi_{xx} = -k_1^2\psi$, and ψ_{II} for the solution to $\psi_{xx} = -k_2^2\psi$, one obtains

$$\psi_I = Ae^{ik_1 x} + Be^{-ik_1 x} \tag{5.188}$$

$$\psi_{II} = Ce^{ik_2 x} + De^{-ik_2 x} \tag{5.189}$$

The term $De^{-ik_2 x}$ in ψ_{II} represents a wave coming from $x = +\infty$, and if there is no such wave, we may take $D = 0$. $Ae^{ik_1 x}$ in ψ_I represents the incident wave (ψ_{inc}), $Be^{-ik_1 x}$ in ψ_I represents the reflected wave (ψ_{ref}), and $Ce^{ik_2 x}$ in ψ_{II} represents the transmitted wave (ψ_{trans}).

The solutions ψ_I and ψ_{II} represent a single solution to the Schrödinger equation for all x, for the step potential and $E > V$. These solutions, both ψ_I and ψ_{II}, and their first derivatives are continuous at the point $x = 0$,

$$\psi_I(0) = \psi_{II}(0) \quad \text{and} \quad \frac{\partial\psi_I(0)}{\partial x} = \frac{\partial\psi_{II}(0)}{\partial x} \tag{5.190}$$

These equations give the relations

$$A + B = C \quad \text{and} \quad A - B = \frac{k_2}{k_1}C \tag{5.191}$$

Solving for C/A and B/A, one obtains

$$\frac{C}{A} = \frac{2}{1 + \frac{k_2}{k_1}} \quad , \quad \frac{B}{A} = \frac{1 - \frac{k_2}{k_1}}{1 + \frac{k_2}{k_1}} \tag{5.192}$$

Substituting these values into

$$T = \left|\frac{C}{A}\right|^2 \frac{k_2}{k_1} \quad \text{and} \quad R = \left|\frac{B}{A}\right|^2 \tag{5.193}$$

gives

$$T = \frac{4\frac{k_2}{k_1}}{\left(1 + \frac{k_2}{k_1}\right)^2} \quad , \quad R = \left|\frac{1 - \frac{k_2}{k_1}}{1 + \frac{k_2}{k_1}}\right|^2 \tag{5.194}$$

The ratio k_2/k_1 can be obtained from

$$\frac{\hbar^2 k_1^2}{2m} = E \quad \text{and} \quad \frac{\hbar^2 k_2^2}{2m} = E - V \tag{5.195}$$

$$k_1^2 = \frac{2mE}{\hbar^2} \quad , \quad k_2^2 = \frac{2m}{\hbar^2}(E - V) \tag{5.196}$$

Hence,

$$\left(\frac{k_2}{k_1}\right)^2 = 1 - \frac{V}{E} \tag{5.197}$$

In this model $E \geq V$, so $0 \leq \frac{k_2}{k_1} \leq 1$.

For $E \gg V$,

$$\frac{k_2}{k_1} \to 1 \ , \ \text{and} \ T \to 1 \ , \ R \to 0$$

There is total transmission.

For $E = V$,

$$\frac{k_2}{k_1} = 0 \ , \ \text{and} \ T = 0 \ , \ R = 1$$

There is total reflection and zero transmission.

For all values of $\frac{k_2}{k_1}$ we see that $T + R = 1$; see Figure 5.16.

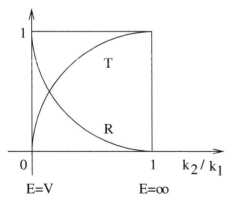

Fig. 5.16. Variation of T and R with respect to $k_2/k_1 = \sqrt{1 - V/E}$.

Let us now consider the simple step for the case $E < V$; see Figure 5.17.

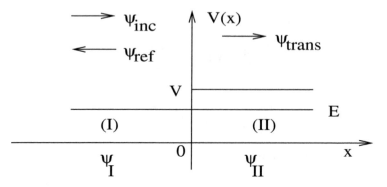

Fig. 5.17. Simple step potential in 1D; $E < V$.

In region I the Schrödinger equation becomes

$$\frac{\partial^2 \psi}{\partial x^2} = -k_1^2 \psi \quad \text{where} \quad \frac{\hbar^2 k_1^2}{2m} = E \tag{5.198}$$

In region II the Schrödinger equation becomes

$$\frac{\partial^2 \psi}{\partial x^2} = \kappa^2 \psi \quad \text{where} \quad \frac{\hbar^2 \kappa^2}{2m} = V - E > 0 \tag{5.199}$$

The kinetic energy in this region (region II) is negative $(-\hbar^2 \kappa^2 / 2m)$. Region II is a classically forbidden region. In quantum mechanics, however, it is possible for particles to penetrate the barrier.

Again calling the solution to $\psi_{xx} = -k_1^2 \psi$ as ψ_I, and the solution to $\psi_{xx} = \kappa^2 \psi$ as ψ_{II}, we obtain

$$\psi_I = Ae^{ik_1 x} + Be^{-k_1 x} \tag{5.200}$$
$$\psi_{II} = Ce^{-\kappa x} \tag{5.201}$$

Continuity of ψ and ψ_x at $t = 0$ gives

$$A + B = C \quad \text{and} \quad A - B = i\frac{\kappa}{k_1} C \tag{5.202}$$

Solving for C/A and B/A one obtains

$$\frac{C}{A} = \frac{2}{1 + i\frac{\kappa}{k_1}} \quad , \quad \frac{B}{A} = \frac{1 - i\frac{\kappa}{k_1}}{1 + i\frac{\kappa}{k_1}} \tag{5.203}$$

Since $|B/A| = 1$,

$$R = \left|\frac{B}{A}\right|^2 = 1 \quad , \quad T = 0$$

There is total reflection, hence the transmission must be zero. If we calculate the transmission current

$$J_{trans} = \frac{\hbar}{2mi} \left(\psi_{II}^* \frac{\partial \psi_{II}}{\partial x} - \psi_{II} \frac{\partial \psi_{II}^*}{\partial x} \right) \tag{5.204}$$

$$= \frac{\hbar}{2mi} |C|^2 \left(e^{-\kappa x} \frac{\partial}{\partial x} e^{-\kappa x} - e^{-\kappa x} \frac{\partial}{\partial x} e^{-\kappa x} \right) = 0$$

Therefore,

$$T = \left|\frac{J_{trans}}{J_{inc}}\right| = 0 \tag{5.205}$$

The rectangular barrier:

Let us consider the rectangular barrier for the case $E > V$ (see Figure 5.18).

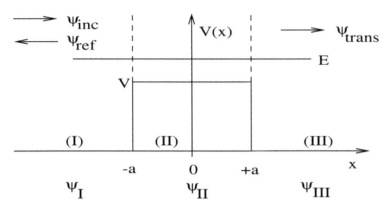

Fig. 5.18. Rectangular potential barrier in 1D; $E > V$.

In this case there are three regions. The potential function $V(x)$ is

$$V(x) = \begin{cases} 0 & \text{for } x < -a \\ V > 0 & \text{for } -a \le x \le a \\ 0 & \text{for } x > a \end{cases} \tag{5.206}$$

The solutions to the time–independent Schrödinger equation in each of the three regions are

$$\psi_I = Ae^{ik_1x} + Be^{-ik_1x} \quad , \quad \frac{\hbar^2 k_1^2}{2m} = E \tag{5.207}$$

$$\psi_{II} = Ce^{ik_2x} + De^{-ik_2x} \quad , \quad \frac{\hbar^2 k_2^2}{2m} = E - V \tag{5.208}$$

$$\psi_{III} = Fe^{ik_1x} \quad , \quad \frac{\hbar^2 k_1^2}{2m} = E \tag{5.209}$$

From the equalities of

$$\frac{\hbar^2 k_1^2}{2m} = E \quad \text{and} \quad \frac{\hbar^2 k_2^2}{2m} + V = E$$

we may write

$$\frac{\hbar^2 k_1^2}{2m} = \frac{\hbar^2 k_2^2}{2m} + V \quad \text{or} \quad k_1^2 - k_2^2 = \frac{2m}{\hbar^2} V \tag{5.210}$$

or

$$(ak_1)^2 - (ak_2)^2 = \frac{2ma^2 V}{\hbar^2} = \left(\frac{g}{2}\right)^2 \tag{5.211}$$

This last equation represents the conservation of energy. The parameter g contains all the potential properties. For rectangular barrier scattering with $E \ge V$, ak_1 and ak_2 lie on a hyperbola, $ak_1 > ak_2 > 0$ (see Figure 5.19).

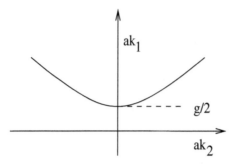

Fig. 5.19. Dependence of k_1 on k_2 in a rectangular barrier potential in 1D for $E \geq V$.

The permitted values of k_1 (and therefore E) comprise a positive unbounded continuum. For each such eigen-k_1 value, there is a corresponding eigenstate (ψ_I, ψ_{II}, ψ_{III}) which is determined in terms of the coefficients B/A, C/A, D/A, F/A. The scattering parameters are obtained from these coefficients,

$$T = \left|\frac{F}{A}\right|^2 \quad , \quad R = \left|\frac{B}{A}\right|^2 \tag{5.212}$$

The coefficients are determined from the boundary conditions at $x = a$ and $x = -a$,

$$\psi_I(-a) = \psi_{II}(-a) \tag{5.213}$$

or

$$e^{-ik_1 a} + \left(\frac{B}{A}\right)e^{ik_1 a} = \left(\frac{C}{A}\right)e^{-ik_2 a} + \left(\frac{D}{A}\right)e^{ik_2 a} \tag{5.214}$$

$$\frac{\partial\psi_I(-a)}{\partial x} = \frac{\partial\psi_{II}(-a)}{\partial x} \tag{5.215}$$

or

$$k_1\left[e^{-ik_1 a} - \left(\frac{B}{A}\right)e^{ik_1 a}\right] = k_2\left[\left(\frac{C}{A}\right)e^{-ik_2 a} - \left(\frac{D}{A}\right)e^{ik_2 a}\right] \tag{5.216}$$

$$\psi_{II}(a) = \psi_{III}(a) \tag{5.217}$$

or

$$\left(\frac{C}{A}\right)e^{ik_1 a} + \left(\frac{D}{A}\right)e^{-ik_2 a} = \left(\frac{F}{A}\right)e^{ik_1 a} \tag{5.218}$$

$$\frac{\partial\psi_{II}(a)}{\partial x} = \frac{\partial\psi_{III}(a)}{\partial x} \tag{5.219}$$

or

$$k_2\left[\left(\frac{C}{A}\right)e^{ik_2 a} - \left(\frac{D}{A}\right)e^{-ik_2 a}\right] = k_1\left(\frac{F}{A}\right)e^{ik_1 a} \tag{5.220}$$

Here there are four equations and four unknowns: $B/A, C/A, D/A, F/A$. Eliminating C/A and D/A in the last two equations one can write the following equalities:

$$\frac{F}{A} = e^{-2ik_1 a}\left[\cos(2k_2 a) - \frac{i}{2}\left(\frac{k_1^2 + k_2^2}{k_1 k_2}\right)\sin(2k_2 a)\right]^{-1} \tag{5.221}$$

$$\frac{B}{A} = \frac{F}{A}\frac{i(k_2^2 - k_1^2)}{2k_1 k_2}\sin(2k_2 a) \tag{5.222}$$

The transmission coefficient is

$$T = \left| \frac{F}{A} \right|^2 \tag{5.223}$$

One can also write

$$T + R = \left| \frac{F}{A} \right|^2 + \left| \frac{B}{A} \right|^2 = 1 \tag{5.224}$$

The transmission coefficient can be simplified as

$$\frac{1}{T} = \left| \frac{A}{F} \right|^2 = 1 + \frac{1}{4} \left(\frac{k_1^2 - k_2^2}{k_1 k_2} \right)^2 \sin^2(2k_2 a) \tag{5.225}$$

Expressing k_1 and k_2 in terms of E and V, one obtains

$$\frac{1}{T} = 1 + \frac{1}{4} \frac{V^2}{E(E-V)} \sin^2(2k_2 a) \quad , \quad E > V \tag{5.226}$$

Using the relation

$$k_2^2 = \frac{2m}{\hbar^2}(E - V)$$

and taking

$$\frac{E}{V} = \bar{E} > 1$$

we can write the relation

$$2k_2 a = 2a\sqrt{\frac{2m}{\hbar^2}(E-V)} = \sqrt{\frac{8ma^2 V}{\hbar^2}(\bar{E}-1)} = \sqrt{g^2(\bar{E}-1)} \tag{5.227}$$

The transmission coefficient then takes the form

$$\frac{1}{T} = 1 + \frac{1}{4}\frac{1}{\bar{E}(\bar{E}-1)} \sin^2\left[g\sqrt{(\bar{E}-1)}\right] \tag{5.228}$$

or

$$T = \frac{1}{1 + \frac{\sin^2\left[g\sqrt{(\bar{E}-1)}\right]}{4\bar{E}(\bar{E}-1)}} \tag{5.229}$$

The reflection coefficient is $R = 1 - T$. Here $T = 1$ when $\sin^2(2k_2 a) = 0$, equivalently

$$2ak_2 = n\pi \quad , \quad n = 1, 2, 3, \cdots \tag{5.230}$$

Setting $k_2 = 2\pi/\lambda$, we write $2a = n(\lambda/2)$. When the barrier width $2a$ is an integral number of $\lambda/2$, $n(\lambda/2)$, the barrier becomes transparent to the incident beam.

The requirement for perfect transmission ($2ak_2 = n\pi$) becomes

$$k_2 = \frac{n\pi}{2a} \quad , \quad k_2^2 = \frac{n^2\pi^2}{4a^2} \tag{5.231}$$

The energy expression then takes the form

$$\frac{\hbar^2 k_2^2}{2m} = \frac{n^2\pi^2\hbar^2}{8ma^2} = E - V \tag{5.232}$$

or

$$E - V = n^2\left(\frac{\pi^2\hbar^2}{8ma^2}\right) = n^2 E_1 \tag{5.233}$$

where E_1 is the lowest energy of a one–dimensional box of width $2a$.

Let us now consider the rectangular barrier for the case $E < V$ (see Figure 5.20).

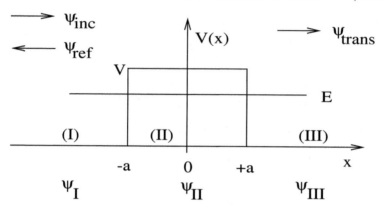

Fig. 5.20. Rectangular potential barrier in 1D; $E < V$.

The solutions to the time–independent Schrödinger equation in each of the three regions are

$$\psi_I(x) = Ae^{ik_1x} + Be^{-ik_1x} \quad , \quad \frac{\hbar^2 k_1^2}{2m} = E \tag{5.234}$$

$$\psi_{II}(x) = Ce^{\kappa x} + De^{-\kappa x} \quad , \quad \frac{\hbar^2 \kappa^2}{2m} = V - E > 0 \tag{5.235}$$

$$\psi_{III}(x) = Fe^{ik_1x} \quad , \quad \frac{\hbar^2 k_1^2}{2m} = E \tag{5.236}$$

The conservation of energy equation becomes

$$\frac{\hbar^2 k_1^2}{2m} = V - \frac{\hbar^2 \kappa^2}{2m} \quad \text{or} \quad \frac{\hbar^2 k_1^2}{2m} + \frac{\hbar^2 \kappa^2}{2m} = V \tag{5.237}$$

or

$$k_1^2 + \kappa^2 = \frac{2m}{\hbar^2} V \tag{5.238}$$

or

$$(ak_1)^2 + (a\kappa)^2 = \frac{2ma^2 V}{\hbar^2} = \left(\frac{g}{2}\right)^2 \tag{5.239}$$

This last equation indicates that for rectangular barrier scattering with $E < V$, ak_1 and $a\kappa$ lie on a circle of radius $g/2$ with $ak_1 \geq 0$, $a\kappa \geq 0$ (see Figure 5.21). The energy spectrum consists of a bounded continuum.

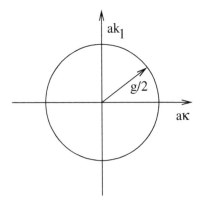

Fig. 5.21. Dependence of k_1 on κ in a rectangular barrier potential in 1D for $E < V$.

The wave functions in each of the three regions for a rectangular barrier with $E < V$ look like as shown in Figure 5.22.

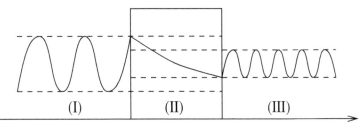

Fig. 5.22. Wave functions in three regions in a rectangular barrier potential in 1D for $E < V$.

The coefficients are determined following the same procedure done for the case $E > V$. In the case of $E < V$, we just replace $ik_2 \rightarrow \kappa$, taking

$$\frac{\hbar^2 \kappa^2}{2m} = V - E > 0 \tag{5.240}$$

So that the transmission coefficient for this case is obtained by making the substitution $ik_2 \rightarrow \kappa$ into

$$\frac{1}{T} = 1 + \frac{1}{4}\left(\frac{k_1^2 - k_2^2}{k_1 k_2}\right)^2 \sin^2(2k_2 a)$$

and using the relation, $\sin(iz) = i\sinh(z)$, the transmission coefficient in the rectangular barrier for the case $E < V$ becomes

$$\frac{1}{T} = 1 + \frac{1}{4}\left(\frac{k_1^2 - \kappa^2}{k_1 \kappa}\right)^2 \sinh^2(2\kappa a) \tag{5.241}$$

Expressing in terms of E and V gives

$$\frac{1}{T} = 1 + \frac{1}{4}\frac{V^2}{E(V - E)}\sinh^2(2\kappa a) \tag{5.242}$$

or

$$T = \frac{1}{1 + \frac{1}{4}\frac{V^2}{E(V-E)}\sinh^2(2\kappa a)} \tag{5.243}$$

Using the relation

$$\frac{\hbar^2\kappa^2}{2m} = V - E \quad \text{or} \quad \kappa = \sqrt{\frac{2m}{\hbar^2}(V-E)} \tag{5.244}$$

and taking $E/V = \bar{E} < 1$, we can write

$$T = \frac{1}{1 + \frac{1}{4}\dfrac{\sinh^2\sqrt{\frac{8ma^2V}{\hbar^2}(1-\bar{E})}}{\bar{E}(1-\bar{E})}} \tag{5.245}$$

$$= \frac{1}{1 + \dfrac{\sinh^2\sqrt{g^2(1-\bar{E})}}{4\bar{E}(1-\bar{E})}} = \frac{1}{1 + \dfrac{\sinh^2(g\sqrt{1-\bar{E}})}{4\bar{E}(1-\bar{E})}}$$

Here the parameter $g^2 = 8ma^2V/\hbar^2$ contains all the information about the potential barrier.

A typical plot of T, for a given g, versus $\bar{E}$ looks like as shown in Figure 5.23.

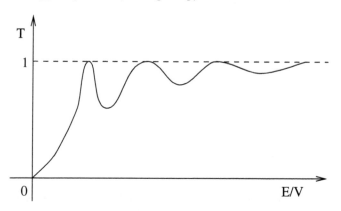

Fig. 5.23. Variation of T with respect to $\bar{E}$ ($= E/V$) for a given g in a rectangular barrier potential in 1D for $E < V$.

For scattering from a potential barrier, the transmission is less than unity at $E = V$. As $E \to V$ or $\bar{E} \to 1$, $T \to 1/(1 + g^2/4)$. The fact that T does not vanish for $E < V$ is a purely quantum mechanical result. This phenomenon of particles passing through barriers higher than their own incident energy is known as tunneling. It allows emission of α–particles from a nucleus and field emission of electrons from a metal surface in the presence of a strong electric field.

Kinematics of a wave packet scattered from a potential barrier:

A wave packet centered at $x = 0$ at $t = 0$ is in the form

$$\psi(x,t) = \frac{1}{\sqrt{2\pi}}\int_{-\infty}^{\infty} b(k)e^{i(kx-wt)}dk \tag{5.246}$$

The same wave packet centered at $x = -X$ at $t = 0$ is in the form

$$\psi(x,t) = \frac{1}{\sqrt{2\pi}} \int_{-\infty}^{\infty} b(k) e^{ik(x+X)} e^{-iwt} dk \qquad (5.247)$$

$$= \frac{1}{\sqrt{2\pi}} \int_{-\infty}^{\infty} b(k) e^{ikX} e^{i(kx-wt)} dk$$

For a chopped pulse, with length L, containing particles moving with momentum $\hbar k_0$, the coefficient $b(k)$ is given by

$$b(k) = \sqrt{\frac{2}{\pi L}} \frac{\sin\left[\frac{(k-k_0)L}{2}\right]}{(k-k_0)} \qquad (5.248)$$

The group velocity of this packet is $v_0 = \hbar k_0/m$. Let us represent the wave packet centered at $x = -X$ at $t = 0$ by ψ_{inc}. This packet, in general, is a superposition of plane wave states of the form

$$\psi_{inc} \sim e^{i(kx-wt)} \qquad (5.249)$$

Each such incident k–component plane wave is reflected and transmitted (see Figure 5.24),

$$\psi = Ae^{ikx} + Be^{-ikx} \quad , \quad \psi = Fe^{ikx}$$

We may write the incident, reflected, and transmitted waves in the following form:

$$\psi_{inc} = \frac{1}{\sqrt{2\pi}} \int_{-\infty}^{\infty} b(k) e^{ikX} e^{i(kx-wt)} dk \quad ; \quad x < -a \qquad (5.250)$$

$$\psi_{ref} = \frac{1}{\sqrt{2\pi}} \int_{-\infty}^{\infty} \sqrt{R} e^{i\phi_R} b(k) e^{ikX} e^{-i(kx+wt)} dk \quad ; \quad x < -a \qquad (5.251)$$

$$\psi_{trans} = \frac{1}{\sqrt{2\pi}} \int_{-\infty}^{\infty} \sqrt{T} e^{i\phi_T} b(k) e^{ikX} e^{i(kx-wt)} dk \quad ; \quad x > +a \qquad (5.252)$$

here

$$\sqrt{R} e^{i\phi_R} = \frac{B}{A} \quad , \quad \sqrt{T} e^{i\phi_T} = \frac{F}{A} \qquad (5.253)$$

$$\phi_T = \phi_R - n\left(\frac{\pi}{2}\right) \quad ; \quad n = 1, 2, 3, \cdots$$

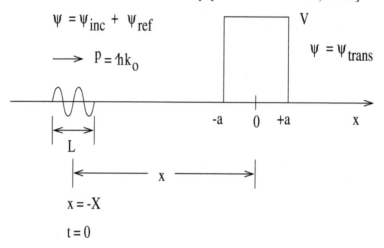

Fig. 5.24. Motion of a chopped pulse (wave packet) through a potential barrier in 1D.

The phase of the Fourier integral for ψ_{ref} vanishes, because the major contribution in a Fourier integral is due to the k component with stationary phase,

$$\frac{\partial}{\partial k}(\phi_R + kX - kx - wt) = 0 \tag{5.254}$$

This gives the trajectory of the reflected packet,

$$x = -\frac{\hbar k_0}{m}t + X + \left(\frac{\partial \phi_R}{\partial k}\right)_{k_0} \quad ; \quad x < -a \tag{5.255}$$

Similarly, for the incident and transmitted packets, respectively, we obtain

$$x = \frac{\hbar k_0}{m}t - X \quad ; \quad x < -a \tag{5.256}$$

$$x = \frac{\hbar k_0}{m}t - X - \left(\frac{\partial \phi_R}{\partial k}\right)_{k_0} \quad ; \quad x > +a \tag{5.257}$$

We see that there is a delay for both the transmitted and reflected packets. The transmitted pulse arrives at any plane $x > a$, $(\partial \phi_T/\partial k)/v_0$ seconds after the free pulse. The reflected pulse arrives at any plane $x < -a$, $(\partial \phi_R/\partial k)/v_0$ seconds after the free pulse would be reflected from an impenetrable wall at the $x = 0$ plane.

5.5　The finite potential well

Unbound states:

Let us consider a finite potential well, a rectangular well. The configuration for this case is shown in Figure 5.25.

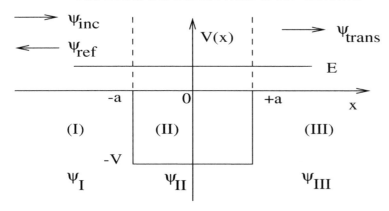

Fig. 5.25. Rectangular potential well in 1D; $E > 0$.

The solutions to the time–independent Schrödinger equation in each of the three regions are

$$\psi_I = A e^{ik_1 x} + B e^{-ik_1 x} \quad , \quad \frac{\hbar^2 k_1^2}{2m} = E \tag{5.258}$$

$$\psi_{II} = C e^{ik_2 x} + D e^{-ik_2 x} \quad ; \quad \frac{\hbar^2 k_2^2}{2m} = E - V = E + |V| \tag{5.259}$$

$$\psi_{III} = F e^{ik_1 x} \quad , \quad \frac{\hbar^2 k_1^2}{2m} = E \tag{5.260}$$

From the equality of

$$\frac{\hbar^2 k_1^2}{2m} = E \quad \text{and} \quad \frac{\hbar^2 k_1^2}{2m} - |V| = E$$

we may write

$$\frac{\hbar^2 k_1^2}{2m} = \frac{\hbar^2 k_1^2}{2m} - |V| \tag{5.261}$$

or

$$k_1^2 - k_2^2 = \frac{2m}{\hbar^2}(-|V|) = -\frac{2m|V|}{\hbar^2} \tag{5.262}$$

or

$$(ak_1)^2 - (ak_2)^2 = -\frac{2ma^2|V|}{\hbar^2} = -\left(\frac{g}{2}\right)^2 \tag{5.263}$$

or

$$(ak_2)^2 - (ak_1)^2 = \left(\frac{g}{2}\right)^2 \tag{5.264}$$

For rectangular well scattering with $E > 0$, ak_2 and ak_1 lie on a hyperbola, $ak_2 > ak_1 > 0$ (see Figure 5.26).

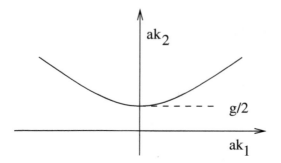

Fig. 5.26. Dependence of k_2 on k_1 in a rectangular well potential in 1D for $E > 0$.

The permitted values of k_2 (and therefore E) comprise a positive unbounded continuum.

The scattering parameters are obtained from the coefficients, B/A and F/A. The coefficients are determined from the boundary conditions at $x = -a$ and $x = a$.

$$\psi_I(-a) = \psi_{II}(-a) \tag{5.265}$$

$$\left.\frac{\partial \psi_I}{\partial x}\right|_{x=-a} = \left.\frac{\partial \psi_{II}}{\partial x}\right|_{x=-a} \tag{5.266}$$

$$\psi_{II}(a) = \psi_{III}(a) \tag{5.267}$$

$$\left.\frac{\partial \psi_{II}}{\partial x}\right|_{x=a} = \left.\frac{\partial \psi_{III}}{\partial x}\right|_{x=a} \tag{5.268}$$

The transmission and the reflection coefficients are obtained from

$$T = \left|\frac{F}{A}\right|^2 \quad , \quad R = \left|\frac{B}{A}\right|^2 \tag{5.269}$$

After solving the coefficients, the transmission coefficient becomes, for $E \geq 0$

$$\frac{1}{T} = 1 + \frac{1}{4}\frac{V^2}{E(E+|V|)}\sin^2(2k_2 a) \tag{5.270}$$

Using the relations

$$k_2^2 = \frac{2m}{\hbar^2}(E+|V|) \quad , \quad g^2 = \frac{8ma^2|V|}{\hbar^2} \quad , \quad \frac{E}{|V|} = \bar{E} > 0$$

we can write

$$\frac{1}{T} = 1 + \frac{1}{4}\frac{\sin^2\left(\sqrt{g^2(\bar{E}+1)}\right)}{\bar{E}(\bar{E}+1)} \tag{5.271}$$

or

$$T = \frac{1}{1 + \frac{\sin^2(g\sqrt{\bar{E}+1})}{4\bar{E}(\bar{E}+1)}} \quad , \quad \bar{E} > 0 \tag{5.272}$$

The perfect transmission $(T = 1)$ is possible when

$$2k_2 a = n\pi \quad , \quad n = 1, 2, 3, \cdots \tag{5.273}$$

On the other hand, $T \to 1$ with increasing incident energy. At $E = 0$, $T = 0$. The plot of T versus $\bar{E}$ looks the same as obtained for the potential barrier. Considering the requirement for perfect transmission

$$2k_2 a = n\pi \quad , \quad k_2 = \frac{n\pi}{2a} \quad , \quad k_2^2 = \frac{n^2\pi^2}{4a^2}$$

the energy expression becomes

$$\frac{\hbar^2 k_2^2}{2m} = \frac{\hbar^2 \pi^2 n^2}{8ma^2} = n^2 \left(\frac{\hbar^2 \pi^2}{8ma^2} \right) = n^2 E_1 \tag{5.274}$$

$$\frac{\hbar^2 k_2^2}{2m} = E + |V| = n^2 E_1 \tag{5.275}$$

Here

$$E_1 = \frac{\hbar^2 \pi^2}{8ma^2}$$

corresponds to the lowest eigen energy of a particle in a one–dimensional box of width $2a$.

The scattering of a beam of particles by a potential well has been used as a model for the scattering of low–energy electrons from atoms. The attractive well represents the field of the nucleus, whose positive charge becomes evident when the scattering electrons penetrate the shell structure of the atomic electrons. The reflection coefficient is a measure of the scattering cross–section. Experiments in which this cross–section is measured (for rear gas atoms) detect a low–energy minimum which is consistent with the first maximum that T goes through for typical values of well depth and width according to the model discussed here,

$$T = \frac{1}{1 + \frac{\sin^2(g\sqrt{\bar{E}+1})}{4\bar{E}(\bar{E}+1)}} \quad , \quad \bar{E} > 0 \tag{5.276}$$

This transparency to low–energy electrons of rare gas atoms is known as the **Ramsauer effect**.

In the models discussed here, the Hamiltonians are of the form

$$H = \frac{p^2}{2m} + V(x) \tag{5.277}$$

In each case considered, the spectrum of energy is a continuum, $E = \hbar^2 k^2/2m$. For each value of k, a corresponding set of coefficient ratios (B/A, C/A for the simple step, and B/A, C/A, D/A, F/A for the rectangular potential) are determined. The coefficient A is fixed by the data on the incident beam. These coefficients then determine the wave function, which is an eigenfunction of the Hamiltonian above. All such scattering eigenstates are unbound states. A continuous spectrum is characteristic of unbound states, while a discrete spectrum is characteristic of bound states (e.g., particle in a box, harmonic oscillator).

Bound states:

In the previous discussion of finite potential well, total energy of the system was positive and the states were unbound corresponding to a continuum of eigen energies,

$$E_k = \frac{\hbar^2 k^2}{2m} \quad , \quad E_k > 0$$

If we look for the solutions of the Schrödinger equation for negative energies, $E < 0$, only a finite, discrete number of eigenstates are found (see Figure 5.27).

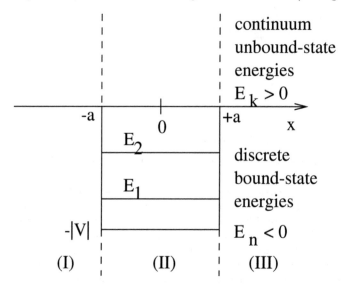

Fig. 5.27. Rectangular well potential in 1D. $E_k > 0$ correspond to unbound–state energies. $E_n < 0$ correspond to bound–state energies.

There are three regions; the Schrödinger equation and the corresponding solutions are (for $|E| < |V|$, $E < 0$, $V < 0$):

Region I: $x < -a$

$$-\frac{\hbar^2}{2m}\frac{\partial^2 \psi}{\partial x^2} = -|E|\psi \quad \text{or} \quad \frac{\partial^2 \psi}{\partial x^2} = \kappa^2 \psi \tag{5.278}$$

Kinetic energy is negative, and the eigenfunction decays,

$$\psi_I(x) = Ae^{\kappa x} \quad , \quad \frac{\hbar^2 \kappa^2}{2m} = |E| > 0 \tag{5.279}$$

Region II: $-a \le x \le a$

$$-\frac{\hbar^2}{2m}\frac{\partial^2 \psi}{\partial x^2} = (|V| - |E|)\psi \quad \text{or} \quad \frac{\partial^2 \psi}{\partial x^2} = -k^2 \psi \tag{5.280}$$

Kinetic energy is positive, and the eigen function oscillates,

$$\psi_{II}(x) = Be^{ikx} + Ce^{-ikx} \quad , \quad \frac{\hbar^2 k^2}{2m} = |V| - |E| > 0 \tag{5.281}$$

We may also choose the superposition of real functions as eigenfunction

$$\psi_{II}(x) = B\cos(kx) + C\sin(kx) \tag{5.282}$$

It is convenient to use this form of ψ_{II} in the calculations.

Region III: $x > +a$

$$-\frac{\hbar^2}{2m}\frac{\partial^2\psi}{\partial x^2} = -|E|\psi \quad \text{or} \quad \frac{\partial^2\psi}{\partial x^2} = \kappa^2\psi \tag{5.283}$$

Kinetic energy is negative, and the eigenfunction decays,

$$\psi_{III}(x) = De^{-\kappa x} \quad , \quad \frac{\hbar^2\kappa^2}{2m} = |E| > 0 \tag{5.284}$$

In these three solutions k and κ obey the constraint (see Figure 5.28),

$$k^2 + \kappa^2 = \frac{2m|V|}{\hbar^2} \tag{5.285}$$

or

$$(ka)^2 + (\kappa a)^2 = \frac{2ma^2|V|}{\hbar^2} = \left(\frac{g}{2}\right)^2 = \rho^2 \tag{5.286}$$

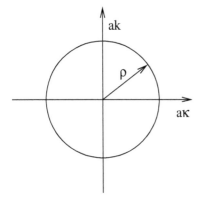

Fig. 5.28. Dependence of k on κ in rectangular well potential in 1D for $E < 0$.

The coefficients A, B, C, D determine the eigenstate corresponding to the eigen energy $\hbar^2\kappa^2/2m$. These coefficients are determined by the continuity conditions at $x = -a$, and $x = a$.

$$\psi_I(-a) = \psi_{II}(-a) \quad , \quad \frac{\partial\psi_I(x)}{\partial x}\Big|_{x=-a} = \frac{\partial\psi_{II}(x)}{\partial x}\Big|_{x=-a} \tag{5.287}$$

$$\psi_{II}(a) = \psi_{III}(a) \quad , \quad \frac{\partial\psi_{II}(x)}{\partial x}\Big|_{x=a} = \frac{\partial\psi_{III}(x)}{\partial x}\Big|_{x=a} \tag{5.288}$$

The corresponding equations are:

$$Ae^{-\kappa a} = B\cos(ka) - C\sin(ka) \tag{5.289}$$

$$\kappa Ae^{-\kappa a} = k(B\sin(ka) + C\cos(ka)) \tag{5.290}$$

$$B\cos(ka) + C\sin(ka) = De^{-\kappa a} \tag{5.291}$$

$$-k(B\sin(ka) - C\cos(ka)) = -\kappa De^{-\kappa a} \tag{5.292}$$

There are four equations and four unknowns (A, B, C, D). We may write this set of simultaneous linear homogeneous equations in the matrix form

$$\begin{pmatrix} e^{-\kappa a} & -\cos(ka) & \sin(ka) & 0 \\ \kappa e^{-\kappa a} & -k\sin(ka) & -k\cos(ka) & 0 \\ 0 & -\cos(ka) & -\sin(ka) & e^{-\kappa a} \\ 0 & k\sin(ka) & -k\cos(ka) & -\kappa e^{-\kappa a} \end{pmatrix} \begin{pmatrix} A \\ B \\ C \\ D \end{pmatrix} = 0 \tag{5.293}$$

To get nontrivial solutions the determinant of the coefficient matrix must vanish. We may rearrange the determinant of the equations as the following:

$$\begin{vmatrix} \cos(ka) & -\sin(ka) & -e^{-\kappa a} & 0 \\ k\sin(ka) & k\cos(ka) & -\kappa e^{-\kappa a} & 0 \\ \cos(ka) & \sin(ka) & 0 & -e^{-\kappa a} \\ -k\sin(ka) & k\cos(ka) & 0 & \kappa e^{-\kappa a} \end{vmatrix} = 0 \tag{5.294}$$

This leads to the equation

$$\left(\tan(ka) - \frac{\kappa}{k}\right)\left(\tan(ka) + \frac{k}{\kappa}\right) = 0 \tag{5.295}$$

Since k and κ are functions of the energy E, Eq. (5.295) imposes restrictions on the values of energy E that permit a solution for A, B, C, D; this means that the energy is quantized and the system has a discrete energy spectrum.

Here there are two types of solution, one obtained from

$$\tan(ka) - \frac{\kappa}{k} = 0 \quad \text{or} \quad \tan(ka) = \frac{\kappa}{k} \tag{5.296}$$

the other obtained from

$$\tan(ka) + \frac{k}{\kappa} = 0 \quad \text{or} \quad \cot(ka) = -\frac{\kappa}{k} \tag{5.297}$$

Let us consider first the solution $\tan(ka) = \kappa/k$:

From the equations Eqs. (5.289) and (5.290) we may write

$$\kappa A e^{-\kappa a} = \kappa\cos(ka)(B - C\tan(ka)) = \kappa\cos(ka)(B - C\frac{\kappa}{k}) \tag{5.298}$$

and

$$\kappa A e^{-\kappa a} = k\cos(ka)(B\tan(ka) + C) = k\cos(ka)(B\frac{\kappa}{k} + C) \tag{5.299}$$

It follows that C must be zero. Thus in this case, the interior solution is

$$\psi(x) = B\cos(kx) \tag{5.300}$$

where B has to be determined from normalization. This solution is an even function of x. The eigenvalue condition, that is Eq. (5.296), can be solved graphically as follows: Define

$$\rho^2 = \frac{2ma^2|V|}{\hbar^2} \quad \text{and} \quad \xi = ka \tag{5.301}$$

Then the eigenvalue condition, Eq. (5.296), can be written as

$$\frac{\sqrt{\rho^2 - \xi^2}}{\xi} = \tan(\xi) \tag{5.302}$$

Then the eigenvalues can be obtained as the points of intersection of the two functions $\tan(\xi)$ and $\sqrt{\rho^2 - \xi^2}/\xi$ plotted as a function of ξ (see Figure 5.29).

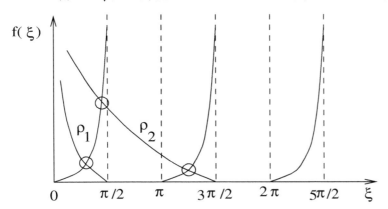

Fig. 5.29. Plot of $f(\xi) = \tan(\xi)$, and $f(\xi) = \sqrt{\rho^2 - \xi^2}/\xi$ for given ρ, versus ξ.

We may conclude that:

- The eigenvalues are discrete.
- The number of eigenvalues increases as ρ^2 increases.
- There is at least one bound state, no matter how shallow or how narrow the well is. This is a special characteristic of the even solution of one–dimensional attractive potential.

Let us now consider the solution $\cot(ka) = -\kappa/k$: From the equations (5.291) and (5.292) we may write

$$\kappa D e^{-\kappa a} = \kappa \sin(ka)(B\cot(ka) + C) = \kappa \sin(ka)\left(-B\frac{\kappa}{k} + C\right) \tag{5.303}$$

and

$$\kappa D e^{-\kappa a} = k \sin(ka)(B - C\cot(ka)) = k \sin(ka)\left(B + C\frac{\kappa}{k}\right) \tag{5.304}$$

It follows that B must be zero. Thus, in this case, the interior solution is

$$\psi(x) = C\sin(kx) \tag{5.305}$$

where C has to be determined from normalization. This solution is an odd function of x. The eigenvalue condition, that is Eq. (5.297), for this solution can be written as

$$\frac{\sqrt{\rho^2 - \xi^2}}{\xi} = -\cot(\xi) \tag{5.306}$$

Similarly, the eigenvalues can be obtained as the points of intersection of the two functions $-\cot(\xi)$ and $\sqrt{\rho^2 - \xi^2}/\xi$ plotted as a function of ξ (see Figure 5.30).

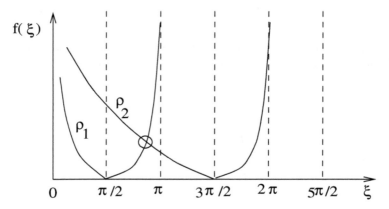

Fig. 5.30. Plot of $f(\xi) = -\cot(\xi)$, and $f(\xi) = \sqrt{\rho^2 - \xi^2}/\xi$ for given ρ, versus ξ.

We may conclude that:

- The eigenvalues are discrete.

- The number of eigenvalues increases as ρ^2 increases.

- There is a lower limit for ρ^2 to have a bound state.

We find that there will be no intersection of the two curves unless $\rho^2 \geq \pi^2/4$; that is

$$\frac{2ma^2|V|}{\hbar^2} \geq \frac{\pi^2}{4} \tag{5.307}$$

This is the range–depth condition for the existence of a bound state of the odd solution of one–dimensional attractive potential.

It is also possible to obtain the eigenvalues following a different procedure: For the even states, $\psi(x) = B\cos(kx)$, we may write the eigenvalue condition, $k\tan(ka) = \kappa$, in terms of nondimensional wavenumbers, $\xi = ka$ and $\eta = \kappa a$

$$\xi \tan(\xi) = \eta \tag{5.308}$$

$$\xi^2 + \eta^2 = \frac{2ma^2|V|}{\hbar^2} = \rho^2 \tag{5.309}$$

The intersection of these equations determines the eigenenergies corresponding to the even eigenstates (see Figure 5.31).

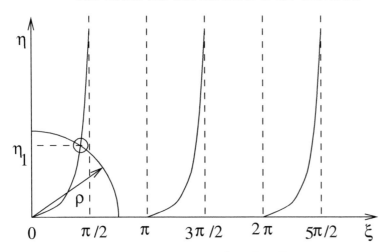

Fig. 5.31. Plot of $\eta = \xi \tan(\xi)$, and $\xi^2 + \eta^2 = \rho^2$ for given ρ, versus ξ.

For given potential parameters, a, $|V|$, and particle mass, m, one can first obtain the radius of the circle, ρ, and then the intersection point on the graph, η_1.

$$\rho = \sqrt{\frac{2ma^2|V|}{\hbar^2}} \quad , \quad E_1 = \left(\frac{\eta_1}{\rho}\right)^2 |V| \qquad (5.310)$$

$$\eta^2 = a^2 \kappa^2 = \frac{2ma^2|E|}{\hbar^2} = \frac{2ma^2|V||E|}{\hbar^2|V|} = \rho^2 \frac{|E|}{|V|}$$

The position of the eigen energy obtained in this way is relative to the $V = 0$ point (see Figure 5.32).

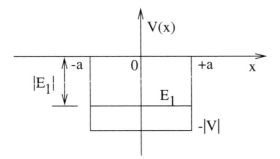

Fig. 5.32. Position of E_1 with respect to $V = 0$ point.

The eigen energies of the odd eigenstates, $\psi(x) = C \sin(kx)$, are the intersections of the two curves (see Figure 5.33).

$$\xi \cot(\xi) = -\eta \qquad (5.311)$$

$$\xi^2 + \eta^2 = \rho^2 \qquad (5.312)$$

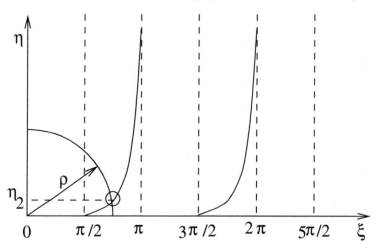

Fig. 5.33. Plot of $\eta = -\xi \cot(\xi)$, and $\xi^2 + \eta^2 = \rho^2$ for given ρ, versus ξ.

Here again for given potential parameters, a, $|V|$, and particle mass, m, one can first obtain the radius of the circle, ρ, and then the intersection point on the graph, η_2.

$$\rho = \sqrt{\frac{2ma^2|V|}{\hbar^2}} \quad , \quad E_2 = \left(\frac{\eta_2}{\rho}\right)^2 |V| \tag{5.313}$$

The position of the eigenenergy obtained in this way is relative to the $V = 0$ point (see Figure 5.34).

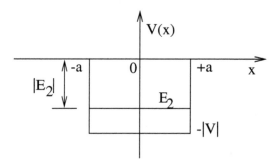

Fig. 5.34. Position of E_2 with respect to $V = 0$ point.

If we obtain the eigenenergies using ξ, instead of η, then the position of the eigenenergy values are defined with respect to the bottom of the potential well (see Figure 5.35). The energies, E', measured with respect to the bottom of the well are directly obtained from k or, equivalently, ξ :

$$E' = |V| - |E| = \frac{\hbar^2 k^2}{2m} = \frac{\hbar^2 \xi^2}{2ma^2} > 0 \tag{5.314}$$

Energy measured from the top of the well, $|E| \propto \eta^2$, represents binding energy.

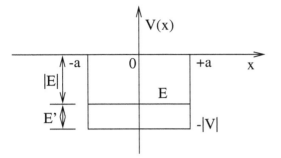

Fig. 5.35. Position of a bound–state energy level, E, with respect to the top, $|E|$, or with respect to the bottom, E', in a rectangular potential well in 1D.

$$|E| = \frac{\hbar^2 \kappa^2}{2m} = \frac{\hbar^2 \kappa^2 a^2}{2ma^2} = \frac{\hbar^2 \eta^2}{2ma^2} \tag{5.315}$$

$$E' = |V| - |E| = \frac{\hbar^2 k^2}{2m} = \frac{\hbar^2 k^2 a^2}{2ma^2} = \frac{\hbar^2 \xi^2}{2ma^2} \tag{5.316}$$

Therefore, one can easily obtain the position of a bound state energy either with respect to the top of the potential well (from η) or with respect to the bottom of the potential well (from ξ) (see Figure 5.36).

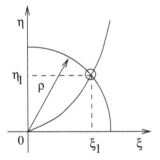

Fig. 5.36. Determination of bound–state energy position in a rectangular potential well in 1D.

$$|E_1| = \left(\frac{\eta_1}{\rho}\right)^2 |V| \quad , \quad E_1' = \left(\frac{\xi_1}{\rho}\right)^2 V \tag{5.317}$$

The time–dependent form of the bound eigenstates appears as

$$\Psi_n(x,t) = \psi_n(x) e^{-\frac{i}{\hbar} E_n t} \quad , \quad E_n < 0 \tag{5.318}$$

For positive energy, the unbound time–dependent eigenstates form a continuum,

$$\Psi_k(x,t) = \psi_k(x) e^{-\frac{i}{\hbar} E_k t} \quad , \quad E_k > 0 \tag{5.319}$$

5.6 Worked examples

Example - 5.1 : Write down the Schrödinger equation for an oscillator in the momentum representation. Find the wave function in the momentum space.

Solution : The Hamiltonian operator in coordinate representation:

$$H = \frac{p^2}{2m} + \frac{1}{2}mw^2x^2$$

The Hamiltonian operator in momentum representation:

$$H = \frac{p^2}{2m} - \frac{1}{2}mw^2\hbar^2\frac{d^2}{dp^2}$$

Thus the Schrödinger equation in momentum representation is

$$\left(\frac{p^2}{2m} - \frac{\hbar^2}{2}mw^2\frac{d^2}{dp^2}\right)a(\mathbf{p}) = Ea(\mathbf{p})$$

The method of solving this is similar to that of the equation in the coordinate representation:

$$a_n(p) = \frac{1}{2^n n!\sqrt{\pi mw\hbar}}e^{-\frac{p^2}{2mw\hbar}}H_n(p/\sqrt{mw\hbar})$$

Example - 5.2 : (a) Using the relation $[x, p] = i\hbar$, show that $[a, a^+] = 1$. (b) Using the result of (a), show that $[H, a] = -\hbar wa$ and $[H, a^+] = \hbar wa^+$.

Solution : (a)

$$[x, p] = i\hbar = \left[\frac{a + a^+}{\sqrt{2}\beta}, \frac{mw_0}{i}\left(\frac{a - a^+}{\sqrt{2}\beta}\right)\right]$$

$$2\beta^2 i\hbar = \frac{mw_0}{i}\{(a + a^+)(a - a^+) - (a - a^+)(a + a^+)\}$$

$$-2\frac{\beta^2\hbar}{mw_0} = -2aa^+ + 2a^+a$$

$$\frac{\beta^2\hbar}{mw_0} = aa^+ - a^+a = [a, a^+]$$

since $\beta^2 = mw_0/\hbar$, $[a, a^+] = 1$

(b)

$$[H, a] = [\hbar w(a^+a + \frac{1}{2}), a] = \hbar w[a^+a + \frac{1}{2}, a]$$

$$= \hbar w\{[a^+a, a] + [\frac{1}{2}), a]\} = \hbar w[a^+a, a]$$

$$= \hbar w\{a^+[a, a] + [a^+, a]a\} = -\hbar wa$$

$$[H, a^+] = [\hbar w(a^+a + \frac{1}{2}), a^+] = \hbar w[a^+a + \frac{1}{2}, a^+]$$

$$= \hbar w\{[a^+a, a^+] + [\frac{1}{2}), a^+]\} = \hbar w[a^+a, a^+]$$

$$= \hbar w\{a^+[a, a^+] + [a^+, a^+]a\} = \hbar wa^+$$

Example - 5.3 : Find $< x >$ for a harmonic oscillator in the superposition state

$$\psi(x,t) = \frac{1}{\sqrt{2}}[\psi_0(x,t) + \psi_1(x,t)]$$

Solution :

$$\psi_0(x,t) = \phi_0(x)e^{-\frac{i}{\hbar}E_0 t} \quad , \quad \psi_1(x,t) = \phi_1(x)e^{-\frac{i}{\hbar}E_1 t}$$

$$< x >=< \psi(x,t)|x|\psi(x,t) > \quad , \quad x = \frac{1}{\sqrt{2\beta}}(a + a^+)$$

$$< x >= \frac{1}{2\sqrt{2\beta}} < \psi_0 + \psi_1|a + a^+|\psi_0 + \psi_1 >$$

Let $\psi' = (a + a^+)\psi = (a + a^+)(\psi_0 + \psi_1)$. The operators a and a^+ act on ϕ_n only, therefore

$$(a + a^+)(\psi_0 + \psi_1) = a\psi_0 + a\psi_1 + a^+\psi_0 + a^+\psi_1$$

$$= a\phi_0(x)e^{-\frac{i}{\hbar}E_0 t} + a\phi_1(x)e^{-\frac{i}{\hbar}E_1 t} + a^+\phi_0(x)e^{-\frac{i}{\hbar}E_0 t} + a^+\phi_1(x)e^{-\frac{i}{\hbar}E_1 t}$$

$$= 0 + \phi_0 e^{-\frac{i}{\hbar}E_1 t} + \phi_1 e^{-\frac{i}{\hbar}E_0 t} + \sqrt{2}\phi_2 e^{-\frac{i}{\hbar}E_1 t}$$

therefore

$$< x >= \frac{1}{2\sqrt{2\beta}}\left(< \phi_0 e^{-\frac{i}{\hbar}E_0 t} + \phi_1 e^{-\frac{i}{\hbar}E_1 t}|\right)$$

$$\times \left(|\phi_0 e^{-\frac{i}{\hbar}E_1 t} + \phi_1 e^{-\frac{i}{\hbar}E_0 t} + \sqrt{2}\phi_2 e^{-\frac{i}{\hbar}E_1 t} >\right)$$

$$= \frac{1}{2\sqrt{2\beta}}\left\{e^{-\frac{i}{\hbar}(E_1 - E_0)t} + e^{-\frac{i}{\hbar}(E_0 - E_1)t}\right\}$$

Using $E_1 - E_0 = \hbar w_0(1 + 1/2) - \hbar w_0(1/2) = \hbar w_0$ we get

$$< x >= \frac{1}{2\sqrt{2\beta}}\left(e^{-iw_0 t} + e^{iw_0 t}\right) = \frac{2}{2\sqrt{2\beta}}\cos(w_0 t) = \frac{1}{\sqrt{2\beta}}\cos(w_0 t)$$

Example - 5.4 : An electron is confined to a potential well of finite depth and width, 10 Å. The eigenstate of highest energy of this system corresponds to the value $\xi = 3.2$. (a) How many bound states does this system have? (b) Estimate the energy of the highest state with respect to the zero energy line at the bottom of the well.

Solution : (a)

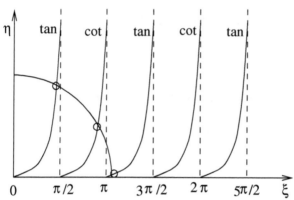

There are 3 bound states, since $3.2 > \pi$.

(b)

$$E' = \left(\frac{\xi}{\rho}\right)^2 V = \frac{\xi^2 \hbar^2}{2ma^2}$$

$$= \frac{(3.2)^2 (1.05 \times 10^{-34})^2}{2(9.1 \times 10^{-31})(5 \times 10^{-10})^2} \; Joule \cong \frac{1}{45} \times 10^{-17} \; J$$

$$= \frac{10^{-17}}{45 \times 1.6 \times 10^{-19}} \; eV \cong 1.34 \; eV$$

Example - 5.5 : An electron beam is incident on a barrier of height $10 \; eV$. At $E = 10 \; eV$, the transmission coefficient has the value $T = 3.37 \times 10^{-3}$. What is the width of the barrier?

Solution : For rectangular barriers we have the relation:

$$\text{as} \quad \frac{V - E}{V} = \frac{\hbar^2 \kappa^2}{2mV} \equiv \epsilon \quad \rightarrow \quad 0$$

$$T \quad \rightarrow \quad \frac{1}{1 + \frac{g^2}{4}} \;, \quad \text{where} \; g^2 = \frac{2m(2a)^2 V}{\hbar^2}$$

taking $2a = L$, and substituting in T, we get

$$T = \frac{\hbar^2}{\hbar^2 + 2mL^2 V} \quad \rightarrow \quad L^2 = \frac{\hbar^2}{2mV}\left(\frac{1}{T} - 1\right)$$

or

$$L^2 = \frac{(1.1 \times 10^{-34})^2}{2 \times 9.1 \times 10^{-31} \times 10 \times 1.6 \times 10^{-3}}\left(\frac{1}{3.37 \times 10^{-3}} - 1\right)$$

$$\cong 123 \times 10^{-20} \; m^2$$

Therefore, $L \cong 11 \times 10^{-10} \; m = 11 \; Å$.

Example - 5.6 : Find the reflection and the transmission coefficients for a repulsive delta function $V(x) = \alpha \delta(x)$, $\alpha > 0$.

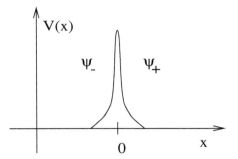

Solution : The Schrödinger equation for this system is the folloowing:

$$-\frac{\hbar^2}{2m}\frac{\partial^2\psi}{\partial x^2} + \alpha\delta(x)\psi = E\psi$$

or

$$\frac{\partial^2\psi}{\partial x^2} + \frac{2mE}{\hbar^2}\psi - \frac{2m\alpha}{\hbar^2}\delta(x)\psi = 0$$

or

$$\frac{\partial^2\psi}{\partial x^2} + k^2\psi - \beta^2\psi = 0$$

where

$$k^2 = \frac{2mE}{\hbar^2} \quad , \quad \beta^2 = \frac{2m\alpha}{\hbar^2}$$

Continuity equation for the first derivative of wave functions is given for the delta function as

$$\frac{\partial\psi_-}{\partial x}\Big|_{x=0} = \frac{\partial\psi_+}{\partial x}\Big|_{x=0} - \frac{2m\alpha}{\hbar^2}\psi_+(0)$$

Therefore,

$$\psi_-(0) = \psi_+(0)$$

$$\frac{\partial\psi_-}{\partial x}\Big|_{x=0} = \frac{\partial\psi_+}{\partial x}\Big|_{x=0} - \beta^2\psi_+(0)$$

The wave functions are

$$\psi_-(x) = Ae^{ikx} + Be^{-ikx} \quad , \quad \psi_+(x) = Ce^{ikx}$$

Continuity equations give:

$$A + B = C \quad \text{and} \quad ikA - ikB = ikC - \beta^2 C = (ik - \beta^2)C$$

or

$$ik(A - B) = (ik - \beta^2)C$$

$$\frac{ik(A - B)}{(ik - \beta^2)} = C \quad ; \quad A + B = \frac{ik(A - B)}{(ik - \beta^2)}$$

$$(ik - \beta^2)(A + B) = ik(A - B) \quad \text{or} \quad (2ik - \beta^2)B = \beta^2 A$$

The reflection coefficient, R, can be calculated from

$$\frac{B}{A} = \frac{\beta^2}{2ik - \beta^2} \quad \rightarrow \quad R = \left|\frac{B}{A}\right|^2 = \frac{\beta^4}{4k^2 + \beta^4}$$

From $C = A + B$ we can write

$$\frac{C}{A} = 1 + \frac{B}{A} = 1 + \frac{\beta^2}{2ik - \beta^2} = \frac{2ik}{2ik - \beta^2}$$

The transmission coefficient, T, can be calculated from

$$T = \left|\frac{C}{A}\right|^2 = \frac{4k^2}{4k^2 + \beta^4}$$

$$R + T = 1 \quad \text{is satisfied.}$$

Example - 5.7 : Obtain the energy spectrum of a particle in the periodic potential shown in the figure.

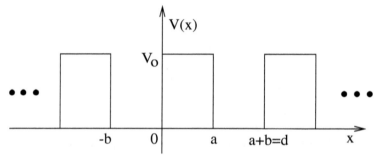

Solution : The periodic condition in potential is $V(x+d) = V(x)$. The solutions of the Schrödinger equation

$$\frac{d^2\psi}{dx^2} + \frac{2m}{\hbar^2}(E - V(x))\psi = 0$$

will also be periodic, $\psi(x+d) = C\psi(x)$, where C is a constant factor. This equality may be generalized as $\psi(x + nd) = C^n\psi(x)$. If $|C| \neq 1$, then along one direction of the x–axis, $\psi(x)$ increases (or decreases) without bound. Therefore, physically meaningful solutions can be obtained only if C is a phase factor, that is, $C = e^{i\phi}$, where ϕ is real. Then $|C| = 1$ and

$$\psi(x + d) = e^{i\phi}\psi(x)$$

In the period $-b < x < a$, we may write the wave functions as

$$\psi(x) = \begin{cases} Ae^{i\beta x} + Be^{-i\beta x} & , -b < x < 0 \\ De^{i\alpha x} + Fe^{-i\alpha x} & , 0 < x < a \end{cases}$$

where

$$\beta = \sqrt{\frac{2m}{\hbar^2}(E - V_0)} \quad , \quad \alpha = \sqrt{\frac{2m}{\hbar^2}E}$$

In the period $a < x < a + d$, the wave functions may be expressed as

$$\psi(x) = e^{i\phi} \begin{cases} Ae^{i\beta(x-d)} + Be^{-i\beta(x-d)} & , a < x < d \\ De^{i\alpha(x-d)} + Fe^{-i\alpha(x-d)} & , d < x < a + d \end{cases}$$

From the continuity conditions at $x = 0$ and at $x = a$ the following system of equations is obtained

$$A + B = D + F$$

$$\beta(A - B) = \alpha(D - F)$$

$$e^{i\phi}(Ae^{-i\beta b} + Be^{i\beta b}) = De^{i\alpha a} + Fe^{-i\alpha a}$$

$$\beta e^{i\phi}(Ae^{-i\beta b} - Be^{i\beta b}) = \alpha(De^{i\alpha a} - Fe^{-i\alpha a})$$

Nontrivial solutions for the coefficients A, B, D, F are obtained only if the determinant of this system of equations vanishes, which gives the condition

$$\cos\phi = \cos(a\alpha)\cosh(b\delta) - \frac{\alpha^2 - \delta^2}{2\alpha\delta}\sin(a\alpha)\sinh(b\delta) \quad \text{for} \quad 0 < E < V_0$$

$$\cos\phi = \cos(a\alpha)\cos(b\beta) - \frac{\alpha^2 + \beta^2}{2\alpha\beta}\sin(a\alpha)\sin(b\beta) \quad \text{for} \quad E > V_0$$

where

$$\delta = \sqrt{\frac{2m}{\hbar^2}(V_0 - E)}$$

Now the energy E, which appears in $\cos\phi$ equations through α, β, and δ, has a value such that $-1 \leq \cos\phi \leq +1$. We now consider two cases:

(1) $0 < E < V_0$:

To get a picture of the structure of the energy spectrum we look at the particular case for which $b \to 0$, $V_0 \to \infty$ in such a manner that $b\delta^2$ remains finite. With the notation

$$\lim_{b \to 0, \delta \to \infty} \frac{ab\delta^2}{2} = P$$

the condition $-1 \leq \cos\phi \leq +1$ becomes $-1 \leq g(E) \leq +1$, where

$$g(E) = P\frac{\sin(a\alpha)}{a\alpha} + \cos(a\alpha)$$

The plot of the function $g(E)$ gives the energy spectrum.

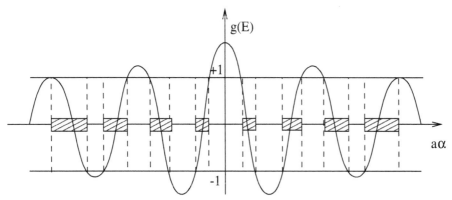

As seen from the figure, the energy spectrum consists of a series of separate regions, inside each of which the energy of the particle can vary continuously. These regions are called the allowed bands, and the ones between them the forbidden bands. It can be seen that the width of the allowed bands increases as the energy increases.

(2) $E > V_0$:

The possible values of the energy are determined in this case by the condition $-1 \leq f(E) \leq +1$, where

$$f(E) = \cos(a\alpha)\cos(b\beta) - \frac{\alpha^2 + \beta^2}{2\alpha\beta}\sin(a\alpha)\sin(b\beta)$$

$$= \cos(a\alpha + b\beta) - \frac{(\alpha - \beta)^2}{2\alpha\beta}\sin(a\alpha)\sin(b\beta)$$

The energy spectrum will also, in this case, have a band structure.

5.7   Problems

Problem - 5.1 : At time zero a linear harmonic oscillator is in a state that is described by the normalized wave function:

$$\psi(x,0) = \frac{1}{\sqrt{5}}u_0(x) + \frac{1}{\sqrt{2}}u_2(x) + c_3 u_3(x)$$

where $u_n(x)$ is the n th time independent eigenfunction for the oscillator. (a) Determine the numerical value of c_3 assuming it to be real and positive. (b) Write out the wave function at time t. (c) What is the expectation value of the energy of the oscillator at $t = 0$? At $t = 1$ s?

Problem - 5.2 : Show that in the n th eigenstate of the simple harmonic oscillator in one dimension the Virial theorem is satisfied.

Problem - 5.3 : Find $\Delta x \Delta p$ for 1D SHO in any state.

Problem - 5.4 : Determine the current density J for the wave functions (a)

$$\psi(x,t) = Ae^{\frac{i}{\hbar}(\sqrt{2mE}x - Et)}$$

(b)

$$\psi(x,t) = Ae^{\frac{i}{\hbar}(\sqrt{2mE}x - Et)} + Be^{\frac{i}{\hbar}(-\sqrt{2mE}x - Et)}$$

Problem - 5.5 : Calculate the possible energy values of a particle in the potential given by $V(x) = \infty$ if $x \leq 0$, and $V(x) = \frac{1}{2}mw^2x^2$ if $x > 0$.

Problem - 5.6 : Find the energies of the bound states of a particle in the potential well given by $V(x) = +\infty$ if $x < 0$, $V(x) = -V_0$ if $0 < x < a$, and $V(x) = 0$ if $x > a$.

PART – II
Three–Dimensional Systems

Chapter 6

N–Particle Systems

6.1   The Schrödinger equation for N–particle systems

After knowing the single particle analysis, one can easily extend the similar discussions for many–particle systems. Consider an N–particle system. The N–particles are described by a wave function $\psi(x_1, \cdots, x_N)$ that is normalized such that

$$\int \cdots \int |\psi(x_1, \cdots, x_N)|^2 dx_1 \cdots dx_N = 1 \tag{6.1}$$

The interpretation of the square of the N–particle wave function is that it yields the probability density for finding particle 1 at x_1, particle 2 at $x_2, \cdots$, particle N at x_N. The time development of such a wave function is given by the solution of the differential equation

$$i\hbar \frac{\partial}{\partial t} \psi(x_1, \cdots, x_N; t) = H\psi(x_1, \cdots, x_N; t) \tag{6.2}$$

where the Hamiltonian H also represents an N–particle system

$$H = \sum_{i=1}^{N} \frac{p_i^2}{2m_i} + V(x_1, \cdots, x_N) \tag{6.3}$$

or

$$H = -\hbar^2 \left(\frac{1}{2m_1} \frac{\partial^2}{\partial x_1^2} + \cdots + \frac{1}{2m_N} \frac{\partial^2}{\partial x_N^2} \right) + V(x_1, \cdots, x_N) \tag{6.4}$$

Operators describing single particle observables commute when they refer to different particles, for example,

$$[p_i, x_j] = -i\hbar \delta_{ij} \tag{6.5}$$

If there are no external fields, then the potential energy can only depend on the relative position of the particles, that is,

$$V = V(x_{12}, \cdots, x_{N-1,N}) \quad ; \quad x_{ij} = x_i - x_j \tag{6.6}$$

Such a potential is invariant under the transformation

$$x_i \to x_i + a \tag{6.7}$$

A special case of this potential is the case of two–body forces, in which case

$$V = \sum_{i<j} V(x_{ij}) \tag{6.8}$$

When there are no external forces, then in classical mechanics the total momentum is conserved. This follows from the equations of motion

$$m_i \frac{d^2 x_i}{dt^2} = -\frac{\partial}{\partial x_i} V(x_{12}, \cdots, x_{N-1,N}) \qquad (6.9)$$

or

$$\frac{d}{dt} \left(\sum_i m_i \frac{dx_i}{dt} \right) = -\sum_i \frac{\partial}{\partial x_i} V(x_{12}, \cdots, x_{N-1,N}) = 0 \qquad (6.10)$$

Total momentum

$$P = \sum_i m_i \frac{dx_i}{dt} \qquad (6.11)$$

is a constant of the motion.

In quantum mechanics the same conclusion holds. We can show this by using the invariance of the Hamiltonian under the transformation $x_i \rightarrow x_i + a$. The invariance implies that both

$$H u_E(x_1, \cdots, x_N) = E u_E(x_1, \cdots, x_N) \qquad (6.12)$$

and

$$H u_E(x_1 + a, \cdots, x_N + a) = E u_E(x_1 + a, \cdots, x_N + a) \qquad (6.13)$$

hold. For an infinitesimal a, the terms of $\mathcal{O}(a^2)$ may be neglected. Then

$$u(x_1 + a, \cdots, x_N + a) \cong u(x_1, \cdots, x_N) + \qquad (6.14)$$

$$a \frac{\partial}{\partial x_1} u(x_1, \cdots, x_N) + \cdots + a \frac{\partial}{\partial x_N} u(x_1, \cdots, x_N)$$

$$\cong u(x_1, \cdots, x_N) + a \sum_i \frac{\partial}{\partial x_i} u(x_1, \cdots, x_N)$$

Subtracting Eq.(2.12) from Eq.(2.13), we can write

$$aH \left(\sum_i \frac{\partial}{\partial x_i} \right) u_E(x_1, \cdots, x_N) = aE \left(\sum_i \frac{\partial}{\partial x_i} \right) u_E(x_1, \cdots, x_N) \qquad (6.15)$$

$$= a \left(\sum_i \frac{\partial}{\partial x_i} \right) E u_E(x_1, \cdots, x_N) = a \left(\sum_i \frac{\partial}{\partial x_i} \right) H u_E(x_1, \cdots, x_N)$$

Defining

$$P = -i\hbar \sum_i \frac{\partial}{\partial x_i} \equiv \sum_i p_i \qquad (6.16)$$

we see that

$$(HP - PH) u_E(x_1, \cdots, x_N) = 0 \qquad (6.17)$$

Since the energy eigenstates for N–particles presumably form a complete set of states, we can rewrite Eq.(6.17) as

$$[H, P] \psi(x_1, \cdots, x_N) = 0 \quad \rightarrow \quad [H, P] = 0 \qquad (6.18)$$

This implies that P, the total momentum of the system, is a constant of the motion.

Consider a two–particle system. For two noninteracting particles we have the simple Hamiltonian

$$H = \frac{p_1^2}{2m_1} + \frac{p_2^2}{2m_2} \tag{6.19}$$

Since the two particles are totaly uncorrelated, the probability of finding one at x_1 and the other at x_2 is the product of two independent probabilities

$$P(x_1, x_2) = P(x_1)P(x_2) \tag{6.20}$$

Thus we expect that the solution of

$$\left(-\frac{\hbar^2}{2m_1} \frac{\partial^2}{\partial x_1^2} - \frac{\hbar^2}{2m_2} \frac{\partial^2}{\partial x_2^2} \right) = Eu(x_1, x_2) \tag{6.21}$$

should be separable into

$$u(x_1, x_2) = \phi(x_1)\phi(x_2) \tag{6.22}$$

Substituting this in Eq.(2.21) we can write

$$-\frac{\hbar^2}{2m_1} \frac{1}{\phi_1(x_1)} \frac{\partial^2 \phi_1(x_1)}{\partial x_1^2} - \frac{\hbar^2}{2m_2} \frac{1}{\phi_2(x_2)} \frac{\partial^2 \phi_2(x_2)}{\partial x_2^2} = E \tag{6.23}$$

The two terms in the equation depend on different variables, and that is why we set both of them equal to the constants E_1 and E_2, respectively,

$$E = E_1 + E_2 \tag{6.24}$$

$$-\frac{\hbar^2}{2m_1} \frac{\partial^2 \phi_1(x_1)}{\partial x_1^2} = E_1\phi_1(x_1) \quad , \quad -\frac{\hbar^2}{2m_2} \frac{\partial^2 \phi_2(x_2)}{\partial x_2^2} = E_2\phi_2(x_2) \tag{6.25}$$

These two equations are easily solved, and we get

$$u(x_1, x_2) = Ce^{ik_1 x_1 + ik_2 x_2} \tag{6.26}$$

with

$$k_1^2 = \frac{2m_1 E_1}{\hbar^2} \quad , \quad k_2^2 = \frac{2m_2 E_2}{\hbar^2} \tag{6.27}$$

Let us now rewrite the solution using the coordinates

$$x = x_1 + x_2 \quad , \quad X = \frac{m_1 x_1 + m_2 x_2}{m_1 + m_2} \quad \text{(C.M. coordinates)} \tag{6.28}$$

If we write

$$k_1 x_1 + k_2 x_2 = \alpha(x_1 - x_2) + \beta \frac{m_1 x_1 + m_2 x_2}{m_1 + m_2} \tag{6.29}$$

we find that

$$\beta = k_1 + k_2 \equiv K \quad , \quad \alpha = \frac{m_2 k_1 - m_1 k_2}{m_1 + m_2} \equiv k \tag{6.30}$$

so that the solution has the form

$$u(x_1, x_2) = Ce^{iKX + ikx} \tag{6.31}$$

where $K = k_1 + k_2$ is the wave number corresponding to the total momentum, and k is the wave number corresponding to the relative momentum. The factor e^{iKX}

represents the motion of the C.M., and the factor e^{ikx} represents the **internal** wave function. The energy may be written as

$$E = \frac{\hbar^2 K^2}{2(m_1 + m_2)} + \frac{\hbar^2 k^2}{2}\left(\frac{1}{m_1} + \frac{1}{m_2}\right) \tag{6.32}$$

The first term in Eq.(6.32) is the energy of the two–particle system, with mass $m_1 + m_2$ moving freely with the total momentum; the second term is the internal energy. If we introduce the reduced mass μ,

$$\frac{1}{\mu} = \frac{1}{m_1} + \frac{1}{m_2} \tag{6.33}$$

then the term $\hbar^2 k^2 / 2\mu$ is effectively a one–particle energy, namely, that of a free particle with mass μ and momentum $\hbar k$. When the Hamiltonian, Eq.(6.19), is altered by the addition of a potential that depends on $x_{12} = x_1 - x_2$ only, then we have

$$\left(-\frac{\hbar^2}{2m_1}\frac{\partial^2}{\partial x_1^2} - \frac{\hbar^2}{2m_2}\frac{\partial^2}{\partial x_2^2}\right) + V(x_{12})u(x_1, x_2) = Eu(x_1, x_2) \tag{6.34}$$

Using the coordinates

$$x = x_{12} = x_1 - x_2 \quad , \quad X = \frac{m_1 x_1 + m_2 x_2}{m_1 + m_2} = \frac{\mu}{m_2}x_1 + \frac{\mu}{m_1}x_2 \tag{6.35}$$

so that

$$x_1 = X + \frac{\mu}{m_1}x \quad , \quad x_2 = X - \frac{\mu}{m_2}x \tag{6.36}$$

The wave equation then takes the form

$$\left[-\frac{\hbar^2}{2(m_1 + m_2)}\frac{\partial^2}{\partial X^2} - \frac{\hbar^2}{2\mu}\frac{\partial^2}{\partial x^2} + V(x)\right]u(x, X) = Eu(x, X) \tag{6.37}$$

If we write

$$u(x, X) = e^{iKX}\phi(x) \tag{6.38}$$

we find that the equation for $\phi(x)$ is

$$-\frac{\hbar^2}{2\mu}\frac{d^2\phi(x)}{dx^2} + V(x)\phi(x) = \epsilon\phi(x) \tag{6.39}$$

that is, a one–particle Schrödinger equation with reduced mass, and energy

$$\epsilon = E - \frac{\hbar^2 K^2}{2(m_1 + m_2)} \tag{6.40}$$

6.2 Identical particles

Electrons are indistinguishable. If this were not so, then the spectrum of an atom, say helium, would vary from experiment to experiment. Similarly, nuclear spectra are always the same, indicating that protons are indistinguishable, as are neutrons. This is a purely quantum–mechanical property.

Electrons are characterized by an internal quantum number, called the spin. This has a further effect on the consequences of indistinguishability.

A Hamiltonian for indistinguishable particles must be completely symmetric in the coordinates of the particles. For a two–particle system, if there is no dependence on the spin labels, the Hamiltonian is

$$H = \frac{p_1^2}{2m} + \frac{p_2^2}{2m} + V(x_1, x_2) \tag{6.41}$$

with

$$V(x_1, x_2) = V(x_2, x_1) \tag{6.42}$$

or

$$H(1, 2) = H(2, 1) \tag{6.43}$$

Here the labels 1 and 2 include all possible quantum numbers, for example $(n_1, l_1, m_{l_1}, s_1, m_{s_1})$. A wave function for an N–particle system, with all the particles identical, will be denoted by $\psi(1, \cdots, N)$, and this stands for the more explicit $\psi(x_1, \sigma_1; \cdots; x_N, \sigma_N)$ where the σ_i's describe the spin states.

For a two–particle system the energy eigenvalue equation reads

$$H(1, 2)u_E(1, 2) = Eu_E(1, 2) \tag{6.44}$$

or $H(2, 1)u_E(2, 1) = Eu_E(2, 1)$, or $H(1, 2)u_E(2, 1) = Eu_E(2, 1)$.

Let us introduce an exchange operator P_{12}, which, acting on a state, interchanges all coordinates (space and time) of particles 1 and 2. The definition of P_{12} implies that

$$P_{12}\psi(1, 2) = \psi(2, 1) \tag{6.45}$$

We may write Eq.(2.43) as follows:

$$HP_{12}u_E(1, 2) = Eu_E(1, 2) = EP_{12}u_E(1, 2) \tag{6.46}$$
$$= P_{12}Eu_E(1, 2) = P_{12}Hu_E(1, 2)$$

This implies the relation

$$[H, P_{12}] = 0 \tag{6.47}$$

Thus, P_{12} is a constant of the motion. Also

$$(P_{12})^2\psi(1, 2) = \psi(1, 2) \tag{6.48}$$

so that the eigenvalues of P_{12} are ∓ 1. Therefore, the exchange operator P_{12} has similar properties with the parity operator P.

$$P\psi(x) = \psi(-x) \ ; \ P\psi^{(+)}(x) = \psi^{(+)}(x) \ ; \tag{6.49}$$
$$P\psi^{(-)}(x) = -\psi^{(-)}(x)$$

Here $\psi(x)$ is any wave function, $\psi^{(+)}(x)$ is an even function, and $\psi^{(-)}(x)$ is an odd function for the parity operator.

The eigenstates are the symmetric and antisymmetric combinations

$$\psi^{(S)}(1, 2) = \frac{1}{\sqrt{2}}[\psi(1, 2) + \psi(2, 1)] \tag{6.50}$$

$$\psi^{(A)}(1, 2) = \frac{1}{\sqrt{2}}[\psi(1, 2) - \psi(2, 1)] \tag{6.51}$$

The fact that P_{12} is a constant of the motion implies that a state that is symmetric at an initial time will always be symmetric, and an antisymmetric state will always be antisymmetric.

6.3　The Pauli principle; fermions and bosons

It is an important law of nature that the symmetry or antisymmetry under the interchange of two particles is a characteristic of the particles, and not something that can be arranged in the preparation of the initial state. The law, which was discovered by Pauli, states that

1. Systems consisting of identical particles of half–odd–integer spin (i.e., spin $1/2, 3/2, \cdots$) are described by antisymmetric wave functions. Such particles are called **fermions**, and are said to obey Fermi-Dirac statistics.

2. Systems consisting of identical particles of integer spin (i.e., spin $0, 1, 2, \cdots$) are described by symmetric wave functions. Such particles are called **bosons**, and are said to obey Bose–Einstein statistics.

The law extends to N–particle states. For a system of N identical fermions, the wave function is antisymmetric under the interchange of any pair of particles. For example, a three–particle wave function, properly antisymmetrized, has the form

$$\psi^{(A)}(1, 2, 3) = \frac{1}{\sqrt{6}}[\psi(1, 2, 3) - \psi(2, 1, 3) + \psi(2, 3, 1) \tag{6.52}$$

$$-\psi(3, 2, 1) + \psi(3, 1, 2) - \psi(1, 3, 2)]$$

whereas the three identical boson wave function has the form

$$\psi^{(S)}(1, 2, 3) = \frac{1}{\sqrt{6}}[\psi(1, 2, 3) + \psi(2, 1, 3) + \psi(2, 3, 1) \tag{6.53}$$

$$+\psi(3, 2, 1) + \psi(3, 1, 2) + \psi(1, 3, 2)]$$

Let us consider a special case, in which N fermions do not interact with each other, but do interact with a common potential. In this case

$$H = \sum_{i=1}^{N} H_i \tag{6.54}$$

where

$$H_i = \frac{p_i^2}{2m} + V(x_i) \tag{6.55}$$

The eigenstates of the one–particle potential are denoted by $u_{E_k}(x)$ where

$$H_k u_{E_k}(x) = E_k u_{E_k}(x) \tag{6.56}$$

A solution of

$$H u_E(1, \cdots, N) = E u_E(1, \cdots, N) \tag{6.57}$$

is

$$u_E(1, \cdots, N) = u_{E_1}(x_1) \cdots u_{E_N}(x_N) \tag{6.58}$$

where

$$E_1 + \cdots + E_N = E \tag{6.59}$$

If there are only two particles, the antisymmetrization of $u_E(1, \cdots, N)$ takes the form

$$u^A(1,2) = \frac{1}{\sqrt{2}}[u_{E_1}(x_1)u_{E_2}(x_2) - u_{E_1}(x_2)u_{E_2}(x_1)] \tag{6.60}$$

With three particles, the form is

$$u^A(1,2,3) = \frac{1}{\sqrt{6}}[u_{E_1}(x_1)u_{E_2}(x_2)u_{E_3}(x_3) - u_{E_1}(x_2)u_{E_2}(x_1)u_{E_3}(x_3) \tag{6.61}$$

$$+u_{E_1}(x_2)u_{E_2}(x_3)u_{E_3}(x_1) - u_{E_1}(x_3)u_{E_2}(x_2)u_{E_3}(x_1)$$

$$+u_{E_1}(x_3)u_{E_2}(x_1)u_{E_3}(x_2) - u_{E_1}(x_1)u_{E_2}(x_3)u_{E_3}(x_2)]$$

For N particles (fermions), the antisymmetrized wave function may be expressed in a determinantal form, the so–called **Slater determinant**:

$$u^A(1, \cdots, N) = \frac{1}{\sqrt{N!}} \begin{vmatrix} u_{E_1}(x_1) & \cdots & u_{E_1}(x_N) \\ \vdots & & \vdots \\ u_{E_N}(x_1) & \cdots & u_{E_N}(x_N) \end{vmatrix} \tag{6.62}$$

The product of the diagonal elements of the Slater determinant is called as the **Hartree product**. As a short hand for an N–fermion (say, electron) wave function, it is often expressed as a Hartree product,

$$u_N = u(1, \cdots, N) = \prod_{i=1}^{N} u_{E_i}(x_i) \tag{6.63}$$

The determinantal wave function can also be written in the symbolic form

$$det(u_N) = \mathcal{A}u_N \tag{6.64}$$

where $\mathcal{A}$ is the antisymmetrization operator. For an N–fermion wave function in the one–particle eigenstate approximation $\mathcal{A}$ has the general form

$$\mathcal{A} = \frac{1}{\sqrt{N!}} \sum_P (-1)^P \mathcal{P} \tag{6.65}$$

This operator converts a product of N doubly indexed elements into a determinantal form. The operator carries out $N!$ different permutations of the fermion coordinates to form the components of an antisymmetric wave function.

$$\sum_P (-1)^P \mathcal{P} = 1 - \sum_{ij} \mathcal{P}_{ij} + \sum_{ijk} \mathcal{P}_{ijk} - \cdots \tag{6.66}$$

where $\mathcal{P}_{ij}$, $\mathcal{P}_{ijk}$ are the two–particle and three–particle permutation operators, respectively. Permutations are carried out only upon the subscripts of the eigenstates.

Some properties of $\mathcal{A}$:

The permutation operators are unitary,

$$\mathcal{P}^{+} = \mathcal{P}^{-1} \tag{6.67}$$

The antisymmetrization operator is self–adjoint,

$$\mathcal{A}^{+} = \mathcal{A} \tag{6.68}$$

$$\mathcal{A}^{2} = \sqrt{N!}\,\mathcal{A} \tag{6.69}$$

If H is a spin–free N–electron Hamiltonian

$$[H, \mathcal{A}] = 0 \tag{6.70}$$

that is H commutes with $\mathcal{A}$.

The wave function for N identical bosons is totally symmetric, and the general form is obtained by expanding the Slater determinant and making all the signs positive.

The interchange of two particles involves the interchange of two columns in the determinant, and this changes the sign.

If two electrons are in the same energy eigenstate, for example, $E_1 = E_2$, and if they are in the same spin state, that is, the spin labels are the same $\sigma_1 = \sigma_2$, then the determinant vanishes when $x_1 = x_2$, that is, the electrons cannot be at the same place. Thus the requirement of antisymmetry introduces an effective interaction between two fermions: Two particles in the same state tend to stay away from each other, since the joint wave function vanishes when their separation goes to zero. Thus even noninteracting particles behave as if there were a repulsive interaction between them.

A complete set of commuting observables for electrons includes an additional two–valued observable associated with the spin. Thus a state of given energy, angular momentum, parity, and so on, can be occupied by two electrons (of opposite spin variables), but by no more than two electrons. This is another version of the Pauli exclusion principle.

Suppose we consider a box of width L, and suppose the particles are localized in $0 \le x \le L/4$ and $3L/4 \le x \le L$, respectively. Then the momentum of the particles can be determined with an accuracy that is resrtricted by the uncertainty principle. The possible values of the energy are given by

$$E_n = \frac{\hbar^2 \pi^2 n^2}{2ML^2} \tag{6.71}$$

the possible values of the momentum are

$$p = \frac{\hbar \pi n}{L} \tag{6.72}$$

Measurements of the momenta of particles are restricted by the uncertainty relation

$$\Delta p \cong \frac{n\hbar}{\Delta x} \cong \frac{n\hbar}{L} \tag{6.73}$$

and hence their energies can only be determined with an accuracy

$$\Delta E \cong \frac{p\Delta p}{M} \cong \frac{\hbar^2 \pi n^2}{ML^2} \tag{6.74}$$

This, however, is larger than

$$E_n - E_{n-1} \cong \frac{\hbar^2 \pi^2 n}{ML^2} \tag{6.75}$$

This analysis shows that there is no possibility that in a macroscopic situation there will be conflict with classical intuition.

If the two atoms are labeled A and B, the question is whether there is a difference between using the wave function

$$\psi_A(x_1)\psi_B(x_2) \tag{6.76}$$

and

$$\frac{1}{\sqrt{2}}[\psi_A(x_1)\psi_B(x_2) - \psi_A(x_2)\psi_B(x_1)] \tag{6.77}$$

to describe the two electron state. If these two atoms form a molecule, the overlap between wave functions falls off exponentially with the distance between the two atoms (see Figure 6.1).

$$S = \int \psi_A(x_1)\psi_B(x_2)dv \tag{6.78}$$

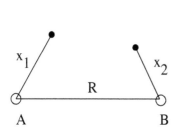
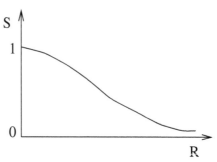

Fig. 6.1. Overlap integral between two atoms.

When the atoms are far apart, it does not make any difference which wave function we use. When the atoms are close, as in a H_2 molecule, the wave functions do overlap and it does make a difference whether one uses the uncorrelated wave functions or the antisymmetrized wave functions. Experimentation tells us that it is the latter that should be used.

An interesting consequence of the Pauli exclusion principle is that the ground state for N electrons in a potential is very different from the ground state for N bosons or N distinguishable particles. Consider, for example, the infinite potential box,

$$\begin{aligned} V(x) &= \infty & x < 0 \\ &= 0 & 0 < x < b \\ &= \infty & b < x \end{aligned} \tag{6.79}$$

The solution of the Schrödinger equation that vanishes at $x = 0$ and $x = b$ is given by

$$u_n(x) = \sin\left(\frac{n\pi x}{b}\right) \tag{6.80}$$

with $n = 1, 2, 3, \cdots$, and the energy eigenvalues are

$$E_n = \frac{\hbar^2 \pi^2 n^2}{2mb^2} \tag{6.81}$$

For N noninteracting bosons, the ground state has all the particles in the $n = 1$ state, and thus the energy is given by

$$E = N \frac{\hbar^2 \pi^2}{2mb^2} \tag{6.82}$$

so that the energy per particle is

$$\frac{E}{N} = \frac{\hbar^2 \pi^2}{2mb^2} \tag{6.83}$$

In general, if one takes the potential well as shown in Figure 6.2., there are two possible solution sets for the boundaries of x: $-b < x < b$

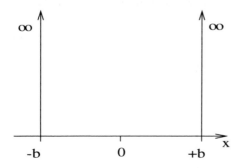

Fig. 6.2. Infinite potential box in 1D with boundaries $-b < x < b$.

Even solutions,
$$u_n(x) = \cos kx \quad , \quad kb = \frac{\pi}{2}, 3\frac{\pi}{2}, 5\frac{\pi}{2}, \cdots$$

Odd solutions,
$$u_n(x) = \sin kx \quad , \quad kb = \pi, 2\pi, 3\pi, \cdots$$

The possible k values from these two solutions are

$$kb = n\frac{\pi}{2} \quad \text{so} \quad E_n = \frac{\pi^2 \hbar^2 n^2}{8mb^2} \quad \text{in both cases.}$$

For N noninteracting fermions the situation is quite different. Only two electrons can go into each of the states $n = 1, 2, 3, \cdots$, so that $N/2$ states are filled. Thus the total energy is given by

$$E = 2 \sum_{n=1}^{N/2} \frac{\hbar^2 \pi^2 n^2}{2mb^2} = \frac{\hbar^2 \pi^2}{mb^2} \frac{N^3}{24} \tag{6.84}$$

Here we used the approximation for large N

$$\sum_{n=1}^{N/2} n^2 \cong \int_1^{N/2} n^2 dn \cong \frac{1}{3}\left(\frac{N}{2}\right)^3 \tag{6.85}$$

Thus the energy per particle

$$\frac{E}{N} = \left(\frac{\hbar^2 \pi^2}{24mb^2}\right) N^2 \tag{6.86}$$

grows with N^2. Equivalently, for a given energy, the number of bosons filling the well is proportional to E, while the number of fermions filling the well is proportional to $E^{1/3}$. The highest level to be filled in the fermion case is the one for which $n = N/2$, and its energy is

$$E_F = \frac{\hbar^2 \pi^2 N^2}{8mb^2} \tag{6.87}$$

This energy is called the **Fermi energy**. We may write in terms of the density of fermions, which in the one–dimensional problem is

$$\frac{N}{b} \equiv \rho \tag{6.88}$$

as

$$E_F = \frac{\hbar^2 \pi^2}{8m}\rho^2 \tag{6.89}$$

The total energy can be expressed in terms of E_F as

$$E = \frac{1}{3}E_F N \tag{6.90}$$

The exclusion principle plays an extremely important role in the structure of atoms.

6.4   Worked examples

Example - 6.1 : For a system of N identical particles, if the symmetrizing and antisymmetrizing operators are given by

$$\mathcal{S} = \frac{1}{\sqrt{N!}}\sum_{\mathcal{P}} \mathcal{P} \quad , \quad \mathcal{A} = \frac{1}{\sqrt{N!}}\sum_{\mathcal{P}} \delta_{\mathcal{P}} \mathcal{P}$$

where $\mathcal{P}$ is the permutation operator and $\delta_{\mathcal{P}} = +1$ for $\mathcal{P}$ even and -1 for $\mathcal{P}$ odd, obtain the eigenvalue of $\mathcal{S}$ and $\mathcal{A}$.

Solution : Let

$$\mathcal{S}\psi = \frac{1}{\sqrt{N!}}\sum_{\mathcal{P}} \mathcal{P}\psi = \psi_s$$

$$\mathcal{S}\psi_s = \frac{1}{\sqrt{N!}}\sum_{\mathcal{P}} \mathcal{P}\frac{1}{\sqrt{N!}}\sum_{\mathcal{P}'} \mathcal{P}'\psi = \frac{1}{N!}\sum_{\mathcal{P}}\sum_{\mathcal{P}'} \mathcal{P}\mathcal{P}'\psi$$

$$= \frac{1}{N!} \sum_{\mathcal{P}} \sum_{\mathcal{P}''} \mathcal{P}'' \psi = \sum_{\mathcal{P}''} \mathcal{P}'' \psi = \sqrt{N!} \mathcal{S} \psi$$

Thus

$$(\mathcal{S}^2 - \sqrt{N!} \mathcal{S}) \psi = 0$$

The only eigenvalues of $\mathcal{S}$ are 0 and $\sqrt{N!}$. Similarly

$$\mathcal{A} \psi = \psi_a$$

$$\mathcal{A} \psi_a = \frac{1}{\sqrt{N!}} \sum_{\mathcal{P}} \delta_{\mathcal{P}} \mathcal{P} \frac{1}{\sqrt{N!}} \sum_{\mathcal{P}'} \delta_{\mathcal{P}'} \mathcal{P}' \psi$$

$$= \frac{1}{N!} \sum_{\mathcal{P}} \sum_{\mathcal{P}'} \delta_{\mathcal{P}} \delta_{\mathcal{P}'} \mathcal{P} \mathcal{P}' \psi = \frac{1}{N!} \sum_{\mathcal{P}} \sum_{\mathcal{P}''} \delta_{\mathcal{P}''} \mathcal{P}'' \psi$$

$$= \sum_{\mathcal{P}''} \delta_{\mathcal{P}''} \mathcal{P}'' \psi = \sqrt{N!} \mathcal{A} \psi$$

Thus

$$(\mathcal{A}^2 - \sqrt{N!} \mathcal{A}) \psi = 0$$

The only eigenvalues of $\mathcal{A}$ are 0 and $\sqrt{N!}$. Here we used $\mathcal{P} \mathcal{P}'$ as another permutation operator and $\delta_{\mathcal{P}} \delta_{\mathcal{P}'} = \delta_{\mathcal{P} \mathcal{P}'} = \delta_{\mathcal{P}''}$.

Example - 6.2 : (a) Show that for a system of two identical particles, each of which can be in one of n quantum states, there are $n(n+1)/2$ symmetric and $n(n-1)/2$ antisymmetric states of the system. (b) Show that, if the particles have spin S, the ratio of symmetric to antisymmetric spin states is $(S+1):S$.

Solution : (a) If there are n single–particle states, there are n symmetric states in which both particles are in the same state. The number of symmetric states in which the particles are in different states is equal to the number of ways we can select two objects from n, that is, $\frac{1}{2}n(n-1)$. So the total number of symmetric states is

$$n + \frac{1}{2}n(n-1) = \frac{1}{2}n(n+1)$$

The number of antisymmetric states is $\frac{1}{2}n(n-1)$, because states in which both particles are in the same state are excluded.

(b) If the particles have spin S, there are $n = 2S + 1$ single–particle states, corresponding to the $2S + 1$ different m values that give the component of spin angular momentum in any direction in space. Therefore the number of symmmetric states is $n_s = (S+1)(2S+1)$, and the number of antisymmetric states is $n_a = S(2S+1)$. The ratio of symmetric to antisymmetric states is thus $(S+1)/S$.

Example - 6.3 : Consider a system consisting of two identical particles, each of which has one–particle states represented in coordinate representation by the wave functions $\psi_\alpha(\mathbf{r})$, $\psi_\beta(\mathbf{r})$. Let us define the antisymmetric and the symmetric wave functions of the system as follows:

$$\psi^{(A)}(\mathbf{r}_1, \mathbf{r}_2) = \frac{1}{\sqrt{2}} [\psi_\alpha(\mathbf{r}_1)\psi_\beta(\mathbf{r}_2) - \psi_\alpha(\mathbf{r}_2)\psi_\beta(\mathbf{r}_1)]$$

$$\psi^{(S)}(\mathbf{r}_1, \mathbf{r}_2) = \frac{1}{\sqrt{2}}[\psi_\alpha(\mathbf{r}_1)\psi_\beta(\mathbf{r}_2) + \psi_\alpha(\mathbf{r}_2)\psi_\beta(\mathbf{r}_1)]$$

If the symmetrization rule for identical particles were ignored, the system would in general have the following wave function

$$\Psi = \lambda\psi^{(A)} + \mu\psi^{(S)} \quad , \quad |\lambda|^2 + |\mu|^2 = 1$$

Show that in this case the probability per unit volume of finding a particle at $\mathbf{r}_1$ and another at $\mathbf{r}_2$ depends on λ and on μ.

Solution : The probability per unit volume of finding a particle at $\mathbf{r}_1$ and another at $\mathbf{r}_2$ is a sum of two quantities

$$P(\mathbf{r}_1, \mathbf{r}_2) = |\psi(\mathbf{r}_1, \mathbf{r}_2)|^2 + |\psi(\mathbf{r}_2, \mathbf{r}_1)|^2$$

$$= 2[|\lambda|^2|\psi^{(A)}(\mathbf{r}_1, \mathbf{r}_2)|^2 + |\mu|^2|\psi^{(S)}(\mathbf{r}_1, \mathbf{r}_2)|^2]$$

and hence it depends on λ and μ; the probability density, $P(\mathbf{r}_1, \mathbf{r}_2)$, depends on these parameters. To avoid this difficulty we could postulate that

$$|\psi^{(A)}(\mathbf{r}_1, \mathbf{r}_2)|^2 = |\psi^{(S)}(\mathbf{r}_1, \mathbf{r}_2)|^2 \quad (1)$$

for all $\mathbf{r}_1$, $\mathbf{r}_2$. But this is not possible since we can write

$$|\psi^{(S)}(\mathbf{r}, \mathbf{r})| = \sqrt{2}|\psi_\alpha(\mathbf{r})\psi_\beta(\mathbf{r})| \neq 0 \quad , \quad \psi^{(A)}(\mathbf{r}, \mathbf{r}) = 0$$

which contradicts (1). The difficulty can be avoided only by resorting to the symmetrization rule according to which the values of the parameters λ and μ are fixed at $\lambda = 0$, $\mu = 1$, or at $\lambda = 1$, $\mu = 0$, according to the type of particle being considered.

Example - 6.4 : Show that the symmetrization and the antisymmetrization operators are orthogonal projection operators, that is, $S^2 = S$, $A^2 = A$, $SA = AS = 0$.

Solution : The symmetrization and the antisymmetrization operators are defined as

$$S = \frac{1}{N!}\sum P \quad , \quad A = \frac{1}{N!}\sum(-1)^p P \quad (1)$$

where the summation is to be taken over all $N!$ permutation operators. Let us suppose that these operators have been placed in some particular order and let us multiply them on the right, or on the left, by a given permutation operator P'. This changes the order in the arrangement of the $N!$ permutation operators. Owing to the summation in (1), we may write

$$P'S = SP' = S \quad , \quad P'A = AP' = (-1)^{p'}A \quad (2)$$

From (1) and (2) the relations stated in the problem can be obtained.

6.5   Problems

Problem - 6.1 : An antisymmetric linear combination of product functions of
the type

$$\psi = \prod_{i=1}^{N} u_i(i)$$

can be constructed by operating on ψ with the antisymmetrizer

$$\mathcal{A} = \frac{1}{\sqrt{N!}} \sum_{\mathcal{P}} (-1)^{\mathcal{P}} \mathcal{P} \quad (1)$$

where $\mathcal{P}$ is a permutation which may be regarded as the result of P interchanges of
pairs of orbital indices or electron labels but not both. The summation is over all
permutations. Use $\mathcal{A}$ to generate the normalized, antisymmetric functions for the
$1s\bar{1}s2s$ configuration of Li, and thus show that this function can be expressed as the
Slater determinant:

$$\frac{1}{\sqrt{3!}} \begin{vmatrix} 1s\alpha(1) & 1s\alpha(2) & 1s\alpha(3) \\ 1s\beta(1) & 1s\beta(2) & 1s\beta(3) \\ 2s\alpha(1) & 2s\alpha(2) & 2s\alpha(3) \end{vmatrix} \quad (2)$$

Problem - 6.2 : (a) Suppose you have solved the Schrödinger equation for the
singly–ionized helium atom and found a set of eigenfunctions $\psi_N(\mathbf{r})$.
(1) How do the $\phi_N(\mathbf{r})$ compare with the hydrogen atom wave functions?
(2) If we include a spin part σ^+ (or σ^-) for spin up (or spin down), how do you
combine the ϕ's and σ's to form an eigenfunction of definite spin?
(b) Now consider the helium atom to have two electrons, but ignore the electromagnetic interactions between them.
(1) Write down a typical two-electron wave function, in terms of the ϕ's and σ's, of
definite spin. Do not choose the ground state.
(2) What is the total spin in your example?
(3) Demonstrate that your example is consistent with the Pauli exclusion principle.
(4) Demonstrate that your example is antisymmetric with respect to electron interchange.

Chapter 7

The Schrödinger Equation in Three–Dimensions

7.1 The two–body systems

Consider an isolated two–body system (see Figure 7.1). The Hamiltonian representing the system is

$$H = \frac{p_1^2}{2m_1} + \frac{p_2^2}{2m_2} + V(|\mathbf{r}_1 - \mathbf{r}_2|) \tag{7.1}$$

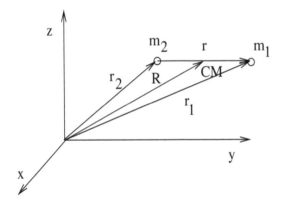

Fig. 7.1. A two–body system.

The interaction potential V is seen to be invariant under translations and rotations. We introduce center–of–mass and relative coordinates by

$$\mathbf{R} = \frac{1}{M}(m_1\mathbf{r}_1 + m_2\mathbf{r}_2) \quad , \quad \mathbf{r} = \mathbf{r}_1 - \mathbf{r}_2 \tag{7.2}$$

where $M = m_1 + m_2$. The total momentum operator

$$\mathbf{P} = \mathbf{p}_1 + \mathbf{p}_2 \tag{7.3}$$

has the coordinate representation

$$\mathbf{P} \equiv -i\hbar \left(\frac{\partial}{\partial \mathbf{r}_1} + \frac{\partial}{\partial \mathbf{r}_2}\right) = -i\hbar \frac{\partial}{\partial \mathbf{R}} \tag{7.4}$$

Furthermore, the relative momentum

$$\mathbf{p} = \frac{1}{M}(m_2\mathbf{p}_1 - m_1\mathbf{p}_2) \tag{7.5}$$

is represented by

$$\mathbf{p} \equiv -i\hbar\frac{1}{M}\left(m_2\frac{\partial}{\partial\mathbf{r}_1} - m_1\frac{\partial}{\partial\mathbf{r}_2}\right) = -i\hbar\frac{\partial}{\partial\mathbf{r}} \tag{7.6}$$

The nonzero commutators involving the new variables are

$$[R_i, P_j] = i\hbar\delta_{ij} \quad , \quad [r_i, p_j] = i\hbar\delta_{ij} \tag{7.7}$$

The utility of the variables $(\mathbf{R}, \mathbf{P}, \mathbf{r}, \mathbf{p})$ is that in terms of them the Hamiltonian can be written as

$$H = \frac{P^2}{2M} + \frac{p^2}{2\mu} + V(r) \tag{7.8}$$

where $\mu = m_1 m_2/M$ is the reduced mass. The Hamiltonian therefore breaks up into two commuting pieces. The first, $P^2/2M$, is just the kinetic energy for the center–of–mass motion. The second portion is the Hamiltonian of a single particle of mass μ moving in the static potential $V(r)$.

The total angular momentum

$$\mathbf{L} = \mathbf{r}_1 \times \mathbf{p}_1 + \mathbf{r}_2 \times \mathbf{p}_2 \tag{7.9}$$

can also be split into operators relating separately to the center–of–mass and relative motion

$$\mathbf{L} = \mathbf{L}_R + \mathbf{L}_r \tag{7.10}$$

where

$$\mathbf{L}_R = \mathbf{R} \times \mathbf{P} \quad , \quad \mathbf{L}_r = \mathbf{r} \times \mathbf{p} \tag{7.11}$$

The observables which possess time–independent expectation values are those which commute with H:

$$H \ , \ \frac{P^2}{2M} \ , \ \frac{p^2}{2m} + V \ , \ \mathbf{P} \ , \ \mathbf{L}_R \ , \ \mathbf{L}_r$$

In classical mechanics all of these quantities are also constants of the motion.

7.2   Separation of variables in the two–body systems

With a slight change of notation we write the Hamiltonian

$$\mathbf{H} = \frac{P^2}{2M} + \frac{p^2}{2\mu} + V(\mathbf{r}) \tag{7.12}$$

as

$$\mathcal{H} = \frac{P^2}{2M} + H \tag{7.13}$$

where H is the Hamiltonian in the center–of–mass system,

$$H = \frac{p^2}{2\mu} + V(\mathbf{r}) \tag{7.14}$$

Separation of variables:

The center–of–mass and relative coordinates can be separated by introducing a wave function of the form

$$\Psi(\mathbf{r}, \mathbf{R}, t) = \exp\left(\frac{i}{\hbar}\mathbf{P} \cdot \mathbf{R}\right)\psi_E(\mathbf{r}) \exp\left[-\frac{i}{\hbar}\left(\frac{P^2}{2M} + E\right)t\right] \tag{7.15}$$

This will be a stationary state solution of

$$i\hbar\frac{\partial\Psi}{\partial t} = \mathcal{H}\Psi \tag{7.16}$$

provided

$$H\psi_E(\mathbf{r}) = E\psi_E(\mathbf{r}) \tag{7.17}$$

Clearly Ψ is an eigenfunction of the total momentum operator with eigenvalue $\mathbf{P}$, and also an eigenfunction of $\mathcal{H}$ with eigenvalue $E + P^2/2M$. When ψ_E describes a bound state one can construct localized two–body states (or wave packets) by superposing solutions with different values of $\mathbf{P}$. We confine ourselves to the important case of a central field $V(r)$. The relative angular momentum is then conserved,

$$[\mathbf{L}_r, H] = 0 \tag{7.18}$$

We will delete the subscript r from $\mathbf{L}_r$ henceforth, because we need only concern ourselves with the motion in the center–of–mass system. The only complete set of compatible observables pertaining to the center–of–mass motion that contains H consists of H, L^2 and L_z. Hence we may choose ψ_E to be a simultaneous eigenfunction of L^2 and L_z, as well as of H,

$$L^2\psi_{Elm}(\mathbf{r}) = l(l+1)\psi_{Elm}(\mathbf{r}) \tag{7.19}$$

$$L_z\psi_{Elm}(\mathbf{r}) = m\psi_{Elm}(\mathbf{r}) \tag{7.20}$$

For our purpose we require the angular momentum operators in spherical coordinates (r, θ, ϕ), with the z–direction as the polar axis. One can write that

$$L_z = -i\left(x\frac{\partial}{\partial y} - y\frac{\partial}{\partial x}\right) \tag{7.21}$$

is simply

$$L_z = -i\frac{\partial}{\partial\phi} \tag{7.22}$$

A lengthier calculation results in

$$L_\pm = \pm e^{\pm i\phi}\left(\frac{\partial}{\partial\theta} \pm i\cot\theta\frac{\partial}{\partial\phi}\right) \tag{7.23}$$

where L_+ and L_- are the non–Hermitian operators

$$L_\pm = L_x \pm iL_y \tag{7.24}$$

These non–Hermitian operators have somewhat simpler properties than L_x and L_y. In terms of L_z, L_+, and L_-, the operator L^2 is

$$L^2 = L_z^2 + \frac{1}{2}(L_+L_- + L_-L_+) \tag{7.25}$$

Direct substitution then reveals that

$$L^2 = -\left(\frac{1}{\sin\theta}\frac{\partial}{\partial\theta}\sin\theta\frac{\partial}{\partial\theta} + \frac{1}{\sin^2\theta}\frac{\partial^2}{\partial\theta^2}\right) \tag{7.26}$$

In spherical coordinates the Laplacian is

$$\nabla^2 = \frac{1}{r^2}\frac{\partial}{\partial r}r^2\frac{\partial}{\partial r} - \frac{L^2}{r^2} \tag{7.27}$$

and the kinetic energy is therefore

$$\frac{p^2}{2\mu} = \frac{1}{2\mu}\left(p_r^2 + \frac{\hbar^2 L^2}{r^2}\right) \tag{7.28}$$

where

$$p_r = -i\hbar\left(\frac{1}{r} + \frac{\partial}{\partial r}\right) \tag{7.29}$$

This equation, Eq.(7.28), has the same form as a familiar expression for the kinetic energy in classical mechanics. The differential operator p_r has been arranged to be Hermitian in the interval $0 \le r < \infty$, i.e., it satisfies

$$\int_0^\infty r^2 dr f^*(r)p_r g(r) = \int_0^\infty r^2 dr [p_r f(r)]^* g(r) \tag{7.30}$$

The Schrödinger equation, Eq.(3.17), becomes

$$\left[\frac{1}{2\mu}\left(p_r^2 + \frac{\hbar^2 l(l+1)}{r^2}\right) + V(r)\right]\psi_{Elm} = E\psi_{Elm} \tag{7.31}$$

This differential operator depends only on the single variable r, while L^2 and L_z depend only on θ and ϕ. We can therefore separate variables. In other words, a solution of the system of differential equations can be written as a product of two functions,

$$\psi_{Elm}(\mathbf{r}) = R_{El}(r)Y_{lm}(\theta, \phi) \tag{7.32}$$

provided $R_{El}(r)$ satisfies the ordinary differential equation

$$\left[\frac{\hbar^2}{2\mu}\left(-\frac{1}{r^2}\frac{d}{dr}r^2\frac{d}{dr} + \frac{l(l+1)}{r^2}\right) + V(r)\right]R_{El}(r) = 0 \tag{7.33}$$

and $Y_{lm}(\theta, \phi)$ satisfies the differential equations

$$\left[L^2 - l(l+1)\right]Y_{lm}(\theta, \phi) = 0 \tag{7.34}$$

$$(L_z - m)Y_{lm}(\theta, \phi) = 0 \tag{7.35}$$

Here all the angular momenta are measured in units of $\hbar$.

7.3   Rotational invariance

The energy eigenvalue E, the potential V, and the mass μ only appear in the radial equation, Eq.(7.33). This means that the angular momentum eigenfunctions $Y_{lm}(\theta, \phi)$, and the associated eigenvalues $l(l + 1)$ and m, do not depend on the particular dynamical system that we are considering. Equations (7.34) and (7.35) are completely free of any physical parameters or constants of nature, and the spectrum and eigenfunctions of the angular momentum observables therefore involve only pure numbers. This is one facet of the fact that all information concerning the angular momentum observables can be deduced from symmetry considerations. The radial equation also demonstrates that the energy eigenvalue E can only depend on l, because the eigenvalue m does not appear in Eq.(7.33). This result is a consequence of the rotational invariance of H. Under the same circumstances, the orientation of a classical orbit does not affect its energy, whereas the magnitude of the angular momentum does.

Here the number m specifies the projection of the angular momentum along the z–axis, but as there is no preferred direction in space, the energy cannot depend on m.

The rotational invariance of the Hamiltonian has several other consequences of great importance. These can be deduced by applying the infinitesimal rotation operators to the energy and angular momentum eigenvalue equations. A function $f(\mathbf{r})$ can be rotated through the infinitesimal angle $\delta\omega$ about the axis $\hat{n}$ as follows:

$$(1 - i\delta\vec{\omega} \cdot \mathbf{L})f(\mathbf{r}) = f(\mathbf{r} - \delta\vec{\omega} \times \mathbf{r}) \tag{7.36}$$

where

$$\delta\vec{\omega} = \delta\omega\hat{n} \tag{7.37}$$

Taking advantage of

$$[\mathbf{L}, H] = [\mathbf{L}, L^2] = 0 \tag{7.38}$$

we see that

$$(H - E)(1 - i\delta\vec{\omega} \cdot \mathbf{L})\psi_{Elm}(\mathbf{r}) = 0 \tag{7.39}$$

$$[L^2 - l(l+1)](1 - i\delta\vec{\omega} \cdot \mathbf{L})\psi_{Elm}(\mathbf{r}) = 0 \tag{7.40}$$

Hence the rotated function $(1 - i\delta\vec{\omega} \cdot \mathbf{L})\psi_{Elm}(\mathbf{r})$ is also a simultaneous eigenfunction of the energy and L^2 with eigenvalue E and $l(l + 1)$, respectively.

On the other hand, if the rotation axis $\hat{n}$ does not coincide with the z–direction, L_z will fail to commute with $(1 - i\delta\vec{\omega} \cdot \mathbf{L})$, and $(1 - i\delta\vec{\omega} \cdot \mathbf{L})\psi_{Elm}$ will not, in general, be an eigenfunction of L_z with eigenvalue m.

When $\psi_{Elm}(\mathbf{r})$ is not a spherically symmetric function (i.e., not a function of $|\mathbf{r}|$ alone), $(1 - i\delta\vec{\omega} \cdot \mathbf{L})\psi_{Elm}$ and ψ_{Elm} are obviously different functions, and therefore linearly independent. But they are both energy eigenfunctions with the same eigenvalue E, and the energy level is therefore degenerate.

If a stationary state of a spherically symmetric Hamiltonian is not itself spherically symmetric, the energy level in question is necessarily degenerate.

Since the degeneracies of the energy spectrum are a consequence of the symmetries of the Hamiltonian, they can be determined experimentally by destroying this symmetry. In the laboratory this can often be achieved by allowing the system under study to interact with forces arranged so as to introduce a preferred spatial direction. For example, the radiation emitted by an atom in a weak externally applied magnetic field **B** reveals groups of energy levels (multiplets) that are nearly degenerate. This is known as the Zeeman effect.

7.4 The Schrödinger equation for noncentral potentials

We have concentrated on the reduction of the three–dimensional energy eigenvalue equation in spherical coordinates so far, since central potentials, for which $V = V(r)$, are by far the most interesting ones. One other situation that is of interest to us is the case when the potential is of the form

$$V(x, y, z) = V_1(x) + V_2(y) + V_3(z) \tag{7.41}$$

The corresponding Schrödinger equation is in the form

$$\left[-\frac{\hbar^2}{2m} \left(\frac{\partial^2}{\partial x^2} + \frac{\partial^2}{\partial y^2} + \frac{\partial^2}{\partial z^2} \right) + V_1(x) + V_2(y) + V_3(z) \right] \psi_E(x, y, z) \tag{7.42}$$

$$= E\psi_E(x, y, z)$$

The solution to this equation is in the form

$$\psi_E(x, y, z) = \psi_{E_1}(x)\psi_{E_2}(y)\psi_{E_3}(z) \tag{7.43}$$

where the functions on the right hand side are solutions of

$$\left[-\frac{\hbar^2}{2m} \frac{\partial^2}{\partial x^2} + V_1(x) \right] \psi_{E_1}(x) = E_1 \psi_{E_1}(x) \tag{7.44}$$

$$\left[-\frac{\hbar^2}{2m} \frac{\partial^2}{\partial y^2} + V_2(y) \right] \psi_{E_2}(y) = E_2 \psi_{E_2}(y) \tag{7.45}$$

$$\left[-\frac{\hbar^2}{2m} \frac{\partial^2}{\partial z^2} + V_3(z) \right] \psi_{E_3}(z) = E_3 \psi_{E_3}(z) \tag{7.46}$$

and

$$E = E_1 + E_2 + E_3 \tag{7.47}$$

Particle in a three–dimensional box:

A particularly interesting example is the three–dimensional generalization of the potential hole with infinite walls. If the three–dimensional box is cubical in shape, with sides L_1, L_2, and L_3, then $V = 0$ inside the box and $V = \infty$ outside the box.

The general solution is

$$\psi_E(x, y, z) = \sqrt{\frac{8}{L_1 L_2 L_3}} \sin\left(\frac{n_1 \pi x}{L_1}\right) \sin\left(\frac{n_2 \pi y}{L_2}\right) \sin\left(\frac{n_3 \pi z}{L_3}\right) \tag{7.48}$$

and

$$E = \frac{\hbar^2 \pi^2}{2m} \left(\frac{n_1^2}{L_1^2} + \frac{n_2^2}{L_2^2} + \frac{n_3^2}{L_3^2} \right) \tag{7.49}$$

if $L_1 = L_2 = L_3 = L$ then

$$E = \frac{\hbar^2 \pi^2}{2mL^2} (n_1^2 + n_2^2 + n_3^2) \tag{7.50}$$

In this form there is a lot of degeneracy in the energy. There are as many solutions for a given E as there are sets of integers $\{n_1 n_2 n_3\}$ that satisfy E.

The degeneracy is usually associated with the existence of mutually commuting operators. Here these operators are H_x, H_y, and H_z defined by

$$H_x = \frac{p_x^2}{2m} + V_1(x) \quad , \quad H_y = \frac{p_y^2}{2m} + V_2(y) \quad , \quad H_z = \frac{p_z^2}{2m} + V_3(z) \tag{7.51}$$

so that

$$H = H_x + H_y + H_z \tag{7.52}$$

It is interesting to calculate the ground state energy of N noninteracting electrons in the box of volume L^3. For each set of integers $\{n_1 n_2 n_3\}$, two electrons can be accommodated. Each set forms a lattice point in a three–dimensional space, and if there are very many of them, then it is a very good approximation to say that they must lie inside a sphere of radius R given by

$$n_1^2 + n_2^2 + n_3^2 = R^2 = \frac{2mE_F}{\hbar^2 \pi^2} L^2 \tag{7.53}$$

The number of lattice points is

$$\frac{1}{8} \left(\frac{4\pi}{3} R^3 \right) = \frac{1}{8} \frac{4\pi}{3} \left(\frac{2mE_F L^2}{\hbar^2 \pi^2} \right)^{3/2} = \frac{\pi}{6} \left(\frac{2mE_F}{\hbar^2 \pi^2} \right)^{3/2} L^3 \tag{7.54}$$

and hence the number of electrons with energy less than the Fermi energy E_F is twice that, that is,

$$N = \frac{\pi}{3} \left(\frac{2mE_F}{\hbar^2 \pi^2} \right)^{3/2} L^3 = \frac{\pi}{3} R^3 \tag{7.55}$$

The density of electrons is given by

$$n = \frac{N}{L^3} = \frac{\pi}{3} \left(\frac{2mE_F}{\hbar^2 \pi^2} \right)^{3/2} \tag{7.56}$$

Therefore the Fermi energy is obtained as

$$E_F = \frac{\hbar^2 \pi^2}{2m} \left(\frac{3n}{\pi} \right)^{2/3} \tag{7.57}$$

At each lattice point the energy is given by

$$E = \frac{\hbar^2 \pi^2}{2mL^2} n^2 \tag{7.58}$$

so that the total energy is obtained from

$$E_{tot} = 2 \cdot \frac{\hbar^2 \pi^2}{2mL^2} \frac{1}{8} \int n^2 d^3 n = 2 \cdot \frac{\hbar^2 \pi^2}{2mL^2} \frac{1}{8} 4\pi \int_0^R n^4 dn \tag{7.59}$$

$$= \frac{\hbar^2 \pi^3}{10mL^2} R^5$$

Since R is related to the number of electrons, we finally get

$$E_{tot} = \frac{\hbar^2 \pi^3}{10mL^2} \left(\frac{3N}{\pi}\right)^{5/3} = \frac{\hbar^2 \pi^3}{10m} \left(\frac{3n}{\pi}\right)^{5/3} L^3 \tag{7.60}$$

The total energy can be expressed in terms of E_F as

$$E_{tot} = \frac{3}{5} E_F N \tag{7.61}$$

Three–dimensional harmonic oscillator:

A system described by the Hamiltonian

$$H = -\frac{\hbar^2}{2m} \nabla^2 + \frac{1}{2} m(\omega_1^2 x^2 + \omega_2^2 y^2 + \omega_3^2 z^2) \tag{7.62}$$

is called an **anisotropic harmonic oscillator**. Since

$$V(x, y, z) = V_1(x) + V_2(y) + V_3(z) \tag{7.63}$$

this problem reduces to the problem of three independent harmonic oscillators of frequencies ω_1, ω_2, ω_3, along the axes x, y, z, respectively. Therefore

$$\psi_{n_1 n_2 n_3}(x, y, z) = \left(\frac{m^3 \omega_1 \omega_2 \omega_3}{\hbar^3 \pi^3}\right)^{1/4} \left(\frac{2^{-(n_1+n_2+n_3)}}{n_1! n_2! n_3!}\right)^{1/2} \tag{7.64}$$

$$\times H_1(\xi_1) H_2(\xi_2) H_3(\xi_3) e^{-\frac{1}{2}(\xi_1^2 + \xi_2^2 + \xi_3^2)}$$

where

$$\xi_1 = \left(\frac{m\omega_1}{\hbar}\right)^{1/2} x \ , \quad \xi_2 = \left(\frac{m\omega_2}{\hbar}\right)^{1/2} y \ , \quad \xi_3 = \left(\frac{m\omega_3}{\hbar}\right)^{1/2} z \tag{7.65}$$

and $n_1, n_2, n_3 = 0, 1, 2, \cdots$. If the ratios of the eigenfunctions are irrational, the energy levels are nondegenerate; otherwise they may be degenerate. The ground state E_{000} (or E_0) is always nondegenerate. For the **isotropic harmonic oscillators** ($\omega_1 = \omega_2 = \omega_3 = \omega$)

$$E_n = \hbar\omega(n + \frac{3}{2}) \ , \quad n = n_1 + n_2 + n_3 \tag{7.66}$$

In this case all the energy levels with the exception of E_0 are degenerate. To calculate the degeneracy of the level of energy E_n, consider for the moment a particular value of the quantum number n_1. n_2 can then have any of the values $0, 1, \cdots, n - n_1$, and the sum $n = n_1 + n_2 + n_3$ for given n and n_1 can be obtained in $n - n_1 + 1$ ways. Since $n_1 = 0, 1, 2, \cdots, n$, the degeneracy of E_n will be

$$g = \sum_{n_1=0}^{n} (n - n_1 + 1) = \frac{1}{2}(n + 1)(n + 2) \tag{7.67}$$

7.5 Worked examples

Example - 7.1 : For a conservative (nonrelativistic) system having $3N$ degrees of freedom and N particles of mass m the Hamiltonian function is

$$\sum_{n=1}^{3N} \frac{p_i^2}{2m} + V(x_1, y_1, z_1, \cdots, x_N, y_N, z_N) = E$$

where $p_1 = p_{x_1}$, $p_2 = p_{y_1}$, etc. (a) Obtain the Hamiltonian operator H for this system in coordinate representation. (b) Assume that $\psi(x,t) = \phi(t)u(x)$ in the time–dependent Schrödinger equation and obtain the time–independent Schrödinger equation for $u(x)$. Here x represents all necessary spatial coordinates.

Solution : (a) The coordinate representation (sometimes called Schrödinger's representation) is obtained by the substitution $p_i \rightarrow -i\hbar\frac{\partial}{\partial q_i}$ or in three–dimensions $\mathbf{p} \rightarrow -i\hbar\nabla$. Since $\nabla \cdot \nabla = \nabla^2$ substitution into the Hamiltonian function gives

$$H = \sum_{i=1}^{N} \left(-\frac{\hbar^2}{2m}\right)\nabla_i^2 + V(\mathbf{r}_1, \mathbf{r}_2, \cdots, \mathbf{r}_N)$$

In Cartesian coordinates the Laplacian is

$$\nabla^2 = \frac{\partial^2}{\partial x^2} + \frac{\partial^2}{\partial y^2} + \frac{\partial^2}{\partial z^2}$$

(b) We are given $H\psi = i\hbar\frac{\partial\psi}{\partial t}$. Substituting $\psi(x,t) = \phi(t)u(x)$ and dividing by ψ we find

$$\frac{Hu}{u} = \frac{i\hbar}{\phi}\frac{d\phi}{dt} \quad (1)$$

This is possible since H operates only on the spatial coordinates. The two sides of (1) depend on separate independent variables. For the equality to hold in general each side must equal a constant. This constant turns out to be the energy (E).

$$Hu = Eu \quad \text{and} \quad \frac{d\phi}{dt} = -\frac{iE}{\hbar}\phi$$

The separation constant E must be an eigenvalue of H. We must therefore solve the time–independent Schrödinger equation:

$$\left[\sum_{i=1}^{N}\left(-\frac{\hbar^2}{2m}\right)\nabla_i^2 + V(\mathbf{r}_1, \cdots, \mathbf{r}_N)\right]u(\mathbf{r}_1, \cdots, \mathbf{r}_N) = Eu(\mathbf{r}_1, \cdots, \mathbf{r}_N)$$

Example - 7.2 : Calculate the ground state energy of N noninteracting particles (both for bosons and fermions) in one–dimensional (with size L) and three–dimensional (with size $L \times L \times L$) boxes. $V = 0$ inside the boxes, and $V = \infty$ outside the boxes. Compare average energy per particle for each case.

Solution : One–dimensional box, one–particle case:

$$E_n = \frac{\hbar^2 \pi^2 n^2}{2mL^2} \quad , \quad n = 1, 2, 3, \cdots \quad , \quad E_{gs} = \frac{\hbar^2 \pi^2}{2mL^2}$$

N–particle case: Bosons,

$$E_{gs} = N \frac{\hbar^2 \pi^2}{2mL^2} \quad , \quad E_{gs}^{av} = \frac{E_{gs}}{N} = \frac{\hbar^2 \pi^2}{2mL^2}$$

Fermions:

$$E_{gs} = 2 \sum_{n=1}^{\frac{N}{2}} \frac{\hbar^2 \pi^2 n^2}{2mL^2} = 2 \frac{\hbar^2 \pi^2}{2mL^2} \sum_{n=1}^{\frac{N}{2}} n^2 \cong 2 \frac{\hbar^2 \pi^2}{2mL^2} \frac{1}{3} \left(\frac{N}{2}\right)^3$$

$$= \frac{\hbar^2 \pi^2 N^3}{24mL^2} \quad ; \quad E_{gs}^{av} = \frac{E_{gs}}{N} = \left(\frac{\hbar^2 \pi^2}{24mL^2}\right) N^2$$

Fermi energy, E_F:

$$E_F = \frac{\hbar^2 \pi^2}{2mL^2} \left(\frac{N}{2}\right)^2 = \frac{\hbar^2 \pi^2 N^2}{8mL^2}$$

$$E_{gs} = \frac{1}{3} E_F N \quad , \quad E_{gs}^{av} = \frac{1}{3} E_F$$

Three–dimensional box, one–particle case:

$$E_n = \frac{\hbar^2 \pi^2 n^2}{2mL^2} \quad ; \quad n^2 = n_x^2 + n_y^2 + n_z^2 \quad ; \quad n_x, n_y, n_z = 1, 2, 3, \cdots$$

N–particle case: Bosons,

$$E_{gs} = N \frac{\hbar^2 \pi^2}{2mL^2} (n_x^2 + n_y^2 + n_z^2) = N \frac{\hbar^2 \pi^2 3}{2mL^2} \quad ; \quad E_{gs}^{av} = \frac{E_{gs}}{N} = \frac{\hbar^2 \pi^2 3}{2mL^2}$$

Fermions:

$$E = \frac{\hbar^2 \pi^2}{2mL^2} (n_x^2 + n_y^2 + n_z^2) \quad ; \quad n_x^2 + n_y^2 + n_z^2 = R^2 = \frac{2mE_F}{\hbar^2 \pi^2} L^2$$

R is the radius of sphere in which all particles lie inside.

$$\text{number of points} = \frac{1}{8} \frac{4}{3} \pi R^3 \quad , \quad N = 2 \frac{1}{8} \frac{4}{3} \pi R^3 = \frac{\pi}{3} R^3$$

$$R^3 = \frac{3N}{\pi} \quad \rightarrow \quad R = \left(\frac{3N}{\pi}\right)^{1/3}$$

$$E_{gs} = 2 \frac{\hbar^2 \pi^2}{2mL^2} \frac{1}{8} \int_{n \leq R} n^2 d^3 n = 2 \frac{\hbar^2 \pi^2}{2mL^2} \frac{1}{8} \int_0^R n^2 4\pi n^2 dn$$

$$= \frac{\hbar^2 \pi^3}{2mL^2} \frac{1}{5} R^5 = \frac{\hbar^2 \pi^3 R^5}{10mL^2} = \frac{\hbar^2 \pi^3}{10mL^2} \left(\frac{3N}{\pi}\right)^{5/3}$$

$$= \frac{\hbar^2 \pi^3}{10mL^2} \left(\frac{3N}{\pi}\right) \left(\frac{3N}{\pi}\right)^{2/3} = \frac{\pi}{5} E_F \left(\frac{3N}{\pi}\right) = \frac{3}{5} E_F N$$

$$E_{gs}^{av} = \frac{E_{gs}}{N} = \frac{3}{5}E_F$$

Example - 7.3 : (a) Find the total, average, and Fermi energy of N noninteracting electrons moving in the field of one–dimensional simple harmonic oscillator potential. (b) Obtain the same quantities for the three–dimensional harmonic oscillator potential.

Solution : (a) For the 1D SHO where $E_n = (n + 1/2)\hbar w$, the state and energy counting is:

$$N = 2\int_0^{N_s} dn = 2N_s \quad \text{so that} \quad N_s = \frac{N}{2}$$

and

$$E_{tot} \cong 2\int_0^{N_s} (n\hbar w)dn = \hbar w N_s^2 = \frac{\hbar w N^2}{4}$$

giving

$$E^{av} = \frac{E_{tot}}{N} = \frac{\hbar w}{4}N \quad \text{and} \quad E_F = E(n = N_s) = \frac{\hbar w}{2}N$$

so that $E^{av} = E_F/2$.

(b) For the 3D case, we fill up allowed energy levels, two electrons to a point in $n = (n_x, n_y, n_z)$ space, subject to the constraints

$$n_x, n_y, n_z \geq 0 \quad , \quad n_x + n_y + n_z \leq N_s$$

until we have used up all N electrons, which corresponds to the region in the first quadrant of n–space below the plane form by $n_x + n_y + n_z = N_s$. The counting then is given by the integral approximation

$$N = 2\int d\mathbf{n} = 2\int_0^{N_s} dn_z \int_0^{N_s - n_z} dn_y \int_0^{N_s - n_z - n_y} dn_x = \frac{N_s^3}{3}$$

so we integrate out to a value $N_s = (3N)^{1/3}$. The total energy is then given by

$$E_{tot} = 2\int (n_x + n_y + n_z)\hbar w d\mathbf{n}$$

$$= 2\int_0^{N_s} dn_z \int_0^{N_s - n_z} dn_y \int_0^{N_s - n_z - n_y} dn_x (n_x + n_y + n_z)\hbar w$$

$$= \hbar w \frac{N_s^4}{4} = \frac{\hbar w}{4}(3N)^{4/3}$$

This total energy grows more slowly with N than does the similar result for the infinite well due to the different basic dependence of the quantized energies on the quantum number (n for the SHO, n^2 for the box). We also have

$$E^{av} = \frac{E_{tot}}{N} = \left(\frac{3^{4/3}}{4}\right)\hbar w N^{1/3} \quad , \quad E_F = \hbar w (3N)^{1/3} \quad , \quad E^{av} = \frac{3}{4}E_F$$

Example - 7.4 : Consider a two–dimensional system of electrons confined to a box of area $A = L^2$. Calculate the total energy of N electrons, as well as the average and Fermi energies.

Solution :

$$E = \frac{\hbar^2 \pi^2}{2mL^2}(n_x^2 + n_y^2) \quad , \quad R_n = \sqrt{\frac{2N}{\pi}} \quad , \quad E_{tot} = \frac{\hbar^2 \pi N^2}{2mL^2}$$

$$E^{av} = \frac{E_{tot}}{N} = \frac{\hbar^2 \pi}{2mL^2}N \quad ; \quad E_F = \frac{\hbar^2 \pi}{2mL^2}(2N) \quad ; \quad E^{av} = \frac{1}{2}E_F$$

In d–dimensions $N \sim R^d$, $R \sim N^{1/d}$, $E_{tot} \sim R^{d+2} \sim N^{(d+2)/d}$

7.6   Problems

Problem - 7.1 : A particle of mass m moves freely in a rectangular box with impenetrable walls. (a) If the dimensions of the box are $2a_x$, $2a_y$, $2a_z$, derive expressions for the solutions of the Schrödinger equation and the corresponding energies. (b) What are the parities of the wave functions? (c) If $a_x = a_y = a_z = a$, what are the degeneracies of the two lowest values of the energy?

Problem - 7.2 : Find the allowed energy levels and the total energy of N noninteracting electrons confined in a box of dimensions $L_x \times L_y \times L_z$

Problem - 7.3 : A quark is confined in a cubical box with sides of length 2 fermis. Find the excitation energy from the ground state to the first excited state in MeV.

Problem - 7.4 : A NaCl crystal has some negative ion vacancies, each containing one electron. Treat these electrons as moving freely inside a volume whose dimensions are on the order of the lattice constant. The crystal is at room temperature. Give a numerical estimate for the longest wavelength of electromagnetic radiation absorbed strongly by these electrons.

Problem - 7.5 : An electron is confined to the interior of a hollow spherical cavity of radius R with impenetrable walls. Find an expression for the pressure exerted on the walls of the cavity by the electron in its ground state.

Problem - 7.6 : A particle of mass m is constrained to move between two concentric impermeable spheres of radii $r = a$ and $r = b$, where the potential in this region is zero. Find the ground state energy and normalized wave function of the particle.

Chapter 8

Angular Momentum

8.1 Commutation relations

When a classical particle moves in a central field the energy and the three components of angular momentum are all constants of the motion. The motion is an orbit whose orientation in space is fixed. To discuss the corresponding situation in quantum mechanics we must examine the commutation properties of the associated operators: If they all commute, then the quantities which they represent may all simultaneously be constants of the motion; if not, we must find as many mutually commuting operators as we can and obtain a maximal set of constants of the motion.

To obtain the commutation properties we may start from the classical angular momentum components (in Cartesian form)

$$\mathbf{L} = \mathbf{r} \times \mathbf{p} \tag{8.1}$$

$$L_x = yp_z - zp_y \quad , \quad L_y = zp_x - xp_z \quad , \quad L_z = xp_y - yp_x \tag{8.2}$$

and set up angular momentum operators using the linear momentum operators

$$p_x = -i\hbar \frac{\partial}{\partial x} \quad , \quad p_y = -i\hbar \frac{\partial}{\partial y} \quad , \quad p_z = -i\hbar \frac{\partial}{\partial z} \tag{8.3}$$

The desired properties are completely determined by the basic commutation relation for position and momentum operators, namely

$$[x, p_x] = i\hbar \quad , \quad [y, p_y] = i\hbar \quad , \quad [z, p_z] = i\hbar \tag{8.4}$$

Angular momentum is a constant of motion, that is the three components of the angular momentum operators commute with the Hamiltonian. This parallels the classical result that central forces imply conservation of the angular momentum.

$$[H, L_x] = [H, L_y] = [H, L_z] = 0 \tag{8.5}$$

However, H, L_x, L_y, L_z do not form a complete set of commuting variables. The commutation relations among the components of the angular momentum are

$$[L_x, L_y] = i\hbar L_z \quad , \quad [L_z, L_x] = i\hbar L_y \quad , \quad [L_y, L_z] = i\hbar L_x \tag{8.6}$$

in general,

$$[L_\alpha, L_\beta] = i\hbar \epsilon_{\alpha\beta\gamma} L_\gamma \tag{8.7}$$

where $\epsilon_{\alpha\beta\gamma}$ is the **Levi–Civita density**;

$$\epsilon_{\alpha\beta\gamma} = +1 \quad , \quad \epsilon_{\alpha\gamma\beta} = -1 \quad , \quad \epsilon_{\alpha\beta\beta} = \epsilon_{\beta\beta\gamma} = \epsilon_{\alpha\alpha\gamma} = \epsilon_{\alpha\gamma\gamma} = 0 \tag{8.8}$$

Only one component of **L** may be chosen with H to form the commuting set of observables. On the other hand, L^2 commutes with all three components of **L**:

$$[L_x, L^2] = [L_y, L^2] = [L_z, L^2] = 0 \tag{8.9}$$

or in general

$$[\mathbf{L}, L^2] = 0 \tag{8.10}$$

Therefore, one may choose as the complete set of commuting observables the operators H, L_z, L^2. One can also include the parity, since H is invariant under $x \rightarrow -x$, $y \rightarrow -y$, $z \rightarrow -z$; specification of L^2 determines the parity.

Let us express the operator **L** in spherical coordinates. Using the following relations (spherical polar coordinates)

$$x = r \sin\theta \cos\phi \ , \quad y = r \sin\theta \sin\phi \ , \quad z = r \cos\theta \tag{8.11}$$

one can obtain

$$L_z = -i\hbar \left(x\frac{\partial}{\partial y} - y\frac{\partial}{\partial x} \right) = -i\hbar \frac{\partial}{\partial\phi} \tag{8.12}$$

$$L_y = -i\hbar \left(z\frac{\partial}{\partial x} - x\frac{\partial}{\partial z} \right) = -i\hbar \left(\cos\phi\frac{\partial}{\partial\theta} - \sin\phi\cot\theta\frac{\partial}{\partial\phi} \right) \tag{8.13}$$

$$L_x = -i\hbar \left(y\frac{\partial}{\partial z} - z\frac{\partial}{\partial y} \right) = -i\hbar \left(-\sin\phi\frac{\partial}{\partial\theta} - \cos\phi\cot\theta\frac{\partial}{\partial\phi} \right) \tag{8.14}$$

8.2   Raising and lowering operators

We will now find the eigenvalues and eigenfunctions of the operators L_z and L^2. Since the angular momentum has the dimensions of $\hbar$, we may write the eigenvalue equations in the form

$$L_z Y_{lm} = m\hbar Y_{lm} \tag{8.15}$$

$$L^2 Y_{lm} = l(l+1)\hbar^2 Y_{lm} \tag{8.16}$$

where m and $l(l+1)$ are real numbers.

The possible eigenvalues of the angular momentum operators L_z and L^2 are completely determined by the commutation rules. To find these eigenvalues it is convenient to introduce two new operators

$$L_\pm = L_x \pm iL_y \tag{8.17}$$

Then one can show that

$$L_\pm = \hbar e^{\pm i\phi} \left(\pm\frac{\partial}{\partial\theta} + i\cot\theta\frac{\partial}{\partial\phi} \right) \tag{8.18}$$

One can then construct the L^2 operator using the relation

$$L_+L_- = (L_x + iL_y)(L_x - iL_y) = L_x^2 + L_y^2 - i[L_x, L_y] \tag{8.19}$$

so that

$$L^2 = L_z^2 + L_+L_- + i[L_x, L_y] = L_+L_- + L_z^2 - \hbar L_z \tag{8.20}$$

or

$$L^2 = L_-L_+ + L_z^2 + \hbar L_z \tag{8.21}$$

The operators $L_\pm$ play the role of raising and lowering operators. The commutation relations are

$$[L_+, L_-] = 2\hbar L_z \quad , \quad [L_+, L_z] = -\hbar L_+ \quad , \quad [L_-, L_z] = \hbar L_- \tag{8.22}$$

From the fact that $[L^2, \mathbf{L}] = 0$, it follows that

$$[L^2, L_\pm] = 0 \quad , \quad [L^2, L_z] = 0 \tag{8.23}$$

This implies that

$$L^2 L_\pm Y_{lm} = L_\pm L^2 Y_{lm} = l(l+1)\hbar^2 L_\pm Y_{lm} \tag{8.24}$$

that is, $L_\pm Y_{lm}$ are also eigenfunctions of L^2 with the eigenvalue $l(l+1)\hbar^2$. On the other hand,

$$L_z L_+ Y_{lm} = (L_+L_z + \hbar L_+)Y_{lm} \tag{8.25}$$
$$= m\hbar L_+Y_{lm} + \hbar L_+Y_{lm} = \hbar(m+1)L_+Y_{lm}$$

so that L_+Y_{lm} is also an eigenfunction of L_z, but with m–value increased by unity. Similarly one can show that

$$L_z L_- Y_{lm} = \hbar(m-1)L_-Y_{lm} \tag{8.26}$$

so that L_-Y_{lm} is an eigenfunction of L_z with m–value lowered by unity. Thus we call $L_\pm$ **raising** and **lowering** operators, respectively.

8.3   Eigensolutions of angular momentum operators

Consider the eigenvalue equation

$$L_z Y_{lm} = m\hbar Y_{lm} \tag{8.27}$$

using

$$L_z = -i\hbar \frac{\partial}{\partial \phi} \tag{8.28}$$

We rewrite this equation

$$\frac{\partial}{\partial \phi} Y_{lm}(\theta, \phi) = im Y_{lm}(\theta, \phi) \tag{8.29}$$

so that the solution is of the form

$$Y_{lm}(\theta, \phi) = \Theta_{lm}(\theta)\Phi_m(\phi) \tag{8.30}$$

where

$$\frac{d\Phi_m(\phi)}{d\phi} = im\Phi_m(\phi) \tag{8.31}$$

The solution to this, with the normalization

$$\int_0^{2\pi} |\Phi_m|^2 d\phi = 1 \tag{8.32}$$

is

$$\Phi_m(\phi) = \frac{1}{\sqrt{2\pi}} e^{im\phi} \tag{8.33}$$

Since a rotation through 2π, or a transformation $\phi \to \phi + 2\pi$, leaves the system invariant, it is necessary that

$$e^{2\pi im} = 1 \tag{8.34}$$

so that m is an integer. The eigenfunctions of the Hermitian operators L_z and L^2 will be orthogonal, if the eigenvalues are different, and with proper normalization, we will write

$$< Y_{l'm'}|Y_{lm} > = \delta_{ll'}\delta_{mm'} \tag{8.35}$$

Since

$$< Y_{lm}|(L_x^2 + L_y^2 + L_z^2)Y_{lm} > = \tag{8.36}$$
$$< L_x Y_{lm}|L_x Y_{lm} > + < L_y Y_{lm}|L_y Y_{lm} > + m^2\hbar^2 \geq 0$$

it follows that

$$l(l+1) \geq 0 \tag{8.37}$$

We may write

$$L_{\pm} Y_{lm} = C_{\pm}(l,m) Y_{l,m\pm1} \tag{8.38}$$

From the hermiticity of L_x and L_y operators, that is

$$L_{\pm}^{\dagger} = (L_x \pm iL_y)^+ = L_x \mp iL_y = L_{\mp} \tag{8.39}$$

Therefore, one can write

$$< L_{\pm} Y_{lm}|L_{\pm} Y_{lm} > \geq 0 \quad or \quad < Y_{lm}|L_{\mp} L_{\pm} Y_{lm} > \geq 0 \tag{8.40}$$

and

$$< Y_{lm}|(L^2 - L_z^2 \pm \hbar L_z)Y_{lm} > \geq 0 \tag{8.41}$$

that is

$$l(l+1) \geq m^2 + m \quad , \quad l(l+1) \geq m^2 - m \tag{8.42}$$

Since $l(l+1) \geq 0$, we can take $l \geq 0$. Then we have

$$-l \leq m \leq l \tag{8.43}$$

If there is a minimum value of m (m_-) then for the corresponding eigenstate

$$L_- Y_{lm_-} = 0 \tag{8.44}$$

Then we may calculate m_- by using $L^2 = L_+ L_- + L_z^2 - \hbar L_z$ and applying it to Y_{lm_-}, we get

$$l(l+1)\hbar^2 = m_-^2 \hbar^2 - m_- \hbar^2 \tag{8.45}$$

Similarly, if there is a maximum value of m (m_+) then

$$L_+ Y_{lm_+} = 0 \tag{8.46}$$

An application of $L^2 = L_- L_+ + L_z^2 + \hbar L_z$ to the maximum eigenstate Y_{lm_+} gives

$$l(l+1)\hbar^2 = m_+^2 \hbar^2 + m_+ \hbar^2 \tag{8.47}$$

Hence,

$$m_- = -l \quad , \quad m_+ = +l \tag{8.48}$$

Since the maximum value is to be reached from the minimum value by unit steps (repeated application of L_+), we find

a) that there are $(2l+1)$ states $(2l+1$ is an integer), and

b) that m can take on the values

$$m = -l, -l+1, -l+2, \cdots, 0, \cdots, l-1, l \tag{8.49}$$

We may also calculate the coefficients $C_\pm(l, m)$ defined in

$$L_\pm Y_{lm} = C_\pm(l, m) Y_{l, m\pm 1} \tag{8.50}$$

We have

$$|C_\pm(l, m)|^2 < Y_{l,m\pm 1} | Y_{l,m\pm 1} > = < L_\pm Y_{lm} | L_\pm Y_{lm} > \tag{8.51}$$
$$= < Y_{lm} | L_\mp L_\pm Y_{lm} > = < Y_{lm} | (L^2 - L_z^2 \mp \hbar L_z) Y_{lm} >$$
$$= \hbar^2 [l(l+1) - m(m \pm 1)]$$

so that

$$C_\pm(l, m) = \hbar \sqrt{l(l+1) - m(m \pm 1)} = \hbar \sqrt{(l \mp m)(l \pm m + 1)} \tag{8.52}$$

Using the function

$$Y_{lm}(\theta, \phi) = \Theta_{lm}(\theta) e^{im\phi} \tag{8.53}$$

the condition

$$L_+ Y_{lm_+} = 0 \tag{8.54}$$

gives

$$\hbar e^{i\phi} \left(\frac{\partial}{\partial \theta} + i \cot \theta \frac{\partial}{\partial \phi} \right) \Theta_{ll}(\theta) e^{il\phi} = \tag{8.55}$$

$$\hbar e^{i(l+1)\phi} \left(\frac{\partial}{\partial \theta} - l \cot \theta \right) \Theta_{ll}(\theta) = 0$$

The solution to this equation is found to be

$$\Theta_{ll}(\theta) = (\sin \theta)^l \tag{8.56}$$

An arbitrary state is obtained by the lowering procedure

$$Y_{lm}(\theta, \phi) = C(L_-)^{l-m} (\sin \theta)^l e^{il\phi} \tag{8.57}$$

here C is a constant, which may be determined from the normalization condition. Consider first

$$L_-Y_{ll}(\theta,\phi) = \hbar e^{-i\phi}\left(-\frac{\partial}{\partial\theta} + i\cot\theta\frac{\partial}{\partial\phi}\right)(\sin\theta)^l e^{il\phi} \qquad (8.58)$$

$$= \hbar e^{i(l-1)\phi}\left(-\frac{\partial}{\partial\theta} - l\cot\theta\right)(\sin\theta)^l$$

Using the relation for any function $f(\theta)$

$$\left(\frac{d}{d\theta} + l\cot\theta\right)f(\theta) = \frac{1}{(\sin\theta)^l}\frac{d}{d\theta}\left[(\sin\theta)^l f(\theta)\right] \qquad (8.59)$$

one gets

$$Y_{l,l-1} = C'\frac{e^{i(l-1)\phi}}{(\sin\theta)^l}\left(-\frac{d}{d\theta}\right)\left[(\sin\theta)^l(\sin\theta)^l\right] \qquad (8.60)$$

Going one step further, one obtains

$$Y_{l,l-2} = C''\frac{e^{i(l-2)\phi}}{(\sin\theta)^{l-1}}\left(-\frac{d}{d\theta}\right)\left[(\sin\theta)^{l-1}\frac{1}{(\sin\theta)^l}\left(-\frac{d}{d\theta}\right)(\sin\theta)^{2l}\right] \qquad (8.61)$$

$$= C''(-1)^2\frac{e^{i(l-2)\phi}}{(\sin\theta)^{l-1}}\frac{d}{d\theta}\left[\frac{1}{\sin\theta}\frac{d}{d\theta}(\sin\theta)^{2l}\right]$$

In terms of the variable $u = \cos\theta$, $d/d\theta = -(1/\sin\theta)d/d\theta$, one can rewrite $Y_{l,l-1}$ and $Y_{l,l-2}$ as

$$Y_{l,l-1} = C'\frac{e^{i(l-1)\phi}}{(\sin\theta)^{l-1}}\frac{d}{du}\left[(1-u^2)^l\right] \qquad (8.62)$$

$$Y_{l,l-2} = C''\frac{e^{i(l-2)\phi}}{(\sin\theta)^{l-2}}\frac{d^2}{du^2}\left[(1-u^2)^l\right] \qquad (8.63)$$

The general form is

$$Y_{lm} = C\frac{e^{im\phi}}{(\sin\theta)^m}\left(\frac{d}{du}\right)^{l-m}\left[(1-u^2)^l\right] \qquad (8.64)$$

Using the normalization condition, one can determine the constant C,

$$< Y_{lm}|Y_{lm} >= 1 \qquad (8.65)$$

$$= \int_0^{2\pi}d\phi\int_{-1}^1 du|C|^2\left[\frac{1}{(1-u^2)^{m/2}}\left(\frac{d}{du}\right)^{l-m}(1-u^2)^l\right]^2$$

After some algebra, one gets the explicit form of the spherical harmonics as

$$Y_{lm}(\theta,\phi) = (-1)^m\left[\frac{2l+1}{4\pi}\frac{(l-m)!}{(l+m)!}\right]^{1/2}P_l^m(\cos\theta)e^{im\phi} \qquad (8.66)$$

for negative m

$$Y_{l,-m} = (-1)^m Y_{lm}^* \qquad (8.67)$$

The associated Legendre polynomials are given by

$$P_l^m(u) = (-1)^{l+m}\frac{(l+m)!}{(l-m)!}\frac{(1-u^2)^{-m/2}}{2^l l!}\left(\frac{d}{du}\right)^{l-m}(1-u^2)^l \qquad (8.68)$$

for negative m

$$P_l^{-m}(u) = (-1)^m\frac{(l-m)!}{(l+m)!}P_l^m(u) \qquad (8.69)$$

Some of the spherical harmonics :

$$Y_{0,0} = \frac{1}{\sqrt{4\pi}}$$

$$Y_{1,0} = \sqrt{\frac{3}{4\pi}} \cos\theta \qquad\qquad = \sqrt{\frac{3}{4\pi}} \frac{z}{r}$$

$$Y_{1,\mp 1} = \pm\sqrt{\frac{3}{8\pi}} \sin\theta e^{\mp i\phi} \qquad = \pm\sqrt{\frac{3}{8\pi}} \frac{x\mp iy}{r}$$

$$Y_{2,0} = \sqrt{\frac{5}{16\pi}}(3\cos^2\theta - 1) \qquad = \sqrt{\frac{5}{16\pi}} \frac{(2z^2 - x^2 - y^2)}{r^2}$$

$$Y_{2,\mp 1} = \pm\sqrt{\frac{15}{8\pi}} \sin\theta \cos\theta e^{\mp i\phi} = \pm\sqrt{\frac{15}{8\pi}} \frac{(x\mp iy)z}{r^2}$$

$$Y_{2,\mp 2} = \sqrt{\frac{15}{32\pi}} \sin^2\theta e^{\mp 2i\phi} \qquad = \sqrt{\frac{15}{32\pi}} \frac{(x\mp iy)^2}{r^2}$$

With the knowledge of L^2, we can now write the radial differential equation that determines the energy eigenvalues and eigenfunctions.

$$-\frac{\hbar^2}{2\mu}\left[\frac{1}{r}\frac{d}{dr}\left(r\frac{d}{dr}\right) + \frac{1}{r}\frac{d}{dr} - \frac{l(l+1)}{r^2}\right] R_{Elm}(r) \qquad (8.70)$$

$$+V(r)R_{Elm}(r) = ER_{Elm}(r)$$

Note that there is no dependence on m in the equation. Thus, for a given l, there will always be a $(2l + 1)$–fold degeneracy, since all the possible m–values will have the same energy.

Remark on the eigenvalues and eigenfunctions of L_z and L^2:

The wave equation

$$\left[-\frac{\hbar^2}{2\mu}\nabla^2 + V(\mathbf{r})\right]\psi(\mathbf{r}) = E\psi(\mathbf{r}) \qquad (8.71)$$

with a spherically symmetric potential energy may be written in spherical coordinates:

$$-\frac{\hbar^2}{2\mu}\left[\frac{1}{r^2}\frac{\partial}{\partial r}\left(r^2\frac{\partial}{\partial r}\right) + \frac{1}{r^2\sin\theta}\frac{\partial}{\partial\theta}\left(\sin\theta\frac{\partial}{\partial\theta}\right) + \frac{1}{r^2\sin^2\theta}\frac{\partial^2}{\partial\phi^2}\right]\psi \qquad (8.72)$$

$$+V(r)\psi = E\psi$$

We first separate the radial and the angular parts by substituting

$$\psi(r,\theta,\phi) = R(r)Y(\theta,\phi) \qquad (8.73)$$

into Eq.(4.72) and dividing through by ψ

$$\frac{1}{R}\frac{d}{dr}\left(r^2\frac{dR}{dr}\right) + \frac{2\mu r^2}{\hbar^2}[E - V(r)] = \qquad (8.74)$$

$$-\frac{1}{Y}\left[\frac{1}{\sin\theta}\frac{\partial}{\partial\theta}\left(\sin\theta\frac{\partial Y}{\partial\theta}\right)+\frac{1}{\sin^2\theta}\frac{\partial^2 Y}{\partial\phi^2}\right]$$

Since the left side of Eq.(8.74) depends only on r, and the right side depends only on θ and ϕ, both sides must be equal to a constant that we call λ. Thus Eq.(8.74) gives us a radial equation

$$\frac{1}{r^2}\frac{d}{dr}\left(r^2\frac{dR}{dr}\right)+\left[\frac{2\mu}{\hbar^2}(E-V(r))-\frac{\lambda}{r^2}\right]R=0 \tag{8.75}$$

and the angular equation

$$\frac{1}{\sin\theta}\frac{\partial}{\partial\theta}\left(\sin\theta\frac{\partial Y}{\partial\theta}\right)+\frac{1}{\sin^2\theta}\frac{\partial^2 Y}{\partial\phi^2}+\lambda Y=0 \tag{8.76}$$

The angular equation, Eq.(8.76), can be further separated by substituting

$$Y(\theta,\phi)=\Theta(\theta)\Phi(\phi) \tag{8.77}$$

into it and following the same procedure to obtain

$$\frac{d^2\Phi}{d\phi^2}+\nu\Phi=0 \tag{8.78}$$

$$\frac{1}{\sin\theta}\frac{d}{d\theta}\left(\sin\theta\frac{d\Theta}{d\theta}\right)+\left(\lambda-\frac{\nu}{\sin^2\theta}\right)\Theta=0 \tag{8.79}$$

The ϕ equation, Eq.(8.78), can be solved at once; its general solution may be written

$$\Phi(\phi)=Ae^{i\sqrt{\nu}\phi}+Be^{-i\sqrt{\nu}\phi} \quad \nu\neq0 \tag{8.80}$$

$$\Phi(\phi)=A+B\phi \quad \nu=0$$

The requirement that Φ and $d\Phi/d\phi$ be continuous throughout the domain 0 to 2π of ϕ demands that ν be chosen equal to the square of an integer. We thus replace Eq.(4.80) by

$$\Phi_m(\phi)=\frac{1}{\sqrt{2\pi}}e^{im\phi} \tag{8.81}$$

where now all physical meaningful solutions are included if m is allowed to be a positive or negative integer or zero. The multiplying constant is chosen equal to $1/\sqrt{2\pi}$ in order that Φ be normalized to unity over the range of ϕ.

Unless $V(r)$ is specified, the farthest we can carry our treatment is the solution of the θ equation, Eq.(8.79), where now $\nu=m^2$. It is convenient to substitute $u=\cos\theta$ for θ and put

$$\Theta(\theta)=P(u) \tag{8.82}$$

then Eq.(4.79) becomes

$$\frac{d}{du}\left[(1-u^2)\frac{dP}{du}\right]+\left(\lambda-\frac{m^2}{1-u^2}\right)P=0 \tag{8.83}$$

Since the domain of θ is 0 to π, the domain of u is 1 to -1. Since Eq.(8.83) is a second–order differential equation, it has two linearly independent solutions. Except for a particular value of λ, both of these are infinite at $u=\mp1$ and are not physically acceptable. If, however, $\lambda = l(l+1)$, where l is a positive integer or zero, one of the solutions is finite at $u = \mp 1$ (the other is not); this finite solution has the form

$$(1-u^2)^{\frac{1}{2}|m|} \tag{8.84}$$

times a polynomial of order $(l - |m|)$ in u and has the parity of $(l - |m|)$. The physically acceptable solutions of Eq.(8.83) when $m = 0$ are called the Legendre polynomials $P_l(u)$. For m not necessarily equal to zero, Eq.(8.83) has physically acceptable solutions if $\lambda = l(l+1)$ and $|m| \leq l$. These solutions, which are called associated Legendre functions, are expressible in terms of the Legendre polynomials:

$$P_l^m(u) = (1-u^2)^{\frac{1}{2}|m|} \frac{d^{|m|}}{du^{|m|}} P_l(u) \tag{8.85}$$

The angular part $Y_{lm}(\theta, \phi)$ of the complete wave function, which is a solution of Eq.(8.76) when $\lambda = l(l+1)$, is called a spherical harmonic. It is apparent that

$$Y_{lm}(\theta, \phi) = N_{lm} P_l^m(\cos\theta) \Phi_m(\phi) \tag{8.86}$$

N_{lm} is the normalization constant for the associated Legendre function. The solutions of Eq.(4.76) corresponding to different eigenvalues λ or l are orthogonal. The eigenvalue l is, however, $(2l+1)$–fold degenerate.

$$< Y_{lm}(\theta, \phi)|Y_{l'm'}(\theta, \phi) >= \int_0^\pi \int_0^{2\pi} Y_{lm}^*(\theta, \phi) Y_{l'm'}(\theta, \phi) \sin\theta d\theta d\phi \tag{8.87}$$

$$= \int_{-1}^1 \int_0^{2\pi} Y_{lm}^* Y_{l'm'} du d\phi = \delta_{ll'}\delta_{mm'}$$

The normalized spherical harmonics are given by

$$Y_{lm}(\theta, \phi) = \epsilon \left[\frac{2l+1}{4\pi} \frac{(l-|m|)!}{(l+|m|)!} \right]^{1/2} P_l^m(\cos\theta) e^{im\phi} \tag{8.88}$$

where $\epsilon = (-1)^m$ for $m > 0$, and $\epsilon = 1$ for $m \leq 1$.

Suppose that the position coordinate $\mathbf{r}$ is reflected through the origin so that $\mathbf{r}$ is replaced by $-\mathbf{r}$; this corresponds to replacing

$$x \rightarrow -x \quad, \quad y \rightarrow -y \quad, \quad z \rightarrow -z$$

or

$$r \rightarrow r \quad, \quad \theta \rightarrow \pi - \theta \quad, \quad \phi \rightarrow \phi + \pi$$

The only change in the wave equation, Eq.(8.71), is that $\psi(r, \theta, \phi)$ is replaced by $\psi(r, \pi - \theta, \phi + \pi)$, the rest of the equation being unaffected. This shows that orthogonal linear combinations of degenerate eigenfunctions can be found that have definite parities and that a nondegenerate eigenfunction must have a definite parity.

The energy levels for a spherically symmetric potential are degenerate at least with respect to the quantum number m, for $l > 0$. In this case, the degenerate eigenfunctions all have the same parity. When $\mathbf{r}$ is reflected through the origin, the radial part $R(r)$ of the solution is unchanged, the ϕ part $\Phi(\phi)$ has the parity of $|m|$, and the θ part $P_l^m(\cos\theta)$ has the parity of $(l - |m|)$. Thus $Y_{lm}(\theta, \phi)$, and hence $\psi(\mathbf{r})$, has the parity of l.

The operator that represents the square of the total angular momentum is

$$L^2 = L_x^2 + L_y^2 + L_z^2 = -\hbar^2 \left[\frac{1}{\sin\theta} \frac{\partial}{\partial\theta} \left(\sin\theta \frac{\partial}{\partial\theta} \right) + \frac{1}{\sin^2\theta} \frac{\partial^2}{\partial\phi^2} \right] \qquad (8.89)$$

Comparison of Eq.(8.89) and Eq.(8.76) shows that $Y_{lm}(\theta,\phi)$ is an eigenfunction of L^2 with the eigenvalue $l(l+1)\hbar^2$:

$$L^2 Y_{lm}(\theta,\phi) = l(l+1)\hbar^2 Y_{lm}(\theta,\phi) \qquad (8.90)$$

In similar fashion, it follows from the structure of Eq.(8.81) and

$$L_z = -i\hbar \frac{\partial}{\partial\phi} \qquad (8.91)$$

that $\Phi_m(\phi)$, and hence also $Y_{lm}(\theta,\phi)$, is an eigenfunction of L_z with the eigenvalue $m\hbar$:

$$L_z Y_{lm}(\theta,\phi) = m\hbar Y_{lm}(\theta,\phi) \qquad (8.92)$$

Thus the separation of the wave equation in spherical polar coordinates results in wave functions that are eigenfunctions of both the total angular momentum and the component of angular momentum along the polar axis. The quantum number l is called the azimuthal or orbital–angular–momentum quantum number. The quantum number m is called the magnetic quantum number, which involves the component of angular momentum along the polar axis (z–axis) (which is usually taken as the direction of external field, particularly magnetic field).

It should be noted that in general the wave equation cannot be separated in this way and angular momentum eigenfunctions obtained if the potential energy $V(\mathbf{r})$ is not spherically symmetric. This corresponds to the classical result that the angular momentum is a constant of the motion only for a central field of force, which is describable by a spherically symmetric potential. There is, however, the characteristic difference between classical and quantum theory that all three components of $\mathbf{L}$ can be precisely specified at once in the classical theory, whereas only L_z and L^2 can, in general, be precisely specified at once in the quantum theory, since $Y_{lm}(\theta,\phi)$ is not an eigenfunction of L_x and L_y (except for the case $l=0$).

8.4   Kinetic energy and angular momentum

The operator L^2 is useful in showing that the kinetic energy commutes with $\mathbf{L}$. From classical mechanics we can write

$$L^2 = (\mathbf{r} \times \mathbf{p})^2 = r^2 p^2 - (\mathbf{r} \cdot \mathbf{p})^2 \qquad (8.93)$$

or

$$p^2 = \frac{L^2}{r^2} + \left(\frac{\mathbf{r} \cdot \mathbf{p}}{r} \right)^2 \qquad (8.94)$$

expressing the kinetic energy in terms of a constant of the motion, L^2, and radial component of momentum. This expression, Eq.(8.94), cannot be taken over into quantum mechanics, because the operator $\mathbf{r} \cdot \mathbf{p}$ is not Hermitian. However, the correct relation can be worked out readily from the operator

$$L^2 = (\mathbf{r} \times \mathbf{p}) \cdot (\mathbf{r} \times \mathbf{p}) = r^2 p^2 - \mathbf{r}(\mathbf{r} \cdot \mathbf{p}) \cdot \mathbf{p} + 2i\hbar \mathbf{r} \cdot \mathbf{p} \tag{8.95}$$

where the middle term is to be understood to mean

$$\mathbf{r}(\mathbf{r} \cdot \mathbf{p}) \cdot \mathbf{p} = \sum_{i=1}^{3} \sum_{k=1}^{3} r_i r_k p_k p_i \tag{8.96}$$

The component of the gradient ∇f in the direction of $\mathbf{r}$ is $\partial f/\partial r$; hence

$$\mathbf{r} \cdot \mathbf{p} = -i\hbar r \frac{\partial}{\partial r} \tag{8.97}$$

and consequently

$$L^2 = r^2 p^2 + \hbar^2 r^2 \frac{\partial^2}{\partial r^2} + 2\hbar^2 r \frac{\partial}{\partial r} \tag{8.98}$$

or more compactly

$$L^2 = r^2 p^2 + \hbar^2 \frac{\partial}{\partial r} \left(r^2 \frac{\partial}{\partial r} \right) \tag{8.99}$$

Since $\mathbf{L}$ and therefore also L^2 commutes with any function of r, we may write

$$T = \frac{p^2}{2\mu} = \frac{L^2}{2\mu r^2} - \frac{\hbar^2}{2\mu r^2} \frac{\partial}{\partial r} \left(r^2 \frac{\partial}{\partial r} \right) \tag{8.100}$$

$\mathbf{L}$ commutes with any radial derivative. Hence $\mathbf{L}$ and T commute, and thus for central forces

$$[H, \mathbf{L}] = 0 \tag{8.101}$$

This commutation relation insures that angular momentum is a constant of the motion.

8.5 Worked examples

Example - 8.1 : Show that for a system with orbital angular momentum $\mathbf{L}$, the operator for the rotation of the system about the z–axis through an angle ϕ is given by

$$U_{R_z}(\phi) = e^{-\frac{i}{\hbar} L_z \phi}$$

Solution : Define the effect of an operator R on any function $f(r)$ by $Rf(r) = f(R^{-1}r)$ where R^{-1} is the inverse of R. For a rotation through an angle θ about the z–axis

$$R = \begin{pmatrix} \cos\theta & \sin\theta & 0 \\ -\sin\theta & \cos\theta & 0 \\ 0 & 0 & 1 \end{pmatrix}$$

If $\psi(r)$ is the state function for a system

$$R\psi(r) = \psi(R^{-1}r) = \psi(x\cos\theta - y\sin\theta, x\sin\theta + y\cos\theta, z)$$

If θ is an infinitesimal angle, the right hand side of the above equation becomes

$$\psi(x - y\theta, x\theta + y, z) = \psi(x, y, z) - \theta\left(y\frac{\partial}{\partial x} - x\frac{\partial}{\partial y}\right)\psi(x, y, z)$$

$$= \left(1 - \frac{i}{\hbar}\theta L_z\right)\psi$$

Thus one says that the angular momentum is the generator for infinitesimal rotation. If $U_{R_z}(\phi)$ is the rotation operator for a finite rotation

$$U_{R_z}(\phi + d\phi) = \left(1 - \frac{i}{\hbar}d\phi L_z\right)U_{R_z}(\phi)$$

$$U_{R_z}(\phi) + d\phi\frac{dU_{R_z}}{d\phi} = \left(1 - \frac{i}{\hbar}d\phi L_z\right)U_{R_z}(\phi)$$

Thus

$$U_{R_z}(\phi) = e^{-\frac{i}{\hbar}\phi L_z}$$

Example - 8.2 : Consider the special case in which the Hamiltonian H for a system is invariant to any rotation $R(\hat{n}, \theta)$, i.e. $R^{-1}HR = H$. Show that $[H, L_\alpha^n] = [H, L^2] = 0$, where $L_\alpha = \mathbf{L} \cdot \hat{n}$. Here θ is the angle of rotation and $\hat{n}$ specifies the rotation axis.

Solution : The invariance of H implies that

$$e^{-iL_\alpha\theta}He^{+iL_\alpha\theta} = H \quad \text{or} \quad He^{+iL_\alpha\theta} - e^{+iL_\alpha\theta}H = 0$$

Expanding the exponential terms gives:

$$H\left(1 + iL_\alpha\theta + (iL_\alpha)^2\frac{\theta^2}{2!} + \cdots\right) - \left(1 + iL_\alpha\theta + (iL_\alpha)^2\frac{\theta^2}{2!} + \cdots\right)H = 0$$

For this equation to hold for all values of θ the coefficient of each power of θ must vanish. Therefore,

$$[H, L_\alpha] = [H, L_\alpha^2] = \cdots = [H, L_\alpha^n] = 0$$

Since $[H, L_\alpha^2] = 0$, we have $[H, L_x^2 + L_y^2 + L_z^2] = 0$ or $[H, L^2] = 0$. For such a system H, L^2, and one component of $\mathbf{L}$ can have a complete set of simultaneous eigenfunctions.

Example - 8.3 : Show that for a system in the eigenstate $|lm >$ of the operator L_z, the mean value of the component of angular momentum along a direction z', which makes an angle θ with the z–axis, is equal to $m\cos\theta$.

Solution : Using the vector model of angular momentum, we have that

$$L_{z'} = L_x\cos(x, z') + L_y\cos(y, z') + L_z\cos\theta$$

Since $|lm>$ is an eigenvector of L_z, we obtain

$$< lm|L_x|lm >=< lm|L_y|lm >= 0$$

and

$$< lm|L_{z'}|lm >=< lm|L_z|lm > \cos\theta = m\cos\theta$$

Example - 8.4 : Show that for orbital angular momentum

$$[L_x, L_y] = iL_z \quad , \quad [L_y, L_z] = iL_x \quad , \quad [L_z, L_x] = iL_y$$

Use these results to show that $\mathbf{L} \times \mathbf{L} = i\mathbf{L}$.

Solution : We start with the equations

$$\mathbf{L} = \mathbf{r} \times \mathbf{p} \quad ; \quad [x_i, p_j] = i\hbar\delta_{ij}$$

$$\hbar L_x = yp_z - zp_y \quad ; \quad \hbar L_y = zp_x - xp_z \quad ; \quad \hbar L_z = xp_y - yp_x$$

Therefore

$$\hbar^2[L_x, L_y] = [(yp_z - zp_y), (zp_x - xp_z)]$$
$$= [yp_z, zp_x] + [zp_y, xp_z] - [yp_z, xp_z] - [zp_y, zp_x]$$

where

$$[yp_z, zp_x] = yp_z zp_x - zp_x yp_z = yp_x[p_z, z] = -i\hbar yp_x$$
$$[zp_y, xp_z] = zp_y xp_z - xp_z zp_y = p_y x[z, p_z] = i\hbar xp_y$$
$$[yp_z, xp_z] = [zp_y, zp_x] = 0$$

Therefore

$$[L_x, L_y] = i\hbar^{-1}(-yp_x + xp_y) = iL_z$$

Similarly

$$[L_y, L_z] = iL_x \quad , \quad [L_z, L_x] = iL_y$$
$$\mathbf{L} \times \mathbf{L} = \mathbf{i}(L_y L_z - L_z L_y) + \mathbf{j}(L_z L_x - L_x L_z) + \mathbf{k}(L_x L_y - L_y L_x)$$
$$= \mathbf{i}[L_y, L_z] + \mathbf{j}[L_z, L_x] + \mathbf{k}[L_x, L_y] = i(\mathbf{i}L_x + \mathbf{j}L_y + \mathbf{k}L_z) = i\mathbf{L}$$

Example - 8.5 : Show that, between the simultaneous eigenvectors of the operators L^2 and L_z, there exist the following relations

$$|l, \pm m >= \sqrt{\frac{(l+m)!}{(2l)!(l-m)!}}(L_\mp)^{l-m}|l, \pm l >$$

$$|l, \pm l >= \sqrt{\frac{(l+m)!}{(2l)!(l-m)!}}(L_\pm)^{l-m}|l, \pm m >$$

Solution : Let us consider only the relation

$$|lm >= \sqrt{\frac{(l+m)!}{(2l)!(l-m)!}}(L_-)^{l-m}|ll > \quad (1)$$

since the other three relations can be derived by similar reasoning. We have that

$$L_-|l, m+1> = \sqrt{l(l+1) - m(m+1)}|lm>$$
$$= \sqrt{(l-m)(l+m+1)}|lm>$$

and hence

$$(L_-)^2|l, m+2> = L_-\sqrt{(l-m-1)(l+m+2)}|l, m+1>$$
$$= \sqrt{(l-m-1)(l-m)(l+m+1)(l+m+2)}|lm>$$

and, in general,

$$(L_-)^{l-m}|ll> = \sqrt{1\cdot 2\cdots(l-m)(l+m+1)\cdots(2l-1)(2l)}|lm>$$

In order to obtain (1) it is sufficient to observe that the product under the square root sign can be written in the form $[(2l)!(l-m)!/(l+m)!]$.

Example - 8.6 : Show that (a) $L_\mp|\alpha m> = C_\mp(\alpha, m)|\alpha, m\mp 1>$ where $C_\mp$ are some complex numbers; (b) $\alpha = l(l+1)$ where l is the highest value of m for given value of α; (c) $C_\mp(\alpha, m) = \sqrt{(l\pm m)(l\mp m+1)}$.

Solution : (a)

$$L_z L_+|\alpha, m> = (L_+ + L_+L_z)|\alpha, m> = (m+1)L_+|\alpha, m>$$
$$L_z L_-|\alpha, m> = (L_-L_z - L_-)|\alpha, m> = (m-1)L_-|\alpha, m>$$
$$L^2 L_\pm|\alpha, m> = \alpha L_\pm|\alpha, m>$$

Thus if $|\alpha, m>$ is an eigenket of L^2 and L_z corresponding to the eigenvalues α and m, respectively, then $L_\pm|\alpha, m>$ is an eigenket of L^2 and L_z with eigenvalues α and $m\pm 1$, respectively. Thus $L_\pm$ are the angular momentum raising and lowering operators. Thus

$$L_\pm|\alpha, m> = C_\mp(\alpha, m)|\alpha, m\pm 1>$$

(b)

$$L_+|\alpha, l> = 0 \quad , \quad L_-|\alpha, l'> = 0$$

where l' is the least value of m for a given α. From these two equations one sees:

$$L_-L_+|\alpha, l> = (L^2 - L_z^2 - L_z)|\alpha, l> = (\alpha - l^2 - l)|\alpha, l> = 0$$

which gives $\alpha = l(l+1)$

$$L_+L_-|\alpha, l'> = (L^2 - L_z^2 + L_z)|\alpha, l'> = (\alpha - l'^2 + l')|\alpha, l'> = 0$$

which gives $\alpha = l'(l'-1)$. Therefore $l^2 + l = l'^2 - l'$. Since l' is the least and l the highest value of m, we get from the above $l' = -l$.

(c)

$$L_+|\alpha, m> = C_+(\alpha, m)|\alpha, m+1>$$

The norm of $L_+|\alpha, m>$ is $<\alpha, m|L_-L_+|\alpha, m>$, since L_- is the Hermitian adjoint of L_+, i.e.,

$$<\alpha, m|L_-L_+|\alpha, m> = |C_+(\alpha, m)|^2$$

Thus to within an arbitrary phase factor

$$C_+ = < \alpha, m|L^2 - L_z^2 - L_z|\alpha, m >$$

$$= \sqrt{l(l+1) - m^2 - m} = \sqrt{(l-m)(l+m+1)}$$

Similarly by considering the norm of $L_-|\alpha, m >$ one can write

$$C_- = \sqrt{(l+m)(l-m+1)}$$

Example - 8.7 : The unitary operator $R(\hat{n}, \theta)$ can be used to relate the wave function ψ' after a rotation of the coordinate system to the wave function ψ before the rotation. Hence $\psi' = R(\hat{n}, \theta)\psi$, where $\hat{n}$ defines the axis of rotation and θ is the rotation angle. The angular momentum $\mathbf{L}$ is defined as follows:

$$R\psi = e^{-i\theta(\hat{n} \cdot \mathbf{L})}\psi$$

(a) Show that the infinitesimal rotations $e^{-id\theta_y L_y}$ and $e^{-id\theta_x L_x}$ do not commute by expanding each to second order. (b) If

$$R_1 = e^{-id\theta_y L_y} e^{-id\theta_x L_x} \quad , \quad R_2 = e^{-id\theta_x L_x} e^{-id\theta_y L_y}$$

the difference $R_2 - R_1$ is found to correspond to a rotation of magnitude $d\theta_x d\theta_y$ about the z–direction, i.e.,

$$(R_2 - R_1)\psi = (R - 1)\psi \quad \text{where} \quad R = e^{-id\theta_x d\theta_y L_z}$$

Use this result with the conclusion of part (a) to show that $[L_x, L_y] = iL_z$.

Solution : (a) The commutator of the operators R_x and R_y is:

$$\left[e^{-id\theta_x L_x}, e^{-id\theta_y L_y} \right]$$

$$\cong \left(1 - id\theta_x L_x - \left(\frac{d\theta_x}{2}\right)^2 L_x^2 \right) \left(1 - id\theta_y L_y - \left(\frac{d\theta_y}{2}\right)^2 L_y^2 \right)$$

$$- \left(1 - id\theta_y L_y - \left(\frac{d\theta_y}{2}\right)^2 L_y^2 \right) \left(1 - id\theta_x L_x - \left(\frac{d\theta_x}{2}\right)^2 L_x^2 \right)$$

$$\cong -d\theta_x d\theta_y [L_x, L_y]$$

The linear terms cancel but terms of second and higher degree in $d\theta$ do not.

(b) We are given:

$$R_2 - R_1 = e^{-id\theta_x d\theta_y L_z} - 1$$

Expanding the right side to second order terms and using the results of part (a) gives

$$-d\theta_x d\theta_y [L_x, L_y] = 1 - id\theta_x d\theta_y L_z - 1$$

or

$$[L_x, L_y] = iL_z$$

The definition of angular momentum given here reveals that the components of $\mathbf{L}$ can be determined by investigating the transformation properties of a system.

8.6   Problems

Problem - 8.1 : Show that if a system is in an eigenstate of L_z, then the mean values of the operators L_x and L_y vanish.

Problem - 8.2 : Since the components of the angular momentum operator do not commute, their simultaneous measurement is not possible. Show that in a state $|lm>$ the greatest accuracy of measurement of the components L_x and L_y is obtained when $|m| = l$.

Problem - 8.3 : Show that $[L^2, L_z] = 0$ by considering the commutators of L_z with L_x^2, L_y^2, L_z^2 separately. You may use $\mathbf{L} \times \mathbf{L} = i\mathbf{L}$.

Problem - 8.4 : Let $\mathbf{L}$ be the orbital angular momentum operator, (r, θ, ϕ) a set of polar coordinates, and P the parity operator of a particle, all referred to the same origin. P performs a reflection about the origin, so that its action on any function of position of the form $F(r, \theta, \phi)$ is given by the relation $PF(r, \theta, \phi) = F(r, \pi - \theta, \phi + \pi)$. Show that $[P, \mathbf{L}] = 0$ and, starting from this fact, prove that each of the spherical harmonics has a well–defined parity, which depends only on the quantum number l. Find this parity as a function of l.

Problem - 8.5 : For what value of L_z will the root–mean–square deviation $(\Delta L_x^2)^{1/2}$ be minimum if the state has a definite value of L_z.

Problem - 8.6 : A certain state $|\psi>$ is an eigenstate of $\mathbf{L}^2$ and L_z:

$$\mathbf{L}^2|\psi> = l(l+1)\hbar^2|\psi> \quad , \quad L_z|\psi> = m\hbar|\psi> .$$

For this state calculate $< L_x >$ and $< L_x^2 >$.

Problem - 8.7 : Suppose an electron is in a state described by the wave function

$$\psi = \frac{1}{\sqrt{4\pi}}(e^{i\phi} \sin\theta + \cos\theta)g(r) \quad \text{where} \quad \int_0^\infty |g(r)|^2 r^2 dr = 1,$$

and θ, ϕ are polar and azimuth angles, respectively. (a) What are the possible results of a measurement of the z–component L_z of the angular momentum of the electron in this state? (b) What is the probability of obtaining each of the possible results in part (a)? (c) What is the expectation value of L_z?

Chapter 9

The Radial Equation for Free and Bound Particles

9.1   The radial Schrödinger equation

The radial Schrödinger equation expressed in the previous chapter

$$-\frac{\hbar^2}{2\mu}\left[\frac{1}{r}\frac{d}{dr}\left(r\frac{d}{dr}\right)+\frac{1}{r}\frac{d}{dr}-\frac{l(l+1)}{r^2}\right]R_{nlm}(r) \tag{9.1}$$

$$+V(r)R_{nlm}(r)=ER_{nlm}(r)$$

may be written as

$$\left(\frac{d^2}{dr^2}+\frac{2}{r}\frac{d}{dr}\right)R_{nlm}(r)-\frac{2\mu}{\hbar^2}\left[V(r)+\frac{l(l+1)\hbar^2}{2\mu r^2}\right]R_{nlm}(r) \tag{9.2}$$

$$+\frac{2\mu E}{\hbar^2}R_{nlm}(r)=0$$

We will examine the solutions to this equation for a variety of potentials restricted by the condition that they go to zero at infinity, and we will also assume that the potentials are not as singular as $1/r^2$ at the origin, so that

$$\lim_{r\to 0} r^2 V(r)=0 \tag{9.3}$$

It is sometimes convenient to introduce the function

$$u_{nlm}(r)=rR_{nlm}(r) \tag{9.4}$$

Using the property

$$\left(\frac{d^2}{dr^2}+\frac{2}{r}\frac{d}{dr}\right)\frac{u_{nlm}(r)}{r}=\frac{1}{r}\frac{d^2 u_{nlm}(r)}{dr^2} \tag{9.5}$$

the radial Schrödinger equation, Eq.(5.2), takes the form

$$\frac{d^2 u_{nlm}(r)}{dr^2}+\frac{2\mu}{\hbar^2}\left[E-V(r)-\frac{l(l+1)\hbar^2}{2\mu r^2}\right]u_{nlm}(r)=0 \tag{9.6}$$

This looks like a one–dimensional equation, except that

a) the potential $V(r)$ is altered by the addition of a repulsive centrifugal barrier

$$V(r)\quad\to\quad V(r)+\frac{l(l+1)\hbar^2}{2\mu r^2}=V_{eff}(r) \tag{9.7}$$

which is usually called as the effective potential, b) the definition of $u_{nlm}(r)$ and the finiteness of the wave function at the origin require that

$$u_{nlm}(0) = 0 \tag{9.8}$$

which makes it more like the one–dimensional problem for which $V = +\infty$ in the left–hand region.

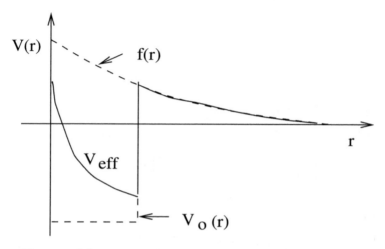

Fig. 9.1. Effective potential acting in radial equation for $u(r) = rR(r)$ when the real potential is a square well, here $f(r) = l(l+1)\hbar^2/2\mu r^2$.

First we consider the radial equation near the origin. As $r \to 0$, the leading terms in the radial equation are

$$\frac{d^2u}{dr^2} - \frac{l(l+1)}{r^2}u \approx 0 \tag{9.9}$$

because the potential does not contribute for small enough r when the condition in Eq.(5.3) is satisfied. If we make the Ansatz

$$u(r) \sim r^s \tag{9.10}$$

we find that the equation will be satisfied, provided that

$$s(s+1) - l(l+1) = 0 \tag{9.11}$$

that is, $s = l+1$ or $s = -l$. The solution that satisfies the condition $u(0) = 0$, that is, the solution that behaves like r^{l+1}, is called the **regular solution**; the solution that behaves like r^{-l} is the **irregular solution**.

For large r we can drop the potential terms, and the radial equation becomes

$$\frac{d^2u}{dr^2} + \frac{2\mu E}{\hbar^2} \approx 0 \tag{9.12}$$

The square integrability condition implies that

$$1 = \int |\psi(r)|^2 dv \tag{9.13}$$

$$= \int_0^\infty r^2 dr \int d\Omega |R_{nlm}(r)Y_{lm}(\theta,\phi)|^2 = \int_0^\infty r^2 dr |R_{nlm}(r)|^2$$

or

$$\int_0^\infty |u_{nlm}(r)|^2 dr = 1 \qquad (9.14)$$

so that the wave function should vanish at ∞.

If $E < 0$, so that

$$\frac{2\mu E}{\hbar^2} = -\alpha^2 \qquad (9.15)$$

the asymptotic solution is

$$u(r) \cong e^{-\alpha r} \qquad (9.16)$$

If $E > 0$, we have solutions that are only normalizable in a box. With

$$\frac{2\mu E}{\hbar^2} = k^2 \qquad (9.17)$$

the solution will be a linear combination of e^{ikr} and e^{-ikr}.

Laplacian in various coordinate systems:

See Figure 9.2.

In Cartesian coordinates: (x, y, z)

$$\nabla^2 = \frac{\partial^2}{\partial x^2} + \frac{\partial^2}{\partial y^2} + \frac{\partial^2}{\partial z^2}$$

In spherical polar coordinates: (r, θ, ϕ)

$$\nabla^2 = \frac{1}{r^2}\frac{\partial}{\partial r}r^2\frac{\partial}{\partial r} + \frac{1}{r^2 \sin\theta}\frac{\partial}{\partial \theta}\sin\theta\frac{\partial}{\partial \theta} + \frac{1}{r^2 \sin^2\theta}\frac{\partial^2}{\partial \phi^2}$$

or

$$\nabla^2 = \frac{1}{r^2}\left[\frac{\partial}{\partial r}r^2\frac{\partial}{\partial r} - L^2\right] \quad , \quad L^2 = -\frac{1}{\sin\theta}\frac{\partial}{\partial \theta}\sin\theta\frac{\partial}{\partial \theta} - \frac{1}{\sin^2\theta}\frac{\partial^2}{\partial \phi^2}$$

In cylindrical coordinates: (r, θ, z)

$$\nabla^2 = \frac{1}{r}\frac{\partial}{\partial r}r\frac{\partial}{\partial r} + \frac{1}{r^2}\frac{\partial^2}{\partial \theta^2} + \frac{\partial^2}{\partial z^2}$$

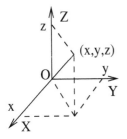
Cartesian coordinates

Spherical polar coordinates

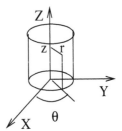
Cylindrical coordinates

Fig. 9.2. Coordinate systems.

9.2   The free particle

In this case $V(r) = 0$, but there is still a centrifugal barrier present. The radial equation takes the form

$$\left[\frac{d^2}{dr^2} + \frac{2}{r}\frac{d}{dr} - \frac{l(l+1)}{r^2} \right] R(r) + k^2 R(r) = 0 \tag{9.18}$$

If we introduce the variable $\rho = kr$, we get

$$\frac{d^2 R}{d\rho^2} + \frac{2}{\rho}\frac{dR}{d\rho} - \frac{l(l+1)}{\rho^2} R + R = 0 \tag{9.19}$$

The solutions to this equation are the following:

The regular solution is (spherical Bessel function)

$$j_l(\rho) = (-\rho)^l \left(\frac{1}{\rho}\frac{d}{d\rho} \right)^l \left(\frac{\sin\rho}{\rho} \right) \tag{9.20}$$

The irregular solution is (spherical Neumann function)

$$n_l(\rho) = -(-\rho)^l \left(\frac{1}{\rho}\frac{d}{d\rho} \right)^l \left(\frac{\cos\rho}{\rho} \right) \tag{9.21}$$

The first few functions are:

$$j_0(\rho) = \frac{\sin\rho}{\rho} \quad , \quad j_1(\rho) = \frac{\sin\rho}{\rho^2} - \frac{\cos\rho}{\rho}$$

$$j_2(\rho) = \left(\frac{3}{\rho^3} - \frac{1}{\rho} \right) \sin\rho - \frac{3}{\rho^3}\cos\rho$$

$$n_0(\rho) = \frac{\cos\rho}{\rho} \quad , \quad n_1(\rho) = -\frac{\cos\rho}{\rho^2} - \frac{\sin\rho}{\rho}$$

$$n_2(\rho) = -\left(\frac{3}{\rho^3} - \frac{1}{\rho} \right) \cos\rho - \frac{3}{\rho^3}\sin\rho$$

The combinations that will be of interest for large ρ are the spherical Hankel functions (first and second kind)

$$h_l^{(1)}(\rho) = j_l(\rho) + i n_l(\rho) \tag{9.22}$$

$$h_l^{(2)}(\rho) = \left[h_l^{(1)}(\rho) \right]^* = j_l(\rho) - i n_l(\rho) \tag{9.23}$$

The first few spherical Hankel functions are:

$$h_0^{(1)}(\rho) = \frac{e^{i\rho}}{i\rho} \quad , \quad h_1^{(1)}(\rho) = -\frac{e^{i\rho}}{\rho}\left(1 + \frac{i}{\rho} \right)$$

$$h_2^{(1)}(\rho) = \frac{ie^{i\rho}}{\rho}\left(1 + \frac{3i}{\rho} - \frac{3}{\rho^2}\right)$$

Special cases of $j_l(\rho)$ and $n_l(\rho)$ are:

a) The behavior near the origin; for $\rho \ll l$, it turns out that

$$j_l(\rho) \cong \frac{\rho^l}{(2l+1)!!} \tag{9.24}$$

and

$$n_l(\rho) \cong -\frac{(2l-1)!!}{\rho^l} \tag{9.25}$$

where the double factorial is defined as $(2l+1)!! = 1 \cdot 3 \cdot 5 \cdots (2l+1)$.

b) For $\rho \gg l$, we have the asymptotic expressions

$$j_l(\rho) \cong \frac{1}{\rho}\sin(\rho - \frac{\pi}{2}l) \tag{9.26}$$

and

$$n_l(\rho) \cong -\frac{1}{\rho}\cos(\rho - \frac{\pi}{2}l) \tag{9.27}$$

so that

$$h_l^{(1)}(\rho) \cong -\frac{i}{\rho}\exp\left[i(\rho - \frac{\pi}{2}l)\right] \tag{9.28}$$

Some properties of the j's and n's are:

$$\int j_0^2(\rho)\rho^2\,d\rho = \frac{1}{2}\rho^3[j_0^2(\rho) + n_0(\rho)j_1(\rho)]$$

$$\int n_0^2(\rho)\rho^2\,d\rho = \frac{1}{2}\rho^3[n_0^2(\rho) - j_0(\rho)n_1(\rho)]$$

$$n_{l-1}(\rho)j_l(\rho) - n_l(\rho)j_{l-1}(\rho) = \frac{1}{\rho^2} \quad , \quad l > 0$$

$$j_l(\rho)\frac{dn_l(\rho)}{d\rho} - n_l(\rho)\frac{dj_l(\rho)}{d\rho} = \frac{1}{\rho^2}$$

The following are the properties of both the j's and the n's:

$$j_{l-1}(\rho) + j_{l+1}(\rho) = \frac{2l+1}{\rho}j_l(\rho) \quad , \quad l > 0$$

$$\frac{dj_l(\rho)}{d\rho} = \frac{1}{2l+1}[lj_{l-1}(\rho) - (l+1)j_{l+1}(\rho)] = j_{l-1}(\rho) - \frac{l+1}{\rho}j_l(\rho)$$

$$\frac{d}{d\rho}\left(\frac{j_l(\rho)}{\rho^l}\right) = -\frac{j_{l+1}(\rho)}{\rho^l}$$

$$\int j_1(\rho)\,d\rho = -j_0(\rho) \quad , \quad \int j_0(\rho)\rho^2\,d\rho = -\rho^2 j_1(\rho)$$

$$\int j_l^2(\rho)\rho^2\,d\rho = \frac{1}{2}\rho^3[j_l^2(\rho) - j_{l-1}(\rho)j_{l+1}(\rho)] \quad , \quad l > 0$$

The solution that is regular at the origin is

$$R_l(r) = j_l(kr) \tag{9.29}$$

Its asymptotic form is

$$R_l(r) \cong -\frac{1}{2ikr}\left(\exp\left[-i(kr - \frac{\pi}{2}l)\right] - \exp\left[i(kr - \frac{\pi}{2}l)\right]\right) \tag{9.30}$$

We describe this as a sum of an **incoming** and **outgoing** spherical wave. In general, any solution (also for $V(r) \neq 0$) for large r is in the form

$$R_l(r) \cong -\frac{1}{2ikr}\left(\exp\left[-i(kr - \frac{\pi}{2}l)\right] - S_l(k)\exp\left[i(kr - \frac{\pi}{2}l)\right]\right) \tag{9.31}$$

where $|S_l(k)|^2 = 1$, and

$$S_l(k) = e^{2i\delta_l(k)} \tag{9.32}$$

here the real function $\delta_l(k)$ is called the **phase shift**. The radial function in the asymptotic region may be rewritten as

$$R_l(r) \cong e^{i\delta_l(k)}\frac{1}{kr}\sin[kr - \frac{\pi}{2}l + \delta_l(k)] \tag{9.33}$$

S is called the **scattering matrix**.

   Free particle wave equation may be solved in two ways. One solution is obtained as a superposition of separable solutions $R_l(r) = j_l(kr)$ multiplied by the appropriate spherical harmonics $Y_{lm}(\theta, \phi)$:

$$\psi(\mathbf{r}) = \sum_{l=0}^{\infty}\sum_{m=-l}^{l} A_{lm}j_l(kr)Y_{lm}(\theta, \phi) \tag{9.34}$$

Another solution of the free particle equation, which reads

$$(\nabla^2 + k^2)\psi(\mathbf{r}) = 0 \tag{9.35}$$

is the plane wave

$$\psi(\mathbf{r}) = e^{i\mathbf{k}\cdot\mathbf{r}} \tag{9.36}$$

We may therefore find A_{lm} such that

$$\sum_{l=0}^{\infty}\sum_{m=-l}^{l} A_{lm}j_l(kr)Y_{lm}(\theta, \phi) = e^{i\mathbf{k}\cdot\mathbf{r}} \tag{9.37}$$

The spherical angles (θ, ϕ) are the coordinates of the vector $\mathbf{r}$ relative to some arbitrarily chosen z–axis. If we define the z–axis by the direction of $\mathbf{k}$, then

$$e^{i\mathbf{k}\cdot\mathbf{r}} = e^{ikr\cos\theta} \tag{9.38}$$

Thus the right side of Eq.(9.37) has no azimuthal angle, ϕ, dependence, and thus on the left side only terms with $m = 0$ can appear; hence, using

$$Y_{l0}(\theta, \phi) = \left(\frac{2l+1}{4\pi}\right)^{1/2} P_l(\cos\theta) \tag{9.39}$$

are the Legendre polynomials, we get the relation

$$e^{ikr\cos\theta} = \sum_{l=0}^{\infty} \left(\frac{2l+1}{4\pi}\right)^{1/2} A_{lm} j_l(kr) P_l(\cos\theta) \tag{9.40}$$

Using the relation

$$\frac{1}{2} \int_{-1}^{1} d(\cos\theta) P_l(\cos\theta) P_{l'}(\cos\theta) = \frac{\delta_{ll'}}{2l+1} \tag{9.41}$$

one obtains

$$A_{lm} j_l(kr) = \frac{1}{2} [4\pi(2l+1)]^{1/2} \int_{-1}^{1} dz P_l(z) e^{ikrz} \tag{9.42}$$

Therefore the plane wave can be expressed as an expansion

$$e^{ikr\cos\theta} = \sum_{l=0}^{\infty} (2l+1) i^l j_l(kr) P_l(\cos\theta) \tag{9.43}$$

This expansion form of plane wave is usually used in scattering theory.

The general solution to the free particle equation, Eq.(9.35), sometimes expressed as the linear combination of plane waves expressed in Eq.(5.36),

$$\psi(\mathbf{r}) = A e^{i\mathbf{k}\cdot\mathbf{r}} + B e^{-i\mathbf{k}\cdot\mathbf{r}} \tag{9.44}$$

The trouble with these wave functions (plane waves) is that they are not square integrable, since

$$\int_{-\infty}^{\infty} |\psi(\mathbf{r})|^2 d\mathbf{r}$$

diverges for all values of A and B. One way of avoiding the normalization difficulty is to deal with the probability current, or **flux**

$$\mathbf{j}(\mathbf{r}) = \frac{\hbar}{2im} [\psi^*(\mathbf{r})\nabla\psi(\mathbf{r}) - \nabla\psi^*(\mathbf{r})\psi(\mathbf{r})] = \mathbf{v}(A^2 - B^2) \tag{9.45}$$

9.3 Three–dimensional square well potential

Consider the square well potential of finite depth, for which $V(r) = -V_0$, $r < a$; $V(r) = 0$, $r > a$, where V_0 is positive (see Figure 9.3).

A spherical region of this type in which the potential is less than that of the surroundings serves to attract a particle just as in the one–dimensional case.

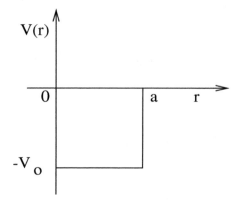

Fig. 9.3. Square well potential in 3D.

Solutions for zero angular momentum:

When $l = 0$, it is easier to solve the wave equation in the form

$$-\frac{\hbar^2}{2m}\frac{d^2u}{dr^2} + \left[V(r) + \frac{l(l+1)\hbar^2}{2mr^2}\right]u = Eu \tag{9.46}$$

than in the form

$$\frac{1}{r^2}\frac{d}{dr}\left(r^2\frac{dR}{dr}\right) + \left(\frac{2m}{\hbar^2}[E - V(r)] - \frac{\lambda}{r^2}\right)R = 0 \tag{9.47}$$

where $\lambda = l(l+1)$. In this case, $u(r) = rR(r)$, and the wave equation is

$$-\frac{\hbar^2}{2m}\frac{d^2u}{dr^2} - V_0 u = Eu \quad , \quad r < a \tag{9.48}$$

$$-\frac{\hbar^2}{2m}\frac{d^2u}{dr^2} = Eu \quad , \quad r > a \tag{9.49}$$

The solutions of these equations are

$$u(r) = A\sin(\alpha r) + B\cos(\alpha r) \quad , \quad \alpha = \left[\frac{2m}{\hbar^2}(V_0 - |E|)\right]^{1/2} \quad , \quad r < a \tag{9.50}$$

$$u(r) = Ce^{-\beta r} \quad , \quad \beta = \left(\frac{2m}{\hbar^2}|E|\right)^{1/2} \quad , \quad r > a \tag{9.51}$$

where we are interested in bound–state energy levels for which $E < 0$. The requirement that $R(r)$ be finite at $r = 0$ demands that we set $B = 0$. Thus the solution has the form of the odd parity solution of the one–dimensional problem.

The energy levels are obtained by equating the two values of $\frac{1}{u}\frac{du}{dr}$ at $r = a$ (this is equivalent to making $\frac{1}{R}\frac{dR}{dr}$ continuous there — logarithmic derivative) and are given by solving

$$\alpha\cot(\alpha a) = -\beta \tag{9.52}$$

Therefore, there is no energy level unless

$$a^2 V_0 > \frac{\pi^2\hbar^2}{8m}$$

there is one bound state if

$$\frac{\pi^2\hbar^2}{8m} < a^2 V_0 \leq \frac{9\pi^2\hbar^2}{8m}$$

etc.

The shape of the lowest three wave functions are shown in Figure 9.4.

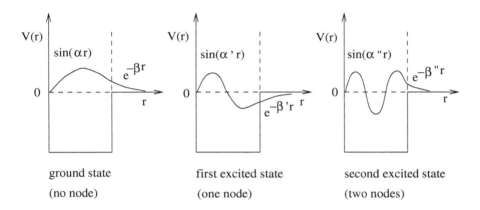

ground state   first excited state   second excited state

(no node)   (one node)   (two nodes)

Fig. 9.4. The wave functions for the lowest three states of a particle in 3D square well potential.

Interior solutions for arbitrary l:

For nonzero values of l, it is more convenient to work with the original radial equation for $R(r)$, Eq.(9.47), than with the equation for $u(r)$, Eq.(9.46). If we put $\rho = \alpha r$, where α is defined as

$$\alpha = \left[\frac{2m}{\hbar^2}(V_0 - |E|)\right]^{1/2}$$

the wave equation, Eq.(9.47), for $r < a$ becomes

$$\frac{d^2 R}{d\rho^2} + \frac{2}{\rho}\frac{dR}{d\rho} + \left[1 - \frac{l(l+1)}{\rho^2}\right]R = 0 \tag{9.53}$$

This differential equation looks like Bessel's equation, hence $R(r)$ can be expressed in terms of Bessel functions. The spherical Bessel function $j_l(\rho)$ that is regular at $\rho = 0$ is defined by

$$j_l(\rho) = \left(\frac{\pi}{2\rho}\right)^{1/2} J_{l+\frac{1}{2}}(\rho) \tag{9.54}$$

where J is an ordinary Bessel function of half–odd–integer order. $j_l(\rho)$ satisfies Eq.(9.53).

The spherical Neumann function is defined by

$$n_l(\rho) = (-1)^{l+1} \left(\frac{\pi}{2\rho} \right)^{1/2} J_{-l-\frac{1}{2}}(\rho) \tag{9.55}$$

$J_{l+\frac{1}{2}}(\rho)$, where l is a positive or negative integer or zero, is expressible as a sum of products of $\sin \rho$ and $\cos \rho$ with polynomials of odd order in $\rho^{-1/2}$.

Since $R(r)$ must be finite for $r = 0$, the desired solution for $r < a$ is

$$R(r) = A j_l(\alpha r) \tag{9.56}$$

Exterior solutions for arbitrary l:

The wave equation for $r > a$ can be put in the form

$$\frac{d^2 R}{d\rho^2} + \frac{2}{\rho} \frac{dR}{d\rho} + \left[1 - \frac{l(l+1)}{\rho^2} \right] R = 0 \tag{9.57}$$

if we redefine ρ to be $i\beta r$, where

$$\beta = \left(\frac{2m}{\hbar^2} |E| \right)^{1/2}$$

Since the domain of ρ does not now extend into zero, there is no reason why n_l cannot appear in the solution. The linear combination of j_l and n_l to be selected will be determined by the asymptotic form, which must fall off exponentially for large r.

The desired solution for $r > a$ is then

$$R(r) = B h_l^{(1)}(i\beta r) = B \left[j_l(i\beta r) + i n_l(i\beta r) \right] \tag{9.58}$$

Summary of the solutions:

$l = 0$ case: $u(r) = A \sin(\alpha r)$, $r < a$; $u(r) = C e^{-\beta r}$, $r > a$

$l \neq 0$ case: $R(r) = A j_l(\alpha r)$, $r < a$; $R(r) = B h_l^{(1)}(i\beta r)$, $r > a$

Energy levels:

The energy levels are obtained by requiring that

$$\frac{1}{R} \frac{dR}{dr}$$

be continuous at $r = a$. When this condition is applied to the interior solution $R(r) = A j_l(\alpha r)$ and the exterior solution $R(r) = B h_l^{(1)}(i\beta r)$ with $l = 0$, we obtain

$$\alpha \cot(\alpha a) = -\beta \tag{9.59}$$

This may be written as

$$\xi \cot \xi = -\eta \quad , \quad \xi^2 + \eta^2 = \frac{2m}{\hbar^2} a^2 V_0 \tag{9.60}$$

where we have put $\xi = \alpha a$ and $\eta = \beta a$. The same condition applied to the solution for $l = 1$ reduces to

$$\frac{\cot \xi}{\xi} - \frac{1}{\xi^2} = \frac{1}{\eta} + \frac{1}{\eta^2} \quad , \quad \xi^2 + \eta^2 = \frac{2m}{\hbar^2} a^2 V_0 \tag{9.61}$$

these equations may be solved numerically or graphically as described in Chapter 5. In general, there is no degeneracy between the eigenvalues obtained from the solution of equations like Eq.(9.60) and Eq.(9.61) for various values of l.

9.4 The hydrogen atom

The hydrogen atom is the simplest atom, since it contains only one electron. Thus the Schrödinger equation becomes a one–particle equation after the C.M. motion is separated out. We will deal with hydrogenlike atoms, that is, atoms containing one electron only, but allowing for a nucleus more complicated than a single proton. The potential then is

$$V(r) = -\frac{Ze^2}{r} \tag{9.62}$$

and the radial Schrödinger equation is

$$\left(\frac{d^2}{dr^2} + \frac{2}{r} \frac{d}{dr} \right) R + \frac{2\mu}{\hbar^2} \left[E + \frac{Ze^2}{r} - \frac{l(l+1)\hbar^2}{2\mu r^2} \right] R = 0 \tag{9.63}$$

We will concentrate on the bound states, that is, $E < 0$ solutions.

It is convenient to make a change of variables,

$$\rho = \left(\frac{8\mu}{\hbar^2} |E| \right)^{1/2} r \tag{9.64}$$

The equation then reads

$$\frac{d^2 R}{d\rho^2} + \frac{2}{\rho} \frac{dR}{d\rho} - \frac{l(l+1)}{\rho^2} R + \left(\frac{\lambda}{\rho} - \frac{1}{4} \right) R = 0 \tag{9.65}$$

where we have introduced the dimensionless parameter

$$\lambda = \frac{Ze^2}{\hbar} \left(\frac{\mu}{2|E|} \right)^{1/2} = Z\alpha \left(\frac{\mu c^2}{2|E|} \right)^{1/2} \tag{9.66}$$

The second form makes it easier to compute with it, since $\alpha \cong 1/137$ (known as the fine structure constant) and the energy is expressed in units of the rest mass; and it is a nonrelativistic equation.

$$\frac{\lambda^2}{Z^2 \alpha^2} = \frac{\mu c^2}{2|E|} \quad \rightarrow \quad |E| = \frac{1}{2} \mu c^2 \left(\frac{Z\alpha}{\lambda} \right)^2$$

We try to solve the Eq.(9.65). First, we extract the large ρ behavior. For large ρ, the only terms that remain in the equation are

$$\frac{d^2 R}{d\rho^2} - \frac{1}{4}R \sim 0 \tag{9.67}$$

and the solution, which behaves properly at infinity, is

$$R \sim e^{-\rho/2} \tag{9.68}$$

We take

$$R(\rho) = e^{-\rho/2}G(\rho) \tag{9.69}$$

substitute this into Eq.(9.65), and obtain the equation for $G(\rho)$. Then the radial equation takes the form

$$\frac{d^2 G}{d\rho^2} - \left(1 - \frac{2}{\rho}\right)\frac{dG}{d\rho} + \left[\frac{\lambda - 1}{\rho} - \frac{l(l+1)}{\rho^2}\right]G = 0 \tag{9.70}$$

We now write a power expansion for $G(\rho)$. This takes the form

$$G(\rho) = \rho^l \sum_{n=0}^{\infty} a_n \rho^n \tag{9.71}$$

When this series form is substituted into Eq.(9.70), we find a relation between various coefficients a_n. The recursion relation is obtained from the differential equation obeyed by

$$H(\rho) = \sum_{n=0}^{\infty} a_n \rho^n \tag{9.72}$$

which is

$$\frac{d^2 H}{d\rho^2} + \left(\frac{2l+2}{\rho} - 1\right)\frac{dH}{d\rho} + \frac{\lambda - 1 - l}{\rho}H = 0 \tag{9.73}$$

One can obtain Eq.(9.73) by taking

$$G(\rho) = \rho^l H(\rho) \tag{9.74}$$

in Eq.(9.70). We then have

$$\sum_{n=0}^{\infty}\left[n(n-1)a_n\rho^{n-2} + na_n\rho^{n-1}\left(\frac{2l+2}{\rho} - 1\right) + (\lambda - 1 - l)a_n\rho^{n-1}\right] = 0 \tag{9.75}$$

that is,

$$\sum_{n=0}^{\infty}\left[(n+1)\left[na_{n+1} + (2l+2)a_{n+1}\right] + (\lambda - 1 - l - n)a_n\right]\rho^{n-1} = 0 \tag{9.76}$$

or

$$\sum_{n=0}^{\infty}\left[(n+1)(n+2l+2)a_{n+1} + (\lambda - 1 - l - n)a_n\right]\rho^{n-1} = 0$$

Since this must vanish term by term, we get the recursion relation

$$\frac{a_{n+1}}{a_n} = \frac{n+l+1-\lambda}{(n+1)(n+2l+2)} \tag{9.77}$$

For large n this ratio is

$$\frac{a_{n+1}}{a_n} \approx \frac{1}{n} \tag{9.78}$$

We do not get a solution $R(\rho)$ that is well behaved at infinity, unless the series

$$\sum_{n=0}^{\infty} a_n \rho^n$$

terminates. This means that for a given l, for some $n = n_r$ we must have $\lambda = n_r + l + 1$. Let us introduce the **principal quantum number** n defined by

$$n = n_r + l + 1 \tag{9.79}$$

Then, it follows from the fact that $n_r \geq 0$, that $n \geq l + 1$, n is an integer, the relation $\lambda = n$ implies that

$$E = -\frac{1}{2} \mu c^2 \frac{(Z\alpha)^2}{n^2} \tag{9.80}$$

or

$$E_n = -\frac{Z^2 e^2}{2 a_\mu n^2}$$

The presence of the reduced mass $\mu = mM/(m+M)$, where m is the electron mass, and M the mass of the nucleus, means that the frequencies

$$\omega_{ij} = \frac{E_i - E_j}{\hbar} = \frac{\mu c^2}{2\hbar} (Z\alpha)^2 \left(\frac{1}{n_j^2} - \frac{1}{n_i^2} \right) \tag{9.81}$$

$$= \frac{mc^2/2\hbar}{(1 + \frac{m}{M})} (Z\alpha)^2 \left(\frac{1}{n_j^2} - \frac{1}{n_i^2} \right)$$

differ slightly for different hydrogenlike atoms. In particular, the difference between the hydrogen spectrum and the deuterium spectrum — where M, the nuclear mass, is very close to being twice the proton mass — was responsible for the discovery of deuterium.

The energy does not depend on l, that is, for a given n the energies of all the states such that $l + 1 \leq n$ are degenerate. There is a $(2l + 1)$–fold degeneracy of the energy states for a given l, since the radial equation does not depend on m. Although the radial equation does depend on l, there is an additional degeneracy. In the first approximation, for a given n, the possible values of $l = 0, 1, 2, \cdots, (n-1)$, and for each l there is the $(2l + 1)$ degeneracy. Thus the total degeneracy is

$$\sum_{l=0}^{n-1} (2l + 1) = n^2 \tag{9.82}$$

There are two possible states for the electron because of its spin, so that the true degeneracy is really $2n^2$.

Let us now return to the differential equation, Eq.(9.73). If we set $\lambda = n$ in the recursion relation, Eq.(9.76), so that

$$a_{k+1} = \frac{k+l+1-n}{(k+1)(k+2l+2)} a_k \tag{9.83}$$

we find that

$$a_{k+1} = (-1)^{k+1} \frac{n-(k+l+1)}{(k+1)(k+2l+2)} \cdot \frac{n-(k+l)}{k(k+2l+1)} \cdots \frac{n-(l+1)}{1 \cdot (2l+2)} a_0 \tag{9.84}$$

With the help of this we can obtain the power series expansion for $H(\rho)$. Equivalently, we observe that the equation for $H(\rho)$ is that for the associated Laguerre polynomials:

$$H(\rho) = L_{n-l-1}^{(2l+1)}(\rho) \tag{9.85}$$

Associated Laguerre polynomials are the solutions for the Laguerre's associated equation

$$x\frac{d^2y}{dx^2} + (k+1-x)\frac{dy}{dx} + ny = 0 \tag{9.86}$$

The associated Laguerre polynomials are defined by

$$L_n^k(x) = (-1)^k \frac{d^k}{dx^k} L_{n+k}(x) = \sum_{r=0}^{n} (-1)^r \frac{(n+k)!}{(n-r)!(k+r)!r!} x^r \tag{9.87}$$

Some properties of the associated Laguerre polynomials:

Generating function:

$$\exp\left[-\left(\frac{xt}{1-t}\right)\right](1-t)^{-k-1} = \sum_{n=0}^{\infty} L_n^k(x)t^n \tag{9.88}$$

$$L_n^k(x) = \frac{e^x x^{-k}}{n!} \frac{d^n}{dx^n}\left(e^{-x}x^{n+k}\right) \tag{9.89}$$

Orthogonality property:

$$\int_0^{\infty} e^{-x} x^k L_n^k(x) L_m^k(x) dx = \frac{(n+k)!}{n!} \delta_{nm} \tag{9.90}$$

Recurrence relations:

$$L_{n-1}^k(x) + L_n^{k-1}(x) = L_n^k(x) \tag{9.91}$$

$$(n+1)L_{n+1}^k(x) = (2n+k+1-x)L_n^k(x) - (n+k)L_{n-1}^k(x) \tag{9.92}$$

$$xL_n^{k'}(x) = nL_n^k(x) - (n+k)L_{n-1}^k(x) \tag{9.93}$$

$$L_n^{k'}(x) = -\sum_{r=0}^{n-1} L_r^k(x) = -L_{n-1}^{k+1}(x) \tag{9.94}$$

$$L_n^{k+1}(x) = \sum_{r=0}^{n} L_r^k(x) \tag{9.95}$$

$$L_n^{\alpha+\beta+1}(x+y) = \sum_{r=0}^{n} L_r^{\alpha}(x) L_{n-r}^{\beta}(y) \tag{9.96}$$

Summary of the radial solution in step–by–step procedure:

$$R(r) \rightarrow R(\rho) \rightarrow e^{-\rho/2} G(\rho) \rightarrow e^{-\rho/2} \rho^l H(\rho)$$

$$\rightarrow e^{-\rho/2} \rho^l \sum_{n=0}^{\infty} a_n \rho^n \rightarrow e^{-\rho/2} \rho^l \sum_{k=0}^{n_r} a_k \rho^k \rightarrow e^{-\rho/2} \rho^l L_{n-l-1}^{2l+1}(\rho)$$

$$E_n = -\frac{1}{2} \mu c^2 \frac{(Z\alpha)^2}{n^2} = -\frac{Z^2 e^2}{2 a_\mu n^2} \quad ; \quad n_r = n - l - 1$$

After conversion back to the radial coordinate r and after normalization, the first few radial functions can be computed.

$$R_{10}(r) = 2 \left(\frac{Z}{a_0}\right)^{3/2} e^{-Zr/a_0} \quad , \quad a_0 = \frac{\hbar}{\mu c \alpha} \left(= \frac{\hbar^2}{m e^2}\right)$$

$$R_{20}(r) = 2 \left(\frac{Z}{2a_0}\right)^{3/2} \left(1 - \frac{Zr}{2a_0}\right) e^{-Zr/2a_0}$$

$$R_{21}(r) = \frac{1}{\sqrt{3}} \left(\frac{Z}{2a_0}\right)^{3/2} \frac{Zr}{a_0} e^{-Zr/2a_0}$$

$$R_{30}(r) = 2 \left(\frac{Z}{3a_0}\right)^{3/2} \left[1 - \frac{2Zr}{3a_0} + \frac{2(Zr)^2}{27a_0^2}\right] e^{-Zr/3a_0}$$

$$R_{31}(r) = \frac{4\sqrt{2}}{9} \left(\frac{Z}{3a_0}\right)^{3/2} \frac{Zr}{a_0} \left(1 - \frac{Zr}{6a_0}\right) e^{-Zr/3a_0}$$

$$R_{32}(r) = \frac{2\sqrt{2}}{27\sqrt{5}} \left(\frac{Z}{3a_0}\right)^{3/2} \left(\frac{Zr}{a_0}\right)^2 e^{-Zr/3a_0}$$

Some properties of the radial wave function:

a) The behavior of r^l for small r, which forces the wave function to stay small for a range of radii that increases with l, is a consequence of the centrifugal repulsive barrier that keeps the electron from coming close to the nucleus.

b) The recursion relation shows that $H(\rho)$ is a polynomial of degree $n_r + n - l - 1$, and thus it has n_r radial nodes (zeros). n_r is called the **radial quantum number**. There will be $n - l$ bumps in the probability density distribution

$$P(r) = r^2 [R_{nl}(r)]^2 \tag{9.97}$$

When, for a given n, l has its largest value $l = n - 1$, then there is only one bump.

$$R_{n,n-1}(r) \propto r^{n-1} e^{-Zr/a_0 n}$$

Hence

$$P(r) \propto r^{2n} e^{-2Zr/a_0 n}$$

will peak at a value of r determined by

$$\frac{dP(r)}{dr} = \left(2nr^{2n-1} - \frac{2Z}{a_0 n}r^{2n}\right)e^{-2Zr/a_0 n} = 0$$

that is, at $r = n^2 a_0/Z$ which is the Bohr atom value for circular orbits. Smaller values of l give probability distributions with more bumps.

c) Plots of the radial probability density $P(r)$ for finding the electron at a distance r from the origin can be constructed with the help of the wave functions. Wave function also has an angular part, whose absolute square is $[P_l^m(\cos\theta)]^2$. As m increases, the probability density is seen to shift from the z–axis toward the equatorial plane. When $|m| = l$, then $|P_l^l(\cos\theta)|^2 \propto \sin^{2l}\theta$. As l increases, the width of the peak decreases like $l^{-1/2}$, and thus for large quantum numbers we get the classical picture of planar orbits.

d) Given the wave functions, we can calculate

$$< r^k > = \int_0^\infty dr\, r^{2+k}[R_{nl}(r)]^2 \tag{9.98}$$

Some useful expectation values:

$$< \tfrac{1}{r} > = \frac{Z}{a_0 n^2}$$

$$< \tfrac{1}{r^2} > = \frac{Z^2}{a_0^2 n^3(l+\frac{1}{2})}$$

$$< r > = \frac{a_0 n^2}{Z}\left[1 + \tfrac{1}{2}\left(1 - \frac{l(l+1)}{n^2}\right)\right]$$

$$< r^2 > = \frac{a_0^2 n^4}{Z^2}\left[1 + \tfrac{3}{2}\left(1 - \frac{l(l+1)-\frac{1}{3}}{n^2}\right)\right]$$

As a summary one can write the bound state wave function for a one–electron atom in general

$$\psi_{nlm}(r,\theta,\phi) = R_{nl}(r)Y_{lm}(\theta,\phi)$$

where the radial wave function $R_{nl}(r)$ is in the form

$$R_{nl}(r) = -\left[\left(\frac{2Z}{na_\mu}\right)^3 \frac{(n-l-1)!}{2n[(n+l)!]^3}\right]^{1/2} e^{-\rho/2}\rho^l L_{n+l}^{2l+1}(\rho)$$

$$\rho = \frac{2Z}{na_\mu}r \quad , \quad a_\mu = \frac{\hbar^2}{e^2\mu}$$

μ is reduced mass. The explicit expression for the Laguerre polynomials $L_{n+l}^{2l+1}(\rho)$ is given by

$$L_{n+l}^{2l+1}(\rho) = \sum_{k=0}^{n_r}(-1)^{k+1}\frac{[(n+l)!]^2 \rho^k}{(n_r-k)!(2l+1+k)!k!}$$

$n_r = n - l - 1$. The spherical harmonics $Y_{lm}(\theta, \phi)$ are given by

$$Y_{lm}(\theta, \phi) = \Theta_{lm}(\theta)\Phi_m(\phi)$$

$$= (-1)^m \left[\frac{(2l+1)(l-m)!}{4\pi(l+m)!}\right]^{1/2} P_l^m(\cos\theta)e^{-im\phi} \quad , \quad m \geq 0$$

with the properties

$$Y_{l,-m}(\theta, \phi) = (-1)^m Y_{lm}^*(\theta, \phi) \quad , \quad |Y_{lm}(\theta, \phi)|^2 = \frac{1}{2\pi}|\Theta_{lm}(\theta)|^2$$

θ dependent function $\Theta_{lm}(\theta)$ can be expressed in terms of the associated Legendre functions P_l^m by

$$\Theta_{lm}(\theta) = (-1)^m \left[\frac{(2l+1)(l-m)!}{2(l+m)!}\right]^{1/2} P_l^m(\cos\theta) \quad , \quad m \geq 0$$

$$= (-1)^{|m|}\Theta_{l|m|}(\theta) \quad , \quad m < 0$$

ϕ dependent function $\Phi_m(\phi)$ is given by

$$\Phi_m(\phi) = \frac{1}{\sqrt{2\pi}}e^{-im\phi}$$

Shapes of the hydrogenic wave functions:

$$\psi_{nlm}(r, \theta, \phi) = R_{nl}(r)Y_{lm}(\theta, \phi) = R_{nl}(r)\Theta_{lm}(\theta)\Phi_m(\phi)$$

$n = 1, 2, 3, \cdots$ principal quantum number

$n - 1 \equiv$ total number of nodes (both in radial and angular parts)

$l = 0, 1, 2, \cdots, n-1$ azimuthal quantum number

$l \equiv$ number of nodes in angular part

$n_r = n - 1 - l$ radial quantum number

$n_r \equiv$ number of nodes in radial part

    The plot of the radial wave functions, R_{nl}, and the corresponding radial probability distributions, $r^2 R_{nl}^2$, are shown in Figure 9.5., and the plot of the angular probability density functions, $[\Theta_{lm}(\theta)]^2$, is shown in Figure 9.6.

9.5   The Spectra of hydrogenic atoms

    The energy eigenvalues of the hydrogenic atoms are given by

$$E_n = -\frac{1}{2n^2}\left(\frac{Ze^2}{\hbar}\right)^2 \mu = -\frac{Z^2 e^2 \mu}{2a_0 n^2 m} = -frac{Z^2 e^2}{2a_\mu n^2} \quad\quad (9.99)$$

where $a_0 = \frac{\hbar^2}{me^2}$ is the Bohr radius, and $a_\mu = a_0 \frac{m}{\mu}$ is the modified Bohr radius. We may also write

$$E_n = -\frac{1}{2}\mu c^2 \frac{(Z\alpha)^2}{n^2} \tag{9.100}$$

where $\alpha = \frac{e^2}{\hbar c} \cong \frac{1}{137}$ is the fine structure constant. In *a.u.* we also have

$$E_n = -\frac{Z^2}{2n^2}\left(\frac{\mu}{m}\right) \tag{9.101}$$

or simply

$$E_n \cong -\frac{Z^2}{2n^2} \tag{9.102}$$

We note from this result that since n may take on all integral values from 1 to $+\infty$, the energy spectrum corresponding to the Coulomb potential $V(r) = -\frac{Ze^2}{r}$ contains an infinite number of discrete energy levels extending from $-Z^2/2$ to zero. This is due to the fact that the magnitude of the Coulomb potential falls off slowly at large r.

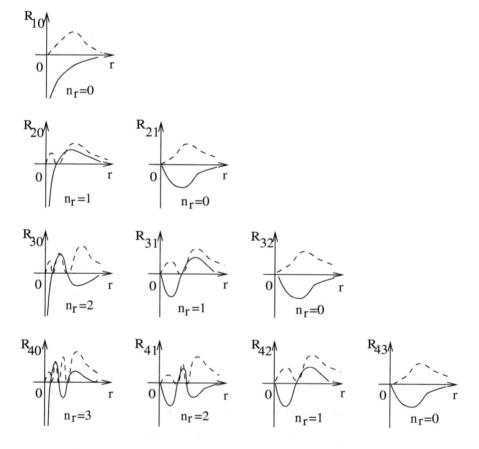

Fig. 9.5. Shapes of the hydrogenic radial wave functions (solid lines), $R_{nl}(r)$, and the corresponding radial probability distributions (dashed lines), $r^2 R_{nl}^2(r)$.

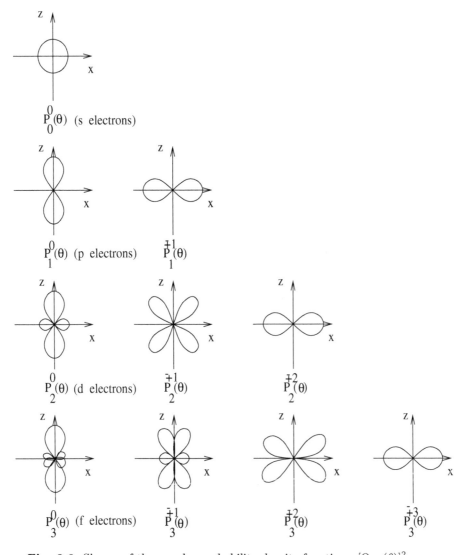

Fig. 9.6. Shapes of the angular probability density functions, $[\Theta_{lm}(\theta)]^2$.

Having obtained the energy levels of one–electron atoms within the framework of the Schrödinger nonrelativistic theory, we may ask about the spectral lines corresponding to transitions from one level to another. In particular, for the most common transitions, the so–called electric dipole transitions, that obey the selection rules

$$\Delta l = l - l' = \mp 1 \quad ; \quad \Delta m = m - m' = 0, \mp 1 \tag{9.103}$$

while $\Delta n = n - n'$ is arbitrary.

The Bohr frequency rule

$$\Delta E = E_n - E_{n'} = h\nu \tag{9.104}$$

can be applied to obtain the frequencies of the spectral lines corresponding to transitions between the energy levels. Thus, using atomic units, we have

$$\nu_{n'n} = Z^2 R(M) \left(\frac{1}{n'^2} - \frac{1}{n^2} \right) \tag{9.105}$$

where

$$R(M) = \frac{\mu}{m} R(\infty) = \frac{\mu}{m} \left(\frac{me^4}{4\pi\hbar^3} \right) \tag{9.106}$$

is the Rydberg constant, and $n > n'$.

Depending on the initial and final energy levels, there are several series of transitions in the hydrogenic atoms. These series of transitions for atomic hydrogen are:

$$\tilde{\nu}_{n'n} = \frac{\nu_{n'n}}{c} = \frac{1}{\lambda} = \frac{R(\infty)}{c} \left(\frac{1}{n'^2} - \frac{1}{n^2} \right) = \tilde{R}(\infty) \left(\frac{1}{n'^2} - \frac{1}{n^2} \right) \tag{9.107}$$

$$\tilde{R}(\infty) = \frac{R(\infty)}{c} = \frac{me^4}{4\pi c\hbar^3} = 109737 \ cm^{-1} \tag{9.108}$$

$$\tilde{R}_{exp} = 109677.58 \ cm^{-1} = R_H \tag{9.109}$$

$$\tilde{R}(M) = \frac{\mu}{m} \tilde{R}(\infty) = \frac{m_p}{m + m_p} \tilde{R}(\infty) \tag{9.110}$$

$$\cong \left(\frac{1836.15}{1837.15} \right) \tilde{R}(\infty) \cong 0.999455678 \tilde{R}(\infty)$$

$$\tilde{R}(M) \cong 0.999455678 \times 109737 \cong 109677.27 \ cm^{-1}$$

The first few hydrogen series:

n'	n	name of the series
1	$2, 3, 4, \cdots$	Lyman
2	$3, 4, 5, \cdots$	Balmer
3	$4, 5, 6, \cdots$	Paschen
4	$5, 6, 7, \cdots$	Brackett
5	$6, 7, 8, \cdots$	Pfund
$\vdots$	$\vdots$	$\vdots$

Ground state wave function of the hydrogenic atoms:

$$\psi_{nlm_l}(r, \theta, \phi) = R_{nl}(r) Y_{lm_l}(\theta, \phi)$$

$$\psi_{100}(r, \theta, \phi) = R_{10}(r) Y_{00}(\theta, \phi)$$

$$= 2 \left(\frac{Z}{a_0} \right)^{3/2} e^{-\frac{Z}{a_0} r} \frac{1}{\sqrt{4\pi}}$$

in *a.u.*

$$\psi_{100}(r, \theta, \phi) = \left(\frac{Z^3}{\pi}\right)^{1/2} e^{-Zr}$$

for $Z = 1$,

$$\psi_{100}(r) = \frac{1}{\sqrt{\pi}} e^{-r}$$

Numerical estimates for various quantities in hydrogen atom:

Particle masses: (m: electron mass , M: proton mass)

$$mc^2 \approx 0.5 \ MeV \quad (0.511 \ MeV) \quad ; \quad Mc^2 \approx 1000 \ MeV \quad (938.3 \ MeV)$$

$$\frac{m}{M} \approx \frac{1}{2000} \ (\frac{1}{1836}) \quad ; \quad \mu = \frac{mM}{m+M} \cong \frac{mM}{M} = m$$

Bohr radius:

$$a_0 = \frac{\hbar^2}{me^2} \quad ; \quad \hbar = 1.054 \times 10^{-27} \ erg \cdot s$$

$$\hbar c \approx 2000 \ eV \cdot \mathring{A} \quad (1973.3 \ eV \cdot \mathring{A})$$

Fine structure constant (dimensionless):

$$\alpha = \frac{e^2}{\hbar c} \approx \frac{1}{137} \ (\frac{1}{137.04})$$

$$a_0 = \frac{\hbar^2}{me^2} = \frac{\hbar c}{mc^2} \left(\frac{\hbar c}{e^2}\right)$$

$$= \frac{\hbar c}{mc^2 \alpha} \approx \frac{2000 \times 137}{0.5 \times 10^6} \ \mathring{A} \cong 0.55 \ \mathring{A}$$

$$a_0 = 0.529177 \ \mathring{A} \approx 0.53 \ \mathring{A} \approx \frac{1}{2} \ \mathring{A}$$

Energy levels:

$$E_n = -\frac{Ry}{n^2}$$

$$Ry = \frac{me^4}{2\hbar^2} = \frac{mc^2}{2}\left(\frac{e^2}{\hbar c}\right)^2 = \frac{1}{2}mc^2\alpha^2$$

$$\cong \frac{1}{2} \times 0.5 \times 10^6 \times \frac{1}{(137)^2} \ eV \cong 13.3 \ eV \quad (13.6 \ eV)$$

so

$$E_n = -\frac{13.6}{n^2} \ eV$$

Some length scales related to a_0:

$$a_0\alpha = \frac{\hbar^2}{me^2} \cdot \frac{e^2}{\hbar c} = \frac{\hbar}{mc} \equiv \lambda_e$$

This quantity (λ_e) is known as the Compton wavelength, which may be defined as the lower limit on how well a particle can be localized. In classical mechanics (nonrelativistic theory) we consider the lower limit is zero, but in reality, as we try to locate the particle better and better, we use more and more energetic probes, photons to be specific. To locate it to some Δx, we need a photon of momentum $\Delta p \sim \frac{\hbar}{\Delta x}$ (from uncertainty principle). The corresponding energy is $\Delta E \sim \frac{\hbar c}{\Delta x}$. Relativistic energy of a particle is $E^2 = c^2 p^2 + m^2 c^4$. If this energy exceeds twice the rest energy of the particle, relativity allows the production of a particle–antiparticle pair in the measurement process. So we require $\Delta E \le 2mc^2$

$$\frac{\hbar c}{\Delta x} \le 2mc^2 \quad \rightarrow \quad \Delta x \ge \frac{\hbar}{2mc} \approx \frac{\hbar}{mc} \equiv \lambda_e$$

If we attempt to localize the particle any better, we will see pair creation and we will have three (or more) particles instead of the one we started.

In our analysis of the hydrogen atom, we treated the electron as a localized point particle. The preceding analysis shows that this is not strictly correct, but it is a fair approximation, since the size of the electron is α times smaller than the size of the atom, a_0

$$\frac{\hbar/mc}{a_0} = \alpha \cong \frac{1}{137}$$

$$\lambda_e = \alpha a_0 \approx \frac{1}{137} \times 0.5 \text{ Å} \approx \frac{1}{250} \text{ Å} = 4 \times 10^{-3} \text{ Å}$$

If we multiply λ_e by α we get another length, called the classical radius of the electron:

$$r_e = \lambda_e \alpha = \frac{\hbar}{mc} \cdot \frac{e^2}{mc^2} \cong 3 \times 10^{-5} \text{ Å}$$

If we imagine the electron to be a spherical charge distribution, the Coulomb energy of the distribution (the energy it takes to assemble it) will be of the order e^2/r_e, where r_e is the radius of the sphere. If we attribute the rest energy of the electron to this Coulomb energy, we arrive at the classical radius.

In summary:

$$a_0\left(\frac{1}{2}\text{Å}\right) \overset{\alpha}{\rightarrow} \lambda_e\left(\frac{\alpha}{2}\text{Å}\right) \overset{\alpha}{\rightarrow} r_e\left(\frac{\alpha^2}{2}\text{Å}\right)$$

The hydrogenic energies may also be obtained as follows:

The ground state energy may be written as (from Bohr model)

$$E_1 = -\frac{1}{2}mv^2 = -\frac{1}{2}mc^2\left(\frac{v}{c}\right)^2 = -\frac{1}{2}mc^2\beta^2$$

where $\beta = v/c$ is the velocity of the electron in the ground state of hydrogen measured in units of velocity of light (c). If we take $\beta = \alpha$

$$E_1 = -\frac{1}{2}mc^2\alpha^2 = -\frac{1}{2}mc^2\left(\frac{e^2}{\hbar c}\right)^2 = -\frac{me^4}{2\hbar^2}$$

in general,

$$E_n = \frac{E_1}{n^2} = -\frac{me^4}{2\hbar^2 n^2}$$

9.6 The Virial Theorem

Let us denote by H the Hamiltonian of a physical system and by ψ its state vector, solutions of the time–dependent Schrödinger equation

$$i\hbar \frac{\partial}{\partial t}\psi = H\psi \qquad (9.111)$$

We know that the time rate of change of the expectation value $< A > = < \psi|A|\psi >$ of an operator which does not depend explicitly at t satisfies the equation

$$i\hbar \frac{d}{dt} < \psi|A|\psi > = < \psi|[A, H]|\psi > \qquad (9.112)$$

where $[A, H] = AH - HA$ is the commutator of the operators A and H.

Let us further assume that H is time–independent and denote respectively by E_n and ψ_n its eigenenergies and eigenfunctions. For a stationary state

$$\Psi_n = \psi_n e^{-\frac{i}{\hbar}E_n t} \qquad (9.113)$$

and a time–independent operator A it is clear that the expectation value

$$< \Psi_n|A|\Psi_n > = < \psi_n|A|\psi_n > \qquad (9.114)$$

does not depend on t so that

$$< \psi_n|[A, H]|\psi_n > = 0 \qquad (9.115)$$

It turns out that the commutator of H and A has a zero expectation value in the state ψ, namely,

$$\frac{d}{dt} < A > = \frac{i}{\hbar} < [H, A] > = 0 \qquad (9.116)$$

Proof:

$$< [H, A] > = < \psi|[H, A]|\psi > = < \psi|HA - AH|\psi > = (E - E) < \psi|A|\psi > = 0$$

This relationship is known as the Hypervirial Theorem. It states that the expectation values of time–independent operators do not vary with time in stationary states.

The time average of the time derivative of the quantity $\mathbf{r} \cdot \mathbf{p}$ is zero for a periodic system (in classical mechanics). The analogous quantity in quantum mechanics is the time derivative of the expectation value of $\mathbf{r} \cdot \mathbf{p}$, or the diagonal matrix element of the commutator $[\mathbf{r} \cdot \mathbf{p}, H]$ is also zero. Therefore

$$\frac{d}{dt} < \mathbf{r} \cdot \mathbf{p} > = \frac{1}{i\hbar} < [\mathbf{r} \cdot \mathbf{p}, H] > = 0 \qquad (9.117)$$

We now apply this result to the particular case of the nonrelativistic motion of a particle of mass μ in a potential $V(r)$, the corresponding time–independent Hamiltonian being

$$H = \frac{p^2}{2\mu} + V = -\frac{\hbar^2}{2\mu} + V(r) \qquad (9.118)$$

Moreover, we choose A to be the time–independent operator $\mathbf{r} \cdot \mathbf{p}$. We then have

$$< \psi_n |[\mathbf{r} \cdot \mathbf{p}, H]|\psi_n > \equiv < [\mathbf{r} \cdot \mathbf{p}, H] >= 0 \qquad (9.119)$$

Using the algebraic properties of the commutators, together with the fundamental commutation relations, we find that

$$[\mathbf{r} \cdot \mathbf{p}, H] = [(xp_x + yp_y + zp_z), \frac{1}{2\mu}(p_x^2 + p_y^2 + p_z^2) + V(x, y, z)] \qquad (9.120)$$

$$= \frac{i\hbar}{\mu}(p_x^2 + p_y^2 + p_z^2) - i\hbar(x\frac{\partial V}{\partial x} + y\frac{\partial V}{\partial y} + z\frac{\partial V}{\partial z})$$

$$= 2i\hbar T - i\hbar(\mathbf{r} \cdot \nabla V)$$

We therefore obtain the relation

$$2 < T >=< \mathbf{r} \cdot \nabla V > \qquad (9.121)$$

which is known as the quantum mechanical Virial Theorem for a particle in a spherically symmetric potential. For a system of N particles having position vectors $\{\mathbf{r}_i\}$ and momenta $\{\mathbf{p}_i\}$, the Virial Theorem becomes:

$$2 < T >= \sum_{i=1}^{N} < \mathbf{r}_i \cdot \nabla_i V > \qquad (9.122)$$

If the interaction potential is spherically symmetric and proportional to r^s, and if the expectation value exists, we obtain

$$2 < T >=< r\frac{\partial V}{\partial r} >= s < V > \qquad (9.123)$$

For example the case $s = 2$ corresponds to the harmonic oscillator, for which $< T >=< V >$. On the other hand the case $s = -1$, corresponding to the hydrogenic atom, yields the relation $2 < T >= - < V >$.

The expectation value of the potential energy

$$V(r) = -\frac{Ze^2}{r} \qquad (9.124)$$

is

$$< V >_{nlm}= -Ze^2 < \frac{1}{r} >_{nlm}= -Ze^2(\frac{Z}{a_\mu n^2}) = -\frac{Z^2 e^2}{a_\mu n^2} = 2E_n \qquad (9.125)$$

The expectation value of the kinetic energy is

$$< T >_{nlm}= E_n- < V >_{nlm}= -E_n \qquad (9.126)$$

so that

$$2 < T >_{nlm}= - < V >_{nlm} \qquad (9.127)$$

This result is a particular case of the Virial Theorem; in other words, hydrogenic solutions satisfy the Virial Theorem.

9.7 Worked examples

Example - 9.1 : An electron is trapped in a spherical box defined by the potential $V(r) = \infty$, $r \geq a$; $V(r) = 0$, $r < a$. Determine the normalized radial solution and bound energies for the case $l = 0$.

Solution : The necessary separated radial equation is

$$\frac{1}{r^2}\frac{\partial}{\partial r}\left(r^2\frac{\partial R}{\partial r}\right) + \frac{2m}{\hbar^2}[E - V(r)]R = 0$$

Substituting $P = rR$ and using $V = 0$ inside, we find

$$P'' + \frac{2m}{\hbar^2}EP = 0$$

The solution of this differential equation is

$$P = c_1 \sin(\alpha r) + \cos(\alpha r) \quad , \quad \text{where} \quad \alpha = \sqrt{\frac{2mE}{\hbar^2}}$$

so that

$$R(r) = \frac{c_1}{r}\sin(\alpha r) + \frac{c_2}{r}\cos(\alpha r)$$

For $R(r)$ to be finite at the origin c_2 must vanish. $R(r)$ must vanish at $r = a$, thus

$$R(a) = 0 = \frac{c_1}{a}\sin(\alpha a) \quad \text{and} \quad \alpha a = n\pi \quad , \quad n = 1, 2, 3, \cdots$$

or

$$E = \frac{n^2\pi^2\hbar^2}{2ma^2} = \frac{n^2h^2}{8ma^2}$$

c_1 is found by normalization.

$$c_1^2 \int_0^a \frac{1}{r^2}\sin^2(\alpha r)r^2 dr = \frac{c_1^2}{\alpha}\int_0^{\alpha a} \sin^2(x)dx$$

$$= \frac{c_1^2}{\alpha}\left(\frac{x}{2} - \frac{\sin(2x)}{4}\right)\Big|_0^{\alpha a} = \frac{c_1^2 a}{2} = 1 \quad \text{or} \quad c_1 = \sqrt{\frac{2}{a}}$$

Therefore

$$R(r) = \sqrt{\frac{2}{a}}\frac{1}{r}\sin\left(\frac{n\pi r}{a}\right)$$

Example - 9.2 : Determine the energy levels for a particle enclosed in a sphere of zero potential, with infinitely high potential walls defining its surface of radius R.

Solution : The Schrödinger equation may be factorized so that its solutions become

$$u(r, \theta, \phi) = \frac{1}{r}\chi_l(r)Y_{lm}(\theta, \phi)$$

with radial equations

$$\chi_l'' + \left[k^2 - \frac{l(l+1)}{r^2} \right] \chi_l = 0 \quad (1) \quad , \quad k^2 = \frac{2mE}{\hbar^2}$$

in $0 \le r \le R$, and zero outside. This differential equation, by using $z = kr$ as variable and splitting off a factor $z^{1/2}$, $\chi_l = z^{1/2}\phi(z)$, is reduced to

$$\phi'' + \frac{1}{z}\phi' + \left[1 - \frac{(l+\frac{1}{2})^2}{z^2} \right] \phi = 0$$

i.e., to the Bessel equation whose solutions are $J_{\pm(l+\frac{1}{2})}(z)$. We therefore write the complete solution of (1)

$$\chi_l(r) = \sqrt{\frac{\pi kr}{2}} \left[C_1 J_{l+\frac{1}{2}}(kr) + C_2 J_{-(l+\frac{1}{2})}(kr) \right]$$

For physically acceptable solutions (as $r \to 0$, $\chi_l(r)$ must be finite) $C_2 = 0$. Therefore

$$\chi_l(r) = \sqrt{\frac{\pi kr}{2}} C_1 J_{l+\frac{1}{2}}(kr)$$

using $\quad j_l(z) = \sqrt{\frac{\pi z}{2}} J_{l+\frac{1}{2}}(z) \quad$ or $\quad \chi_l(r) = C j_l(kr)$

From this set of solutions we select the eigenfunctions by the condition

$$j_l(kR) = 0 \quad \text{or} \quad J_{l+\frac{1}{2}}(kR) = 0$$

Since, for each given value of $l + \frac{1}{2}$, the Bessel function has an infinite number of zeros, we find an infinite number of values $k_{n_r,l}$ and of energy levels

$$E_{n_r,l} = \frac{\hbar^2 k_{n_r,l}^2}{2m}$$

for each l, with $n_r = 1, 2, 3, \cdots$ the radial quantum number counting the zeros (nodes of the wave function).

For the lowest l values,

$$j_0(z) = \sin(z) \quad ; \quad j_1(z) = \frac{\sin(z)}{z} - \cos(z)$$

$$j_2(z) = -3\frac{\cos(z)}{z} + (\frac{3}{z^2} - 1)\sin(z)$$

For higher values of l they may be constructed from the recurrence relation

$$j_l(z) = \frac{l}{z} j_{l-1}(z) - j'_{l-1}(z)$$

Their zeros may be determined from simple transcendental equations:

$$j_0(z) = 0 \quad \text{if} \quad \sin(z) = 0 \quad \text{or} \quad z = n_r \pi$$

$$n_r \pi = kR \quad \rightarrow \quad k = \frac{n_r \pi}{R} \; ; \quad E_{n_r,0} = \frac{\hbar^2}{2m} \left(\frac{n_r \pi}{R}\right)^2 = \frac{\hbar^2 \pi^2}{2mR^2} n_r^2$$

$$j_1(z) = 0 \quad \text{if} \quad \tan(z) = z$$

$$j_2(z) = 0 \quad \text{if} \quad \tan(z) = \frac{3z}{3 - z^2} \quad \text{etc.}$$

Example - 9.3 : A particle of mass m interacts in three dimensions with a spherically symmetric potential of the form $V(r) = -c\delta(r - a)$. Here c is a positive constant. Find the minimum value of c for which there is a bound state.

Solution : The result is the radial equation for a modified radial wave function $u(r) = rR(r)$, where $\psi(r, \theta, \phi) = R(r)Y(\theta, \phi)$

$$-\frac{\hbar^2}{2m} \frac{d^2}{dr^2} u(r) + \left[V(r) + \frac{l(l+1)\hbar^2}{2mr^2}\right] u(r) = Eu(r)$$

The condition for the minimum value of c will arise when there is only one bound state, which will be the ground state, with $l = 0$. Since $V \rightarrow 0$ as $r \rightarrow \infty$, the bound state has $E < 0$, so we are left with, for $r \neq a$,

$$\frac{d^2 u}{dr^2} - k^2 u = 0 \quad \text{where} \quad k^2 = \frac{2m|E|}{\hbar^2}$$

Denote the regions $r < a$ and $r > a$ as regions I and II, respectively. The wave function in I is $u_I(r) = A\sinh(kr)$, because $u(0) = 0$, and in II $u_{II}(r) = Be^{-kr}$, so that $u(r) \rightarrow 0$ as $r \rightarrow \infty$.

Since we require the wave function to be continuous at $r = a$,

$$A \sinh(ka) = Be^{-ka} \quad (1)$$

The condition on the derivative of $u(r)$ is found by integrating the radial equation across the delta function:

$$-\frac{\hbar^2}{2m} \int_{a-\epsilon}^{a+\epsilon} \frac{d^2 u(r)}{dr^2} dr - c \int_{a-\epsilon}^{a+\epsilon} \delta(r - a)u(r)dr = \int_{a-\epsilon}^{a+\epsilon} Eu(r)dr$$

$$u'(a + \epsilon) - u'(a - \epsilon) = -\lambda u(a) - \mathcal{O}(\epsilon) \quad , \quad \lambda = \frac{2mc}{\hbar^2}$$

as $\epsilon \rightarrow 0$ $u'_{II}(a) - u'_I(a) = -\lambda u(a)$. Taking the derivatives, using (1), one obtains

$$\coth(ka) = -\left(\frac{\lambda a}{ka} - 1\right)$$

For small x we can write $\coth(x) = \frac{1}{x}[1 + \mathcal{O}(x^2)]$, so that

$$\frac{1}{x} = \frac{\lambda a}{x} - 1 \quad \text{or} \quad x = \lambda a - 1$$

In order for there to be a solution with $x > 0$, we require $\lambda a - 1 > 0$, or

$$c > \frac{\hbar^2}{2ma}$$

Example - 9.4 : A hydrogen–like wave function is shown below with r in units of a_0,

$$\psi = \frac{\sqrt{2}}{81\sqrt{\pi}} Z^{3/2}(6 - Zr)e^{-Zr/3} \cos\theta$$

(a) Determine the values of the quantum numbers n, l, m for ψ by inspection. (b) Generate from ψ another eigenfunction having the same values of n and l but with the magnetic quantum number equal to $m + 1$. (c) Determine the most probable value of r for an electron in the state specified by ψ when $Z = 1$.

Solution : (a) The exponential factor in ψ is of the form $e^{-\sqrt{-E}r}$ and since $E = -Z^2/n^2$ we see that $n = 3$. The azimuthal quantum number l can be determined either by recalling the factor r^l which multiplies the Laguerre polynomial in hydrogen–like wave functions or by carrying out the operation:

$$L^2\psi = L^2 f(r) \cos\theta = f(r)\left[-\frac{1}{\sin\theta}\frac{\partial}{\partial\theta}\left(\sin\theta\frac{\partial}{\partial\theta}\cos\theta\right)\right]$$

$$= f(r)\left[\frac{1}{\sin\theta}\frac{d}{d\theta}(\sin^2\theta)\right] = 2f(r)\cos\theta = l(l+1)\psi$$

Either way $l = 1$. For the magnetic quantum number we operate with L_z.

$$L_z\psi = -i\frac{\partial}{\partial\phi}[f(r)\cos\theta] = 0 = m\psi$$

Thus $m = 0$.

(b)

$$L_+\psi_m = \sqrt{(l-m)(l+m+1)}\psi_{m+1} = \sqrt{2}\psi_{m+1}$$

since $l = 1$ and $m = 0$. Using $L_+ = L_x + iL_y$,

$$L_+ = i(\sin\phi - i\cos\phi)\frac{\partial}{\partial\theta} + i(\cos\phi - i\sin\phi)\cot\theta\frac{\partial}{\partial\phi}$$

$$L_+\psi_{m=0} = e^{i\phi}\frac{\partial}{\partial\theta}f(r)\cos\theta = -e^{i\phi}f(r)\sin\theta = \sqrt{2}\psi_{m+1}$$

Therefore

$$\psi_{m+1} = -\frac{1}{\sqrt{2}}f(r)\sin\theta e^{i\phi}$$

(c) The most probable value of r occurs when $r^2\psi^2$ reaches its max. value.

$$\frac{\partial}{\partial r}(r\psi)^2 = 0 \quad \rightarrow \quad \frac{\partial}{\partial r}(r\psi) = 0 \quad \rightarrow \quad r^2 - 15r + 36 = 0$$

Therefore $r = 12 ; 3 , r = 12a_0$ corresponds to max. and is the most probable value.

Example - 9.5 : The radial equation for hydrogen–like atoms is given by

$$\frac{1}{r^2}\frac{\partial}{\partial r}\left(r^2\frac{\partial R}{\partial r}\right) + \left[\frac{2Z}{r} + E - \frac{l(l+1)}{r^2}\right]R = 0$$

(a) Transform this radial equation into a one–dimentional wave equation by introducing the function $P(r) = rR(r)$. (b) Plot the effective potential function which appears in the equation for $P(r)$ versus r for the first few values of l assuming that $Z = 1$.

Solution : (a) The Schrödinger equation for the electronic energy of the hydrogen atom is in the form

$$-\frac{1}{r^2}\frac{d}{dr}\left(r^2\frac{dR}{dr}\right) + \left[\frac{l(l+1)}{r^2} + V(r)\right]R = ER \quad (1)$$

The transformation $P(r) = rR(r)$ can be carried out easily by working out the first term:

$$\frac{1}{r^2}\frac{d}{dr}\left(r^2\frac{dR}{dr}\right) = \frac{1}{r}\left(r\frac{dR}{dr} + r\frac{d^2R}{dr^2}\right)$$

But

$$\frac{dP}{dr} = r\frac{dR}{dr} + R$$

and

$$\frac{d^2P}{dr^2} = 2\frac{dR}{dr} + r\frac{d^2R}{dr^2}$$

Therefore (1) becomes

$$-\frac{1}{r}\frac{d^2P}{dr^2} + \left[\frac{l(l+1)}{r^2} + V(r)\right]\frac{P}{r} = E\frac{P}{r}$$

or

$$-\frac{d^2P}{dr^2} + \left[\frac{l(l+1)}{r^2} - \frac{2Z}{r}\right]P = EP \quad (E \text{ in } Ryd.)$$

(b)

$$V_{eff}(r) = \frac{l(l+1)}{r^2} - \frac{2}{r}$$

The repulsive term in this potential is associated with the centrifugal force which is experienced in the rotating coordinate system.

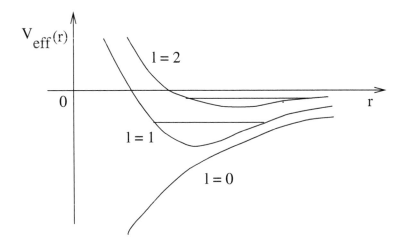

Example - 9.6 : Determine the energy levels of a particle moving in a centrally symmetric field with potential energy

$$V(r) = \frac{A}{r^2} + Br^2$$

where A and B are positive constants.

Solution : There is only a discrete spectrum. Schrödinger's equation is

$$\frac{d^2R}{dr^2} + \frac{2}{r}\frac{dR}{dr} + \frac{2m}{\hbar^2}\left[E - \frac{\hbar^2}{2m}l(l+1)\frac{1}{r^2} - \frac{A}{r^2} - Br^2\right]R = 0$$

Introducing the variable

$$\xi = \left(\frac{2mB}{\hbar^2}\right)^{1/2}r^2$$

and the notation

$$l(l+1) + \frac{2mA}{\hbar^2} = 2s(2s+1)$$

$$\left(\frac{2m}{\hbar^2 B}\right)^{1/2}E = 4(n+s) + 3$$

we obtain the equation

$$\xi R" + \frac{3}{2}R' + \left[n + s + \frac{3}{4} - \frac{1}{4}\xi - s(s+\frac{1}{2})\frac{1}{\xi}\right]R = 0$$

The solution required behaves asymptotically as $e^{-\xi/2}$ when $\xi \to \infty$, while for small ξ it is proportional to ξ^s, where s must be taken as the positive quantity

$$s = \frac{1}{4}\left[-1 + \sqrt{(2l+1)^2 + \frac{8mA}{\hbar^2}}\right]$$

Hence we look for a solution in the form

$$R = \xi^s e^{-\xi/2}w$$

obtaining for w the equation

$$\xi w" + (2s + \frac{3}{2} - \xi)w' + nw = 0$$

where

$$w = F(-n, 2s + \frac{3}{2}, \xi)$$

where n must be a nonnegative integer. We consequently find as the energy levels the infinite set of equidistant values

$$E_n = \hbar\sqrt{\frac{B}{2m}}\left[4n + 2 + \sqrt{(2l+1)^2 + \frac{8mA}{\hbar^2}}\right] \quad , \quad n = 0, 1, 2, \cdots$$

Example - 9.7 : Determine the eigenvalues for the Morse potential

$$V(r) = D\left[1 - e^{-a(r-r_e)}\right]^2$$

This potential function represents an anharmonic oscillator.

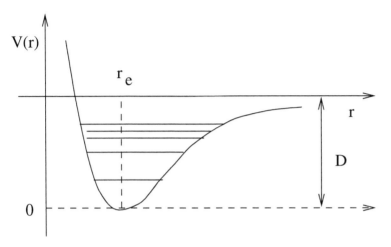

Solution : The radial equation

$$\frac{1}{r^2}\frac{d}{dr}\left(r^2\frac{dR}{dr}\right) + \left[-\frac{l(l+1)}{r^2} + \frac{2m}{\hbar^2}(E - V(r))\right]R = 0$$

may be simplified by the substitution $R(r) = \frac{1}{r}\chi(r)$, which leads to the equation

$$\frac{d^2\chi}{dr^2} + \left[-\frac{l(l+1)}{r^2} + \frac{2m}{\hbar^2}(E - V(r))\right]\chi = 0 \quad (1)$$

With the introduction of the Morse function, the radial equation (1) becomes

$$\frac{d^2\chi}{dr^2} + \left\{-\frac{l(l+1)}{r^2} + \frac{2m}{\hbar^2}\left[E - D - De^{-2a(r-r_e)} + 2De^{-a(r-r_e)}\right]\right\}\chi = 0$$

If we make the substitutions

$$y = e^{-a(r-r_e)} \quad \text{and} \quad A = l(l+1)\frac{\hbar^2}{2mr_e^2}$$

the radial equation becomes

$$\frac{d^2\chi}{dy^2} + \frac{1}{y}\frac{d\chi}{dy} + \frac{2m}{a^2\hbar^2}\left(\frac{E-D}{y^2} + \frac{2D}{y} - D - \frac{Ar_e^2}{y^2r^2}\right)\chi = 0 \quad (2)$$

The quantity r_e^2/r^2 may be expanded in terms of y in the following way (Taylor series):

$$\frac{r_e^2}{r^2} = \left(1 - \frac{\ln y}{ar_e}\right)^{-2} = 1 + \frac{2}{ar_e}(y-1) + \left(-\frac{1}{ar_e} + \frac{3}{a^2r_e^2}\right)(y-1)^2 + \cdots$$

Using the first three terms of this expansion in (2) we obtain

$$\frac{d^2\chi}{dy^2} + \frac{1}{y}\frac{d\chi}{dy} + \frac{2m}{a^2\hbar^2}\left(\frac{E - D - c_0}{y^2} + \frac{2D - c_1}{y} - D - c_2\right)\chi = 0 \quad (3)$$

where

$$c_0 = A\left(1 - \frac{3}{ar_e} + \frac{3}{a^2r_e^2}\right) \quad , \quad c_1 = A\left(\frac{4}{ar_e} - \frac{6}{a^2r_e^2}\right)$$

$$c_2 = A\left(-\frac{1}{ar_e} + \frac{3}{a^2r_e^2}\right)$$

The substitutions

$$\chi(y) = e^{-z/2}z^{b/2}F(z) \quad , \quad z = 2yd$$

$$d^2 = \frac{2m}{a^2\hbar^2}(D + c_2) \quad , \quad b^2 = -\frac{8m}{a^2\hbar^2}(E - D - c_0)$$

simplify (3) considerably, yielding the equation

$$\frac{d^2F}{dz^2} + \left(\frac{b+1}{z} - 1\right)\frac{dF}{dz} + \frac{v}{z}F = 0 \quad (4)$$

in which

$$v = \frac{m}{a^2\hbar^2 d}(2D - c_1) - \frac{1}{2}(b - 1)$$

Equation (4) is closely related to the radial equation of the hydrogen atom

$$\rho L'' + [2(l + 1) - \rho]L' + (\lambda - l - 1)L = 0$$

and may be solved in exactly the same manner. But in this case it is necessary to restrict v to the values $0, 1, 2, \cdots$ in order to obtain a polynomial solution. If we solve for E we obtain the equation

$$E_{l,v} = D + c_0 - \frac{(D - \frac{1}{2}c_1)^2}{D + c_2}$$

$$+ \frac{2a\hbar}{\sqrt{2m}}\frac{(D - \frac{1}{2}c_1)}{\sqrt{D + c_2}}\left(v + \frac{1}{2}\right) - \frac{a^2\hbar^2}{2m}\left(v + \frac{1}{2}\right)^2$$

By expanding in terms of powers of $\frac{c_1}{D}$ and $\frac{c_2}{D}$, this relation may be brought into the form usually employed in the study of observed spectra; namely,

$$\tilde{\nu} = \frac{E_{l,v}}{hc} = \tilde{\nu}_e\left(v + \frac{1}{2}\right) - x_e\tilde{\nu}_e\left(v + \frac{1}{2}\right)^2$$

$$+ l(l + 1)B_e + D_e l^2(l + 1)^2 - \alpha_e\left(v + \frac{1}{2}\right)l(l + 1)$$

where

$$\tilde{\nu}_e = \frac{a}{2\pi c}\sqrt{\frac{2D}{m}} \quad , \quad x_e = \frac{\hbar\tilde{\nu}_e\pi c}{2D} \quad , \quad B_e = \frac{\hbar}{4\pi I_e c}$$

$$D_e = -\frac{\hbar^3}{16\pi^3 m^3 \tilde{\nu}_e^2 c^3 r_e^6} \quad , \quad \alpha_e = \frac{3\hbar^2 \tilde{\nu}_e}{4\pi m r_e^2 D}\left(\frac{1}{a r_e} - \frac{1}{a^2 r_e^2}\right)$$

c is speed of light.

When this potential models the vibrational motion of a diatomic molecule then the parameters in the potential and energy expression and the terms in the energy expressions have the following meaning:

D and a are empirical constants and r_e is the value of r for which $V(r)$ is a minimum, or the equilibrium interatomic distance. The constant D is usually assumed to be the spectroscopic dissociation energy of the molecule. The quantity x_e is an anharmonicity constant depending upon the exact shape of the potential, and α_e is a vibration–rotation coupling constant related to the change in the moment of inertia with vibrational states. The first two terms in $\tilde{\nu}$ are just those of an anharmonic oscillator. The third and fourth terms in $\tilde{\nu}$ represent the energy of a nonrigid rotator in terms of centrifugal stretching. The last term in $\tilde{\nu}$ represents the vibrational–rotational coupling. v is called the vibrational quantum number and l is called the rotational quantum number.

9.8 Problems

Problem - 9.1 : The radial equation for hydrogen–like atoms can be written in the form:

$$-\frac{d^2 P}{dr^2} + \left[\frac{l(l+1)}{r^2} - \frac{2Z}{r}\right]P = EP \quad ; \quad P = rR$$

(a) Verify that $P = r^{-l}$, r^{l+1} are solutions to this equation in the limit that $r \ll 1$. (b) Verify that $P = e^{\pm\sqrt{-E}r}$ are solutions to this equation in the limit that $r \gg 1$.

Problem - 9.2 : A particle of mass μ is confined to a cylindrical potential box. The potential is zero for $0 \le z \le H$ and $x^2 + y^2 \le \rho^2$; otherwise it is infinite. (a) Separate the wave equation for this system into three equations which depend separately on the cylindrical coordinates r, ϕ, and z. (b) Obtain equations for the eigenfunctions on the ϕ, z, and r equations of part (a). (c) Obtain an equation for the allowed energies of this system in terms of the separation constants, and explicitly write out the energy equation for the ground state.

Problem - 9.3 : Assume that the solution for the radial equation

$$-\frac{d^2 P}{dr^2} + \left[\frac{l(l+1)}{r^2} - \frac{2Z}{r}\right]P = EP \quad , \quad P = rR$$

has the form

$$P(r) = r^{l+1}e^{-\sqrt{-E}r}f(r) \quad (1) \quad \text{where} \quad f(r) = \sum_{k=0}^{\infty}A_k r^k$$

Derive a recurrence relation for A_k.

Problem - 9.4 : Consider a two–dimensional hydrogen atom in which an electron is bound to the nucleus by a Coulombic force and is constrained to move in a plane. Determine the eigenfunctions and eigenvalues for this system.

Problem - 9.5 : Determine the energy levels of a particle moving in a centrally symmetric field with potential energy

$$V(r) = \frac{A}{r^2} - \frac{B}{r}$$

where A and B are positive constants.

Problem - 9.6 : An electron in the Coulomb field of a proton is in a state described by the wave function

$$\psi(\mathbf{r}) = N[4\psi_{100}(\mathbf{r}) + 3\psi_{211}(\mathbf{r}) - \psi_{210}(\mathbf{r}) + \sqrt{10}\psi_{21-1}(\mathbf{r})]$$

(a) What is the normalization constant N? (b) What is the expectation value of energy? (c) What is the expectation value of L^2? (d) What is the expectation value of L_z?

Problem - 9.7 : Determine the average kinetic and potential energy for the electron in the $1s$, $2s$, $2p$, and $3p$ states by using the exact hydrogenic orbitals.

Problem - 9.8 : Prove that in any single–particle bound energy eigenstate the following relation is satisfied in nonrelativistic quantum mechanics for a central field of force potential $V(r)$,

$$|\psi(0)|^2 = \frac{m}{2\pi} \left\langle \frac{dV(r)}{dr} \right\rangle - \frac{1}{2\pi} \left\langle \frac{L^2}{r^3} \right\rangle$$

where $\psi(0)$ is the wave function at the origin, m is the particle mass, and L^2 is the square of the orbital angular momentum operator (take $|hbar = 1$). Give a classical interpretation of this equation for the case of a state with angular momentum $\neq 0$.

Chapter 10

Interaction of Electrons with Electromagnetic Field

10.1 Maxwell's equations and gauge transformations

Hydrogen atom may be modeled as the interaction of an electron with the static Coulomb field due to a point charge. One may generalize this to the interaction with an external magnetic or electric field. Before going into the details it is better to review the classical theory of electromagnetism.

Maxwell's equations in Gaussian units, in the vacuum:

$$\nabla \cdot \mathbf{B}(\mathbf{r}, t) = 0 \tag{10.1}$$

$$\nabla \cdot \mathbf{E}(\mathbf{r}, t) = 4\pi\rho(\mathbf{r}, t) \tag{10.2}$$

$$\nabla \times \mathbf{E}(\mathbf{r}, t) + \frac{1}{c}\frac{\partial \mathbf{B}(\mathbf{r}, t)}{\partial t} = 0 \tag{10.3}$$

$$\nabla \times \mathbf{B}(\mathbf{r}, t) - \frac{1}{c}\frac{\partial \mathbf{E}(\mathbf{r}, t)}{\partial t} = \frac{4\pi}{c}\mathbf{j}(\mathbf{r}, t) \tag{10.4}$$

where $\rho(\mathbf{r}, t)$ and $\mathbf{j}(\mathbf{r}, t)$ are the charge and current densities that are the sources of the electromagnetic fields $\mathbf{E}(\mathbf{r}, t)$ and $\mathbf{B}(\mathbf{r}, t)$. The conservation of charge equation

$$\frac{\partial \rho(\mathbf{r}, t)}{\partial t} + \nabla \cdot \mathbf{j}(\mathbf{r}, t) = 0 \tag{10.5}$$

is automatically satisfied. The Eqs. (10.1) and (10.3) may be satisfied by expressing the fields in terms of a scalar potential $\phi(\mathbf{r}, t)$ and a vector potential $\mathbf{A}(\mathbf{r}, t)$

$$\mathbf{B}(\mathbf{r}, t) = \nabla \times \mathbf{A}(\mathbf{r}, t) \tag{10.6}$$

$$\mathbf{E}(\mathbf{r}, t) = -\frac{1}{c}\frac{\partial \mathbf{A}(\mathbf{r}, t)}{\partial t} - \nabla\phi(\mathbf{r}, t) \tag{10.7}$$

The fields $\mathbf{E}$ and $\mathbf{B}$ do not determine ϕ and $\mathbf{A}$ uniquely. New potentials, given by

$$\mathbf{A}'(\mathbf{r}, t) = \mathbf{A}(\mathbf{r}, t) - \nabla f(\mathbf{r}, t) \tag{10.8}$$

$$\phi'(\mathbf{r}, t) = \phi(\mathbf{r}, t) + \frac{1}{c}\frac{\partial f(\mathbf{r}, t)}{\partial t} \tag{10.9}$$

also satisfy the same $\mathbf{E}$ and $\mathbf{B}$ fields. The transformation from the set $(\mathbf{A}, \phi)$ to $(\mathbf{A}', \phi')$ is known as a **gauge transformation**.

Some definitions:

Position probability density:

$$P(\mathbf{r}, t) = \psi^*(\mathbf{r}, t)\psi(\mathbf{r}, t) = |\psi(\mathbf{r}, t)|^2 \qquad (10.10)$$

Probability current density:

$$\mathbf{S}(\mathbf{r}, t) = \frac{\hbar}{i2m}[\psi^*\nabla\psi - (\nabla\psi^*)\psi] \qquad (10.11)$$

Velocity operator:

$$\frac{\hbar}{im}\nabla$$

Continuity equation:

$$\frac{\partial P(\mathbf{r}, t)}{\partial t} + \nabla \cdot \mathbf{S}(\mathbf{r}, t) = 0 \qquad (10.12)$$

Charge density:

$$\rho(\mathbf{r}, t) = eP(\mathbf{r}, t) \qquad (10.13)$$

Current density:

$$\mathbf{j}(\mathbf{r}, t) = e\mathbf{S}(\mathbf{r}, t) \qquad (10.14)$$

Conservation of charge equation:

$$\frac{\partial \rho(\mathbf{r}, t)}{\partial t} + \nabla \cdot \mathbf{j}(\mathbf{r}, t) = 0 \qquad (10.15)$$

With substitutions

$$i\hbar\frac{\partial\psi}{\partial t} \quad \rightarrow \quad i\hbar\frac{\partial\psi}{\partial t} - e\phi\psi \qquad (10.16)$$

and

$$-i\hbar\nabla\psi \quad \rightarrow \quad -i\hbar\nabla\psi - \frac{e}{c}\mathbf{A}\psi \qquad (10.17)$$

The current density becomes

$$\mathbf{j}(\mathbf{r}, t) = \frac{e\hbar}{i2m}[\psi^*\nabla\psi - (\nabla\psi^*)\psi] - \frac{e^2}{mc}\mathbf{A}\psi^*\psi \qquad (10.18)$$

The invariance of $\mathbf{E}$ and $\mathbf{B}$ allows us to choose the arbitrary function $f(\mathbf{r}, t)$ in the most convenient way. The source–dependent equations, Eqs.(10.2) and (10.4), now read

$$-\nabla^2\phi(\mathbf{r}, t) - \frac{1}{c}\frac{\partial}{\partial t}[\nabla \cdot \mathbf{A}(\mathbf{r}, t)] = 4\pi\rho(\mathbf{r}, t) \qquad (10.19)$$

and

$$\nabla \times [\nabla \times \mathbf{A}(\mathbf{r}, t)] + \frac{1}{c^2}\frac{\partial^2\mathbf{A}(\mathbf{r}, t)}{\partial t^2} + \frac{1}{c}\frac{\partial}{\partial t}(\nabla\phi) = \frac{4\pi}{c}\mathbf{j}(\mathbf{r}, t) \qquad (10.20)$$

which may be rewritten as

$$-\nabla^2\mathbf{A}(\mathbf{r}, t) + \frac{1}{c^2}\frac{\partial^2\mathbf{A}(\mathbf{r}, t)}{\partial t^2} \qquad (10.21)$$

$$+\nabla\left(\nabla \cdot \mathbf{A}(\mathbf{r}, t) + \frac{1}{c}\frac{\partial\phi(\mathbf{r}, t)}{\partial t}\right) = \frac{4\pi}{c}\mathbf{j}(\mathbf{r}, t)$$

Here the following relation is used,

$$\nabla \times (\nabla \times \mathbf{A}) = -\nabla^2 \mathbf{A} + \nabla(\nabla \cdot \mathbf{A})$$

If the charge distribution is static, that is, $\rho(\mathbf{r})$ is independent of time, it is convenient to choose the gauge such that

$$\nabla \cdot \mathbf{A}(\mathbf{r}, t) = 0 \qquad (10.22)$$

This choice of $f(\mathbf{r}, t)$ is given the name of **Coulomb gauge**. In this case we have

$$-\nabla^2 \phi(\mathbf{r}) = 4\pi\rho(\mathbf{r}) \qquad (10.23)$$

that is, we have a time–independent scalar potential, and then the equation for $\mathbf{A}(\mathbf{r}, t)$ reads

$$-\nabla^2 \mathbf{A}(\mathbf{r}, t) + \frac{1}{c^2} \frac{\partial^2 \mathbf{A}(\mathbf{r}, t)}{\partial t^2} = \frac{4\pi}{c} \mathbf{j}(\mathbf{r}, t) \qquad (10.24)$$

When the charge distribution is not static, it is more convenient to choose the so–called **Lorentz gauge** for which

$$\nabla \cdot \mathbf{A}(\mathbf{r}, t) + \frac{1}{c} \frac{\partial \phi(\mathbf{r}, t)}{\partial t} = 0 \qquad (10.25)$$

This leaves the equation for the vector potential unaltered, but now the scalar equation also obeys a wave equation.

10.2 Motion of a free electron in a uniform magnetic field

The equation describing the interaction of a point electron of mass μ with an electromagnetic field is the classical Lorentz force equation

$$\mu \frac{d^2 \mathbf{r}}{dt^2} = -e \left[\mathbf{E}(\mathbf{r}, t) + \frac{1}{c} \mathbf{v} \times \mathbf{B}(\mathbf{r}, t) \right] \qquad (10.26)$$

This equation can be obtained from classical Hamiltonian

$$H_0 = \frac{p^2}{2\mu} \qquad (10.27)$$

by making the alteration

$$\mathbf{p} \;\rightarrow\; \mathbf{p} + \frac{e}{c} \mathbf{A}(\mathbf{r}, t) \qquad (10.28)$$

and adding the static scalar potential $e\phi(\mathbf{r})$, so that

$$H = \frac{1}{2\mu} \left[\mathbf{p} + \frac{e}{c} \mathbf{A}(\mathbf{r}, t) \right]^2 + e\phi(\mathbf{r}) \qquad (10.29)$$

The corresponding Schrödinger equation may be written as

$$\frac{1}{2\mu} \left(-i\hbar\nabla + \frac{e}{c} \mathbf{A}(\mathbf{r}, t) \right)^2 \psi(\mathbf{r}, t) = [E - e\phi(\mathbf{r})] \psi(\mathbf{r}, t) \qquad (10.30)$$

The left side is

$$\frac{1}{2\mu}\left(-i\hbar\nabla + \frac{e}{c}\mathbf{A}\right)\cdot\left(-i\hbar\nabla + \frac{e}{c}\mathbf{A}\right)\psi \tag{10.31}$$

$$= \frac{1}{2\mu}\left(-\hbar^2\nabla^2 - i\hbar\frac{e}{c}\nabla\cdot\mathbf{A} - i\hbar\frac{e}{c}\mathbf{A}\cdot\nabla + \frac{e^2}{c^2}A^2\right)\psi$$

$$= \frac{1}{2\mu}\left(-\hbar^2\nabla^2 - 2i\hbar\frac{e}{c}\mathbf{A}\cdot\nabla + \frac{e^2}{c^2}A^2\right)\psi$$

For a constant magnetic field, **B**, we may take

$$\mathbf{A} = -\frac{1}{2}\mathbf{r}\times\mathbf{B} \tag{10.32}$$

This assumption of **A** gives

$$\nabla\times\mathbf{A} = \mathbf{B} \tag{10.33}$$

Hence the second term in Eq.(10.31) becomes

$$i\hbar\frac{e}{2c\mu}\mathbf{r}\times\mathbf{B}\cdot\nabla\psi = -i\hbar\frac{e}{2c\mu}\mathbf{B}\cdot\mathbf{r}\times\nabla\psi \tag{10.34}$$

$$= \frac{e}{2c\mu}\mathbf{B}\cdot\mathbf{r}\times\frac{\hbar}{i}\nabla\psi = \frac{e}{2c\mu}\mathbf{B}\cdot\mathbf{L}\psi$$

and the third term is

$$\frac{e^2}{8\mu c^2}(\mathbf{r}\times\mathbf{B})^2\psi = \frac{e^2}{8\mu c^2}\left[r^2 B^2 - (\mathbf{r}\cdot\mathbf{B})^2\right]\psi \tag{10.35}$$

$$= \frac{e^2 B^2}{8\mu c^2}(x^2 + y^2)\psi$$

if **B** is the direction that defines the z–axis. This is of the form of a two–dimensional harmonic ocsillator potential. The Schrödinger equation may now be written as (for **B**//z–axis)

$$\left(-\frac{\hbar^2}{2\mu}\nabla^2 + \frac{e}{2\mu c}L_z B + \frac{e^2(x^2 + y^2)}{8\mu c^2}B^2\right)\psi \tag{10.36}$$

$$= (E - e\phi)\psi$$

The possible solutions for this equation under certain conditions:

a) B^2 term is negligible and $\phi = 0$ case:

The Schrödinger equation takes the form (which describes the helical motion of a free electron moving along z direction)

$$\left(-\frac{\hbar^2}{2\mu}\nabla^2 + \frac{eBL_z}{2\mu c}\right)\psi = E\psi \tag{10.37}$$

Let

$$\frac{eBL_z}{2\mu c} = H_1 \tag{10.38}$$

The Hamiltonian with $B = 0$ is altered by the addition of H_1. If we define the frequency, called the **Larmor frequency**,

$$\frac{eB}{2\mu c} = \omega_L \tag{10.39}$$

and considering that the energy eigenstates are simultaneously eigenstates of L^2 and L_z, then one can write

$$H_1 \psi_{lm} = \omega_L m \hbar \psi_{lm} \quad ; \quad m = 0, \mp 1, \mp 2, \cdots \tag{10.40}$$

The solution for energy may now be expressed as

$$E = \frac{\hbar^2 k^2}{2\mu} + \omega_L \hbar m \tag{10.41}$$

b) Only $\phi = 0$ case:

The Schrödinger equation now takes the form

$$\left[-\frac{\hbar^2}{2\mu} \nabla^2 + \frac{eB}{2\mu c} L_z + \frac{e^2 B^2}{8\mu c^2} (x^2 + y^2) \right] \psi = E\psi \tag{10.42}$$

The presence of the potential $(x^2 + y^2)$ suggests the use of cylindrical coordinates for the separation of the variables. Writing $x = \rho \cos\phi$, $y = \rho \sin\phi$ the Laplacian in cylindrical coordinates takes the form

$$\nabla^2 = \frac{\partial^2}{\partial z^2} + \frac{\partial^2}{\partial \rho^2} + \frac{1}{\rho} \frac{\partial}{\partial \rho} + \frac{1}{\rho^2} \frac{\partial^2}{\partial \phi^2} \tag{10.43}$$

If we now write

$$\psi(\mathbf{r}) = u_m(\rho) e^{im\phi} e^{ikz} \tag{10.44}$$

we find that the differential equation satisfied by $u_m(\rho)$ is

$$\frac{d^2 u}{d\rho^2} + \frac{1}{\rho} \frac{du}{d\rho} - \frac{m^2}{\rho^2} u - \frac{e^2 B^2}{4\hbar^2 c^2} \rho^2 u$$

$$+ \left(\frac{2\mu E}{\hbar^2} - \frac{eBm}{\hbar c} - k^2 \right) u = 0 \tag{10.45}$$

If we introduce the variable

$$x^2 = \frac{eB}{2\hbar c} \rho^2 \tag{10.46}$$

we can rewrite the equation in the form

$$\frac{d^2 u}{dx^2} + \frac{1}{x} \frac{du}{dx} - \frac{m^2}{x^2} u - x^2 u + \lambda u = 0 \tag{10.47}$$

where

$$\lambda = \frac{4\mu c}{eB\hbar} \left(E - \frac{\hbar^2 k^2}{2\mu} \right) - 2m \quad ; \quad m = 0, \mp 1, \mp 2, \cdots \tag{10.48}$$

The behaviour of $u(x)$ for large x (asymptotic case), at infinity, determined from

$$\frac{d^2 u}{dx^2} - x^2 u \approx 0 \tag{10.49}$$

is
$$u(x) \approx e^{-\frac{1}{2}x^2} \tag{10.50}$$

The behaviour of $u(x)$ for small x, near $x = 0$, determined from

$$\frac{d^2u}{dx^2} + \frac{1}{x}\frac{du}{dx} - \frac{m^2}{x^2}u \approx 0 \tag{10.51}$$

is
$$u(x) \approx x^{|m|} \tag{10.52}$$

We thus write
$$u(x) = x^{|m|}e^{-\frac{1}{2}x^2}G(x) \tag{10.53}$$

and determine the differential equation obeyed by $G(x)$. After some algebra, one obtains

$$\frac{d^2G}{dx^2} + \left(\frac{2|m|+1}{x} - 2x\right)\frac{dG}{dx} + (\lambda - 2 - 2|m|)G = 0 \tag{10.54}$$

This can be brought into the same form as hydrogen–atom solution if we change variables to $y = x^2$. The equation then takes the form

$$\frac{d^2G}{dy^2} + \left(\frac{|m|+1}{y} - 1\right)\frac{dG}{dy} + \frac{\lambda - 2 - 2|m|}{4y}G = 0 \tag{10.55}$$

We must have
$$\frac{\lambda}{4} - \frac{1+|m|}{2} = n_r \tag{10.56}$$

as an eigenvalue condition, with $n_r = 0, 1, 2, \cdots$. This implies that $(E - \hbar^2 k^2/2\mu)$, the energy with the kinetic energy of the free motion in the z–direction subtracted out, is given by

$$E - \frac{\hbar^2 k^2}{2\mu} = \frac{eB\hbar}{2\mu c}(2n_r + 1 + |m| + m) \tag{10.57}$$

and
$$G(y) = L_{n_r}^{|m|}(y) \tag{10.58}$$

or
$$E = \frac{\hbar^2 k^2}{2\mu} + \omega_L\hbar(2n_r + 1 + |m| + m) \tag{10.59}$$

10.3 Motion of a bound electron in a uniform magnetic field

Let us consider a hydrogenic atom with nuclear charge Ze, a single electron bounded by the Coulomb potential, in a uniform external magnetic field $\mathbf{B}$ in the z–direction ($\mathbf{B}//z$–axis). The Hamiltonian which describes the motion of the electron is

$$H = \frac{1}{2\mu}\left[\mathbf{p} + \frac{e}{c}\mathbf{A}(\mathbf{r}, t)\right]^2 + e\phi + V(\mathbf{r}) \tag{10.60}$$

where
$$V(\mathbf{r}) = -\frac{Ze^2}{r} \tag{10.61}$$

The corresponding Schrödinger equation may be expressed (following the same procedure in Section 10.2) as (for $\mathbf{B}//z$–axis)

$$\left(-\frac{\hbar^2}{2\mu}\nabla^2 - \frac{Ze^2}{r} + \frac{eL_z}{2\mu c}B + \frac{e^2(x^2+y^2)}{8\mu c^2}B^2\right)\psi = (E - e\phi)\psi \qquad (10.62)$$

The solution for negligible B^2 term and $\phi = 0$, that is, for the Schrödinger equation

$$\left(-\frac{\hbar^2}{2\mu}\nabla^2 - \frac{Ze^2}{r} + \frac{eB}{2\mu c}L_z\right) = E\psi \qquad (10.63)$$

may be obtained for the energies

$$E = -\frac{1}{2}\mu c^2\left(\frac{Z\alpha}{n}\right)^2 + \omega\hbar m \stackrel{\text{or}}{=} -\frac{Z^2e^2}{2a_0n^2} + \omega\hbar m \qquad (10.64)$$

here m is the z–component of the angular momentum eigenvalue, with $m = 0, \mp1,$ $\mp2, \cdots, \mp l$. Therefore the existing energy levels, with their $(2l+1)$–fold degeneracy, are split into $(2l + 1)$ components that are equally spaced; the size of the splitting is $\omega_L\hbar$:

$$\omega_L\hbar = \frac{eB\hbar}{2\mu c} = \frac{e\hbar}{2\mu c}B = \mu_B B \qquad (10.65)$$

where μ_B is the **Bohr magneton**.

Since there are selection rules according to which only transitions in which the m–value changes by zero or unit are allowed ($\Delta m = 0, \mp1$), it turns out that the single line representing a transition with $B = 0$ splits into three lines. This effect is the **normal Zeeman effect**. The interactions of the electron spin with the magnetic field changes the pattern predicted above. The effect in this case is the **anomalous Zeeman effect**.

The splitting of the energy levels and the possible transitions between $3d$ and $2p$ states are shown in Figure 10.1.

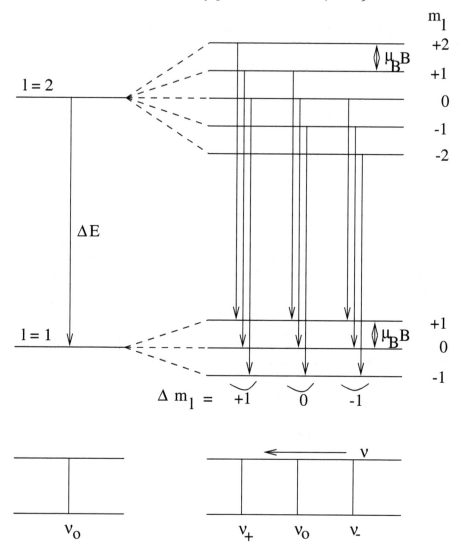

Fig. 10.1. Normal Zeeman effect for $3d \rightarrow 2p$ transitions.

$$\nu_0 = \frac{\Delta E}{h} \quad , \quad \nu_+ = \nu_0 + \frac{\mu_B B}{h} \quad , \quad \nu_- = \nu_0 - \frac{\mu_B B}{h}$$

$$\Delta E = E_{3d} - E_{2p} = E_3 - E_2 = -\frac{Z^2 e^2}{2a_0} \left(\frac{1}{3^2} - \frac{1}{2^2} \right) = \frac{5Z^2 e^2}{72a_0} = \frac{5Z^2}{72} \quad a.u.$$

for hydrogen atom $Z = 1$, $\Delta E = 5/72$ $a.u.$

10.4 The principle of gauge invariance and flux quantization

Another quantum mechanical effect connected with the interaction with a magnetic field is the principle of gauge invariance. The Schrödinger equation

$$\frac{1}{2\mu}\left(-i\hbar\nabla + \frac{e}{c}\mathbf{A}\right)^2 \psi(\mathbf{r},t) = [E - e\phi(\mathbf{r})]\psi(\mathbf{r},t) \tag{10.66}$$

appears to violate the principle of gauge invariance, since it is $\mathbf{A}(\mathbf{r},t)$ that appears in the equation, and under the transformation $\mathbf{A} \rightarrow \mathbf{A}+\nabla f(\mathbf{r},t)$ the Hamiltonian is changed according to

$$\frac{1}{2\mu}\left(-i\hbar\nabla + \frac{e}{c}\mathbf{A}\right)^2 \rightarrow \frac{1}{2\mu}\left(-i\hbar\nabla + \frac{e}{c}\mathbf{A} + \frac{e}{c}\nabla f\right)^2$$

It is possible to save gauge invariance by using the fact that a change of the wave function by a phase factor, which may depend on $\mathbf{r}$, has no physical consequences. Thus if we require that $\mathbf{A} \rightarrow \mathbf{A} + \nabla f$ must be accompanied by the transformation

$$\psi(\mathbf{r},t) \rightarrow e^{i\Lambda(\mathbf{r},t)}\psi(\mathbf{r},t) \tag{10.67}$$

then

$$\frac{1}{2\mu}\left(-i\hbar\nabla + \frac{e}{c}\mathbf{A}\right)^2 \psi(\mathbf{r},t)$$

becomes

$$\frac{1}{2\mu}\left(-i\hbar\nabla + \frac{e}{c}\mathbf{A} + \frac{e}{c}\nabla f\right)\cdot\left(-i\hbar\nabla + \frac{e}{c}\mathbf{A} + \frac{e}{c}\nabla f\right)e^{i\Lambda}\psi \tag{10.68}$$

$$= \frac{1}{2\mu}\left(-i\hbar\nabla + \frac{e}{c}\mathbf{A} + \frac{e}{c}\nabla f\right)$$

$$\cdot\left[e^{i\Lambda}\left(-i\hbar\nabla\psi + \frac{e}{c}\mathbf{A}\psi + \frac{e}{c}\nabla f\psi + \hbar\nabla\Lambda\psi\right)\right]$$

$$= \frac{1}{2\mu}e^{i\Lambda}\left(-i\hbar\nabla + \frac{e}{c}\mathbf{A} + \frac{e}{c}\nabla f + \hbar\nabla\Lambda\right)^2\psi$$

Thus with the choice $\Lambda = -(e/\hbar c)f$, that is, with the transformation law

$$\psi(\mathbf{r},t) \rightarrow \exp\left[-\frac{ie}{\hbar c}f(r,t)\right]\psi(\mathbf{r},t) \tag{10.69}$$

gauge invariance is restored.

In a field free region $(\mathbf{B} = 0 \rightarrow \nabla\times\mathbf{A} = 0, \mathbf{A}$ may be written as $\mathbf{A} = \nabla f)$, we may describe the motion of an electron in two ways: either we do not consider the presence of a field at all, and write

$$\left[\frac{1}{2\mu}(-i\hbar\nabla)^2 + V(\mathbf{r})\right]\psi = E\psi \tag{10.70}$$

for the energy eigenfunction equation, or we write the equation with the vector potential given by ($\mathbf{A} = \nabla f$)

$$\frac{1}{2\mu}\left(-i\hbar\nabla + \frac{e}{c}\mathbf{A}\right)^2 \psi' + V(\mathbf{r})\psi' = E\psi' \qquad (10.71)$$

and take

$$\psi' = e^{-(ie/\hbar c)f}\psi \qquad (10.72)$$

The function $f(\mathbf{r}, t)$ may be written in terms of $\mathbf{A}(\mathbf{r}, t)$

$$f(\mathbf{r}, t) = \int^{\mathbf{r}} \mathbf{A}(\mathbf{r}', t) \cdot d\mathbf{r}' \qquad (10.73)$$

where the path of integration is taken from an arbitrary fixed point, for example, the origin, or infinity, to the point $\mathbf{r}$. The integral only makes sense if $\mathbf{B} = 0$, that is, in a field–free region, since the difference in the integral along two different paths, labeled 1 and 2, is

$$\int_{(1)} \mathbf{A}(\mathbf{r}', t) \cdot d\mathbf{r}' - \int_{(2)} \mathbf{A}(\mathbf{r}', t) \cdot d\mathbf{r}' = \oint \mathbf{A}(\mathbf{r}', t) \cdot d\mathbf{r}' \qquad (10.74)$$

$$= \int_S \nabla' \times \mathbf{A}(\mathbf{r}', t) \cdot d\mathbf{S} = \int_S \mathbf{B} \cdot d\mathbf{S} = \Phi$$

where Φ is the flux of magnetic field through the surface spanned by the two paths (see Figure 10.2).

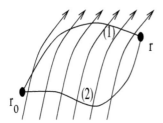

Fig. 10.2. Flux of magnetic field through the surface spanned by the two paths.

Thus only if $\Phi = 0$ will the phase factor in $\psi' = \exp\left[-(ie/\hbar c)f\right]\psi$ be independent of the choice of path in the line integral. Such an independence is required if we insist that the wave function be single–valued.

If the two paths include flux (if $\Phi \neq 0$), then the wave functions of electrons traveling along the two paths will acquire different phases; the phase factor becomes

$$e^{ie\Phi/\hbar c}$$

The requirement that the electron wave function be single–valued, so that the phase factor is unity, implies that the enclosed flux is quantized

$$\Phi = \frac{2\pi\hbar c}{e}n \quad , \quad n = 0, \mp 1, \mp 2, \cdots \qquad (10.75)$$

Such a situation arises in the motion of electrons in a superconducting ring surrounding a region containing flux (see Figure 10.3).

When a superconductor ring is placed in an external magnetic field at a temperature above the critical temperature (T_c), it acts like any other metal, and magnetic field lines (or flux lines) panetrate it. When the temperature is lowered below the T_c the ring becomes superconductor and expels magnetic flux lines and $B = 0$ inside the ring. This is the **Meissner effect**. After cooling below T_c or when the ring becomes superconductor some of these flux lines become trapped inside the ring. It is the trapped flux that is found to be quantized (see Figure 10.3).

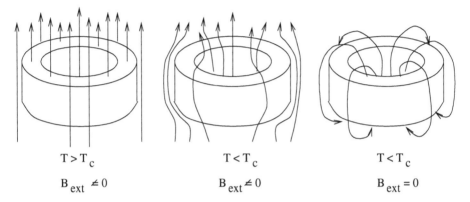

$$T > T_c$$
$$B_{ext} \neq 0$$

$$T < T_c$$
$$B_{ext} \neq 0$$

$$T < T_c$$
$$B_{ext} = 0$$

Fig. 10.3. Flux quantization and the Meissner effect.

Experimental measurement of the trapped flux shows that Eq.(10.75) holds, with the modification that

$$\Phi = \frac{2\pi\hbar c}{2e} n \tag{10.76}$$

This shows that the phenomenon of superconductivity is the result of electron pairing in the materials under certain conditions.

Another important aspect of the dependence of the wave function on the flux can be seen in an interference experiment in which a solenoid confining magnetic flux is placed between the slits in a two–slit experiment. The interference pattern at the screen is due to the superposition of two parts of the wave function

$$\psi = \psi_1 + \psi_2 \tag{10.77}$$

where ψ_1 denotes the part of the wave function that describes the electron following path 1, and ψ_2 the part appropriate to path 2. In the presence of the solenoid we have

$$\psi = \psi_1 \exp\left(\frac{ie}{\hbar c} \int_1 d\mathbf{r} \cdot \mathbf{A}\right) + \psi_2 \exp\left(\frac{ie}{\hbar c} \int_2 d\mathbf{r} \cdot \mathbf{A}\right) \tag{10.78}$$

$$= \left[\psi_1 \exp\left(\frac{ie}{\hbar c} \Phi\right) + \psi_2\right] \exp\left(\frac{ie}{\hbar c} \int_2 d\mathbf{r} \cdot \mathbf{A}\right)$$

The flux thus causes a relative change in phase between ψ_1 and ψ_2, and this will change the interference pattern. This is the **Aharanov–Bohm effect** (see Figure 10.4).

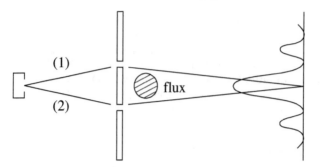

Fig. 10.4. Interference of waves and the Aharanov–Bohm effect.

Remark on Gauge Transformations:

In classical electromagnetic theory, a gauge transformation on the scalar potential $\phi(\mathbf{r},t)$ and the vector potential $\mathbf{A}(\mathbf{r},t)$ is defined by the relations

$$\phi'(\mathbf{r},t) = \phi(\mathbf{r},t) - \frac{\partial}{\partial t} f(\mathbf{r},t) \tag{10.79}$$

$$\mathbf{A}'(\mathbf{r},t) = \mathbf{A}(\mathbf{r},t) + \nabla f(\mathbf{r},t)$$

where $f(\mathbf{r},t)$ is an arbitrary differentiable scalar function of space and time. Since the fields are related to the potentials by

$$\mathbf{E}(\mathbf{r},t) = -\nabla\phi(\mathbf{r},t) - \frac{\partial}{\partial t}\mathbf{A}(\mathbf{r},t) \tag{10.80}$$

$$\mathbf{B}(\mathbf{r},t) = \nabla \times \mathbf{A}(\mathbf{r},t)$$

a gauge transformation of the potentials has no effect on the fields; that is, when ϕ and $\mathbf{A}$ are replaced by ϕ' and $\mathbf{A}'$, the fields are unchanged. Hence, the potentials are not determined uniquely.

It now will be shown that the transformations (Eq.10.79) also may arise in a quantum mechanical context. The fundamental Hamiltonian for the interaction of a nonrelativistic electron with an external electromagnetic field is

$$H(\mathbf{r},t) = \frac{1}{2m}[\mathbf{p} + e\mathbf{A}(\mathbf{r},t)]^2 - e\phi(\mathbf{r},t) \tag{10.81}$$

The unitary transformation

$$\psi'(\mathbf{r},t) = O\psi(\mathbf{r},t) \tag{10.82}$$

transforms the time–dependent Schrödinger equation

$$i\hbar\frac{\partial}{\partial t}\psi(\mathbf{r},t) = H\psi(\mathbf{r},t) \tag{10.83}$$

into

$$i\hbar\frac{\partial}{\partial t}\psi'(\mathbf{r},t) = H'\psi'(\mathbf{r},t) \tag{10.84}$$

where

$$H' = OHO^+ + [\mathbf{p} + e(\mathbf{A} + \nabla f)]^2 \tag{10.85}$$

Now, let

$$O = e^{-\frac{i}{\hbar}ef} \quad , \quad f \equiv f(\mathbf{r}, t) \tag{10.86}$$

With $\mathbf{p} = -i\hbar\nabla$,

$$\mathbf{p}O^+ = eO^+\nabla f + O^+\mathbf{p} \tag{10.87}$$

$$(\mathbf{p} + e\mathbf{A})O^+ = O^+[\mathbf{p} + e(\mathbf{A} + \nabla f)]$$

$$(\mathbf{p} + e\mathbf{A})^2 O^+ = (\mathbf{p} + e\mathbf{A})O^+[\mathbf{p} + e(\mathbf{A} + \nabla f)] \tag{10.88}$$

$$= O^+[\mathbf{p} + e(\mathbf{A} + \nabla f)]^2$$

we have

$$O(\mathbf{p} + e\mathbf{A})^2 O^+ = [\mathbf{p} + e(\mathbf{A} + \nabla f)]^2 \tag{10.89}$$

$$O\phi O^+ = \phi \quad , \quad [\mathbf{p} + e(\mathbf{A} + \nabla f)]^2 = e\frac{\partial f}{\partial t} \tag{10.90}$$

The Hamiltonian, Eq.(10.81), is transformed then into

$$H' = \frac{1}{2m}[\mathbf{p} + e(\mathbf{A} + \nabla f)]^2 - e(\phi - \frac{\partial f}{\partial t}) \tag{10.91}$$

$$= \frac{1}{2m}[\mathbf{p} + e\mathbf{A}']^2 - e\phi'$$

What has been shown is that the operator O defined by Eq.(10.86) acting through the transformations Eq.(10.85) generates a gauge transformation of the type of Eq.(10.79) when the Hamiltonian has the form of Eq.(10.81). This, then, is the quantum mechanical counterpart to the classical gauge transformation. Since O is a unitary operator, its effect on the wave function ψ is to alter the phase but not the absolute value. Consequently, the physical content of quantum mechanics cannot depend on the specific choice of the gauge function $f(\mathbf{r}, t)$.

Let us now consider the case of an electron interacting with an electromagnetic field that has no scalar potential ($\phi = 0$).

The field then, is, described completely by the vector potential alone, as in the case of a radiation field. The Hamiltonian, Eq.(10.81), reduces to

$$H = \frac{1}{2m}[\mathbf{p} + e\mathbf{A}(\mathbf{r}, t)]^2 \tag{10.92}$$

and the transformed Hamiltonian, Eq.(10.91), reduces to

$$H' = \frac{1}{2m}[\mathbf{p} + e(\mathbf{A} + \nabla f)]^2 + e\frac{\partial f}{\partial t} \tag{10.93}$$

We now choose a gauge function in the form

$$f(\mathbf{r}, t) = -\mathbf{A}(t) \cdot \mathbf{r} \tag{10.94}$$

in which the vector potential is independent of spatial coordinates. If we are interested in atomic systems, it is sufficient for $\mathbf{A}$ (or the fields $\mathbf{E}$ and $\mathbf{B}$) not to vary over a distance of a few Angstrom. In other words, the wavelengths associated with $\mathbf{A}$ must be long in comparison with atomic dimensions, or,

$$\mathbf{k}(t) \cdot \mathbf{r} \ll 1 \tag{10.95}$$

In the context of a multipole expansion, this condition is recognized as the approximation whereby all multipoles except the leading one — the dipole term — are ignored. Specifically, it is the electric dipole term that is of primary interest.

With the gauge function, Eq.(10.94), we have

$$\nabla f = -\mathbf{A}(t) \quad , \quad \frac{\partial f}{\partial t} = -\mathbf{r} \cdot \frac{\partial \mathbf{A}}{\partial t} = \mathbf{r} \cdot \mathbf{E}(t) \tag{10.96}$$

and

$$H' = \frac{p^2}{2m} + e\mathbf{r} \cdot \mathbf{E}(t) \tag{10.97}$$

Thus, the effect of the gauge transformation with the gauge function, Eq.(10.94), has been to eliminate the vector potential from the Hamiltonian.

The Hamiltonian, Eq.(10.92), contains the term $e(\mathbf{p} \cdot \mathbf{A} + \mathbf{A} \cdot \mathbf{p})$. If $g(\mathbf{r}, t)$ is an arbitrary function of position and time,

$$\mathbf{p} \cdot \mathbf{A}g(\mathbf{r}, t) = \mathbf{A} \cdot [\mathbf{p}g(\mathbf{r}, t) + g(\mathbf{r}, t)[\mathbf{p} \cdot \mathbf{A}] \tag{10.98}$$

If we are interested in radiation fields, the vector potential is subject to the Coulomb gauge, $\nabla \cdot \mathbf{A} = 0$. Therefore,

$$(\mathbf{p} \cdot \mathbf{A}) = -i\hbar \nabla \cdot \mathbf{A} = 0 \tag{10.99}$$

and

$$\mathbf{p} \cdot \mathbf{A} = \mathbf{A} \cdot \mathbf{p} \tag{10.100}$$

We see, then, that the interaction term in H is

$$\frac{e}{m}\mathbf{p} \cdot \mathbf{A} + \left(\frac{e}{2m}\right)^2 A^2 \tag{10.101}$$

On the other hand, the interaction term in the gauge–transformed Hamiltonian, Eq.(10.97), is

$$e\mathbf{r} \cdot \mathbf{E}(t) = -\mathbf{d} \cdot \mathbf{E}(t) \tag{10.102}$$

where $\mathbf{d} = -e\mathbf{r}$ is the electronic dipole moment operator. Thus, within the dipole approximation, the single term, Eq.(10.102), is equivalent to the two terms, Eq.(10.101).

10.5   Worked examples

Example - 10.1 : Show that the Hamiltonian operator for a particle of mass m and charge e in an electromagnetic field is

$$H = -\frac{\hbar^2}{2m}\nabla^2 + \frac{i\hbar e}{2m}(2\mathbf{A} \cdot \nabla) + \frac{e^2 A^2}{2m} + e\phi + V(\mathbf{r})$$

Assume that the gauge can be chosen so that $\nabla \cdot \mathbf{A} = 0$ and $\phi = 0$ for a radiation field.

Solution : Substitution of $p_x = -i\hbar \frac{\partial}{\partial x}$ into the classical Hamiltonian

$$H = \frac{1}{2m}[(p_x - eA_x)^2 + (p_y - eA_y)^2 + (p_z - eA_z)^2] + e\phi$$

gives the desired result. First we find the components

$$(p_x - eA_x)^2 f = \left(-i\hbar\frac{\partial}{\partial x} - eA_x\right)^2 f$$

$$= \left(i\hbar\frac{\partial}{\partial x} + eA_x\right)\left(i\hbar\frac{\partial}{\partial x} + eA_x\right) f$$

$$= \left(i\hbar\frac{\partial}{\partial x} + eA_x\right)\left(i\hbar f' + eA_x f\right)$$

$$= -\hbar^2\frac{\partial^2}{\partial x^2} f + i\hbar e(A_x f' + f A'_x) + i\hbar e A_x f' + e^2 A_x^2 f$$

$$= \left(-\hbar^2\frac{\partial^2}{\partial x^2} + 2i\hbar e A_x\frac{\partial}{\partial x} + i\hbar e\frac{\partial A_x}{\partial x} + e^2 A_x^2\right) f$$

Similarly equations result for the y and z components so that we have

$$H = \frac{1}{2m}\left\{-\hbar^2\nabla^2 + i\hbar e[(2\mathbf{A} \cdot \nabla) + (\nabla \cdot \mathbf{A})] + e^2 A^2\right\} + e\phi$$

Then choosing the gauge so that $\nabla \cdot \mathbf{A} = 0$ and $\phi = 0$ gives

$$H = \frac{1}{2m}\left[-\hbar^2\nabla^2 + 2ie\hbar\mathbf{A} \cdot \nabla + e^2 A^2\right]$$

and for a many–particle system in a radiation field we have

$$H = \sum_j \frac{1}{2m_j}\left[-\hbar^2\nabla_j^2 + 2ie_j\hbar\mathbf{A}_j \cdot \nabla_j + e_j^2 A_j^2\right] + V(\mathbf{r}_1, \cdots, \mathbf{r}_n)$$

where V is the particle–particle potential energy function.

Example - 10.2 : In spectroscopy the interaction between an electron (neglecting spin) and the radiation field can often be described by the perturbing Hamiltonian:

$$H'(t) = -i\hbar\frac{e}{m}\mathbf{A} \cdot \nabla = \frac{e}{m}\mathbf{A} \cdot \mathbf{p} \quad (1)$$

This equation is appropriate for weak radiation fields where the term $e^2 A^2/2m$ can be neglected. (a) Show that

$$< f|p_x|i >= imw_{fi} < f|x|i > \quad \text{where} \quad w_{fi} = \frac{E_f - E_i}{\hbar}$$

u_f and u_i are eigenfunctions of the unperturbed Hamiltonian. (b) Assume that the vector potential for the radiation field can be expressed as $\mathbf{A}(t) = 2\mathbf{A}_0\cos(wt)$, where $w \sim w_{fi}$ and show that

$$| < f|\frac{e}{m}\mathbf{A}_0 \cdot \mathbf{p}|i > |^2 \approx | < f| - \vec{\mu} \cdot \vec{\mathcal{E}}_0|i > |^2$$

where $\vec{\mathcal{E}}_0$ gives the amplitude and direction of the electric field associated with the radiation. This justifies the use of the classical energy expression for a dipole in an electric field as the perturbing Hamiltonian when the dependence of $\mathbf{A}$ on $\mathbf{r}$ can be neglected, i.e., when $\cos(wt - \mathbf{k} \cdot \mathbf{r}) \sim \cos(wt)$. This is called the long wave approximation. Note that $\vec{\mathcal{E}} = -\partial\mathbf{A}/\partial t$ when $\phi = 0$ and energy $W = \varepsilon_0 < \mathcal{E}^2 >$.

Solution : (a) We have

$$\frac{\partial R}{\partial t} = \frac{i}{\hbar}[H, R] \quad \text{or} \quad <f|\frac{\partial R}{\partial t}|i> = \frac{i}{\hbar} <f|[H, R]|i>$$

Therefore

$$<f|p_x|i> = m<f|\dot{x}|i> = m\frac{i}{\hbar} <f|Hx - xH|i>$$

but

$$<f|Hx|i> = \sum_j <f|H|j><j|x|i> = E_f <f|x|i>$$

and

$$<f|xH|i> = E_i <f|x|i>$$

Therefore

$$<f|p_x|i> = m\frac{i}{\hbar}(E_f - E_i) <f|x|i> = imw_{fi} <f|x|i> \quad (2)$$

(b) Using (1) and (2) we obtain

$$<f|\frac{e}{m}\mathbf{A_0}\cdot\mathbf{p}|i> = iw_{fi}e\mathbf{A_0}\cdot <f|x\mathbf{i} + y\mathbf{j} + z\mathbf{k}|i>$$

$$= iw_{fi}\mathbf{A_0}\cdot <f|\vec{\mu}|i>$$

We take $\mathbf{A_0} = A_0\mathbf{k}$ for convenience and write

$$|<f|\frac{e}{m}\mathbf{A_0}\cdot\mathbf{p}|i>|^2 = w_{fi}^2 A_0^2| <f|\mu_z|i>|^2$$

but

$$\vec{\mathcal{E}} = -\frac{\partial\mathbf{A}}{\partial t} = 2w\mathbf{A_0}\sin(wt) = 2\vec{\mathcal{E}_0}\sin(wt)$$

and

$$W = \varepsilon_0 <\mathcal{E}^2> = 4\varepsilon_0 w^2 A_0^2 <\sin^2(wt)> = 2\varepsilon_0 w^2 A_0^2$$

Therefore, when $w \sim w_{fi}$

$$w_{fi}^2 \frac{W}{2\varepsilon_0 w^2}| <f|\mu_z|i>|^2 \cong \frac{W}{2\varepsilon_0}| <f|\mu_z|i>|^2$$

Starting with $|<f|-\vec{\mu}\cdot\vec{\mathcal{E}_0}|i>|^2$ and choosing $\vec{\mathcal{E}_0}$ to lie along the z–direction:

$$\mathcal{E}_0^2| <f|\mu_z|i>|^2 = w^2 A_0^2| <f|\mu_z|i>|^2 = \frac{W}{2\varepsilon_0}| <f|\mu_z|i>|^2$$

Example - 10.3 : Show that the Schrödinger equation of a system of electrons in a homogeneous magnetic field **B** can be written as

$$i\hbar\frac{\partial\psi}{\partial t} = \left[\frac{1}{2m}\sum_k p_k^2 - \frac{e}{2mc}(\mathbf{L} + 2\mathbf{S})\cdot\mathbf{B} + \frac{e^2}{8mc^2}\sum_k(\mathbf{B}\times\mathbf{r}_k)^2 + U\right]\psi$$

where $\mathbf{L} = \sum_k \mathbf{l}_k$ is the total orbital angular momentum operator of the system, $\mathbf{S} = \sum_k \mathbf{s}_k$ is the total spin operator, and U is the potential energy of all interactions other than those with the field **B**.

Solution : The Schrödinger equation for an electron in a potential V and a magnetic field $\mathbf{B} = \nabla \times \mathbf{A}$ is

$$i\hbar \frac{\partial \psi}{\partial t} = H\psi \quad , \quad H = \frac{1}{2m}(\mathbf{p} - \frac{e}{c}\mathbf{A})^2 + V + V_{spin}$$

where V_{spin} is the operator which represents the potential energy of the interaction between the intrinsic magnetic moment of the electron and the field **B**. The wave function ψ is of course a function of the position and of the spin orientations of the electron.

By analogy with the classical expression for the potential energy of a magnetic dipole in a field **B**, the term V_{spin} is written as

$$V_{spin} = -\left(\frac{e}{mc}\mathbf{s}\right) \cdot \mathbf{B} = -\frac{e}{mc}(\mathbf{s} \cdot \mathbf{B}) \quad , \quad \mathbf{s} = \frac{\hbar}{2}\vec{\sigma}$$

The Schrödinger equation then becomes

$$i\hbar \frac{\partial \psi}{\partial t} = \frac{1}{2m}(\mathbf{p} - \frac{e}{c}\mathbf{A})^2 \psi - \frac{e}{mc}(\mathbf{s} \cdot \mathbf{B})\psi + V\psi$$

which is sometimes called the Pauli equation. Consider now the particular case of a homogeneous magnetic field **B**. The vector potential for such a field can be written in the form

$$\mathbf{A} = \frac{1}{2}(\mathbf{B} \times \mathbf{r}) \quad \text{where} \quad \nabla \cdot \mathbf{A} = 0$$

and hence

$$\mathbf{p} \cdot \mathbf{A} - \mathbf{A} \cdot \mathbf{p} = i\hbar \nabla \cdot \mathbf{A} = 0$$

Schrödinger equation then becomes

$$i\hbar \frac{\partial \psi}{\partial t} = \frac{p^2}{2m}\psi - \frac{e}{2mc}(\mathbf{B} \times \mathbf{r}) \cdot \mathbf{p}\psi - \frac{e}{mc}(\mathbf{s} \cdot \mathbf{B})\psi + \frac{e^2}{8mc^2}(\mathbf{B} \times \mathbf{r})^2 \psi + V\psi$$

But

$$(\mathbf{B} \times \mathbf{r}) \cdot \mathbf{p} = \mathbf{B} \cdot (\mathbf{r} \times \mathbf{p}) = \mathbf{B} \cdot \mathbf{l}$$

where l is the orbital angular momentum operator of the electron. Hence the Schrödinger equation becomes

$$i\hbar \frac{\partial \psi}{\partial t} = \frac{p^2}{2m}\psi - (\vec{\mu}_l \cdot \mathbf{B})\psi - (\vec{\mu}_s \cdot \mathbf{B})\psi + \frac{e^2}{8mc^2}(\mathbf{B} \times \mathbf{r})^2 \psi + V\psi$$

here $\vec{\mu}_l = (e/2mc)\mathbf{l}$ arises as an operator for the effective magnetic moment due to orbital motion (in contrast to μ_s, which is the intrinsic magnetic moment operator). Note that the ratio of the orbital magnetic moment μ_l to the mechanical angular momentum $\mathbf{l}$ is (as in classical physics) equal to $e/2mc$. For the corresponding quantities attributed to the spin, this ratio is twice as large, a fact which was considered very puzzling until explained by Dirac's theory of the electron.

For a system of electrons in a homogeneous field **B**, Pauli's equation takes the form

$$i\hbar \frac{\partial \psi}{\partial t} = \left[\frac{1}{2m}\sum_k p_k^2 - \frac{e}{2mc}(\mathbf{L} + 2\mathbf{S}) \cdot \mathbf{B} + \frac{e^2}{8mc^2}\sum_k (\mathbf{B} \times \mathbf{r}_k)^2 + U\right]\psi$$

where $\mathbf{L} = \sum_k \mathbf{l}_k$ is the total orbital angular momentum operator, $\mathbf{S} = \sum_k \mathbf{s}_k$ is the total spin operator, and U is the potential energy of all interactions other than those with the field $\mathbf{B}$.

10.6   Problems

Problem - 10.1 : The classical Hamiltonian is related to the Lagrangian by the equation

$$H(p, q, t) = \sum_i p_i \dot{q}_i - \mathcal{L}(q, \dot{q}, t)$$

where the momentum conjugate to q_i is defined by $p_i = \partial \mathcal{L}/\partial \dot{q}_i$. Use the Lagrangian

$$\mathcal{L} = \frac{1}{2}m(\mathbf{v} \cdot \mathbf{v}) + e(\mathbf{v} \cdot \mathbf{A}) - e\phi$$

to show that the Hamiltonian for a particle of mass m and charge e in a radiation field is given by

$$H = \frac{1}{2}m\left[(p_x - eA_x)^2 + (p_y - eA_y)^2 + (p_z - eA_z)^2\right] + e\phi$$

Problem - 10.2 : A hydrogen atom is placed in a time–dependent homogeneous electric field given by

$$\mathcal{E}(t) = \left(\frac{B\tau}{e\pi}\right)\frac{1}{\tau^2 + t^2}$$

where B and τ are constants. If, at $t = -\infty$, the atom is in its ground state, calculate the probability that it will be in a $2p$ state at $t = +\infty$.

Problem - 10.3 : Consider a particle of charge e and mass m in constant, crossed $\mathbf{E}$ and $\mathbf{B}$ fields: $\mathbf{E} = (0, 0, E)$, $\mathbf{B} = (0, B, 0)$, $\mathbf{r} = (x, y, z)$. (a) Write the Schrödinger equation in a convenient gauge. (b) Separate variables and reduce it to a one–dimensional problem. (c) Calculate the expectation value of the velocity in the x–direction in any energy eigenstate, sometimes called the drift velocity.

Problem - 10.4 : Consider an electron confined to the interior of a hollow cylindrical shell whose axis coincides with the z–axis. The wave function is required to vanish on the inner and outer walls, $\rho = \rho_a$ and ρ_b, and also at the top and bottom, $z = 0$ and L. (a) Find the eigensolutions, both energy and wave function. (b) Repeat the same problem when there is a uniform magnetic field $\mathbf{B} = B\mathbf{k}$ for $0 < \rho < \rho_a$. (c) Compare the ground state of the $B = 0$ problem with that of the $B \neq 0$ problem. Show that if we require the ground state energy to be unchanged in the presence of B, we obtain flux quantization

$$\pi\rho_a^2 B = \frac{2\pi N\hbar c}{e} \quad, \quad (N = 0, \mp 1, \mp 2, \cdots)$$

Problem - 10.5 : Show that free electrons cannot absorb or emit photons.

Chapter 11

Matrix Representations

11.1 Matrix representations of wave functions and operators

A complete description of the state of a dynamical system is provided by the wave function $\psi(\mathbf{r}_1, \mathbf{r}_2, \cdots)$ for the state. There are various ways of representing the wave function and hence the state. To discuss these representations let us choose a complete orthonormal set of functions $u_j(\mathbf{r}_1, \mathbf{r}_2, \cdots)$; these satisfy the equations

$$< u_j | u_k >= \int u_j^* u_k \, dr_1 dr_2 \cdots = \delta_{jk} \tag{11.1}$$

For simplicity, assume that the set $\{u_j\}$ is discrete and finite. Since the u_j form a complete set, any physically admissible wave function ψ can be expanded in terms of them:

$$\psi = \sum_j^n a_j | u_j > \tag{11.2}$$

where

$$a_j =< u_j > |\psi > \tag{11.3}$$

The set of numbers a_j constitutes a complete description of the state, since the functions u_j are assumed to be given and known. This set of numbers a_j is said to form a **representation** of the wave function ψ.

A type of equation frequently met in quantum–mechanical formalism is of the form

$$Q\psi = \psi' \tag{11.4}$$

where Q is a differential operator. Expressing both ψ and ψ' in terms of u_j, one can write

$$Q \sum_j^n a_j | u_j >= \sum_j^n a_j' | u_j > \tag{11.5}$$

Multiplying both sides by $< u_k |$ leads to

$$\sum_j^n Q_{kj} a_j = a_k' \tag{11.6}$$

where

$$Q_{kj} =< u_k | Q u_j >=< u_k | Q | u_j > \tag{11.7}$$

Q_{kj} is known as a **matrix element** of Q. It is convenient to express Eq.(11.6) in matrix notation; the elements Q_{kj} can be arranged in a square array:

$$\underline{\mathbf{Q}} \equiv \begin{bmatrix} Q_{11} & Q_{12} & \cdots & Q_{1n} \\ Q_{21} & Q_{22} & \cdots & Q_{2n} \\ \vdots & \vdots & & \vdots \\ Q_{n1} & Q_{n2} & \cdots & Q_{nn} \end{bmatrix} \tag{11.8}$$

This array is defined to be the matrix $\mathbf{Q}$.

In a similar way, the numbers a_j and a'_j can be arranged in linear arrays known as column vectors (or column matrices):

$$\mathbf{a} \equiv \begin{bmatrix} a_1 \\ a_2 \\ \vdots \\ a_n \end{bmatrix} \quad , \quad \underline{\mathbf{a}}' \equiv \begin{bmatrix} a'_1 \\ a'_2 \\ \vdots \\ a'_n \end{bmatrix} \tag{11.9}$$

In matrix notation Eq.(11.6) becomes

$$\underline{\mathbf{Q}}\,\mathbf{a} = \underline{\mathbf{a}}' \tag{11.10}$$

11.2 Matrix algebra

Algebraic properties of matrices:

If two matrices have equal dimensions, i.e., the number of rows are equal and the number of columns are equal, it is possible to define matrix addition:

$$\mathbf{R} + \mathbf{S} = \mathbf{T} \tag{11.11}$$

The addition rule is

$$R_{ij} + S_{ij} = T_{ij} \tag{11.12}$$

The general law for the multiplication of two matrices

$$\mathbf{R}\,\mathbf{S} = \mathbf{T} \tag{11.13}$$

is given by

$$\sum_k R_{ik} S_{kj} = T_{ij} \tag{11.14}$$

From this it can be seen that a necessary requirement for matrix multiplication is that the number of rows of the matrix $\mathbf{S}$ be equal to the number of columns of the matrix $\mathbf{R}$. The resultant product matrix will have the number of rows of the matrix $\mathbf{R}$ and the number of columns of the matrix $\mathbf{S}$.

$$\begin{bmatrix} \mathbf{R} \end{bmatrix}_{n \times k} \begin{bmatrix} \mathbf{S} \end{bmatrix}_{k \times m} = \begin{bmatrix} \mathbf{T} \end{bmatrix}_{n \times m} \tag{11.15}$$

The above rules for matrix addition and multiplication imply several general algebraic relations that are often taken to be postulates of matrix algebra:

i) Multiplication is associative

$$\mathbf{A}(\mathbf{BC}) = (\mathbf{AB})\mathbf{C} \tag{11.16}$$

ii) There exists a square identity matrix $\mathbf{I}$ such that

$$\mathbf{IA} = \mathbf{A} \tag{11.17}$$

It is clear that $I_{jk} = \delta_{jk}$.

iii) The distribution law holds

$$\mathbf{A}(\mathbf{B} + \mathbf{C}) = \mathbf{AB} + \mathbf{AC} \tag{11.18}$$

iv) A square matrix may have an inverse

$$\mathbf{AA}^{-1} = \mathbf{I} = \mathbf{A}^{-1}\mathbf{A} \tag{11.19}$$

If it does, it is said to be nonsingular.

v) In general, multiplication is noncommutative

$$\mathbf{AB} \neq \mathbf{BA} \tag{11.20}$$

If $\mathbf{AB} = \mathbf{BA}$, the matrices are said to commute.

11.2.1 Some definitions:

– The transpose of the matrix $\mathbf{A}$ is written $\tilde{\mathbf{A}}$ and has elements

$$\tilde{A}_{ij} \equiv A_{ji} \tag{11.21}$$

– The Hermitian adjoint of a matrix $\mathbf{A}$ is written $\mathbf{A}^+$ and has elements

$$A_{ij}^+ \equiv A_{ji}^* \tag{11.22}$$

– The Hermitian adjoint of the product of two matrices is equal to the product of their Hermitian adjoints taken in inverse order

$$(\mathbf{AB})^+ = \mathbf{B}^+\mathbf{A}^+ \tag{11.23}$$

– If a matrix is equal to its transpose, it is symmetric

$$A_{ij} = \tilde{A}_{ij} = A_{ji} \tag{11.24}$$

– A matrix is Hermitian if it is equal to its Hermitian adjoint

$$A_{ij} = A_{ij}^+ = A_{ji}^* \tag{11.25}$$

– A matrix is unitary if its inverse is equal to its Hermitian adjoint

$$A_{ij}^{-1} = A_{ij}^+ \tag{11.26}$$

– A matrix representation of a Hermitian operator is Hermitian

$$Q_{jk}^* \equiv < u_j|Qu_k >^* = < Qu_K|u_j > = < u_k|Qu_j > \equiv Q_{kj} \tag{11.27}$$

The matrix of the product of two operators is the product of the corresponding matrices. This is shown by making use of the closure relation

$$\sum_j u_j^*(\mathbf{r}_1, \mathbf{r}_2, \cdots) u_j(\mathbf{r}_1', \mathbf{r}_2', \cdots) = \delta(\mathbf{r}_1 - \mathbf{r}_1')\delta(\mathbf{r}_2 - \mathbf{r}_2') \cdots \qquad (11.28)$$

The product of two matrices is given by

$$\sum_k Q_{jk} P_{kl} = \sum_k < u_j|Qu_k >< u_k|Pu_l > \qquad (11.29)$$

$$\equiv \sum_k \left(\int u_j^* Q u_k d\mathbf{r} \right) \left(\int u_k^* P u_l d\mathbf{r}' \right)$$

From the closure relation, this is equal to

$$\sum_k Q_{jk} P_{kl} = \int u_j^* Q \delta(\mathbf{r} - \mathbf{r}') P u_l d\mathbf{r} d\mathbf{r}' \qquad (11.30)$$

$$= \int (Qu_j)^* \delta(\mathbf{r} - \mathbf{r}') P u_l d\mathbf{r} d\mathbf{r}' = \int (Qu_j)^* P u_l d\mathbf{r}$$

$$= \int u_j^* QP u_l d\mathbf{r} \equiv [\mathbf{QP}]_{jl}$$

From this, it follows that the matrices of commuting operators are commuting, and that the matrix of an operator that is inverse to Q is the matrix inverse to Q; the algebraic properties of the differential operators are mirrored in their matrices.

It is generally desirable to take the set of functions u_k used as the base of the matrix representation to be the eigenfunctions of some quantum mechanical operator. For example, the u_k may be the eigenfunction of the Hamiltonian

$$Hu_k = E_k u_k \qquad (11.31)$$

Then

$$H_{ij} \equiv < u_i|Hu_j >=< u_i|E_j u_j >= E_j \delta_{ij} \qquad (11.32)$$

In this case, $\mathbf{H}$ has nonzero elements only along the diagonal of the matrix. Such a matrix is said to be diagonal. If the orthonormal set of base functions is simultaneuously a set of eigenfunctions of several commuting operators, the matrices of all these operators are diagonal.

Remark on the matrix representation of Hamiltonian eigenvalue equation:

Let $H\psi = E\psi$ for the eigenfunction $\psi = \sum_i a_i|u_i >$

a) If the set $\{|u_i >\}$ is not orthonormal:

$$H \sum_i a_i|u_i >= E \sum_i a_i|u_i >$$

Multiplying this equation from left by $< u_j|$ gives

$$\sum_i a_i < u_j|H|u_i >= E \sum_i a_i < u_j|u_i >$$

or

$$\sum_i a_i H_{ij} = E \sum_i a_i S_{ij}$$

where the Hamiltonian matrix element is $H_{ij} =< u_i|H|u_j >$ and the overlap matrix element is $S_{ij} =< u_i|u_j >$.

Secular equations:

$$\sum_i (H_{ij} - ES_{ij})a_i = 0 \quad \rightarrow \quad (\mathbf{H} - E\underline{\mathbf{S}})\underline{\mathbf{a}} = 0$$

or

$$\begin{pmatrix} H_{11} - ES_{11} & \cdots & H_{1n} - ES_{1n} \\ \vdots & & \vdots \\ H_{n1} - ES_{n1} & \cdots & H_{nn} - ES_{nn} \end{pmatrix} \begin{pmatrix} a_1 \\ \vdots \\ a_n \end{pmatrix} = 0$$

The eigenvalues are the roots of the secular determinant:

$$\begin{vmatrix} H_{11} - ES_{11} & \cdots & H_{1n} - ES_{1n} \\ \vdots & & \vdots \\ H_{n1} - ES_{n1} & \cdots & H_{nn} - ES_{nn} \end{vmatrix} = 0$$

b) If the set $\{|u_i >\}$ is orthonormal:

In this case $S_{ij} =< u_i|u_j >= \delta_{ij}$, and the secular equations take the form

$$\sum_i (H_{ij} - E\delta_{ij})a_i = 0 \quad \rightarrow \quad \sum_i (H_{ij} - EI)a_i = 0 \quad \rightarrow \quad (\underline{\mathbf{H}} - E\underline{\mathbf{I}})\underline{\mathbf{a}} = 0$$

or

$$\begin{pmatrix} H_{11} - E & \cdots & H_{1n} \\ \vdots & & \vdots \\ H_{n1} & \cdots & H_{nn} - E \end{pmatrix} \begin{pmatrix} a_1 \\ \vdots \\ a_n \end{pmatrix} = 0$$

the secular determinant is

$$\begin{vmatrix} H_{11} - E & \cdots & H_{1n} \\ \vdots & & \vdots \\ H_{n1} & \cdots & H_{nn} - E \end{vmatrix} = 0$$

11.3   Types of matrix representations

A physical system may change from one instant of time to another. This time dependence may be variously viewed as a change in the state vectors, in the dynamical variables, or in both. In quantum mechanics, the state of a system at any given time is described by a unit vector in a Hilbert space, in which sets of axes can be defined by the eigenvectors of complete sets of observables of the system. Any change with time in the state of the system can be investigated by keeping the axes fixed and allowing the state vector to rotate, or keeping the state vector fixed and allowing the axes to rotate, or by permitting simultaneous rotation of the state vector and of the axes, using in each case the appropriate equations of motion of the vector concerned. The three possibilities described above are called the Schrödinger, the Heisenberg, and the interaction representations (or pictures) respectively. In other words, matrix representations may be classified with respect to the dependence of base functions and operators, particularly of the Hamiltonian to time explicitly.

Schrödinger Representation:

If the orthonormal set of base functions $\{u_n\}$ is time–independent, the Schrödinger equation is left unchanged in form after transforming to a matrix representation. Let the expansion coefficients be represented by $\psi_n(t)$;

$$\psi(\mathbf{r}, t) \equiv \sum_n \psi_n(t) u_n(\mathbf{r}) \tag{11.33}$$

Substituting this expression into the Schrödinger equation yields

$$\mathbf{H}\underset{\sim}{\psi} = i\hbar \frac{d}{dt} \underset{\sim}{\psi} \tag{11.34}$$

The time derivative of the matrix $\underset{\sim}{\psi}$ with elements ψ_n represents the matrix having the elements $\dot{\psi}_n$. A representation of this type, in which the base functions $\{u_n\}$ are time–independent, but the wave vector $\underset{\sim}{\psi}$ time–dependent, is known as the Schrödinger representation. In this representation operator $\mathbf{H}$ is time–independent but the wave function $\underset{\sim}{\psi}$ is time–dependent.

Heisenberg Representation:

Consider a set of functions of both $\mathbf{r}$ and t, $\{u_n(\mathbf{r}, t)\}$, that is orthonormal set at $t = 0$ and satisfies Schrödinger's equation. The set continues to be orthonormal at all times, as is seen below:

$$H u_n = i\hbar \frac{\partial u_n}{\partial t} \tag{11.35}$$

Then

$$< u_m | H u_n > = i\hbar < u_m | \frac{\partial u_n}{\partial t} > \tag{11.36}$$

Alternatively,

$$< u_n | H u_m > = i\hbar < u_n | \frac{\partial u_m}{\partial t} > \tag{11.37}$$

or

$$< H u_m | u_n > = -i\hbar < \frac{\partial u_m}{\partial t} | u_n > \tag{11.38}$$

Since H is Hermitian, Eq.(7.36) can be written as

$$< H u_m | u_n > = i\hbar < u_m | \frac{\partial u_n}{\partial t} > \tag{11.39}$$

Now we have Eqs.(11.38) and (11.39). Subtracting these equations, one gets

$$0 = i\hbar \left[< u_m | \frac{\partial u_n}{\partial t} > + < \frac{\partial u_m}{\partial t} | u_n > \right] = i\hbar \frac{d}{dt} < u_m | u_n > \qquad (11.40)$$

This implies that the orthonormality of the set of functions $\{u_n\}$ does not change with time.

Since $< u_m | u_n > = \delta_{mn}$ for all time, the functions $\{u_n\}$ can be used to obtain a matrix representation. Let

$$\psi = \sum_n \psi_n u_n(\mathbf{r}, t)$$

Each term in the sum satisfies the Schrödinger equation; the sum with constant coefficients ψ_n will therefore also satisfy this equation, and the representation of the wave function ψ, which is a solution to Schrödinger equation, is time–independent. The coefficients ψ_n, which are the representation, are time–independent.

On the other hand, an operator in this representation is usually time–dependent. Consider an operator with matrix elements $Q_{ij} = < u_i | Q | u_j >$. The time derivative of this matrix has elements

$$\dot{Q}_{ij} = < \frac{\partial u_i}{\partial t} | Q | u_j > + < u_i | \frac{\partial Q}{\partial t} | u_j > + < u_i | Q | \frac{\partial u_j}{\partial t} > \qquad (11.41)$$

From equation

$$H u_n = i\hbar \frac{\partial u_n}{\partial t} \qquad (11.42)$$

this can be written as

$$\dot{Q}_{ij} = \frac{i}{\hbar} [< H u_i | Q u_j > - < u_i | Q H u_j >] + < u_i | \frac{\partial Q}{\partial t} | u_j > \qquad (11.43)$$

$$= \frac{i}{\hbar} < u_i | (HQ - QH) | u_j > + < u_i | \frac{\partial Q}{\partial t} | u_j >$$

since H is Hermitian. Thus the matrix relation

$$\dot{\mathbf{Q}} = \frac{i}{\hbar} [\mathbf{H}, \mathbf{Q}] + \frac{\partial \mathbf{Q}}{\partial t} \qquad (11.44)$$

holds. The classical analogue of this equation is

$$\dot{Q} = \{Q, H\} + \frac{\partial Q}{\partial t} \qquad (11.45)$$

In other words, any matrix $\mathbf{Q}$ has a time dependence such that it obeys the classical equation of motion,

$$\frac{dF}{dt} = \{F, H\} + \frac{\partial F}{\partial t} \qquad (11.46)$$

This type of representation is known as the Heisenberg representation. Its time dependence connected only with the operators. Wave function is time–independent.

Interaction Representation:

Assume that the Hamiltonian can be divided into two parts,

$$H = H_0 + H_1 \tag{11.47}$$

Choose an orthonormal set of base functions which satisfy the Schrödinger equation with H_0 as the Hamiltonian:

$$H_0 u_k = i\hbar \frac{\partial u_k}{\partial t} \tag{11.48}$$

If we expand the wave function ψ in terms of the functions u_k,

$$\psi = \sum_k \psi_k u_k \tag{11.49}$$

the complete Schrödinger equation, with $H = H_0 + H_1$, becomes

$$H \sum_k \psi_k u_k = i\hbar \frac{\partial}{\partial t} \sum_k \psi_k u_k \tag{11.50}$$

Since the u_k satisfy Eq.(11.48), this becomes

$$H_1 \sum_k \psi_k u_k + i\hbar \sum_k \psi_k \frac{u_k}{\partial t} = i\hbar \frac{\partial}{\partial t} \sum_k \psi_k u_k \tag{11.51}$$

or

$$H_1 \sum_k \psi_k u_k = i\hbar \sum_k \frac{\partial \psi_k}{\partial t} u_k \tag{11.52}$$

If we multiply each side of this equation on the left by u_m^* and integrate over all space, it becomes the matrix equation

$$\mathbf{H}_1 \underline{\psi} = i\hbar \frac{\partial}{\partial t} \underline{\psi} \tag{11.53}$$

In this representation the equations of motion of any time–independent matrix operator $\mathbf{Q}$ are

$$\dot{\mathbf{Q}} = \frac{i}{\hbar}[\mathbf{H}_0, \mathbf{Q}] \tag{11.54}$$

Both wave function and operator are time–dependent. When $H_1 = 0$, this representation reduces to Heisenberg representation.

As a summary in a Schrödinger representation the base functions are time–independent, leading to time–dependent representations for wave functions. Heisenberg representations, on the other hand, have time–independent wave function representations, since the base functions are chosen to satisfy the time–dependent Schrödinger equation. In the interaction representation, the Hamiltonian is separated into two parts, one generally describing two independent systems and the other being a weak coupling term. The base functions are then chosen to be solutions of the Schrödinger equation, omitting the coupling term.

Summary of the equations of motion of the state vector $|\psi>$ and of any observable Q of the system in each of the three representations:

Representation (or picture)	Equation of motion	Wave functions and observables
Schrödinger	$i\hbar \frac{\partial}{\partial t}\|\psi(t)> = H\|\psi(t)>$	$\|\psi(t)> = U(t)\|\psi(0)>$
Heisenberg	$\frac{\partial}{\partial t}\|\psi_H> = 0$ $i\hbar \frac{dQ_H}{dt} = [Q_H, H_H]$ $+i\hbar U^+ \frac{\partial Q_H}{\partial t} U$	$\|\psi_H> = U^+(t)\|\psi(t)>$ $Q_H(t) = U^+(t)QU(t)$
Interaction	$i\hbar \frac{\partial}{\partial t}\|\psi_I(t)> = H_1\|\psi_I(t)>$ $i\hbar \frac{dQ_I}{dt} = [Q_I, H_0]$ $+i\hbar U_0^+ \frac{\partial Q_I}{\partial t} U_0$	$\|\psi_I(t)> = U_0^+(t)\|\psi(t)>$ $Q_I(t) = U_0^+(t)QU_0(t)$

11.4   Harmonic oscillator in matrix representations

The Hamiltonian operator for the harmonic oscillator is

$$\underline{\mathbf{H}} = \frac{\underline{\mathbf{p}}^2}{2m} + \frac{1}{2}k\underline{\mathbf{x}}^2 \tag{11.55}$$

We are required to find state vectors ψ_E, such that

$$H\psi_E = E\psi_E \tag{11.56}$$

the operators $\underline{\mathbf{x}}$ and $\underline{\mathbf{p}}$ being subject to the commutation rules

$$[\underline{\mathbf{x}}, \underline{\mathbf{p}}] = i\hbar \underline{\mathbf{1}} \tag{11.57}$$

For simplicity let us introduce the units $m = 1$, $k = 1$, $\hbar = 1$. Then we have

$$\underline{\mathbf{H}} = \frac{1}{2}(\underline{\mathbf{p}}^2 + \underline{\mathbf{x}}^2) \tag{11.58}$$

and

$$[\underline{\mathbf{x}}, \underline{\mathbf{p}}] = i\underline{\mathbf{1}} \tag{11.59}$$

It is useful to introduce, in addition to $\underline{\mathbf{x}}$ and $\underline{\mathbf{p}}$, the operators

$$\underline{\mathbf{a}} = \frac{i}{\sqrt{2}}(\underline{\mathbf{p}} - i\underline{\mathbf{x}}) = \frac{1}{\sqrt{2}}(\underline{\mathbf{x}} + i\underline{\mathbf{p}}) \tag{11.60}$$

and

$$a^+ = \frac{1}{\sqrt{2i}}(p + ix) = \frac{1}{\sqrt{2}}(x - ip) \tag{11.61}$$

(a^+ is the Hermitian conjugate of a, because p and x are Hermitian) in terms of which

$$aa^+ = \frac{1}{2}[p^2 - i(xp - px) + x^2] = H + \frac{1}{2}1 \tag{11.62}$$

and

$$a^+a = \frac{1}{2}[p^2 + i(xp - px) + x^2] = H + -\frac{1}{2}1 = N \tag{11.63}$$

where N is called as the number operator. It follows that the commutator of a and a^+ is

$$[a, a^+] = 1 \tag{11.64}$$

and

$$H = \frac{1}{2}(aa^+ + a^+a) \tag{11.65}$$

Also

$$[a, H] = a \quad , \quad [a^+, H] = -a^+ \tag{11.66}$$

From the definition of a scalar product,

$$< \psi_E | a^+ a \psi_E > = < a\psi_E | a\psi_E > \geq 0 \tag{11.67}$$

If we suppose that ψ_E is an eigenvector of H, then we can write

$$< \psi_E | a^+ a \psi_E > = < \psi_E | (H - \frac{1}{2})\psi_E > = (E - \frac{1}{2}) < \psi_E | \psi_E > \geq 0 \tag{11.68}$$

from which it follows that

$$E \geq \frac{1}{2} \tag{11.69}$$

Applying the operator a to each member of $H\psi_E = E\psi_E$ and using Eq.(11.66), we obtain

$$aH\psi_E = (Ha + a)\psi_E = Ea\psi_E \tag{11.70}$$

or

$$Ha\psi_E = (E - 1)a\psi_E \tag{11.71}$$

The vector $a\psi_E$ is an eigenvector of H belonging to the eigenvalue $E - 1$, that is, the lowering operator a generates the vector

$$\psi_{E-1} = a\psi_E$$

from the vector ψ_E. Repeated application of a therefore lowers E indefinitely, so long as none of the ψ is zero:

$$a\psi_E = \psi_{E-1} \quad , \quad a^2\psi_E = \psi_{E-2} \quad , \quad a^3\psi_E = \psi_{E-3} \quad , \quad \text{etc.}$$

However, this result contradicts ($E \geq 1/2$), unless the operator a produces the zero vector at some stage. Hence, there is an eigenvalue $E_0 = E - n$ for which $\psi_{E_0} \neq 0$, but

$$a\psi_{E_0} = 0 \tag{11.72}$$

For this eigenvalue we have

$$\mathbf{a}^+\mathbf{a}\psi_{E_0} = (\mathbf{H} - \frac{1}{2}\mathbf{1})\psi_{E_0} = (E_0 - \frac{1}{2})\psi_{E_0} = 0 \qquad (11.73)$$

since

$$\psi_{E_0} \neq 0 \quad , \quad E_0 = \frac{1}{2}$$

The smallest eigenvalue of $\mathbf{H}$ is $1/2$; therefore,

$$E = n + \frac{1}{2} \quad , \quad n = 0, 1, 2, \cdots \qquad (11.74)$$

which is exactly the result obtained in x–representation. We now relabel the eigenfunctions ψ_E with the index n:

$$\psi_{E=n+\frac{1}{2}} = \text{const.} \times \psi_n$$

$$\mathbf{H}\psi_n = (n + \frac{1}{2})\psi_n$$

$$\mathbf{H} = \mathbf{N} + \frac{1}{2}\mathbf{1}$$

$$\mathbf{H}\psi_n = (\mathbf{N} + \frac{1}{2}\mathbf{1})\psi_n = (n + \frac{1}{2})\psi_n$$

The algebraic properties of these eigenvectors follow easily from the properties of the matrices which represent $\mathbf{a}$ and $\mathbf{a}^+$ in the energy representation. If we assume that the states ψ_n are normalized, so that

$$< \psi_{n'}|\psi_n > = \delta_{nn'} \qquad (11.75)$$

then we have

$$\mathbf{a}\psi_n = \alpha_n\psi_{n-1} \quad , \quad \mathbf{a}^+\psi_n = \beta_n\psi_{n+1} \qquad (11.76)$$

The matrix elements of $\mathbf{a}$ and $\mathbf{a}^+$ are therefore

$$< n'|\mathbf{a}|n > = \alpha_n\delta_{n',n-1} \qquad (11.77)$$

$$< n'|\mathbf{a}^+|n > = \beta_n\delta_{n',n+1} \qquad (11.78)$$

the constants α_n and β_n are related by

$$\alpha_{n+1} = \beta_n^* \qquad (11.79)$$

The matrix element of $\mathbf{a}\mathbf{a}^+$ is

$$< n'|\mathbf{a}\mathbf{a}^+|n > = (n+1)\delta_{nn'} \qquad (11.80)$$

Writing out the matrix product, we obtain

$$< n'|\mathbf{a}\mathbf{a}^+|n > = \sum_{n"} < n'|\mathbf{a}|n" >< n"|\mathbf{a}^+|n > \qquad (11.81)$$

$$= \sum_{n"} \alpha_{n"}\delta_{n',n"-1}\beta_n\delta_{n",n+1} = \alpha_{n+1}\beta_n\delta_{nn'}$$

whence

$$\alpha_{n+1}\beta_n = n+1 \tag{11.82}$$

and

$$|\alpha_{n+1}|^2 = |\beta_n|^2 = n+1 \tag{11.83}$$

Hence, the arbitrary phases of the vectors ψ can be chosen so that

$$\alpha_n = \sqrt{n} \quad , \quad \beta_n = \sqrt{n+1} \tag{11.84}$$

and we have

$$\mathbf{a}\psi_n = \sqrt{n}\psi_{n-1} \tag{11.85}$$

$$\mathbf{a}^+\psi_n = \sqrt{n+1}\psi_{n+1} \tag{11.86}$$

$$\mathbf{N}\psi_n = n\psi_n \tag{11.87}$$

The relations

$$\mathbf{a} + \mathbf{a}^+ = \sqrt{2}\mathbf{x} \quad , \quad \mathbf{a} - \mathbf{a}^+ = \sqrt{2}i\mathbf{p}$$

therefore lead to

$$\sqrt{n}\psi_{n-1} - \sqrt{2}\mathbf{x}\psi_n + \sqrt{n+1}\psi_{n+1} = 0 \tag{11.88}$$

$$\sqrt{n}\psi_{n-1} - \sqrt{2}i\mathbf{p}\psi_n - \sqrt{n+1}\psi_{n+1} = 0 \tag{11.89}$$

These recurrence relations lead to the Schrödinger equation, and we know that in the x–representation, the vector ψ_n has the components

$$\psi_n(x) = \frac{1}{\sqrt{2^n n! \sqrt{\pi}}} e^{-\frac{1}{2}x^2} H_n(x) \tag{11.90}$$

or

$$\psi_n(x) = \frac{1}{\sqrt{n!}}(\mathbf{a}^+)^n \psi_0(x)$$

or

$$|n> = \frac{1}{\sqrt{n!}}(\mathbf{a}^+)^n |0>$$

Summary of matrix representation of harmonic oscillator:

$$\mathbf{a}|n> = \sqrt{n}|n-1> \quad , \quad \mathbf{a}^+|n> = \sqrt{n+1}|n+1>$$

$$\mathbf{a}^+\mathbf{a} = \mathbf{N} \quad , \quad \mathbf{a}\mathbf{a}^+ = \mathbf{H} - \frac{1}{2}\mathbf{1} \quad , \quad \mathbf{H} = \frac{1}{2}(\mathbf{a}\mathbf{a}^+ + \mathbf{a}^+\mathbf{a})$$

$$\mathbf{N}|n> = n|n> \quad , \quad \mathbf{H}|n> = (n + \frac{1}{2})|n>$$

The lowering and raising operators $\mathbf{a}$ and $\mathbf{a}^+$ play a fundamental role in the quantum theory of the electromagnetic field. It is well known that the Maxwell field can be represented as a linear combination of harmonic oscillators of different frequencies in various states of excitation. Consequently, the electromagnetic field can be represented by a Hamiltonian function which describes an infinite set of harmonic oscillators, each of which represents a normal mode of the field oscillations. In quantum theory, each of these oscillators is quantized. The total energy of the field is a sum of terms of the form $(n + 1/2)\hbar\omega$; there is one such term for each normal mode.

In a transition in which a quantum of energy $\hbar\omega$ is absorbed by an atomic system, the corresponding quantum number n is reduced by unity. Similarly, the addition of a photon to the field, by emission of a quantum, corresponds to an increase of n by one.

The operators $\underline{a}$ and $\underline{a}^+$ have exactly this effect, which describe the interaction between the field and the charged particles. When these operators are used in this manner, they are called **destruction** and **creation** operators for photons.

The essential properties of $\underline{a}$ and $\underline{a}^+$, which are required for the radiation theory, are expressed by the equations

$$< n' |\underline{a}| n > = \sqrt{n}\delta_{n',n-1} \quad , \quad < n' |\underline{a}^+| n > = \sqrt{n + 1}\delta_{n',n+1} \tag{11.91}$$

The matrices which represent these operators are therefore:

$$\underline{a} = \begin{bmatrix} 0 & \sqrt{1} & 0 & 0 & \cdots \\ 0 & 0 & \sqrt{2} & 0 & \cdots \\ 0 & 0 & 0 & \sqrt{3} & \cdots \\ 0 & 0 & 0 & 0 & \cdots \\ \vdots & \vdots & \vdots & \vdots & \end{bmatrix} \tag{11.92}$$

$$\underline{a}^+ = \begin{bmatrix} 0 & 0 & 0 & 0 & \cdots \\ \sqrt{1} & 0 & 0 & 0 & \cdots \\ 0 & \sqrt{2} & 0 & 0 & \cdots \\ 0 & 0 & \sqrt{3} & 0 & \cdots \\ \vdots & \vdots & \vdots & \vdots & \end{bmatrix} \tag{11.93}$$

and $\underline{x}$ and $\underline{p}$ are the Hermitian matrices

$$\underline{x} = \frac{1}{\sqrt{2}} \begin{bmatrix} 0 & \sqrt{1} & 0 & 0 & \cdots \\ \sqrt{1} & 0 & \sqrt{2} & 0 & \cdots \\ 0 & \sqrt{2} & 0 & \sqrt{3} & \cdots \\ 0 & 0 & \sqrt{3} & 0 & \cdots \\ \vdots & \vdots & \vdots & \vdots & \end{bmatrix} \tag{11.94}$$

$$\underline{p} = \frac{i}{\sqrt{2}} \begin{bmatrix} 0 & -\sqrt{1} & 0 & 0 & \cdots \\ \sqrt{1} & 0 & -\sqrt{2} & 0 & \cdots \\ 0 & \sqrt{2} & 0 & -\sqrt{3} & \cdots \\ 0 & 0 & \sqrt{3} & 0 & \cdots \\ \vdots & \vdots & \vdots & \vdots & \end{bmatrix} \tag{11.95}$$

Note that the selection rules for the matrix elements $\underline{x}$ and $\underline{p}$, namely $\Delta n = \pm 1$, appear explicitly in p–matrix, in that nonzero elements occur only in two diagonals adjacent to the principal diagonal. The Hamiltonian is represented by the diagonal matrix

$$\mathbf{H} = \begin{bmatrix} \frac{1}{2} & 0 & 0 & 0 & \cdots \\ 0 & \frac{3}{2} & 0 & 0 & \cdots \\ 0 & 0 & \frac{5}{2} & 0 & \cdots \\ 0 & 0 & 0 & \frac{7}{2} & \cdots \\ \vdots & \vdots & \vdots & \vdots & \end{bmatrix} \tag{11.96}$$

The $\mathbf{a}$ and $\mathbf{a}^+$ operators greatly facilitate the calculation of the matrix of elements of the other operators between oscillator eigenstates. Consider for example $< 3|\mathbf{x}^3|2 >$. In the x–basis one would have to carry out the following integral:

$$< 3|\mathbf{x}^3|2 > = \frac{1}{\sqrt{\pi}} \left(\frac{1}{2^3 3!} \cdot \frac{1}{2^2 2!} \right)^{1/2}$$

$$\times \int_{-\infty}^{\infty} e^{-x^2/2} H_3(x) x^3 e^{-x^2/2} H_2(x) dx$$

where as in the $|n >$ basis

$$< 3|\mathbf{x}^3|2 > = \frac{1}{(2)^{3/2}} < 3|(\mathbf{a} + \mathbf{a}^+)^3|2 >$$

$$= \frac{1}{(2)^{3/2}} < 3|(\mathbf{a}^3 + \mathbf{a}^2\mathbf{a}^+ + \mathbf{a}\mathbf{a}^+\mathbf{a} + \mathbf{a}\mathbf{a}^+\mathbf{a}^+$$

$$+ \mathbf{a}^+\mathbf{a}\mathbf{a} + \mathbf{a}^+\mathbf{a}\mathbf{a}^+ + \mathbf{a}^+\mathbf{a}^+\mathbf{a} + \mathbf{a}^+\mathbf{a}^+\mathbf{a}^+)|2 >$$

Since $\mathbf{a}$ lowers n by one unit and $\mathbf{a}^+$ raises it by one unit and we want to go up by one unit from $n = 2$ to $n = 3$, the only nonzero contribution comes from $\mathbf{a}\mathbf{a}^+\mathbf{a}^+$, $\mathbf{a}^+\mathbf{a}\mathbf{a}^+$, and $\mathbf{a}^+\mathbf{a}^+\mathbf{a}$. Now

$$\mathbf{a}\mathbf{a}^+\mathbf{a}^+|2 > = 4\sqrt{3}|3 >$$
$$\mathbf{a}^+\mathbf{a}\mathbf{a}^+|2 > = 3\sqrt{3}|3 >$$
$$\mathbf{a}^+\mathbf{a}^+\mathbf{a}|2 > = 2\sqrt{3}|3 >$$

so that

$$< 3|\mathbf{x}^3|2 > = \frac{1}{(2)^{3/2}} (9\sqrt{3}) = 9\sqrt{\frac{3}{8}}$$

11.5   Matrix representations of angular momentum operators

Let us first develop the matrix formalism for orbital angular momentum operators. The set of functions which are spherical harmonics form an orthonormal set of functions in the sense that

$$< Y_{lm}|Y_{l'm'} > = \int Y_{lm} Y_{l'm'} \, d\phi \sin\theta d\theta = \delta_{ll'} \delta_{mm'} \tag{11.97}$$

Consequently, one can expand any wave function in terms of this set of spherical harmonics:

$$\psi = \sum_{l,m} a_{lm}(r,t)Y_{lm}(\theta,\phi) \tag{11.98}$$

The expansion coefficients are given by

$$< Y_{lm}|\psi > = a_{lm} \tag{11.99}$$

where the integral is taken over angle variables only. The matrix elements for the z–component of the angular momentum operator in this representation are

$$[L_z]_{lm,l'm'} = < \psi_{lm}|L_z|\psi_{l'm'} > = m\hbar\delta_{ll'}\delta_{mm'} \tag{11.100}$$

Similarly for L^2

$$[L^2]_{lm,l'm'} = < \psi_{lm}|L^2|\psi_{l'm'} > = l(l+1)\hbar^2\delta_{ll'}\delta_{mm'} \tag{11.101}$$

Written out in matrix form, the matrices representing L_z and L^2 have the form:

$$\mathbf{L}_z = \hbar \begin{bmatrix} 0 & & & & & & & & \\ & 1 & & & & & & & \\ & & 0 & & & & & & \\ & & & -1 & & & & & \\ & & & & 2 & & & & \\ & & & & & 1 & & & \\ & & & & & & 0 & & \\ & & & & & & & -1 & \\ & & & & & & & & -2 \\ & & & & & & & & & \ddots \end{bmatrix} \tag{11.102}$$

$$\mathbf{L}^2 = \hbar^2 \begin{bmatrix} 0 & & & & & & & & \\ & 2 & & & & & & & \\ & & 2 & & & & & & \\ & & & 2 & & & & & \\ & & & & 6 & & & & \\ & & & & & 6 & & & \\ & & & & & & 6 & & \\ & & & & & & & 6 & \\ & & & & & & & & 6 \\ & & & & & & & & & \ddots \end{bmatrix} \tag{11.103}$$

In these matrices all elements are zero except those on the diagonal and that on the diagonal are the eigenvalues of the corresponding operators. The next problem is that of calculating the matrix elements for the operators L_x and L_y. To do this, use is made of the operators $L_\mp$. We have

$$L_\mp Y_{lm} = \sqrt{(l \pm m)(l \mp m + 1)}\hbar Y_{l,m\mp 1} \tag{11.104}$$

From this, we can evaluate the matrix elements of $\mathbf{L}_-$ and $\mathbf{L}_+$:

$$[\mathbf{L}_-]_{lm,l'm'} = < Y_{lm}|L_-|Y_{l'm'} > = \sqrt{(l'-m')(l'-m'+1)}\hbar\delta_{ll'}\delta_{m,m'-1} \tag{11.105}$$

$$[\mathbf{L}_+]_{lm,l'm'} = <Y_{lm}|L_+|Y_{l'm'}> = \sqrt{(l'-m')(l'+m'+1)}\,\hbar\delta_{ll'}\delta_{m,m'+1} \quad (11.106)$$

Since the operators L_+ and L_- are Hermitian adjoints,

$$[\mathbf{L}_-]_{lm,l'm'} = <L_+Y_{lm}|Y_{l'm'}> = <Y_{l'm'}|L_+Y_{l'm'}>^+ \quad (11.107)$$

Hence the matrix elements of $\mathbf{L}_+$ and $\mathbf{L}_-$ are related to each other by

$$[\mathbf{L}_-]_{lm,l'm'} = [\mathbf{L}_+]^+_{l'm',lm} \quad (11.108)$$

Writing these results in matrix form gives

$$\mathbf{L}_- = \hbar
\begin{bmatrix}
0 & & & & & & & & \\
& 0 & 0 & 0 & & & & & \\
& \sqrt{2} & 0 & 0 & & & & & \\
& 0 & \sqrt{2} & 0 & & & & & \\
& & & & 0 & 0 & 0 & 0 & 0 \\
& & & & 2 & 0 & 0 & 0 & 0 \\
& & & & 0 & \sqrt{6} & 0 & 0 & 0 \\
& & & & 0 & 0 & \sqrt{6} & 0 & 0 \\
& & & & 0 & 0 & 0 & 2 & 0 \\
& & & & & & & & \ddots
\end{bmatrix}
\quad (11.109)$$

$$\mathbf{L}_+ = \hbar
\begin{bmatrix}
0 & & & & & & & & \\
& 0 & \sqrt{2} & 0 & & & & & \\
& 0 & 0 & \sqrt{2} & & & & & \\
& 0 & 0 & 0 & & & & & \\
& & & & 0 & 2 & 0 & 0 & 0 \\
& & & & 0 & 0 & \sqrt{6} & 0 & 0 \\
& & & & 0 & 0 & 0 & \sqrt{6} & 0 \\
& & & & 0 & 0 & 0 & 0 & 2 \\
& & & & 0 & 0 & 0 & 0 & 0 \\
& & & & & & & & \ddots
\end{bmatrix}
\quad (11.110)$$

The matrices L_x and L_y then can be obtained from L_+ and L_- by making use of

$$\mathbf{L}_x = \frac{1}{2}(\mathbf{L}_+ + \mathbf{L}_-) \quad , \quad \mathbf{L}_y = -\frac{i}{2}(\mathbf{L}_+ - \mathbf{L}_-) \quad (11.111)$$

Thus

$$\mathbf{L}_x = \frac{\hbar}{\sqrt{2}}
\begin{bmatrix}
0 & & & & & & \\
& 0 & 1 & 0 & & & \\
& 1 & 0 & 1 & & & \\
& 0 & 1 & 0 & & & \\
& & & & 0 & \sqrt{2} & 0 & 0 & 0 \\
& & & & \sqrt{2} & 0 & \sqrt{3} & 0 & 0 \\
& & & & 0 & \sqrt{3} & 0 & \sqrt{3} & 0 \\
& & & & 0 & 0 & \sqrt{3} & 0 & \sqrt{2} \\
& & & & 0 & 0 & 0 & \sqrt{2} & 0 \\
& & & & & & & & \ddots
\end{bmatrix}
\quad (11.112)$$

$$
L_y = \frac{i\hbar}{\sqrt{2}}
\begin{bmatrix}
0 & & & & & & \\
0 & -1 & 0 & & & & \\
1 & 0 & -1 & & & & \\
0 & 1 & 0 & & & & \\
& & & 0 & -\sqrt{2} & 0 & 0 & 0 \\
& & & \sqrt{2} & 0 & -\sqrt{3} & 0 & 0 \\
& & & 0 & \sqrt{3} & 0 & -\sqrt{3} & 0 \\
& & & 0 & 0 & \sqrt{3} & 0 & -\sqrt{2} \\
& & & 0 & 0 & 0 & \sqrt{2} & 0 \\
& & & & & & & & \ddots
\end{bmatrix}
\tag{11.113}
$$

One can check that the matrices satisfy the commutation relations. For example, for $l = 1$ (with $\hbar = 1$)

$$
[L_+, L_-] =
$$

$$
\begin{bmatrix} 0 & \sqrt{2} & 0 \\ 0 & 0 & \sqrt{2} \\ 0 & 0 & 0 \end{bmatrix}
\begin{bmatrix} 0 & 0 & 0 \\ \sqrt{2} & 0 & 0 \\ 0 & \sqrt{2} & 0 \end{bmatrix}
-
\begin{bmatrix} 0 & 0 & 0 \\ \sqrt{2} & 0 & 0 \\ 0 & \sqrt{2} & 0 \end{bmatrix}
\begin{bmatrix} 0 & \sqrt{2} & 0 \\ 0 & 0 & \sqrt{2} \\ 0 & 0 & 0 \end{bmatrix}
$$

$$
=
\begin{bmatrix} 2 & 0 & 0 \\ 0 & 2 & 0 \\ 0 & 0 & 0 \end{bmatrix}
-
\begin{bmatrix} 0 & 0 & 0 \\ 0 & 2 & 0 \\ 0 & 0 & 2 \end{bmatrix}
= 2
\begin{bmatrix} 1 & 0 & 0 \\ 0 & 0 & 0 \\ 0 & 0 & -1 \end{bmatrix}
= 2L_z
$$

Spin angular momentum in matrix representation:

Let us now consider the spin angular momentum of the particles. If the independent variables which can be measured simultaneously include an internal variable for the particle describing the spin orientation, the wave function can be expanded in terms of the eigenfunctions of this variable:

$$
\psi = \sum_{m_s=-s}^{s} a_{m_s}(r, t)\phi_{m_s}
\tag{11.114}
$$

The spin functions ϕ_{m_s} may be a function of some internal variables of the particles, perhaps including the positions of subparticles out of which the particle is constructed. Fortunately, for problems not involving internal structure, it is not necessary to know these internal variables. However, it is possible to work with the coefficients a_{m_s}. Consider a particle having a spin angular momentum of $\hbar$. In general, the spin of a particle is fixed and only its orientation can change. Consequently, the operator S^2 has a definite quantum number which is simply a constant. For the case of spin one ($s = 1$) under consideration, there are only three possible orientations for the spin, and the wave function can be written as a column matrix (as a vector):

$$
\psi =
\begin{bmatrix} a_1(r, t) \\ a_0(r, t) \\ a_{-1}(r, t) \end{bmatrix}
= \psi(r, t, \phi_{m_s})
\tag{11.115}
$$

where m_s can take on only the values $0, \mp 1$.

The matrices for the three components of the spin angular momentum are:

$$\mathbf{S}_x = \frac{\hbar}{\sqrt{2}} \begin{bmatrix} 0 & 1 & 0 \\ 1 & 0 & 1 \\ 0 & 1 & 0 \end{bmatrix} \quad , \quad \mathbf{S}_y = \frac{i\hbar}{\sqrt{2}} \begin{bmatrix} 0 & -1 & 0 \\ 1 & 0 & -1 \\ 0 & 1 & 0 \end{bmatrix} \tag{11.116}$$

$$\mathbf{S}_z = \hbar \begin{bmatrix} 1 & 0 & 0 \\ 0 & 0 & 0 \\ 0 & 0 & -1 \end{bmatrix} \quad , \quad \mathbf{S}^2 = 2\hbar^2 \begin{bmatrix} 1 & 0 & 0 \\ 0 & 1 & 0 \\ 0 & 0 & 1 \end{bmatrix} = 2\hbar^2 \mathbf{1}$$

The eigenvalue equation for S_z is

$$S_z \psi = m_s \psi \tag{11.117}$$

or in matrix notation

$$(\mathbf{S}_z - m_s \mathbf{1})\psi = 0 \tag{11.118}$$

This corresponds to a set of linear equations in the three unknown components of ψ that is homogeneous and has a nontrivial solution only when the determinant of the coefficients (secular determinant) vanishes:

$$det(\mathbf{S}_z - m_s \mathbf{1}) = |\mathbf{S}_z - m_s \mathbf{1}| = 0 \tag{11.119}$$

If the determinant is expanded, one obtains

$$m_s(m_s^2 - \hbar^2) = 0 \tag{11.120}$$

which has roots $m_s = \hbar$, 0 , $-\hbar$.

In this formalism, $|a_1|^2$ is interpreted as the probability that the z–component of $\mathbf{S}$ will have the value $+\hbar$ when the particle is located at a particular point r. The expectation value of S_z can be written as

$$< S_z >= \frac{\hbar|a_1|^2 + 0 \cdot |a_0|^2 + (-\hbar)|a_{-1}|^2}{|a_1|^2 + |a_0|^2 + |a_{-1}|^2} \tag{11.121}$$

in matrix notation

$$< S_z >= \frac{\psi^+ \mathbf{S}_z \psi}{\psi^+ \psi} \tag{11.122}$$

where

$$\psi^+ = \begin{bmatrix} a_1 \\ a_0 \\ a_{-1} \end{bmatrix}^+ = [a_1^*, a_0^*, a_{-1}^*] \tag{11.123}$$

If wave function is normalized,

$$\psi^+ \psi = |a_1|^2 + |a_0|^2 + |a_{-1}|^2 = 1 \tag{11.124}$$

11.6   Worked examples

Example - 11.1 : Diagonalize the rotation matrix

$$\tilde{R} = \begin{pmatrix} \cos\theta & -\sin\theta \\ \sin\theta & \cos\theta \end{pmatrix}$$

to obtain the eigenvalues and eigenvectors.

Solution :

$$\begin{vmatrix} \cos\theta - \lambda & -\sin\theta \\ \sin\theta & \cos\theta - \lambda \end{vmatrix} = \lambda^2 - 2\lambda\cos\theta + 1 = 0$$

Eigenvalues:

$$\lambda_{\mp} = \cos\theta \mp i\sin\theta \quad \to \quad \lambda_+ = e^{i\theta} \;,\quad \lambda_- = e^{-i\theta}$$

The corresponding eigenvectors:

$$\tilde{R}\tilde{\psi} = \lambda\tilde{\psi} \quad \to \quad (\tilde{R} - \lambda)\tilde{\psi} = 0 \;,\quad \tilde{\psi} = \begin{pmatrix} a_1 \\ a_2 \end{pmatrix}$$

$$\lambda_+ : \quad [\cos\theta - (\cos\theta + i\sin\theta)]a_1 + (-\sin\theta)a_2 = 0$$

$$-i\sin\theta a_1 - \sin\theta a_2 = 0 \quad \to \quad -ia_1 = a_2$$

$$\psi_+ = a_1|1> +a_2|2> = a_1|1> -ia_1|2> = a_1(|1> -i|2>)$$

Therefore

$$\psi_+ = \frac{1}{\sqrt{2}}(|1> -i|2>)$$

$$\lambda_- : \quad [\cos\theta - (\cos\theta - i\sin\theta)]a_1 + (-\sin\theta)a_2 = 0$$

$$i\sin\theta a_1 - \sin\theta a_2 = 0 \quad \to \quad ia_1 = a_2$$

$$\psi_- = a_1|1> +a_2|2> = a_1|1> +ia_1|2> = a_1(|1> +i|2>)$$

Therefore

$$\psi_- = \frac{1}{\sqrt{2}}(|1> +i|2>)$$

Since the eigenvectors do not depend on the angle of rotation, we see that all rotations about the same axis have the same eigenvectors. These rotations thus commute.

Example - 11.2 : Show that the operator

$$S_{12} = 2\left[3\frac{(\tilde{S}\cdot\tilde{r})^2}{r^2} - \tilde{S}\right]$$

regarded as a function of $\tilde{r}$, depends only on the polar angles θ and ϕ, and that this dependence has the form of a spherical harmonic with $l = 2$.

Solution : In spherical polar coordinates,

$$\tilde{r} = (r\sin\theta\cos\phi \;,\; r\sin\theta\sin\phi \;,\; r\cos\theta)$$

we have

$$S_x\cos\phi + S_y\sin\phi = \frac{1}{2}\left(S_+e^{-i\phi} + S_-e^{i\phi}\right)$$

$$\frac{1}{r^2}(\tilde{S}\cdot\tilde{r})^2 = \left[\frac{1}{2}\left(S_+e^{-i\phi} + S_-e^{i\phi}\right)\sin\theta + S_z\cos\theta\right]^2$$

$$= \frac{1}{4}S_+^2 e^{-2i\phi}\sin^2\theta + \frac{1}{4}S_-^2 e^{2i\phi}\sin^2\theta + \frac{1}{2}(S_+S_z + S_zS_+)e^{-i\phi}\sin\theta\cos\theta$$

$$+\frac{1}{2}(S_-S_z + S_zS_-)e^{i\phi}\sin\theta\cos\theta + S_z^2\cos^2\theta + \frac{1}{4}(S_+S_- + S_-S_+)\sin^2\theta$$

Using the relations

$$[S_+, S_-] = 2S_z \quad , \quad S^2 = S_+S_- + S_z^2 - S_z$$

$$S_+S_- + S_-S_+ = 2S_+S_- - 2S_z = 2(S^2 - S_z^2)$$

we get

$$S_{12} = \frac{3}{2}e^{-2i\phi}\sin^2\theta + 3(S_+S_z + S_zS_+)e^{-i\phi}\sin\theta\cos\theta$$

$$-(3S_z^2 - S^2)(1 - 3\cos^2\theta) + 3(S_-S_z + S_zS_-)e^{i\phi}\sin\theta\cos\theta$$

$$+\frac{3}{2}S_-^2 e^{2i\phi}\sin^2\theta$$

In terms of the spherical harmonics with $l = 2$, we find that

$$S_{12} = \sqrt{\frac{24\pi}{5}}\left[S_+^2 Y_2^{-2} + (S_+S_z + S_zS_+)Y_2^{-1}\right]$$

$$+\sqrt{\frac{24\pi}{5}}\left[\sqrt{\frac{2}{3}}(3S_z^2 - S^2)Y_2^0 - (S_-S_z + S_zS_-)Y_2^1 + S_-^2 Y_2^2\right]$$

Example - 11.3 : For conservative systems, show that if, at $t = 0$, the state vector $|\psi(t) >$ is an eigenvector of the observable A with eigenvalue a, then, for $t > 0$, $|\psi(t) >$ will be an eigenvector of the operator $A_H(-t)$ with the same eigenvalue a.

Solution : Taking $t_0 = 0$, and writing for simplicity $U(t,0) = U(t)$, we have from

$$|\psi_H >= U^+(t,t_0)|\psi(t) > \quad , \quad A_H(t) = U^+(t,t_0)AU(t,t_0)$$

that

$$|\psi(t) >= U(t)|\psi(0) > \quad , \quad A_H(t) = U^+(t)AU(t) \quad (1)$$

Also in this problem, $A|\psi(0) >= a|\psi(0) >$. By making the substitution $|\psi(0) >= U^+(t)|\psi(t) >$ and multiplying both sides of the resulting equation from the left by $U(t)$ we have

$$U(t)AU^+(t)|\psi(t) >= a|\psi(t) >$$

By comparing with (1) and taking

$$U(t,t_0) = e^{-\frac{i}{\hbar}H(t-t_0)} \quad , \quad U^{(0)}(t,t_0) = e^{-\frac{i}{\hbar}H_0(t-t_0)}$$

it follows that

$$A_H(-t)|\psi(t) >= a|\psi(t) >$$

Example - 11.4 : Treating the coordinate x as an operator in the Schrödinger representation, determine the corresponding operators x_H in the Heisenberg representation (a) for the free particle and (b) for the harmonic oscillator.

Solution : Since the Hamiltonian is in both cases time–independent, we will use the relations

$$|\psi_H> = U^+(t,t_0)|\psi(t)> \quad , \quad A_H(t) = U^+(t,t_0)AU(t,t_0)$$

and

$$U(t,t_0) = e^{-\frac{i}{\hbar}H(t-t_0)} \quad , \quad U^{(0)}(t,t_0) = e^{-\frac{i}{\hbar}H_0(t-t_0)}$$

and take $t_0 = 0$. (a) For a free particle

$$x_H = e^{\frac{i}{\hbar}Ht}xe^{-\frac{i}{\hbar}Ht} \quad , \quad H = -\frac{\hbar^2}{2m}\frac{\partial^2}{\partial x^2} \quad (1)$$

$$[H,x]\psi(x) = -\frac{\hbar^2}{2m}\frac{\partial\psi(x)}{\partial x}$$

and consequently

$$[H,x] = -\frac{\hbar^2}{2m}\frac{\partial}{\partial x} \quad , \quad [H,[H,x]] = 0 \quad , \quad [H,[H,[H,x]]] = 0 \quad (2)$$

Using relation

$$e^L A e^{-L} = A + [L,A] + \frac{1}{2!}[L,[L,A]] + \frac{1}{3!}[L,[L,[L,A]]] + \cdots$$

we find from (1)

$$x_H(t) = x - \frac{i\hbar}{m}t\frac{\partial}{\partial x} \quad (3)$$

(b) For the oscillator

$$H = -\frac{\hbar^2}{2m}\frac{\partial^2}{\partial x^2} + \frac{1}{2}mw^2x^2$$

Working out the commutators (2) and regrouping the terms, we obtain

$$x_H(t) = x\cos(wt) - \frac{i\hbar}{mw}\sin(wt)\frac{\partial}{\partial x} \quad (4)$$

With the help of the momentum operator $p = -i\hbar\frac{\partial}{\partial x}$, we can write the relations (3) and (4) in the form

$$x_H(t) = x + \frac{p}{m}t \quad , \quad x_H(t) = x\cos(wt) + \frac{p}{mw}\sin(wt)$$

11.7   Problems

Problem - 11.1 : Consider a system of two nucleons. Let $\mathbf{r} = \mathbf{r}_1 - \mathbf{r}_2$ be their relative position vector, and $\frac{1}{2}\tilde{\sigma}_1$, $\frac{1}{2}\tilde{\sigma}_2$ their spin operators. Part of the interaction energy between nucleons is similar in form to the classical interaction between two dipoles, which can be written in the form

$$V = V(r)\left[3\frac{(\tilde{\sigma}_1\cdot\tilde{r})(\tilde{\sigma}_2\cdot\tilde{r})}{r^2} - \tilde{\sigma}_1\cdot\tilde{\sigma}_2\right]$$

This is the so called tensor force, which is not electromagnetic in origin. Show that the operator

$$S_{12} = \left[3 \frac{(\tilde{\sigma}_1 \cdot \tilde{r})(\tilde{\sigma}_2 \cdot \tilde{r})}{r^2} - \tilde{\sigma}_1 \cdot \tilde{\sigma}_2 \right]$$

which expresses the spin dependence of the tensor force, can be written in the form

$$S_{12} = 2 \left[3 \frac{(\tilde{S} \cdot \tilde{r})^2}{r^2} - \tilde{S}^2 \right]$$

where $\tilde{S} = \frac{1}{2}(\tilde{\sigma}_1 + \tilde{\sigma}_2)$ is the total spin operator.

Problem - 11.2 : Find the matrices corresponding to the spin operator components S_x, S_y, S_z for a particle of spin $s = 1$, in the representation in which the operators S^2 and S_z are diagonal.

Problem - 11.3 : Let us denote by

$$S(t, t_0) = U^{(0)+}(t, t_0)U(t, t_0)$$

the transformation operator between the Heisenberg and the interaction pictures. Show that this operator is the solution of the differential equation

$$i\hbar \frac{\partial S(t, t_0)}{\partial t} = H'_I S(t, t_0)$$

with the initial condition $S(t_0, t_0) = 1$.

Problem - 11.4 : Determine explicitly the transformation operator $S(t, 0)$ between the Heisenberg and the interaction pictures for a one–dimensional harmonic oscillator of mass m and charge e placed in a constant uniform electric field $\mathcal{E}$, with

$$H = \frac{p^2}{2m} + \frac{1}{2}mw^2x^2 - e\mathcal{E}x$$

where

$$H_0 = \frac{p^2}{2m} + \frac{1}{2}mw^2x^2 \quad , \quad H' = -e\mathcal{E}x$$

Example - 11.5 : In constructing eigenfunctions of S^2 it is convenient to define the unnormalized projection operator

$$\mathcal{O}_{S_j} = \prod_{i \neq j} [S^2 - S_i(S_i + 1)]$$

which has the effect of annihilating those eigenfunctions of S^2 which have the eigenvalues $S_i(S_i + 1)$ with $i \neq j$. Show that

$$\mathcal{O}_{S=1}[\alpha(1)\beta(2) + \beta(1)\alpha(2)] = \text{const.} \times [\alpha(1)\beta(2) + \beta(1)\alpha(2)]$$

and

$$\mathcal{O}_{S=1}[\alpha(1)\beta(2) - \beta(1)\alpha(2)] = 0$$

Chapter 12

Spin and the Addition of Angular Momenta

12.1 Systems with spin one–half

Systems with spin one–half are of especial interest, since this is the spin encountered in the stable particles: electrons, positrons, protons, and neutrons. (The neutron is stable only in an atomic nucleus.) In this case, the wave function is of the form

$$\psi = \begin{bmatrix} a_{1/2}(r,t) \\ a_{-1/2}(r,t) \end{bmatrix} \tag{12.1}$$

The components of spin satisfy the commutation relation:

$$[S_x, S_y] = \hbar S_z \tag{12.2}$$

The spin angular momentum component operators in matrix form are:

$$\mathbf{S}_x = \frac{\hbar}{2} \begin{bmatrix} 0 & 1 \\ 1 & 0 \end{bmatrix} \quad , \quad \mathbf{S}_x = \frac{i\hbar}{2} \begin{bmatrix} 0 & -1 \\ 1 & 0 \end{bmatrix} \tag{12.3}$$

$$\mathbf{S}_z = \frac{\hbar}{2} \begin{bmatrix} 1 & 0 \\ 0 & -1 \end{bmatrix} \quad , \quad \mathbf{S}^2 = \frac{3\hbar^2}{4} \begin{bmatrix} 1 & 0 \\ 0 & 1 \end{bmatrix}$$

$$\mathbf{S}_x^2 = \frac{\hbar^2}{4}\mathbf{1} \quad , \quad \mathbf{S}_y^2 = \frac{\hbar^2}{4}\mathbf{1} \quad , \quad \mathbf{S}_z^2 = \frac{\hbar^2}{4}\mathbf{1} \quad , \quad \mathbf{S}^2 = \frac{3\hbar^2}{4}\mathbf{1}$$

The operators

$$\boldsymbol{\sigma} = \left(\frac{\hbar}{2}\right)\mathbf{S} \tag{12.4}$$

are known as the **Pauli spin operators**, or **Pauli matrices**.

$$\sigma_x = \begin{pmatrix} 0 & 1 \\ 1 & 0 \end{pmatrix} \quad , \quad \sigma_y = i\begin{pmatrix} 0 & -1 \\ 1 & 0 \end{pmatrix} \quad , \quad \sigma_z = \begin{pmatrix} 1 & 0 \\ 0 & -1 \end{pmatrix} \tag{12.5}$$

$$\sigma_x^2 = \sigma_y^2 = \sigma_z^2 = \begin{pmatrix} 1 & 0 \\ 0 & 1 \end{pmatrix} = \mathbf{1} \tag{12.6}$$

These matrices satisfy the commutation relations

$$[\sigma_x, \sigma_y] = 2i\sigma_z \tag{12.7}$$

For particles with spin one–half, there are only two possible orientations (eigenstates) of the spin with respect to some fixed direction in space, usually taken as the z–axis. These are commonly referred to as the orientations in which the spin is either **parallel** (up) or **antiparallel** (down) to z (see Figure 12.1).

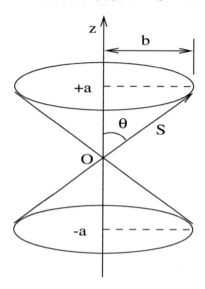

Fig. 12.1. Possible orientations of spin vector in space (vector model of spin). For $s = \frac{1}{2}$: $S = \sqrt{3}\hbar/2$, $a = \hbar/2$, $b = \hbar/\sqrt{2}$, $\cos\theta = 1/\sqrt{3}$, $\theta \cong 55°$.

Average values of spin components:

$$< S_x >=< S_y >= 0 \quad , \quad < S_z >= \frac{\hbar}{2} \tag{12.8}$$

but

$$< S_x^2 >=< S_y^2 >=< S_z^2 >= \frac{\hbar^2}{4} \tag{12.9}$$

The component operators have the following algebraic properties:

$$S_x S_y + S_y S_x = 0 \tag{12.10}$$

S_x and S_y are anticommute,

$$S_x^2 = S_y^2 = S_z^2 = \frac{\hbar^2}{4} \quad , \quad S^2 = \frac{3}{4}\hbar^2 \tag{12.11}$$

or

$$S^2 = s(s+1)\hbar^2 = \frac{1}{2}(\frac{1}{2}+1) = \frac{3}{4}\hbar^2$$

The corresponding Pauli matrices also have these algebraic properties:

$$\sigma_x \sigma_y + \sigma_y \sigma_x = 0 \tag{12.12}$$

Raising and lowering operators (or matrices) can also be defined for spin:

$$S_+ = \hbar \begin{pmatrix} 0 & 1 \\ 0 & 0 \end{pmatrix} \quad , \quad S_- = \hbar \begin{pmatrix} 0 & 0 \\ 1 & 0 \end{pmatrix} \tag{12.13}$$

The eigenstates of S_z are usually called as **spinors**. To find these eigenspinors, we solve

$$S_z \begin{bmatrix} u \\ v \end{bmatrix} = \pm \frac{\hbar}{2} \begin{bmatrix} u \\ v \end{bmatrix} \tag{12.14}$$

or

$$\begin{pmatrix} 1 & 0 \\ 0 & -1 \end{pmatrix} \begin{pmatrix} u \\ v \end{pmatrix} = \pm \begin{pmatrix} u \\ v \end{pmatrix} \quad \rightarrow \quad \begin{pmatrix} u \\ -v \end{pmatrix} = \pm \begin{pmatrix} u \\ v \end{pmatrix} \tag{12.15}$$

The $(+)$ eigensolution has $v = 0$,

$$\mathcal{X}_+ = \begin{pmatrix} 1 \\ 0 \end{pmatrix} \quad for \quad S_z = +\frac{1}{2}\hbar \tag{12.16}$$

and the $(-)$ eigensolution has $u = 0$,

$$\mathcal{X}_- = \begin{pmatrix} 0 \\ 1 \end{pmatrix} \quad for \quad S_z = -\frac{1}{2}\hbar \tag{12.17}$$

An arbitrary spinor can be expanded in this complete set

$$\begin{pmatrix} \alpha_+ \\ \alpha_- \end{pmatrix} = \alpha_+ \begin{pmatrix} 1 \\ 0 \end{pmatrix} + \alpha_- \begin{pmatrix} 0 \\ 1 \end{pmatrix} \tag{12.18}$$

if the expansion is normalized, then

$$|\alpha_+|^2 + |\alpha_-|^2 = 1 \tag{12.19}$$

For example, the eigenstates of the operator $S_x \cos\phi + S_y \sin\phi$ can be found as follows:

$$(S_x \cos\phi + S_y \sin\phi) \begin{pmatrix} u \\ v \end{pmatrix} = \frac{1}{2}\hbar\lambda \begin{pmatrix} u \\ v \end{pmatrix}$$

that is

$$\begin{pmatrix} 0 & \cos\phi - i\sin\phi \\ \cos\phi + i\sin\phi & 0 \end{pmatrix} \begin{pmatrix} u \\ v \end{pmatrix} = \lambda \begin{pmatrix} u \\ v \end{pmatrix}$$

or

$$\begin{pmatrix} 0 & e^{-i\phi} \\ e^{i\phi} & 0 \end{pmatrix} \begin{pmatrix} u \\ v \end{pmatrix} = \lambda \begin{pmatrix} u \\ v \end{pmatrix}$$

This implies that $ve^{-i\phi} = \lambda u$, $ue^{i\phi} = \lambda v$; hence $\lambda = \pm 1$. The eigenvectors corresponding to $\lambda = +1$ and $\lambda = -1$ are, respectively

$$\frac{1}{\sqrt{2}} \begin{pmatrix} e^{-i\phi} \\ 1 \end{pmatrix}, \quad \frac{1}{\sqrt{2}} \begin{pmatrix} -e^{-i\phi} \\ 1 \end{pmatrix}$$

For a given state α, the expectation value of $\mathbf{S}$ may be calculated. We have

$$< \mathbf{S} > = < \alpha|\mathbf{S}|\alpha > = \sum_{i,j} < \alpha|i >< i|\mathbf{S}|j >< j|\alpha > \tag{12.20}$$

or equivalently

$$< \mathbf{S} > = (\alpha_+^*, \alpha_-^*)\mathbf{S} \begin{pmatrix} \alpha_+ \\ \alpha_- \end{pmatrix} \tag{12.21}$$

Therefore

$$<\mathbf{S}_x> = \frac{\hbar}{2}(\alpha_+^* \alpha_- + \alpha_-^* \alpha_+) \tag{12.22}$$

$$<\mathbf{S}_y> = \frac{i\hbar}{2}(-\alpha_+^* \alpha_- + \alpha_-^* \alpha_+) \tag{12.23}$$

$$<\mathbf{S}_z> = \frac{\hbar}{2}(|\alpha_+|^2 - |\alpha_-|^2) \tag{12.24}$$

All these results are real.

$$<\mathbf{S}_x^2> = <\mathbf{S}_y^2> = <\mathbf{S}_z^2> = \frac{\hbar^2}{4} \quad , \quad <\mathbf{S}^2> = \frac{3}{4}\hbar^2 \tag{12.25}$$

Remark on Spin-$\frac{1}{2}$ System:

The Pauli spin operators in their usual matrix representation are

$$\sigma_x = \begin{pmatrix} 0 & 1 \\ 1 & 0 \end{pmatrix} \quad , \quad \sigma_y = \begin{pmatrix} 0 & -i \\ i & 0 \end{pmatrix} \quad , \quad \sigma_z = \begin{pmatrix} 1 & 0 \\ 0 & -1 \end{pmatrix} \tag{12.26}$$

with

$$Tr\{\sigma_i \sigma_j\} = 2\delta_{ij} \quad , \quad \sigma_x^2 = \sigma_y^2 = \sigma_z^2 = I \tag{12.27}$$

where I is 2×2 identity matrix,

$$I = \begin{pmatrix} 1 & 0 \\ 0 & 1 \end{pmatrix}$$

The three spin operators, collectively represented by the vector operator $\vec{\sigma}$, are Hermitian and obey the commutation rules

$$[\sigma_x, \sigma_y] = 2i\sigma_z \quad , \quad [\sigma_y, \sigma_z] = 2i\sigma_x \quad , \quad [\sigma_z, \sigma_x] = 2i\sigma_y \tag{12.28}$$

or, written more compactly,

$$[\sigma_i, \sigma_j] = 2i\sigma_k \quad , \quad (i, j, k \ \ cyclik) \tag{12.29}$$

One can also show that

$$\sigma_i \sigma_j = i\sigma_k \quad , \quad (i, j, k \ \ cyclik) \tag{12.30}$$

We note that

$$\sigma_i \sigma_j \sigma_k = i \quad , \quad Tr\{\sigma_i\} = 0 \quad , \quad det(\sigma_i) = -1 \tag{12.31}$$

The spin states

$$|\alpha> = \begin{pmatrix} 1 \\ 0 \end{pmatrix} \quad , \quad |\beta> = \begin{pmatrix} 0 \\ 1 \end{pmatrix} \tag{12.32}$$

are simultaneously eigenstates of σ_z and $\sigma^2 (= \sigma_x^2 + \sigma_y^2 + \sigma_z^2)$:

$$\sigma_z|\alpha> = |\alpha> \quad , \quad \sigma^2|\alpha> = 3|\alpha> \tag{12.33}$$

$$\sigma_z|\beta> = -|\beta> \quad , \quad \sigma^2|\beta> = 3|\beta>$$

and satisfy

$$<\alpha|\alpha> = <\beta|\beta> = 1 \quad , \quad <\alpha|\beta> = 0 \tag{12.34}$$

The spin states $|\alpha>$ and $|\beta>$ form a complete set in the spin-$\frac{1}{2}$ system.

The density matrix for spin-$\frac{1}{2}$ system may be written

$$\rho = \begin{pmatrix} <\alpha|\rho|\alpha> & <\alpha|\rho|\beta> \\ <\beta|\rho|\alpha> & <\beta|\rho|\beta> \end{pmatrix} \equiv \begin{pmatrix} \rho_{\alpha\alpha} & \rho_{\alpha\beta} \\ \rho_{\beta\alpha} & \rho_{\beta\beta} \end{pmatrix} \qquad (12.35)$$

$$= c_0 I + c_1 \sigma_x + c_2 \sigma_y + c_3 \sigma_z$$

The last expression arising from the fact that any 2×2 matrix can be expressed in terms of the Pauli spin operators together with the 2×2 identity matrix, I. Evaluating the matrix elements one can write

$$\rho = \begin{pmatrix} c_0 + c_3 & c_1 - ic_2 \\ c_1 + ic_2 & c_0 - c_3 \end{pmatrix} \qquad (12.36)$$

But

$$Tr\{\rho\} = 1 = 2c_0 \quad , \quad Tr\{\rho\sigma_x\} \equiv <\sigma_x> = 2c_1 \qquad (12.37)$$

$$Tr\{\rho\sigma_y\} \equiv <\sigma_y> = 2c_2 \quad , \quad Tr\{\rho\sigma_z\} \equiv <\sigma_z> = 2c_3$$

Thus,

$$\rho = \frac{1}{2} \begin{pmatrix} 1 + <\sigma_z> & <\sigma_x> -i<\sigma_y> \\ <\sigma_x> +i<\sigma_y> & 1 - <\sigma_z> \end{pmatrix} \qquad (12.38)$$

$$= \frac{1}{2}(I + <\vec{\sigma}> \cdot \vec{\sigma})$$

If $|\psi>$ is a general, normalized one–particle spin state, it may be expanded in the (complete) basis set consisting of $|\alpha>$ and $|\beta>$.

$$|\psi> = a|\alpha> +b|\beta> \quad , \quad |a|^2 + |b|^2 = 1 \qquad (12.39)$$

We note that the state described by the wave function $|\psi>$ is, by definition, a pure state. It is then readily verified that the density operator acquires the form

$$\rho = |\psi><\psi| = \begin{pmatrix} |a|^2 & ab^* \\ a^*b & |b|^2 \end{pmatrix} \qquad (12.40)$$

Expansions like $|\psi> = a|\alpha> +b|\beta>$, $|a|^2 + |b|^2 = 1$ are regarded as coherent superpositions of basis states and are characterized by nonvanishing off–diagonal matrix elements of the density operator. Comparing all ρ expressions above, one finds

$$<\sigma_x> = Tr\{\rho\sigma_x\} = \rho_{\alpha\beta} + \rho_{\beta\alpha} = b^*a + a^*b \qquad (12.41)$$

$$<\sigma_y> = Tr\{\rho\sigma_y\} = i(\rho_{\alpha\beta} - \rho_{\beta\alpha}) = i(b^*a + a^*b)$$

$$<\sigma_z> = Tr\{\rho\sigma_z\} = \rho_{\alpha\alpha} - \rho_{\beta\beta} = |a|^2 - |b|^2$$

$$<\sigma_x^2> + <\sigma_y^2> + <\sigma_z^2> = (|a|^2 + |b|^2)^2 = 1$$

$$det(\rho) = \rho_{\alpha\alpha}\rho_{\beta\beta} - \rho_{\alpha\beta}\rho_{\beta\alpha} = 0$$

It is observed that $<\sigma_z>$ — the difference between the two diagonal matrix elements of ρ — corresponds to the difference in the occupation probabilities of the two spin states.

We define two new operators:

$$\sigma_+ = \frac{1}{2}(\sigma_x + i\sigma_y) = \begin{pmatrix} 0 & 1 \\ 0 & 0 \end{pmatrix} = \begin{pmatrix} 1 \\ 0 \end{pmatrix} = (0 \ \ 1) = |\alpha >< \beta| \qquad (12.42)$$

$$\sigma_- = \frac{1}{2}(\sigma_x - i\sigma_y) = \begin{pmatrix} 0 & 0 \\ 1 & 0 \end{pmatrix} = \begin{pmatrix} 0 \\ 1 \end{pmatrix} = (1 \ \ 0) = |\beta >< \alpha|$$

In terms of σ_+ and σ_-, the Pauli spin operators are

$$\sigma_x = \sigma_+ + \sigma_- \quad , \quad \sigma_y = i(\sigma_- - \sigma_+) \qquad (12.43)$$

$$\sigma_z = 1 - 2\sigma_-\sigma_+ = 2\sigma_+\sigma_- - 1$$

and the density operator is

$$\rho = \begin{pmatrix} < \sigma_+\sigma_- > & < \sigma_- > \\ < \sigma_+ > & < \sigma_-\sigma_+ > \end{pmatrix} \qquad (12.44)$$

$$= \frac{1}{2}I + (< \sigma_+\sigma_- > -\frac{1}{2})\sigma_z + < \sigma_+ > \sigma_- + < \sigma_- > \sigma_+$$

with

$$< \sigma_+\sigma_- >= Tr\{\rho\sigma_+\sigma_-\} \qquad (12.45)$$

$$< \sigma_+ >= Tr\{\rho\sigma_+\} \quad , \quad < \sigma_- >= Tr\{\rho\sigma_-\}$$

Additional relations among the spin operators:

$$(\sigma_+)^2 = (\sigma_-)^2 = 0 \qquad (12.46)$$

$$\sigma_+\sigma_- = \begin{pmatrix} 1 & 0 \\ 0 & 0 \end{pmatrix} = |\alpha >< \alpha|$$

$$\sigma_-\sigma_+ = \begin{pmatrix} 0 & 0 \\ 0 & 1 \end{pmatrix} = |\beta >< \beta|$$

$$\sigma_+\sigma_- + \sigma_-\sigma_+ = \begin{pmatrix} 1 & 0 \\ 0 & 1 \end{pmatrix} = |\alpha >< \alpha| + |\beta >< \beta| = I$$

$$\sigma_+\sigma_- - \sigma_-\sigma_+ \equiv [\sigma_+, \sigma_-] = \begin{pmatrix} 1 & 0 \\ 0 & -1 \end{pmatrix} = \sigma_z = 2(\sigma_+\sigma_- - \frac{1}{2})$$

For the commutators one gets

$$[\sigma_+, \sigma_x] = \sigma_z \quad , \quad [\sigma_-, \sigma_x] = -\sigma_z \qquad (12.47)$$

$$[\sigma_+, \sigma_y] = i\sigma_z \quad , \quad [\sigma_-, \sigma_y] = i\sigma_z$$

$$[\sigma_+, \sigma_z] = -2\sigma_+ \quad , \quad [\sigma_-, \sigma_z] = 2\sigma_-$$

The spin operators acting on the eigenstates $|\alpha >$ and $|\beta >$ yield:

$$\sigma_x|\alpha >= |\beta > \quad , \quad \sigma_x|\beta >= |\alpha > \qquad (12.48)$$

$$\sigma_y|\alpha >= i|\beta > \quad , \quad \sigma_y|\beta >= -i|\alpha >$$

$$\sigma_z|\alpha >= |\alpha > \quad , \quad \sigma_z|\beta >= -|\beta >$$

$$\sigma_+|\alpha >= 0 \quad , \quad \sigma_+|\beta >= |\alpha >$$

$$\sigma_- |\alpha> = |\beta> \quad , \quad \sigma_- |\beta> = 0$$

In place of the Pauli spin operators, it is often desirable to employ the Hermitian operators defined by

$$\mathbf{s} = \frac{1}{2}\vec{\sigma} \tag{12.49}$$

that satisfy the angular momentum commutation rules

$$[s_x, s_y] = is_z \quad , \quad [s_y, s_z] = is_x \quad , \quad [s_z, s_x] = is_y \tag{12.50}$$

or

$$\mathbf{s} \times \mathbf{s} = i\mathbf{s} \tag{12.51}$$

Since the spin states $|\alpha>$ and $|\beta>$ are simultaneous eigenstates of σ_z and σ^2, they are also simultaneous eigenstates of s_z and s^2:

$$s_z|\alpha> = \frac{1}{2}|\alpha> \quad , \quad s^2|\alpha> = \frac{3}{4}|\alpha> \tag{12.52}$$

$$s_z|\beta> = -\frac{1}{2}|\beta> \quad , \quad s^2|\beta> = \frac{3}{4}|\beta>$$

In a more general notation, the spin states are written $|sm>$ where

$$|\alpha> \equiv |\frac{1}{2},\frac{1}{2}> \quad , \quad |\beta> \equiv |\frac{1}{2},-\frac{1}{2}> \tag{12.53}$$

The above four equations are then represented by

$$s_z|sm> = m|sm> \quad , \quad s^2|sm> = s(s+1)|sm> \tag{12.54}$$

with $s = \frac{1}{2}$, and $m = \mp\frac{1}{2}$.

This formalism may be extended to systems involving N spins. We define

$$\mathbf{S} = \sum_{i=1}^{N} \mathbf{s}_i \tag{12.55}$$

where $\mathbf{s}_i$ is the spin operator for the i th spin. We also have

$$\mathbf{S} \times \mathbf{S} = i\mathbf{S} \tag{12.56}$$

Employing the rules for the coupling of angular momenta, one may construct simultaneous eigenstates of S_z and S^2:

$$S_z|SM> = M|SM> \quad , \quad S^2|SM> = S(S+1)|SM> \tag{12.57}$$

where

$$S = 0, 1, 2, \cdots, \frac{N}{2} \quad ; \quad N \text{ even} \tag{12.58}$$

$$= \frac{1}{2}, \frac{3}{2}, \cdots, \frac{N}{2} \quad ; \quad N \text{ odd}$$

$$M = S, \ S-1, \ \cdots, \ -S+1, \ -S$$

Thus, for two spins, $S = 0$ (singlet) or 1 (triplet). The normalized eigenstates — obtained by the rules for the addition of angular momenta — are: for the singlet

$$|00> = \frac{1}{\sqrt{2}}(|\alpha(1)>|\beta(2)> - |\alpha(2)>|\beta(1)>) \tag{12.59}$$

and for the triplet

$$|11> = |\alpha(1) > |\alpha(2) > \quad , \quad |1-1> = |\beta(1) > |\beta(2) > \qquad (12.60)$$

$$|10> = \frac{1}{\sqrt{2}}(|\alpha(1) > |\beta(2) > +|\alpha(2) > |\beta(1) >)$$

It is observed that the states corresponding to $S = 1$ are symmetric under an interchange of spin (i.e., indices 1 and 2) while the state corresponding to $S = 0$ is antisymmetric.

Spin Magnetic Moment of Electron:

A bound electron has an intrinsic magnetic dipole moment by virtue of its spin, and that magnetic moment is

$$\vec{\mu}_s = -\frac{eg}{2mc}\mathbf{S} \qquad (12.61)$$

where g is the **gyromagnetic ratio**. Its value is very close to 2,

$$g = 2(1 + \frac{\alpha}{2\pi} + \cdots) \cong 2.0023192 \qquad (12.62)$$

For such a bound electron, the Hamiltonian in the presence of an external magnetic field **B** has an additional term which is just the interaction potential energy

$$H_1 = -\vec{\mu}_s \cdot \mathbf{B} = \frac{eg}{2mc}\mathbf{S} \cdot \mathbf{B} \qquad (12.63)$$

The Schrödinger equation for the state

$$\psi(t) = \begin{pmatrix} \alpha_+ \\ \alpha_- \end{pmatrix}$$

is

$$i\hbar\frac{d\psi(t)}{dt} = H_1\psi(t)$$

(in the interaction representation)

$$i\hbar\frac{d\psi(t)}{dt} = \frac{eg}{2mc}\mathbf{S} \cdot \mathbf{B}\psi(t) \qquad (12.64)$$

If **B** is taken parallel to the z–axis, and if we write

$$\psi(t) = e^{-i\omega t}\begin{pmatrix} \alpha_+ \\ \alpha_- \end{pmatrix}$$

then the equation becomes

$$\hbar\omega = \begin{pmatrix} \alpha_+ \\ \alpha_- \end{pmatrix} = \frac{egB}{2mc}\frac{\hbar}{2}\begin{pmatrix} 1 & 0 \\ 0 & -1 \end{pmatrix}\begin{pmatrix} \alpha_+ \\ \alpha_- \end{pmatrix} \qquad (12.65)$$

The solution corresponds to different frequencies ω. We have

$$\text{for} \ \ \omega = \frac{egB}{4mc} \quad , \quad \begin{pmatrix} \alpha_+ \\ \alpha_- \end{pmatrix} = \begin{pmatrix} 1 \\ 0 \end{pmatrix}$$

and

$$for \ \ \omega = -\frac{egB}{4mc} \quad , \quad \begin{pmatrix} \alpha_+ \\ \alpha_- \end{pmatrix} = \begin{pmatrix} 0 \\ 1 \end{pmatrix}$$

Thus, if the initial state is

$$\psi(0) = \begin{pmatrix} a \\ b \end{pmatrix} \tag{12.66}$$

then the state at a later time will be

$$\psi(t) = \begin{pmatrix} ae^{-i\omega t} \\ be^{i\omega t} \end{pmatrix} \quad , \quad \omega = \frac{egB}{4mc} \tag{12.67}$$

Suppose that at $t = 0$ the spin is an eigenstate of S_x with eigenvalue $+\hbar/2$, that is, it points in the x–direction. This means that

$$\frac{\hbar}{2} \begin{pmatrix} 0 & 1 \\ 1 & 0 \end{pmatrix} \begin{pmatrix} a \\ b \end{pmatrix} = \frac{\hbar}{2} \begin{pmatrix} a \\ b \end{pmatrix}$$

that is

$$\begin{pmatrix} a \\ b \end{pmatrix} = \frac{1}{\sqrt{2}} \begin{pmatrix} 1 \\ 1 \end{pmatrix}$$

Then, at a later time

$$< \mathbf{S}_x > = \frac{1}{\sqrt{2}}(e^{i\omega t}, e^{-i\omega t})\frac{\hbar}{2}\begin{pmatrix} 0 & 1 \\ 1 & 0 \end{pmatrix}\frac{1}{\sqrt{2}}\begin{pmatrix} e^{-i\omega t} \\ e^{i\omega t} \end{pmatrix} = \frac{\hbar}{2}\cos(2\omega t)$$

Similarly

$$< \mathbf{S}_y > = \frac{1}{\sqrt{2}}(e^{i\omega t}, e^{-i\omega t})\frac{i\hbar}{2}\begin{pmatrix} 0 & -1 \\ 1 & 0 \end{pmatrix}\frac{1}{\sqrt{2}}\begin{pmatrix} e^{-i\omega t} \\ e^{i\omega t} \end{pmatrix} = \frac{\hbar}{2}\sin(2\omega t)$$

$$< \mathbf{S}_z > = \frac{1}{\sqrt{2}}(e^{i\omega t}, e^{-i\omega t})\frac{\hbar}{2}\begin{pmatrix} 1 & 0 \\ 0 & -1 \end{pmatrix}\frac{1}{\sqrt{2}}\begin{pmatrix} e^{-i\omega t} \\ e^{i\omega t} \end{pmatrix} = 0$$

Thus the spin precesses about the z–axis, the direction of B, with frequency

$$\omega = \frac{egB}{4mc} \cong \frac{eB}{2mc}$$

In a solid the gyromagnetic factor g of an electron is affected by the nature of the forces acting in the solid. A knowledge of g provides very useful constraints on what these forces could be, and it is therefore important to be able to measure g. This can be done by the **paramagnetic resonance method**, which we now describe.

Paramagnetic resonance method:

Consider an electron, whose only degrees of freedom are the spin states, under the influence of a large magnetic field B_0 pointing in the z–direction, and constant in time, and a small oscillating field $B_1 \cos(\omega t)$, pointing in the x–direction. The corresponding Schrödinger equation is

$$i\hbar\frac{d\psi(t)}{dt} = \frac{eg}{2mc}\mathbf{S} \cdot \mathbf{B}\psi(t) \tag{12.68}$$

$$i\hbar \frac{d}{dt}\begin{pmatrix} a(t) \\ b(t) \end{pmatrix} = \frac{egh}{4mc}\begin{pmatrix} B_0 & B_1\cos(\omega t) \\ B_1\cos(\omega t) & -B_0 \end{pmatrix}\begin{pmatrix} a(t) \\ b(t) \end{pmatrix} \tag{12.69}$$

or with

$$\omega_0 = \frac{egB_0}{4mc} \quad , \quad \omega_1 = \frac{egB_1}{4mc}$$

$$i\frac{da(t)}{dt} = \omega_0 a(t) + \omega_1\cos(\omega t) \tag{12.70}$$

$$i\frac{db(t)}{dt} = \omega_1\cos(\omega t)a(t) - \omega_0 b(t)$$

Let

$$A(t) = a(t)e^{i\omega_0 t} \quad , \quad B(t) = b(t)e^{-i\omega_0 t} \tag{12.71}$$

These satisfy the equations

$$i\frac{dA(t)}{dt} = \omega_1\cos(\omega t)B(t)e^{2i\omega_0 t} \approx \frac{1}{2}\omega_1 e^{i(2\omega_0 - \omega)t}B(t) \tag{12.72}$$

$$i\frac{dB(t)}{dt} = \omega_1\cos(\omega t)A(t)e^{-2i\omega_0 t} \approx \frac{1}{2}\omega_1 e^{-i(2\omega_0 - \omega)t}A(t) \tag{12.73}$$

Eliminating $B(t)$ between these two equations, one obtains

$$\frac{d^2 A(t)}{dt^2} - i(2\omega_0 - \omega)\frac{dA(t)}{dt} + \frac{\omega_1^2}{4}A(t) = 0 \tag{12.74}$$

A trial solution

$$A(t) = A(0)e^{i\lambda t} \tag{12.75}$$

gives

$$-\lambda^2 + (2\omega_0 - \omega)\lambda + \frac{\omega_1^2}{4} = 0 \tag{12.76}$$

The roots of this equation are

$$\lambda_\pm = \frac{1}{2}[2\omega_0 - \omega \pm \sqrt{(2\omega_0 - \omega)^2 + \omega_1^2}] \tag{12.77}$$

The most general solution is

$$A(t) = A_+ e^{i\lambda_+ t} + A_- e^{i\lambda_- t} \tag{12.78}$$

and hence

$$B(t) = -\frac{2}{\omega_1}e^{-i(2\omega_0 - \omega)t}(\lambda_+ A_+ e^{i\lambda_+ t} + \lambda_- A_- e^{i\lambda_- t}) \tag{12.79}$$

This finally yields

$$a(t) = e^{-i\omega_0 t}(A_+ e^{i\lambda_+ t} + A_- e^{i\lambda_- t}) \tag{12.80}$$

$$b(t) = -\frac{2}{\omega_1}e^{-i(\omega_0 - \omega)t}(\lambda_+ A_+ e^{i\lambda_+ t} + \lambda_- A_- e^{i\lambda_- t})$$

If at $t = 0$ the electron spin points in the positive z–direction, then $a(0) = 1$ and $b(0) = 0$, that is,

$$A_+ + A_- = 1 \quad , \quad \lambda_+ A_+ + \lambda_- A_- = 0$$

so that

$$A_+ = \frac{\lambda_-}{\lambda_- - \lambda_4} \quad , \quad A_- = -\frac{\lambda_+}{\lambda_- - \lambda_4} \tag{12.81}$$

The probability that at some later time t the spin points in the negative z–direction is $|b(t)|^2$:

$$|b(t)|^2 = \frac{4}{\omega_1^2} \left| \frac{\lambda_+ + \lambda_-}{\lambda_- - \lambda_+} e^{i\lambda_+ t} - \frac{\lambda_+ + \lambda_-}{\lambda_- - \lambda_+} e^{i\lambda_- t} \right|^2 \tag{12.82}$$

$$= \frac{\omega_1^2}{(2\omega_0 - \omega)^2 + \omega_1^2} \left| 1 - e^{-i(\lambda_+ - \lambda_-)t} \right|^2$$

$$= \frac{\omega_1^2}{(2\omega_0 - \omega)^2 + \omega_1^2} \cdot \frac{1}{2} \left(1 - \cos[\sqrt{(2\omega_0 - \omega)^2 + \omega_1^2} t] \right)$$

This quantity is small, since $\omega_1 \ll \omega, \omega_0$. When $\omega_1 \to 2\omega_0$, then

$$|b(t)|^2 \to \frac{1}{2}[1 - \cos(\omega_1 t)]$$

In this case $|b(t)|^2$ takes its maximum value, and the energy difference between the spin up and the spin down states, absorbed from the external field, signals the resonance frequency, so that ω_0, and hence g can be measured with great precision.

12.2 The addition of angular momenta

The combination of the angular momenta associated with two parts of a system (such as the orbital angular momenta of two electrons in an atom, or the spin and orbital angular momenta of the same electron) forms the angular momentum of the whole system. The magnitude of the sum of two angular momentum vectors can have any value ranging from the sum of their magnitudes (parallel case) to the difference of their magnitudes (antiparallel case), by integer steps. This is called the **triangle rule**. For example,

$$\mathbf{J} = \mathbf{J}_1 + \mathbf{J}_2 \tag{12.83}$$

$$j = j_1 + j_2 , \; j_1 + j_2 - 1 , \; \cdots , \; |j_1 - j_2| \tag{12.84}$$

$$|j_1 - j_2| \leq j \leq j_1 + j_2 \tag{12.85}$$

Any number of angular momentum eigenstates can be combined by taking them two at a time in accordance with the methods described below:

Eigenvalues of the total angular momentum:

Let us consider two commuting angular momentum operators $\mathbf{J}_1$ and $\mathbf{J}_2$; all components of $\mathbf{J}_1$ commute with all components of $\mathbf{J}_2$, and $\mathbf{J}_1$ and $\mathbf{J}_2$ separately satisfy the relation

$$\mathbf{J}_i \times \mathbf{J}_i = i\hbar \mathbf{J}_i \tag{12.86}$$

The orthonormal eigenstates of J_1^2 and J_{1_z} are $|j_1 m_1>$; $\mathbf{J}_2$ has no effect on them. Similarly, $|j_2 m_2>$ are orthonormal eigenstates of J_2^2 and J_{2_z}, and $\mathbf{J}_1$ has no effect on them. This representation is specified by the orthonormal set of kets $|j_1 m_1 j_2 m_2>$, each of which is a product of the kets $|j_1 m_1>$ and $|j_2 m_2>$, namely

$$|j_1 m_1 j_2 m_2> = |j_1 m_1> |j_2 m_2> \tag{12.87}$$

Since $\mathbf{J}_1$ and $\mathbf{J}_2$ commute, the total angular momentum $\mathbf{J} = \mathbf{J}_1 + \mathbf{J}_2$ also satisfies the relation $\mathbf{J} \times \mathbf{J} = i\hbar\mathbf{J}$. The orthonormal eigenstates of J^2 and J_z are $|jm>$ and satisfy a second representation. We wish to find the unitary transformation that changes from one of these representations to the other. The kets of the first representation may be denoted simply $|m_1 m_2 >$,

$$|j_1 m_1 j_2 m_2 > \equiv |j_1 j_2 > |m_1 m_2 > \equiv |m_1 m_2 > \qquad (12.88)$$

and our object is to find the unitary transformation

$$< m_1 m_2 | jm >$$

Since

$$J_z = J_{1_z} + J_{2_z} \qquad (12.89)$$

it is apparent that $< m_1 m_2 | jm >$ is zero unless $m = m_1 + m_2$. The largest value of m is $j_1 + j_2$ and this value occurs only once, when $m_1 = j_1$ and $m_2 = j_2$. This shows that the largest value of j is $j_1 + j_2$ and that there is only one such state. The next largest value of m is $j_1 + j_2 - 1$, and this occurs twice: when $m_1 = j_1$ and $m_2 = j_2 - 1$, and when $m_1 = j_1 - 1$ and $m_2 = j_2$ (provided that neither j_1 nor j_2 is zero), and so on. By an extension of this argument we can see that each j value, ranging from $j_1 + j_2$ to $|j_1 - j_2|$ by integer steps, appears just once. This establishes the triangle rule of the vector model. Each j value of the new representation has associated with it $2j + 1$ linearly independent combinations of the original eigenstates. Thus the number of $|jm>$ eigenstates is

$$\sum_{j=|j_1-j_2|}^{j_1+j_2} (2j+1) = (2j_1 + 1)(2j_2 + 1) \qquad (12.90)$$

Clebsch–Gordan coefficients:

The elements of the unitary matrix $< m_1 m_2 | jm >$ are the coefficients of the expansion of the eigenstates $|jm>$ in terms of the eigenstates $|m_1 m_2 >$,

$$|jm> = \sum_{m_1=-j_1}^{j_1} \sum_{m_2=-j_2}^{j_2} |m_1 m_2 >< m_1 m_2 | jm > \qquad (12.91)$$

They are called Clebsch–Gordan, Wigner, or vector–coupling coefficients. The inverse expansion to this is

$$|m_1 m_2> = \sum_{j=|j_1-j_2|}^{j_1+j_2} \sum_{m=-j}^{j} |jm><jm|m_1 m_2 > \qquad (12.92)$$

The unitary character of the transformation matrix is expressed through the relations:

$$\sum_{j,m} < m_1 m_2 | jm >< jm | m_1' m_2' > \qquad (12.93)$$

$$= < m_1 m_2 | m_1' m_2' > = \delta_{m_1 m_1'} \delta_{m_2 m_2'}$$

$$\sum_{m_1,m_2} < jm|m_1m_2 >< m_1m_2|j'm' > \tag{12.94}$$

$$=< jm|j'm' >= \delta_{jj'}\delta_{mm'} >$$
$$< jm|m_1m_2 >=< m_1m_2|jm >^* \tag{12.95}$$

Recursion relations:

We apply the angular momentum raising operator J_+ ($= J_x + iJ_y$) to the left side, and the equal operator $J_{1+} + J_{2+}$ to the right side of Eq.(12.9). Taking $\hbar = 1$, we obtain

$$[j(j+1) - m(m+1)]^{1/2}|j, m+1 >= \tag{12.96}$$

$$\sum_{m_1,m_2} \{[j_1(j_1+1) - m_1(m_1+1)]^{1/2}|m_1+1, m_2 >$$

$$+[j_2(j_2+1) - m_2(m_2+1)]^{1/2}|m_1, m_2+1 >\} < m_1m_2|jm >$$

From this equation one can write the following recursion relation

$$[j(j+1) - m(m+1)]^{1/2} < m_1m_2|j, m+1 >= \tag{12.97}$$

$$[j_1(j_1+1) - m_1(m_1-1)]^{1/2} < m_1-1, m_2|jm >$$
$$+[j_2(j_2+1) - m_2(m_2-1)]^{1/2} < m_1, m_2-1|jm >$$

Similar equation can be obtained for J_-. The result is

$$[j(j+1) - m(m-1)]^{1/2} < m_1m_2|j, m-1 >= \tag{12.98}$$

$$[j_1(j_1+1) - m_1(m_1+1)]^{1/2} < m_1+1, m_2|jm >$$
$$+[j_2(j_2+1) - m_2(m_2+1)]^{1/2} < m_1, m_2+1|jm >$$

Construction procedure:

The matrix $< m_1m_2|jm >$ has $(2j_1 + 1)(2j_2 + 1)$ rows and columns but breaks up into disconnected submatrices in accordance with the value of $m = m_1 + m_2$. Thus there will be a 1×1 submatrix for which $m = j_1 + j_2$ and $j = j_1 + j_2$. Then there will be a 2×2 submatrix for which $m = j_1 + j_2 - 1$ and j is either $j_1 + j_2$ or $j_1 + j_2 - 1$. The rank of these submatrices at first increases by unit from one to the next until a maximum rank is reached and maintained for one or more submatrices; thereafter it decreases by unit until the last 1×1 submatrix has $m = -j_1 - j_2$ and $j = j_1 + j_2$. Each of these submatrices is unitary, so that the first 1×1 submatrix is a number of unit magnitude, which we choose by convention to be $+1$:

$$< j_1j_2|j_1 + j_2, j_1 + j_2 >= 1 \tag{12.99}$$

Some particular coefficients:

For $j_1 = 1/2$, $j_2 = 1/2$: [Number of states $= (2j_1 + 1)(2j_2 + 1) = 4$]

(j,m) (m_1,m_2)	$(1,1)$	$(1,0)$	$(0,0)$	$(1,-1)$
$(\frac{1}{2},\frac{1}{2})$	1			
$(\frac{1}{2},-\frac{1}{2})$		$\frac{1}{\sqrt{2}}$	$\frac{1}{\sqrt{2}}$	
$(-\frac{1}{2},\frac{1}{2})$		$\frac{1}{\sqrt{2}}$	$-\frac{1}{\sqrt{2}}$	
$(-\frac{1}{2},-\frac{1}{2})$				1

For $j_1 = 1$, $j_2 = 1/2$: [Number of states $= (2j_1 + 1)(2j_2 + 1) = 6$]

(j,m) (m_1,m_2)	$(\frac{3}{2},\frac{3}{2})$	$(\frac{3}{2},\frac{1}{2})$	$(\frac{1}{2},\frac{1}{2})$	$(\frac{3}{2},-\frac{1}{2})$	$(\frac{1}{2},-\frac{1}{2})$	$(\frac{3}{2},-\frac{3}{2})$
$(1,\frac{1}{2})$	1					
$(1,-\frac{1}{2})$		$\frac{1}{\sqrt{3}}$	$\sqrt{\frac{2}{3}}$			
$(0,\frac{1}{2})$		$\sqrt{\frac{2}{3}}$	$-\frac{1}{\sqrt{3}}$			
$(0,-\frac{1}{2})$				$\sqrt{\frac{2}{3}}$	$\frac{1}{\sqrt{3}}$	
$(-1,\frac{1}{2})$				$\frac{1}{\sqrt{3}}$	$-\sqrt{\frac{2}{3}}$	
$(-1,-\frac{1}{2})$						1

For $j_1 = 1$, $j_2 = 1$: [Number of states $= (2j_1 + 1)(2j_2 + 1) = 9$]

(j,m) (m_1,m_2)	$(2,2)$	$(2,1)$	$(1,1)$	$(2,0)$	$(1,0)$	$(0,0)$	$(2,-1)$	$(1,-1)$	$(2,-2)$
$(1,1)$	1								
$(1,0)$		$\frac{1}{\sqrt{2}}$	$\frac{1}{\sqrt{2}}$						
$(0,1)$		$\frac{1}{\sqrt{2}}$	$-\frac{1}{\sqrt{2}}$						
$(1,-1)$				$\frac{1}{\sqrt{6}}$	$\frac{1}{\sqrt{2}}$	$\frac{1}{\sqrt{2}}$			
$(0,0)$				$\sqrt{\frac{2}{3}}$		$-\frac{1}{\sqrt{3}}$			
$(-1,1)$				$\frac{1}{\sqrt{6}}$	$-\frac{1}{\sqrt{2}}$	$\frac{1}{\sqrt{3}}$			
$(0,-1)$							$\frac{1}{\sqrt{2}}$	$\frac{1}{\sqrt{2}}$	
$(-1,0)$							$\frac{1}{\sqrt{2}}$	$-\frac{1}{\sqrt{2}}$	
$(-1,-1)$									1

Racah obtained the general formula:

$$< m_1 m_2 | jm >= \delta_{m_1+m_2,m}(\Delta_1 \cdot \Delta_2)^{1/2} \sum_\nu \frac{(-1)^\nu}{\Delta_3} \qquad (12.100)$$

where

$$\Delta_1 = \frac{(j_1 + j_2 - j)!(j_1 + j - j_2)!(j_2 + j - j_1)!}{(j_1 + j_2 + j + 1)!}$$

$$\Delta_2 = (2j + 1)(j_1 + m_1)!(j_1 - m_1)!$$

$$\times (j_2 + m_2)!(j_2 - m_2)!(j + m)!(j - m)!$$

$$\Delta_3 = \nu!(j_1 - m_1 - \nu)!(j - j_2 + m_1 + \nu)!(j_2 + m_2 - \nu)!$$

$$\times (j - j_1 + m_2 + \nu)!(j_1 + j_2 - j - \nu)!$$

and ν runs over all values which do not lead to negative factorials.

12.3 Worked examples

Example - 12.1 : Show that (a) the Pauli matrices anticommute, $\tilde{\sigma}_\alpha\tilde{\sigma}_\beta+\tilde{\sigma}_\beta\tilde{\sigma}_\alpha = 0$, $\alpha \neq \beta$; (b) $\tilde{\sigma}_\alpha\tilde{\sigma}_\beta = i\tilde{\sigma}_\gamma$, $\alpha \neq \beta \neq \gamma$; (c) the Pauli matrices are unitary; (d) $\tilde{\sigma} \times \tilde{\sigma} = 2i\tilde{\sigma}$.

Solution : (a)

$$\tilde{\sigma}_x\tilde{\sigma}_y + \tilde{\sigma}_y\tilde{\sigma}_x = \begin{pmatrix} 0 & 1 \\ 1 & 0 \end{pmatrix}\begin{pmatrix} 0 & -i \\ i & 0 \end{pmatrix} + \begin{pmatrix} 0 & -i \\ i & 0 \end{pmatrix}\begin{pmatrix} 0 & 1 \\ 1 & 0 \end{pmatrix}$$

$$= \begin{pmatrix} i & 0 \\ 0 & -i \end{pmatrix} + \begin{pmatrix} -i & 0 \\ 0 & i \end{pmatrix} = 0$$

Similarly $\tilde{\sigma}_x\tilde{\sigma}_z + \tilde{\sigma}_z\tilde{\sigma}_x = 0$ and $\tilde{\sigma}_y\tilde{\sigma}_z + \tilde{\sigma}_z\tilde{\sigma}_y = 0$

(b)

$$\tilde{\sigma}_x\tilde{\sigma}_y = \begin{pmatrix} 0 & 1 \\ 1 & 0 \end{pmatrix}\begin{pmatrix} 0 & -i \\ i & 0 \end{pmatrix} = \begin{pmatrix} i & 0 \\ 0 & -i \end{pmatrix} = i\begin{pmatrix} 1 & 0 \\ 0 & -1 \end{pmatrix} = i\tilde{\sigma}_z$$

Similarly $\tilde{\sigma}_y\tilde{\sigma}_z = i\tilde{\sigma}_x$ and $\tilde{\sigma}_z\tilde{\sigma}_x = i\tilde{\sigma}_y$

(c) Unitary property of a matrix: $\tilde{A}^+ = \tilde{A}^{-1}$ or $\tilde{A}^+\tilde{A} = \tilde{1}$. $\tilde{\sigma}_\alpha^+\tilde{\sigma}_\alpha = 1$, $\alpha = x, y, z$, thus $\tilde{\sigma}_\alpha^+ = \tilde{\sigma}_\alpha^{-1}$ so

$$\tilde{\sigma}_x^+ = \begin{pmatrix} 0 & 1 \\ 1 & 0 \end{pmatrix} = \tilde{\sigma}_x^{-1} \quad , \quad \tilde{\sigma}_y^+ = \begin{pmatrix} 0 & -i \\ i & 0 \end{pmatrix} = \tilde{\sigma}_y^{-1}$$

$$\tilde{\sigma}_z^+ = \begin{pmatrix} 1 & 0 \\ 0 & -1 \end{pmatrix} = \tilde{\sigma}_z^{-1}$$

so that σ_α's are unitary.

(d)
$$\tilde{\sigma} \times \tilde{\sigma} = \mathbf{i}[\tilde{\sigma}_y\tilde{\sigma}_z - \tilde{\sigma}_z\tilde{\sigma}_y] + \mathbf{j}[\tilde{\sigma}_z\tilde{\sigma}_x - \tilde{\sigma}_x\tilde{\sigma}_z] + \mathbf{k}[\tilde{\sigma}_x\tilde{\sigma}_y - \tilde{\sigma}_y\tilde{\sigma}_x]$$

from part (a) we have $\tilde{\sigma}_\alpha\tilde{\sigma}_\beta = -\tilde{\sigma}_\beta\tilde{\sigma}_\alpha$ so

$$\tilde{\sigma} \times \tilde{\sigma} = 2[\mathbf{i}(\tilde{\sigma}_y\tilde{\sigma}_z) + \mathbf{j}(\tilde{\sigma}_z\tilde{\sigma}_x) + \mathbf{k}(\tilde{\sigma}_x\tilde{\sigma}_y)]$$

from part (b) $\tilde{\sigma}_\alpha\tilde{\sigma}_\beta = i\tilde{\sigma}_\gamma$, therefore

$$\tilde{\sigma} \times \tilde{\sigma} = 2i[\mathbf{i}\tilde{\sigma}_x + \mathbf{j}\tilde{\sigma}_y + \mathbf{k}\tilde{\sigma}_z] = 2i\tilde{\sigma}$$

Example - 12.2 : Show that for fixed values of l_1 and l_2, there are a total of $(2l_1 + 1)(2l_2 + 1)$ product wave functions of the form $|l_1m_1 > |l_2m_2 >$.

Solution : This question is equivalent to show the equality

$$\sum_{J=|l_1-l_2|}^{l_1+l_2} (2J + 1) = (2l_1 + 1)(2l_2 + 1)$$

We can write the summation as follows:

$$\sum \equiv \sum_{J=|l_1-l_2|}^{l_1+l_2} (2J+1) = \sum_{J=0}^{l_1+l_2} (2J+1) - \sum_{J=0}^{|l_1-l_2|-1} (2J+1)$$

Using the general summation result

$$\sum_{j=0}^{J} (2j+1) = 2\left(\frac{J(J+1)}{2}\right) + (J+1) = (J+1)^2$$

we can write

$$\sum = (l_1+l_2+1)^2 - (l_1-l_2)^2 = 4l_1l_2 + 2l_1 + 2l_2 + 1 = (2l_1+1)(2l_2+1)$$

Example - 12.3 : Find two spin-$\frac{1}{2}$ states.

Solution : In this case we have $j_1 = j_2 = \frac{1}{2}$. Total number of states $=$ $(2j_1 + 1)(2j_2 + 1) = (2 \times \frac{1}{2} + 1)(2 \times \frac{1}{2} + 1) = 2 \times 2 = 4$. The four possible states in the $|m_1m_2>$ representation are:

$$|\frac{1}{2},\frac{1}{2}> \ , \ |\frac{1}{2},-\frac{1}{2}> \ , \ |-\frac{1}{2},\frac{1}{2}> \ , \ |-\frac{1}{2},-\frac{1}{2}>$$

In the $|jm>$ representation the four possible states are:

$$|1,1> \ , \ |1,0> \ , \ |1,-1> \ , \ |0,0>$$

The state of highest weight is given by

$$|1,1>=|\frac{1}{2},\frac{1}{2}> \quad (1)$$

The two other states with total $j = 1$ are obtained by applying $J_- = J_{1-} + J_{2-}$ to both sides of Eq. (1). This gives

$$J_-|1,1>= (J_{1-} + J_{2-})|\frac{1}{2},\frac{1}{2}>$$

$$\sqrt{1(1+1) - 1(1-1)}|1,0>$$

$$= \sqrt{\frac{1}{2}(\frac{1}{2}+1) - \frac{1}{2}(\frac{1}{2}-1)}(|-\frac{1}{2},\frac{1}{2}> +|\frac{1}{2},-\frac{1}{2}>)$$

so

$$|1,0>= \frac{1}{\sqrt{2}}(|-\frac{1}{2},\frac{1}{2}> +|\frac{1}{2},-\frac{1}{2}>) \quad (2)$$

Similarly applying J_- once more to Eq. (2) or else realizing that both spins 1/2 must point down to get a total j_z of -1 we find that

$$|1,-1>=|-\frac{1}{2},-\frac{1}{2}>$$

The singlet $(j = 0)$ state must be orthogonal to the above three states. After some calculation one gets

$$|0,0> = \frac{1}{\sqrt{2}}(|-\frac{1}{2},\frac{1}{2}> - |\frac{1}{2},-\frac{1}{2}>)$$

In matrix form

$$
\begin{bmatrix} |1,1> \\ |1,0> \\ |0,0> \\ |1,-1> \end{bmatrix}
=
\begin{bmatrix}
1 & 0 & 0 & 0 \\
0 & \frac{1}{\sqrt{2}} & \frac{1}{\sqrt{2}} & 0 \\
0 & \frac{1}{\sqrt{2}} & -\frac{1}{\sqrt{2}} & 0 \\
0 & 0 & 0 & 1
\end{bmatrix}
\begin{bmatrix} |\frac{1}{2},\frac{1}{2}> \\ |-\frac{1}{2},\frac{1}{2}> \\ |\frac{1}{2},-\frac{1}{2}> \\ |-\frac{1}{2},-\frac{1}{2}> \end{bmatrix}
$$

The 4×4 matrix $< m_1 m_2 | j m >$ is unitary.

Example - 12.4 : Consider the addition of an orbital angular momentum L and a spin angular momentum S for the case $l = 1$, $s = \frac{1}{2}$. The eigenfunctions of the operators L^2, S^2, L_z, S_z are products of the space functions $\phi_1, \phi_0, \phi_{-1}$ and the spin functions α, β. Derive the eigenfunctions Φ_{jm_j} of the operators L^2, S^2, J^2, J_z in terms of the first set of eigenfunctions.

Solution : Denote the lowering operators for the total, orbital, and spin angular momenta by J_-, L_-, S_-.

$$L_-\phi_1 = \sqrt{2}\hbar\phi_0 \quad , \quad L_-\phi_0 = \sqrt{2}\hbar\phi_{-1} \quad , \quad L_-\phi_{-1} = 0$$

$$S_-\alpha = \hbar\beta \quad , \quad S_-\beta = 0$$

For $l = 1$, $s = 1/2$, the possible values of j are 3/2, 1/2. We start with the Φ for the maximum value of j, m_j, namely $j = m_j = 3/2$. The function $\Phi_{3/2,3/2}$ must be represented by the single product eigenfunction with $m_l + m_s = 3/2$. So

$$\Phi_{3/2,3/2} = \phi_1\alpha$$

To obtain $\Phi_{3/2,1/2}$, we use the relation $J_- = L_- + S_-$ and evaluate

$$J_-\Phi_{3/2,3/2} = (L_- + S_-)\phi_1\alpha$$

$$\sqrt{3}\hbar\Phi_{3/2,1/2} = \hbar(\sqrt{2}\phi_0\alpha + \phi_1\beta)$$

therefore

$$\Phi_{3/2,1/2} = \frac{1}{\sqrt{3}}(\sqrt{2}\phi_0\alpha + \phi_1\beta)$$

We obtain $\Phi_{3/2,-1/2}$ by the same procedure

$$J_-\Phi_{3/2,1/2} = (L_- + S_-) = \frac{1}{\sqrt{3}}(\sqrt{2}\phi_0\alpha + \phi_1\beta)$$

which gives

$$\Phi_{3/2,-1/2} = \frac{1}{\sqrt{3}}(\phi_{-1}\alpha + \sqrt{2}\phi_0\beta)$$

It is obvious that

$$\Phi_{3/2,-3/2} = \phi_{-1}\beta$$

The eigenfunctions $\Phi_{1/2,1/2}$ and $\Phi_{1/2,-1/2}$ are obtained from the normalization condition. $\Phi_{1/2,1/2}$ is orthogonal to $\Phi_{3/2,1/2}$. Thus

$$\Phi_{1/2,1/2} = \frac{1}{\sqrt{3}}(\phi_0\alpha - \sqrt{2}\phi_1\beta)$$

Similarly $\Phi_{1/2,1/2}$ is orthogonal to $\Phi_{3/2,-1/2}$. Thus

$$\Phi_{1/2,-1/2} = \frac{1}{\sqrt{3}}(\sqrt{2}\phi_{-1}\alpha - \phi_0\beta)$$

12.4   Problems

Problem - 12.1 : Let s_1 and s_2 be the spin operators of two spin-$\frac{1}{2}$ particles. Find the simultaneous eigenfunctions of the operators s^2 and s_z, where $\mathbf{s} = \mathbf{s}_1 + \mathbf{s}_2$. Show that these are also eigenfunctions of the operator $\mathbf{s}_1 \cdot \mathbf{s}_2$.

Problem - 12.2 : Show that the operator $(\tilde{\sigma}_1 \cdot \tilde{\sigma}_2)^n$, where $\tilde{\sigma}_1$ and $\tilde{\sigma}_1$ are Pauli matrices, depends linearly on the product $(\tilde{\sigma}_1 \cdot \tilde{\sigma}_2)$. Find the explicit form of this dependence.

Problem - 12.3 : Establish the identity

$$(\tilde{\sigma} \cdot \tilde{A})(\tilde{\sigma} \cdot \tilde{B}) = \tilde{A} \cdot \tilde{B} + i\tilde{\sigma}(\tilde{A} \times \tilde{B})$$

where $\tilde{\sigma} = (\sigma_x, \sigma_y, \sigma_z)$ are the Pauli matrices, and $\tilde{A}$ and $\tilde{B}$ are vector operators which commute with $\tilde{\sigma}$, but do not necessarily commute with each other.

Problem - 12.4 : When the angular momentum vectors $\mathbf{j}_1$ and $\mathbf{j}_2$ couple to give the resultant $\mathbf{j}$ we require that $m = m_1 + m_2$. Therefore, $m_{max} = (m_1 + m_2)_{max} = j_1 + j_2$ and we see that $j_{max} = j_1 + j_2$. What is the minimum value of j?

Problem - 12.5 : Consider the spin functions for a three–electron system: (a) Show that the function $\alpha(1)\alpha(3)\alpha(3)$ is an eigenfunction of both S^2 and S_z and determine the corresponding eigenvalues. (b) Use the ladder operator $S_- = S_{1-} + S_{2-} + S_{3-}$ to generate all $2S + 1$ of the eigenfunctions for $S = \frac{3}{2}$.

PART – III
Approximation Methods and
Scattering Theory

Chapter 13

Time–Independent Perturbation Theory

13.1 Nondegenerate perturbation theory

There are few potentials $V(r)$ for which the Schrödinger equation is exactly solvable. We must therefore develop approximation techniques to obtain the eigenvalues and eigenfunctions for potentials that do not lead to exactly soluble equations. One of these methods is the perturbation theory, which was developed both for time independent and time dependent phenomena. We will discuss the time independent perturbation theory. The formulation of the time independent perturbation theory is different for nondegenerate and degenerate states.

We assume that we have found the eigenvalues and the complete set of eigenfunctions for a Hamiltonian H_0,

$$H_0\phi_n = E_n^{(0)}\phi \tag{13.1}$$

and we ask for the eigenvalues and eigenfunctions for the Hamiltonian

$$H = H_0 + \lambda H_1 \tag{13.2}$$

that is, for the solutions of

$$(H_0 + \lambda H_1)\psi_n = E_n\psi_n \tag{13.3}$$

We will express the desired quantities as power series in λ. The convergency of the series is not important, but for small λ, the first few terms do properly describe the physical system. We will assume that as $\lambda \to 0$, $E_n \to E_n^{(0)}$ and $\psi_n \to \phi_n$.

Since the ϕ_i form a complete set, we may expand ψ_n in a series involving all the ϕ_i. We write

$$\psi_n = N(\lambda)\left[\phi_n + \sum_{k\neq n} C_{nk}(\lambda)\phi_k\right] \tag{13.4}$$

The factor $N(\lambda)$ is for the normalization of ψ_n. We assume that the linear coefficients $C_{nk}(\lambda)$ are real and positive. Since we require that $\psi_n \to \phi_n$ as $\lambda \to 0$, we have $N(0) = 1$, $C_{nk}(0) = 0$. More generally, we have

$$C_{nk}(\lambda) = \lambda C_{nk}^{(1)} + \lambda^2 C_{nk}^{(2)} + \cdots \tag{13.5}$$

$$E_n = E_n^{(0)} + \lambda E_n^{(1)} + \lambda^2 E_n^{(1)} + \cdots$$

The Schrödinger equation then reads

$$(H_0 + \lambda H_1) \left[\phi_n + \sum_{k \neq n} \lambda C_{nk}^{(1)} \phi_k + \sum_{k \neq n} \lambda^2 C_{nk}^{(2)} \phi_k + \cdots \right] \tag{13.6}$$

$$= (E_n^{(0)} + \lambda E_n^{(1)} + \lambda^2 E_n^{(1)} + \cdots)$$

$$\times \left[\phi_n + \sum_{k \neq n} \lambda C_{nk}^{(1)} \phi_k + \sum_{k \neq n} \lambda^2 C_{nk}^{(2)} \phi_k + \cdots \right]$$

The normalization factor $N(\lambda)$ does not appear in this linear equation. Identifying powers of λ yields a series of equations.

$$H_0 \phi_n + H_0 \sum_{k \neq n} \lambda C_{nk}^{(1)} \phi_k + H_0 \sum_{k \neq n} \lambda^2 C_{nk}^{(2)} \phi_k + \cdots \tag{13.7}$$

$$+ \lambda H_1 \phi_n + \lambda H_1 \sum_{k \neq n} \lambda C_{nk}^{(1)} \phi_k + \lambda H_1 \sum_{k \neq n} \lambda^2 C_{nk}^{(2)} \phi_k + \cdots$$

$$= E_n^{(0)} \phi_n + E_n^{(0)} \sum_{k \neq n} \lambda C_{nk}^{(1)} \phi_k + E_n^{(0)} \sum_{k \neq n} \lambda^2 C_{nk}^{(2)} \phi_k + \cdots$$

$$+ \lambda E_n^{(1)} \phi_n + \lambda E_n^{(1)} \sum_{k \neq n} \lambda C_{nk}^{(1)} \phi_k + \lambda E_n^{(1)} \sum_{k \neq n} \lambda^2 C_{nk}^{(2)} \phi_k + \cdots$$

$$+ \lambda^2 E_n^{(2)} \phi_n + \lambda^2 E_n^{(2)} \sum_{k \neq n} \lambda C_{nk}^{(1)} \phi_k + \lambda^2 E_n^{(2)} \sum_{k \neq n} \lambda^2 C_{nk}^{(2)} \phi_k + \cdots$$

for λ^0 :

$$H_0 \phi_n = E_n^{(0)} \phi_n \tag{13.8}$$

for λ^1 :

$$H_0 \sum_{k \neq n} C_{nk}^{(1)} \phi_k + H_1 \phi_n = E_n^{(0)} \sum_{k \neq n} C_{nk}^{(1)} \phi_k + E_n^{(1)} \phi_n \tag{13.9}$$

for λ^2 :

$$H_0 \sum_{k \neq n} C_{nk}^{(2)} \phi_k + H_1 \sum_{k \neq n} C_{nk}^{(1)} \phi_k = E_n^{(0)} \sum_{k \neq n} C_{nk}^{(2)} \phi_k \tag{13.10}$$

$$+ E_n^{(1)} + \sum_{k \neq n} C_{nk}^{(1)} \phi_k + E_n^{(2)} \phi_n$$

and so on. Consider the equation for λ^1, Eq.(13.9). Using Eq.(13.8) we obtain

$$E_n^{(1)} \phi_n = H_1 \phi_n + \sum_{k \neq n} (E_k^{(0)} - E_n^{(0)}) C_{nk}^{(1)} \phi_k \tag{13.11}$$

If we now take a scalar product with ϕ_n, and using $< \phi_k | \phi_l >= \delta_{kl}$, we obtain

$$\lambda E_n^{(1)} =< \phi_n | \lambda H_1 | \phi_n > \tag{13.12}$$

or

$$E_n^{(1)} =< \phi_n | H_1 | \phi_n >$$

This expression states that the first order energy shift for a given state is just the expectation value of the perturbing potential in that state.

If we take the scalar product of Eq.(13.11) with ϕ_m, for $m \neq n$, then

$$< \phi_m|H_1|\phi_n > +(E_m^{(0)} - E_n^{(0)})C_{nm}^{(1)} = 0 \qquad (13.13)$$

from this equation we can write

$$\lambda C_{nk}^{(1)} = \frac{< \phi_m|\lambda H_1|\phi_n >}{E_m^{(0)} - E_n^{(0)}} \qquad (13.14)$$

or

$$C_{nk}^{(1)} = \frac{< \phi_m|H_1|\phi_n >}{E_m^{(0)} - E_n^{(0)}}$$

The numerator is the matrix element of H_1 in the basis of states in which H_0 is diagonal.

Now consider the equation for λ^2, Eq.(13.10). Taking the scalar product with ϕ_n yields

$$E_n^{(2)} = \sum_{k \neq n} < \phi_n|H_1|\phi_k > C_{nk}^{(1)} \qquad (13.15)$$

$$= \sum_{k \neq n} \frac{< \phi_n|H_1|\phi_k >< \phi_k|H_1|\phi_n >}{E_n^{(0)} - E_k^{(0)}} = \sum_{k \neq n} \frac{|< \phi_k|H_1|\phi_n >|^2}{E_n^{(0)} - E_k^{(0)}}$$

The last equality follows from the Hermiticity of H_1:

$$< \phi_n|H_1|\phi_k >=< \phi_k|H_1|\phi_n >^*$$

This expression, Eq.(13.15), states that the second order energy shift is the sum of terms, whose strength is given by the square of the matrix element connecting the given state ϕ_n to all other states by the perturbing potential, weighted by the reciprocal of the energy difference between the states.

An expression for $C_{nk}^{(2)}$ may be obtained from the equation for λ^2 by taking the scalar product with ϕ_m, $m \neq n$.

$$\sum_{k \neq n} C_{nk}^{(2)} E_k^{(0)} < \phi_m|\phi_k > + \sum_{k \neq n} C_{nk}^{(1)} < \phi_m|H_1|\phi_k > \qquad (13.16)$$

$$= E_n^{(0)} \sum_{k \neq n} C_{nk}^{(2)} < \phi_m|\phi_k > +E_n^{(1)} \sum_{k \neq n} C_{nk}^{(1)} < \phi_m|\phi_k > +E_n^{(2)} < \phi_m|\phi_n >$$

or

$$C_{nm}^{(2)} E_m^{(0)} + \sum_{k \neq n} \frac{< \phi_m|H_1|\phi_k >< \phi_k|H_1|\phi_n >}{E_n^{(0)} - E_k^{(0)}} \qquad (13.17)$$

$$= E_n^{(0)} C_{nm}^{(2)} + E_n^{(1)} \frac{< \phi_m|H_1|\phi_n >}{E_n^{(0)} - E_m^{(0)}}$$

Taking $C_{nm}^{(2)}$ from this equation reads, for $n \neq m$

$$C_{nm}^{(2)} = \frac{1}{E_n^{(0)} - E_m^{(0)}} \qquad (13.18)$$

$$\times \left[\sum_{k \neq n} \frac{< \phi_m |H_1| \phi_k >< \phi_k |H_1| \phi_n >}{E_n^{(0)} - E_k^{(0)}} - E_n^{(1)} \frac{< \phi_m |H_1| \phi_n >}{E_n^{(0)} - E_m^{(0)}} \right]$$

or

$$C_{nm}^{(2)} = \sum_{k \neq n} \frac{< \phi_m |H_1| \phi_k >< \phi_k |H_1| \phi_n >}{(E_n^{(0)} - E_m^{(0)})(E_n^{(0)} - E_k^{(0)})} - E_n^{(1)} \frac{< \phi_m |H_1| \phi_n >}{(E_n^{(0)} - E_m^{(0)})^2}$$

The normalization factor $N(\lambda)$ can be determined from

$$< \psi_n | \psi_n > = N^2(\lambda) \left[1 + \lambda^2 \sum_{k \neq n} |C_{nk}^{(1)}|^2 + \cdots \right] = 1 \qquad (13.19)$$

$$N(\lambda) = \left[1 + \lambda^2 \sum_{k \neq n} |C_{nk}^{(1)}|^2 + \cdots \right]^{-1/2} \qquad (13.20)$$

It is therefore 1 to first order in λ. Hence, to first order in λ, we may write

$$\psi_n = \phi_n + \lambda \sum_{k \neq n} C_{nk}^{(1)} \phi_k \qquad (13.21)$$

To second order in λ, we may write

$$\psi_n = \left[1 + \lambda^2 \sum_{k \neq n} |C_{nk}^{(1)}|^2 \right]^{-1/2} \qquad (13.22)$$

$$\times \left[\phi_n + \lambda \sum_{k \neq n} C_{nk}^{(1)} \phi_k + \lambda^2 \sum_{k \neq n} C_{nk}^{(2)} \phi_k \right]$$

So far the developed method is valid for nondegenerate states. In order to consider degeneracy a modification is needed.

13.2 Degenerate perturbation theory

In the degenerate case, instead of a unique ϕ_n, there is a finite set of $\phi_n^{(i)}$, all of which have the same energy $E_n^{(0)}$. We choose the set of $\phi_n^{(i)}$ such that $< \phi_m^{(j)} | \phi_n^{(i)} > = \delta_{mn} \delta_{ij}$. The wave function in this case may be expanded as follows:

$$\psi_n = N(\lambda) \left[\sum_i \alpha_i \phi_n^{(i)} + \lambda \sum_{k \neq n} C_{nk}^{(1)} \sum_i \beta_i \phi_k^{(i)} + \cdots \right] \qquad (13.23)$$

The coefficients α_i, β_i, $\cdots$ will have to be determined. When this wave function is used in the Schrödinger equation

$$(H_0 + \lambda H_1) \psi_n = E_n \psi_n \qquad (13.24)$$

we get, to first order in λ:

$$H_0 \sum_{k \neq n} C_{nk}^{(1)} \sum_i \beta_i \phi_k^{(i)} + H_1 \sum_i \alpha_i \phi_n^{(i)} \qquad (13.25)$$

$$= E_n^{(1)} \sum_i \alpha_i \phi_n^{(i)} + E_n^{(0)} \sum_{k \neq n} C_{nk}^{(1)} \sum_i \beta_i \phi_k^{(i)}$$

Taking the scalar product with $\phi_n^{(j)}$ gives the first order shift equation

$$\sum_i \alpha_i < \phi_n^{(j)} |\lambda H_1| \phi_n^{(i)} >= \lambda E_n^{(1)} \alpha_j \qquad (13.26)$$

or

$$\sum_i \alpha_i < \phi_n^{(j)} |H_1| \phi_n^{(i)} >= E_n^{(1)} \alpha_j$$

This is a finite–dimensional eigenvalue problem. For example, if there is a two–fold degeneracy, and if we use the notation

$$< \phi_n^{(j)} |H_1| \phi_n^{(i)} >= h_{ij} \qquad (13.27)$$

this equation reads

$$h_{11}\alpha_1 + h_{12}\alpha_2 = E_n^{(1)} \alpha_1 \qquad (13.28)$$

$$h_{21}\alpha_1 + h_{22}\alpha_2 = E_n^{(1)} \alpha_2$$

or

$$\begin{pmatrix} h_{11} & h_{12} \\ h_{21} & h_{22} \end{pmatrix} \begin{pmatrix} \alpha_1 \\ \alpha_2 \end{pmatrix} = E_n^{(1)} \begin{pmatrix} \alpha_1 \\ \alpha_2 \end{pmatrix} \qquad (13.29)$$

Both the eigenvalues and the α_i, can be determined from this equation, if we add the condition that

$$\sum_i |\alpha_i|^2 = 1 \qquad (13.30)$$

The coefficients are necessary in the second order shift calculations. Here we are just interested in the first order energy eigenvalues in degenerate perturbation theory. In the first order calculations $h_{ij} = 0$ for $i \neq j$. This means that the matrix h_{ij} is diagonal, then the first order shifts are just the diagonal elements of this matrix. This will happen when the perturbation H_1 commutes with the operator whose eigenvalues the "i" labels represent.

For example, in the H–atom, there is a degeneracy associated with the eigenvalues of L_z, that is, all m–values have the same energy. If it happens that

$$[H_1, L_z] = 0 \qquad (13.31)$$

and if we choose our $\phi_n^{(i)}$ to be eigenfunctions of L_z, then h_{ij} will be diagonal. To see this, note that with

$$L_z \phi_n^{(i)} = \hbar m^{(i)} \phi_n^{(i)} \qquad (13.32)$$

$$< \phi_n^{(j)} |[H_1, L_z]| \phi_n^{(i)} >=< \phi_n^{(j)} |H_1 L_z - L_z H_1| \phi_n^{(i)} > \qquad (13.33)$$

$$= \hbar(m^{(i)} - m^{(j)}) h_{ij} = 0$$

that is $h_{ij} = 0$ for $m^{(i)} \neq m^{(j)}$.

Let us consider the Stark effect on hydrogen like atoms (the effect of external electric field on the energy levels). The unperturbed Hamiltonian is

$$H_0 = \frac{p^2}{2\mu} - \frac{e^2}{r} \tag{13.34}$$

whose eigenfunctions we denote by $\phi_{nlm}(\mathbf{r})$. The perturbing potential is

$$\lambda H_1 = e\vec{\mathcal{E}} \cdot \mathbf{r} = e\mathcal{E}z \tag{13.35}$$

where $\mathcal{E}$ is the electric field. The quantity $e\mathcal{E}$ will play the role of the parameter λ. The energy shift of the ground state, which is a nondegenerate, is given by the expression

$$E_{100}^{(1)} = e\mathcal{E} < \phi_{100}|z|\phi_{100} >= e\mathcal{E} \int d^3r |\phi_{100}(r)|^2 z \tag{13.36}$$

This integral vanishes. Thus for the ground state there is no energy shift that is linear in the electric field $\mathcal{E}$. This is known as the linear Stark effect. Classically, a system that has an electric dipole moment $\mathbf{d}$ will experience an energy shift of magnitude $-\mathbf{d} \cdot \vec{\mathcal{E}}$. Thus an atom, in its ground state, has no permanent dipole moment. In general one can say that systems in nondegenerate states cannot have permanent dipole moments.

Let us look at the second–order term. It reads

$$E_{100}^{(2)} = e^2\mathcal{E}^2 \sum_{nlm} \frac{|< \phi_{nlm}|z|\phi_{100} >|^2}{E_1^{(0)} - E_n^{(0)}} \tag{13.37}$$

The matrix element here is

$$< \phi_{nlm}|z|\phi_{100} >= \int d^3r R_{nlm}(r)Y_{lm}^*(\theta,\phi)r\cos\theta R_{100}(r)Y_{00}(\theta,\phi) \tag{13.38}$$

using

$$Y_{00} = \frac{1}{\sqrt{4\pi}} \quad , \quad \cos\theta = \sqrt{\frac{4\pi}{3}}Y_{10}$$

the angular part of this integral becomes

$$\int d\Omega Y_{lm}^*(\theta,\phi)\frac{1}{\sqrt{3}}Y_{10}(\theta,\phi) = \frac{1}{\sqrt{3}}\delta_{l1}\delta_{m0} \tag{13.39}$$

The fact that the m–value must be the same for the two states is one of the selection rules, $\Delta m = 0$. It follows from the fact that

$$[L_z, z] = 0 \tag{13.40}$$

the perturbation commutes with L_z.

The radial part has the following integral

$$R = \int_0^\infty r^2 dr R_{n10}(r)r R_{100}(r) \tag{13.41}$$

This integral can be evaluated. The result is

$$|< \phi_{nlm}|z|\phi_{100} >|^2 = \frac{1}{3}\frac{2^8 n^7 (n-1)^{2n-5}}{(n+1)^{2n+5}}a_0^2 \equiv f(n)a_0^2 \tag{13.42}$$

For the second order shift, this gives

$$E_{100}^{(2)} = -e^2 \mathcal{E}^2 a_0^2 \sum_{n=2}^{\infty} \frac{f(n)}{\frac{1}{2}\mu c^2 \alpha^2 (1 - \frac{1}{n^2})} \tag{13.43}$$

$$= -\frac{2e^2 \mathcal{E}^2 a_0^2}{\mu c^2 \alpha^2} \sum_{n=2}^{\infty} \frac{n^2 f(n)}{n^2 - 1} = -2a_0^3 \mathcal{E}^2 \sum_{n=2}^{\infty} \frac{n^2 f(n)}{n^2 - 1} \cong -2.25 a_0^3 \mathcal{E}^2$$

This is known as the quadratic Stark effect. For any hydrogen like atom with nuclear charge Z, $a_0 \rightarrow a_0/Z$.

If we differentiate the energy shift with respect to $\mathcal{E}$, we get an expression for the dipole moment,

$$d = -\frac{\partial E_{100}^{(2)}}{\partial \mathcal{E}} \cong 4.5 a_0^3 \mathcal{E} \tag{13.44}$$

This is proportional to the electric field strength, that is, the dipole moment is induced. The polarizability, defined by

$$P = \frac{d}{\mathcal{E}} \cong 4.5 a_0^3 \tag{13.45}$$

can thus be calculated.

The relation, which appears in Eq.(13.38)

$$\sum_{nlm} < \phi_{100}|z|\phi_{nlm} >< \phi_{nlm}|z|\phi_{100} > \tag{13.46}$$

$$= \sum_{nlm} | < \phi_{nlm}|z|\phi_{100} > |^2 = < \phi_{100}|z^2|\phi_{100} >$$

is called a sum rule.

Let us now calculate the first order (linear in $\mathcal{E}$) Stark effect for $n = 2$ states of the hydrogen atom, that is the linear Stark effect for the excited states of the hydrogen atom. For the unperturbed system there are four $n = 2$ states that have the same energy. These are

$$(\phi_{200} , \phi_{211} , \phi_{210} , \phi_{21,-1}) \text{ or } (2s , 2p_x , 2p_y , 2p_z)$$

or

$$(2s , 2p_{+1} , 2p_0 , 2p_{-1}) \text{ or } (|200 > , |211 > , |210 > , |21, -1 >)$$

Among these four states, the $l = 0$ state has even parity, and the $l = 1$ states have odd parity. We want to solve an equation like

$$\sum_i \alpha_i < \phi_n^{(j)}|\lambda H_1|\phi_n^{(i)} >= \lambda E_n^{(1)} \alpha_j \quad ; \quad j = 1 - 4 \tag{13.47}$$

(there are four such equations).

Note that the perturbing potential (z) commutes with L_z so that it only connects states with the same m–value, and parity forces us to consider only terms in which the perturbing potential must connect $l = 1$ to $l = 0$ terms, that is,

$$< 21, \pm 1|z|21, \pm 1 >= 0 \tag{13.48}$$

then the matrix in Eq.(13.48) is only a 2×2 matrix. The equation reads

$$e\mathcal{E}\begin{pmatrix} <200|z|200> & <200|z|210> \\ <210|z|200> & <210|z|210> \end{pmatrix}\begin{pmatrix} \alpha_1 \\ \alpha_2 \end{pmatrix} = E^{(1)}\begin{pmatrix} \alpha_1 \\ \alpha_2 \end{pmatrix} \qquad (13.49)$$

The diagonal elements are zero, because of parity, and the off–diagonal elements are equal, since they are complex congugates of each other, and they are real. We have

$$<200|z|210> = \int_0^\infty r^2 dr (2a_0)^{-3} e^{-r/a_0} \frac{2r}{\sqrt{3}a_0}(1 - \frac{r}{2a_0})r \qquad (13.50)$$

$$\times \int d\Omega Y_{00}^*(\sqrt{\frac{4\pi}{3}}Y_{10})Y_{10} = -3a_0$$

and hence Eq.(13.50) becomes

$$\begin{pmatrix} 0 & -3e\mathcal{E}a_0 \\ -3e\mathcal{E}a_0 & 0 \end{pmatrix}\begin{pmatrix} \alpha_1 \\ \alpha_2 \end{pmatrix} = E^{(1)}\begin{pmatrix} \alpha_1 \\ \alpha_2 \end{pmatrix} \qquad (13.51)$$

or

$$\begin{pmatrix} -E^{(1)} & -3e\mathcal{E}a_0 \\ -3e\mathcal{E}a_0 & -E^{(1)} \end{pmatrix}\begin{pmatrix} \alpha_1 \\ \alpha_2 \end{pmatrix} = 0$$

The eigenvalues of this secular equation are

$$E^{(1)} = \pm 3e\mathcal{E}a_0 \qquad (13.52)$$

The corresponding eigenstates are

$$\frac{1}{\sqrt{2}}\begin{pmatrix} 1 \\ -1 \end{pmatrix} \quad , \quad \frac{1}{\sqrt{2}}\begin{pmatrix} 1 \\ -1 \end{pmatrix} \qquad (13.53)$$

or

$$\frac{1}{\sqrt{2}}(|200> -|210>) \quad , \quad \frac{1}{\sqrt{2}}(|200> +|210>) \qquad (13.54)$$

Thus the linear Stark effect for the $n = 2$ states yields a splitting of degenerate levels as shown in Figure 13.1.

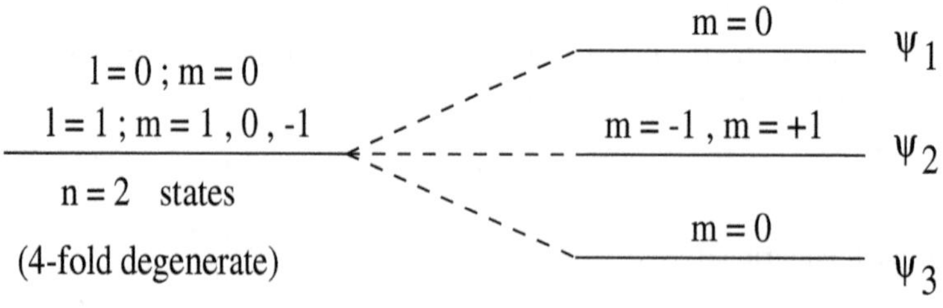

Fig. 13.1. Splitting of $n = 2$ states in electric field (Linear Stark effect).

$\psi_1 = \frac{1}{\sqrt{2}}(|200> -|210>)$, $\psi_2 = |21,\pm 1 >$ (2–fold degenerate), $\psi_3 = \frac{1}{\sqrt{2}}(|200> +|210>)$.

13.3 Worked examples

Example - 13.1 : To the first order of approximation of perturbation theory, calculate the correction to the ground state of a hydrogen–like atom due to the finite spatial extension of the nucleus. For simplicity assume that the nucleus is spherical, of radius R, and that its charge Ze is uniformly distributed throughout its volume.

Solution : The potential energy of the electron is given by

$$V(r) = \begin{cases} -\frac{Ze^2}{R}\left(\frac{3}{2} - \frac{1}{2}\frac{r^2}{R^2}\right) , & 0 \le r \le R \\ -\frac{Ze^2}{r} & , r \ge R \end{cases}$$

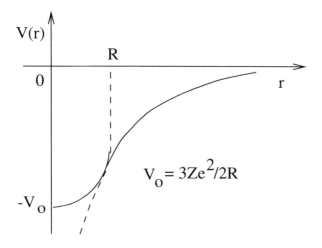

We take the unperturbed Hamiltonian with a point nucleus ($V = -Ze^2/r$), and the perturbation as

$$H_1 = -\frac{Ze^2}{R}\left(\frac{3}{2} - \frac{1}{2}\frac{r^2}{R^2}\right) - \left(-\frac{Ze^2}{r}\right) \quad \text{for} \quad 0 \le r \le R$$

$$H_1 = 0 \quad \text{for} \quad r \ge R$$

The first order correction to the ground state energy is then

$$E_1^{(1)} = <\psi_{100}|H_1|\psi_{100}>$$

$$|\psi_{100}> = \frac{1}{\sqrt{\pi a^3}}e^{-r/a} \quad , \quad a = \frac{\hbar^2}{2me^2}$$

$$E_1^{(1)} = \frac{1}{\pi a^3}\int_0^R \left[\frac{Ze^2}{R} - \frac{Ze^2}{R}\left(\frac{3}{2} - \frac{1}{2}\frac{r^2}{R^2}\right)\right] 4\pi r^2 dr$$

$$= \frac{2}{5} \frac{Ze^2}{a} \left(\frac{R}{a}\right)^2 > 0$$

Since the unperturbed ground state energy is

$$E_1^{(0)} = -\frac{Z^2 m e^4}{2\hbar^2}$$

we have, to first order of perturbation theory

$$E_1 \approx E_1^{(0)} + E_1^{(1)} = E_1^{(0)} \left[1 - \frac{4}{5}\left(\frac{R}{a}\right)^2\right]$$

The nuclear radius $\propto Z^{1/3}$ and the first Bohr orbit $\propto Z^{-1}$, therefore the importance of the perturbation increases as $Z^{8/3}$. The corrections to the energy levels of the atom due to the spatial extension of the nucleus are called isotopic corrections, since they vary from isotope to another of the same element.

Example - 13.2 : Consider an electron in a potential box having the length a. When an electric field $\mathcal{E}$ is turned on in the x–direction, the electron experiences a force equal to $-e\mathcal{E}$ and the potential function has added to it the form $e\mathcal{E}x$. The potential then has the form as shown in the figure.

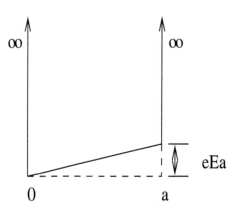

(a) What is the lowest allowed energy (in a first order approximation) for the electron? You may assume that $e\mathcal{E}a$ is much smaller than the ground state energy in the absence of the electric field. (b) Use first order perturbation theory to obtain an approximation to the ground state wave function and evaluate the first term in the correction.

Solution : (a) For this system $H = H_0 + H_1$, with

$$H_0 = -\frac{\hbar^2}{2m}\frac{d^2}{dx^2} \quad \text{and} \quad H_1 = e\mathcal{E}x$$

We know the solutions for $H_0\psi_n^{(0)} = E_n^{(0)}\psi_n^{(0)}$;

$$E_n^{(0)} = \frac{n^2 h^2}{8ma^2} \quad , \quad \psi_n^{(0)} = \sqrt{\frac{2}{a}}\sin\left(\frac{n\pi x}{a}\right)$$

First order correction is : $E_n = E_n^{(0)} + E_n^{(1)}$, where

$$E_n^{(1)} = < \psi_n^{(0)}|H_1|\psi_n^{(0)} >= \int_0^a \psi_n^{(0)} H_1 \psi_n^{(0)} dx$$

$$= e\mathcal{E} \int_0^a \psi_n^{(0)} x \psi_n^{(0)} dx = e\mathcal{E} < x >= \frac{e\mathcal{E}a}{2}$$

Therefore,

$$E_1 = E_1^{(0)} + E_1^{(1)} = \frac{h^2}{8ma^2} + \frac{e\mathcal{E}a}{2}$$

(b)

$$\psi_1 = \psi_1^{(0)} + \sum_{k \neq n} \frac{< k|H_1|k >}{E_n^{(0)} - E_k^{(0)}} \psi_k^{(0)}$$

The first term in the correction is

$$\frac{< 2|H_1|1 >}{E_1^{(0)} - E_2^{(0)}} = \frac{\int_0^a \psi_2^{(0)} e\mathcal{E} x \psi_1^{(0)} dx}{-3 \left(\frac{h^2}{8ma^2} \right)}$$

After evaluating the integral, we obtain

$$\cong 0.889 \left(\frac{2e\mathcal{E}a}{3\pi^2} \right) \left(\frac{8ma^2}{h^2} \right)$$

and finally

$$\psi_1 \cong \psi_1^{(0)} + 0.480 e\mathcal{E}a \left(\frac{ma^2}{h^2} \right) \psi_2^{(0)} + \cdots$$

Example - 13.3 : A particle of mass m is constrained to move in the xy plane so that the Hamiltonian is

$$H = \frac{1}{2m}(p_x^2 + p_y^2) + \frac{1}{2}k(x^2 + y^2) + axy$$

Use the degenerate perturbation theory to determine the energy splitting for the lowest degenerate states and the first order wave functions for these states.

Solution : The energies and degeneracies for the two–dimensional harmonic oscillator (when $a = 0$) are the following:

$$E_{n_x n_y} = (n_x + n_y + 1)h\nu$$

E	n_x	n_y	g
$h\nu$	0	0	1
$2h\nu$	1	0	2
	0	1	
$3h\nu$	1	1	3
	2	0	
	0	2	
$\vdots$	$\vdots$	$\vdots$	$\vdots$

The degenerate functions corresponding to $2h\nu$ are

$$n_x = 1 \ , \ n_y = 0 \quad \rightarrow \quad \phi_1 = \psi_1(x)\psi_0(y) = |1, 0>$$

$$n_x = 0 \ , \ n_y = 1 \quad \rightarrow \quad \phi_2 = \psi_0(x)\psi_1(y) = |0, 1>$$

$$\Psi_2 = c_1\phi_1 + c_2\phi_2$$

When the perturbation $H_1 = axy$ is present the secular equation

$$|(H_1)_{ij} - ES_{ij}| = 0$$

must be solved to obtain the first order corrections to the energy. Here

$$(H_1)_{ij} = <i|axy|j>$$

or

$$= <n_x n_y|axy|n'_x n'_y> = a <n_x|x|n'_x><n_y|y|n'_y>$$

The secular equation becomes

$$\begin{vmatrix} a<1|x|1><0|y|0> -E_2^{(1)} & a<1|x|0><0|y|1> \\ a<0|x|1><1|y|0> & a<0|x|0><1|y|1> -E_2^{(1)} \end{vmatrix} = 0$$

The nonvanishing matrix elements are:

$$<1|x|0> = <0|x|1> = <1|y|0> = <0|y|1> = \frac{1}{\sqrt{2\alpha}}$$

where $\alpha = 4\pi^2 m\nu/h$. We obtain

$$\begin{vmatrix} -E_2^{(1)} & \frac{a}{2\alpha} \\ \frac{a}{2\alpha} & -E_2^{(1)} \end{vmatrix} = 0 \quad \rightarrow \quad E_{2\mp}^{(1)} = \mp\frac{a}{2\alpha}$$

The eigenfunctions are determined as follows:

For $E_{2+} = E_2^{(0)} + E_{2+}^{(1)}$:

$$-\left(\frac{a}{2\alpha}\right)c_1 + \left(\frac{a}{2\alpha}\right)c_2 = 0 \quad \rightarrow \quad c_1 = c_2$$

$$\psi_{2+} = \frac{1}{\sqrt{2}}(|1, 0> +|0, 1>)$$

For $E_{2-} = E_2^{(0)} + E_{2-}^{(1)}$:

$$\left(\frac{a}{2\alpha}\right)c_1 + \left(\frac{a}{2\alpha}\right)c_2 = 0 \quad \rightarrow \quad c_1 = -c_2$$

$$\psi_{2-} = \frac{1}{\sqrt{2}}(|1, 0> -|0, 1>)$$

Example - 13.4 : A plane rigid rotator having a moment of inertia I and an electric dipole moment **d** is placed in a homogeneous electric field **E**. By considering the electric field as a perturbation, determine the first nonvanishing correction to the energy levels of the rotator.

Solution : The Schrödinger equation of a plane rigid rotator is

$$-\frac{\hbar^2}{2I}\frac{d^2\psi}{d\phi^2} = E\psi$$

where ϕ is the angle of rotation about the z–axis. The energies and the normalized wave functions are found to be:

$$E_m^{(0)} = \frac{\hbar^2 m^2}{2I} \quad , \quad \psi_m^{(0)}(\phi) = \frac{1}{\sqrt{2\pi}}e^{im\phi} \quad , \quad m = 0, \mp 1, \mp 2, \cdots$$

the levels for $m \neq 0$ being doubly degenerate, since states with $+m$ and with $-m$ have the same energy.

Treating the electric field as a perturbation, the Hamiltonian of the system becomes

$$H = H_0 + H_1 = -\frac{\hbar^2}{2I}\frac{d^2}{d\phi^2} - Ed\cos\phi$$

Since the parity operator P [defined by $Pf(\phi) = f(-\phi)$] commutes with both H and H_1, the perturbation theory for nondegenerate levels can be used. We have, then

$$< m|H_1|m' >= \int_0^{2\pi} \psi_m^{(0)*} H_1 \psi_{m'}^{(0)} d\phi$$

$$= -\frac{Ed}{2\pi}\int_0^{2\pi} e^{i(m'-m)\phi} \cos\phi d\phi = \begin{cases} 0 & , \ m' \neq m \pm 1 \\ -\frac{1}{2}Ed & , \ m' = m \pm 1 \end{cases}$$

Therefore, $E_m^{(1)} = 0$,

$$E_m^{(2)} = \frac{|< m|H_1|m-1 >|^2}{E_m^{(0)} - E_{m-1}^{(0)}} + \frac{|< m|H_1|m+1 >|^2}{E_m^{(0)} - E_{m+1}^{(0)}} = \frac{IE^2d^2}{\hbar^2(4m^2 - 1)}$$

and hence, to the second order of perturbation, we have

$$E_m = E_m^{(0)} + E_m^{(1)} + E_m^{(2)} = \frac{\hbar^2 m^2}{2I} + \frac{IE^2d^2}{\hbar^2(4m^2 - 1)}$$

Example - 13.5 : A particle of mass m in a harmonic oscillator potential $V(x) = \frac{1}{2}mw^2x^2$ is subject to a small perturbing potential of the same type, namely $V'(x) = \lambda x^2$. The energy spectrum of such a system is given by $E_n' = (n+1/2)\hbar\bar{w}$ where $\bar{w} = w\sqrt{1+2\lambda/mw^2}$. (a) Expand E_n' for small λ to $\mathcal{O}(\lambda)^2$. (b) Evaluate the first– and second–order shifts in energy, and compare the results to the exact answer in part (a).

Solution : (a) The exact energy is expanded as

$$E_n' = (n+\frac{1}{2})\hbar\bar{w} = (n+\frac{1}{2})\hbar w\sqrt{1+\frac{2\lambda}{mw^2}}$$

$$\cong (n+\frac{1}{2})\hbar w\left(1+\frac{\lambda}{mw^2} - \frac{\lambda^2}{2m^2w^4} + \cdots\right)$$

(b) The first order shift is

$$E_n^{(1)} = < n|V'(x)|n > = \lambda < n|x^2|n > = \lambda(n + \frac{1}{2})\frac{\hbar}{mw}$$

The second order term is

$$E_n^{(2)} = \sum_k \frac{|< n|V'(x)|k >|^2}{(E_n^{(0)} - E_k^{(0)})} = \lambda^2 \sum_k \frac{|< n|x^2|k >|^2}{(E_n^{(0)} - E_k^{(0)})}$$

Using the fact that $E_n^{(0)} - E_k^{(0)} = (n - k)\hbar w$ and

$$< |x|k > = \left(\frac{\hbar}{2mw}\right)[(2n + 1)\delta_{n,k} + \sqrt{(n + 1)(n + 2)}\delta_{n+2,k} + \sqrt{n(n - 1)}\delta_{n-2,k}]$$

we can write

$$E_n^{(2)} = \lambda^2 \left(\frac{\hbar^2}{4m^2 w^2}\right)\left[\frac{(n + 1)(n + 2)}{\hbar w(-2)} + \frac{n(n - 1)}{\hbar w(+2)}\right]$$

$$= -\lambda^2 \left(\frac{\hbar}{2m^2 w^3}\right)(n + \frac{1}{2})$$

$E_n^{(1)}$ and $E_n^{(2)}$ values correspond to the second and third terms in E_n', as expected.

13.4   Problems

Problem - 13.1 : Find the energy spectrum of a system whose Hamiltonian is

$$H = H_0 + H_1 = -\frac{\hbar^2}{2m}\frac{d^2}{dx^2} + \frac{1}{2}mw^2 x^2 + ax^3 + bx^4$$

where a and b are small constants (the anharmonic oscillator).

Problem - 13.2 : A linear harmonic oscillator is perturbed by an electric field of strength $\mathcal{E}$. If the oscillating mass has charge $-e$, the perturbing Hamiltonian becomes $H_1 = e\mathcal{E}x$. Determine the perturbation correction to the energy through second order.

Problem - 13.3 : Calculate the first order perturbation energy of a nonrigid rotator.

Problem - 13.4 : A rigid rotator, which is not restricted to move in a plane, having a moment of inertia I and an electric dipole moment $\mathbf{d}$ is placed in a homogeneous electric field $\mathbf{E}$. By considering the electric field as a perturbation, determine the first nonvanishing correction to the energy levels of the rotator.

Problem - 13.5 : A charged particle is bound in a harmonic oscillator potential $V = \frac{1}{2}kx^2$. The system is placed in an external electric field E that is constant in space and time. Calculate the shift of the energy of the ground state to order E^2.

Chapter 14

The Variational Method

14.1 The variational principle

The variational method is useful in obtaining the bound state energies and wave functions of a time–independent Hamiltonian H. The variational method is often the method of choice for studying complex systems such as many–electron atoms and molecules. The overwhelming majority of its practical applications involve numerical computation.

Let us denote by E_n the eigenvalues of the Hamiltonian H and let ψ_n be the corresponding orthonormal eigenfunctions, and assume that H has at least one discrete eigenvalue

$$H\psi_n = E_n\psi_n \quad ; \quad n \geq 1 \tag{14.1}$$

Let ϕ be an arbitrary normalizable function, and let $E[\phi]$ be the functional

$$E[\phi] = \frac{<\phi|H|\phi>}{<\phi|\phi>} \tag{14.2}$$

here the integration is extended over the full range of all the coordinates of the system.

If the function ϕ is identical to one of the exact eigenfunctions ψ_n of H, then $E[\phi]$ will be identical to the corresponding exact eigenvalue E_n. Any function ϕ for which the functional $E[\phi]$ is stationary is an eigenfunction of the discrete spectrum of H. Therefore, if ϕ and ψ_n differ by an arbitrary infinitesimal variation $\delta\phi$,

$$\phi = \psi_n + \delta\phi \tag{14.3}$$

then the corresponding first–order variational of $E[\phi]$ vanishes:

$$\delta E = 0 \tag{14.4}$$

and the eigenfunctions of H are solutions of the variational equation, Eq.(14.4). Varying Eq.(14.2) we get

$$\delta <\phi|\phi> +E <\delta\phi|\phi> +E <\phi|\delta\phi> = <\delta\phi|H|\phi> + <\phi|H|\delta\phi> \tag{14.5}$$

Since $<\phi|\phi>$ is assumed to be finite and nonvanishing, Eq.(14.4) is equivalent to

$$<\delta\phi|(H - E)\phi> + <\phi|(H - E)|\delta\phi> = 0 \tag{14.6}$$

Each term in Eq.(14.6) may be set separately equal to zero. Using the fact that H is Hermitian, $H = H^\dagger$, each term in Eq.(14.6) is separately equivalent to the Schrödinger equation

$$(H - E[\phi])\phi = 0 \tag{14.7}$$

Thus any function $\phi = \psi_n$ for which the functional in Eq.(14.2) is stationary is an eigenfunction of H corresponding to the eigenvalue $E_n = E[\psi_n]$. If ϕ and ψ_n differ by $\delta\phi$, the variational principle, Eq.(14.4), implies that the leading term of the difference between $E[\phi]$ and the true eigenvalue E_n is quadratic in $\delta\phi$. As a result, errors in the approximate energy are of second order in $\delta\phi$ when the energy is calculated from the functional in Eq.(14.2). Therefore, a first order error in the wave function leads to only a second error in the energy.

An important additional property of the functional $E[\phi]$, Eq.(14.2), is that it provides an **upper bound** to the exact ground state energy E_0. To show this let us expand the arbitrary, normalizable function ϕ in the complete set of orthonormal eigenfunctions ψ_n of H. That is

$$\phi = \sum_n a_n \psi_n \tag{14.8}$$

Substituting Eq.(14.8) in Eq.(14.2), we find that

$$E[\phi] = \frac{\sum_n |a_n|^2 E_n}{\sum_n |a_n|^2} \tag{14.9}$$

where we have used the fact that

$$H\psi_n = E_n\psi_n \quad , \quad < \phi|\phi >= \sum_n |a_n|^2$$

If we now subtract E_0, the lowest energy eigenvalue, from both sides of Eq.(14.9) we have

$$E[\phi] - E_0 = \frac{\sum_n |a_n|^2 (E_n - E_0)}{\sum_n |a_n|^2} \tag{14.10}$$

Since $E_n \geq E_0$, the right hand side of Eq.(14.10) is nonnegative, so that

$$E_0 \leq E[\phi] \tag{14.11}$$

and the functional $E[\phi]$ gives an upper bound — or in other words a minimum principle for the ground state energy. This property, Eq.(14.11), constitutes the basis of the **Rayleigh–Ritz variational method** for the approximate calculation of E_0, and the inequality in Eq.(14.11) represents the **variational theorem**. This method consists in evaluating the quantity $E[\phi]$ by using trial functions ϕ which depend on a certain number of variational parameters, and then to minimize $E[\phi]$ with respect to these parameters in order to obtain the best approximation of E_0 allowed by the form chosen for ϕ.

The Rayleigh–Ritz variational method can also be used to obtain an upper bound for the energy of an excited state, provided that the trial function ϕ is made orthogonal to all the energy eigenfunctions corresponding to states having a lower energy than the energy level considered.

14.2   Linear variation functions

In practice the trial function ϕ is constructed by choosing a certain number (N) of linearly independent functions $\{\psi_n\} \; ; \; n = 1, \cdots, N$

$$\phi = \sum_{n=1}^{N} c_n \psi_n \tag{14.12}$$

where the coefficients $\{c_n\}$ are linear variational parameters which must be determined by minimising the functional $E[\phi]$ in order to obtain the best approximation to E_0. Substituting Eq.(14.12) in Eq.(14.2) we find that

$$E[\phi] = \frac{\sum_{n=1}^{N} \sum_{n'=1}^{N} c_n^* c_{n'} H_{nn'}}{\sum_{n=1}^{N} \sum_{n'=1}^{N} c_n^* c_{n'} S_{nn'}} \tag{14.13}$$

where we have set

$$H_{nn'} = <\psi_n|H|\psi_{n'}> \quad , \quad S_{nn'} = <\psi_n|\psi_{n'}> \tag{14.14}$$

If the functions ψ_n are orthonormal, then $S_{nn'} = \delta_{nn'}$.

In order to find the values of the variational parameters $\{c_n\}$ which minimise $E[\phi]$, we apply the variational principle, Eq.(14.4), to $E[\phi]$, Eq.(14.13). This is done by differentiating $E[\phi]$ with respect to each c_n or c_n^*, expressing that

$$\frac{\partial E}{\partial c_n} = 0 \quad or \quad \frac{\partial E}{\partial c_n^*} = 0 \tag{14.15}$$

This procedure results in a system of N linearly and homogeneous equations in the variables $\{c_n\}$, namely

$$\sum_{n=1}^{N} c_n (H_{nn'} - E S_{nn'}) = 0 \quad ; \quad n' = 1, 2, \cdots, N \tag{14.16}$$

This set of linear equations is known as **secular equations**. The necessary and sufficient condition for this system to have a nontrivial solution is that the determinant of the coefficients vanishes. That is,

$$|H_{nn'} - E S_{nn'}| = 0 \tag{14.17}$$

This determinant is known as **secular determinant**.

Let $\{E_n^{(N)}\} \; ; \; n = 1, 2, \cdots, N$ be the N roots of this determinant; the lowest root is an upper bound to the ground state energy E_0. The other roots are upper bounds to excited state energies of the system. Substituting the calculated energies from Eq.(14.17) into the secular equations, Eq.(14.16), and solving for the coefficients $\{c_n\}$ we then obtain the corresponding *optimum* approximation to the wave function.

If we construct a new trial function ϕ' containing an additional basis function ψ_{N+1}, namely

$$\phi' = \sum_{n=1}^{N+1} c_n \psi_n \tag{14.18}$$

the new $(N + 1)$ roots of the determinantal equation are separated by the old (N) roots. This property, which is illustrated in Figure 14.1., is known as the **Hyleraas–Undheim theorem**.

Each root $E_i^{(N)}$ of the determinantal equation, Eq.(14.17), is an upper bound to the corresponding exact eigenvalue E_i.

The variational principle is extremly powerful, and very easy to use. To find the ground state energy of a system one has to write down a trial wave function with a large number of adjustable parameters, calculate $< H >$, and optimize the parameters to get the lowest possible value. Even if ψ has no relation to the true wave function, one often gets miraculously accurate values for ground state energy, E_0. Naturally, if we have some way of guessing a realistic ψ, so much the better. The only trouble with the method is that we never know for sure how close we are to the target. All we can be certain of is that we have got an upper bound.

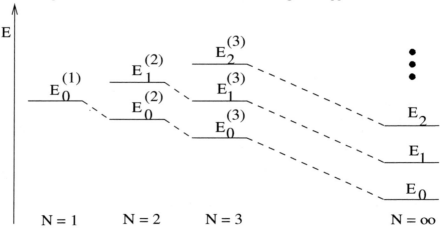

Fig. 14.1. Approximate eigenvalues given by the Rayleigh–Ritz variational method with linear trial functions.

14.3   Worked examples

Example - 14.1 : Assume that the lowest eigenfunction of the simple harmonic oscillator is approximated by $u_0 = Ne^{-cx^2}$ and use the variational method to determine c. Also, calculate the energy associated with this eigenfunction.

Solution : First we determine N from the normalization condition,

$$\int_{-\infty}^{\infty} N^2 e^{-2cx^2}\, dx = N^2 \sqrt{\frac{\pi}{2c}} = 1 \quad \rightarrow \quad N^2 = \sqrt{\frac{2c}{\pi}}$$

From the harmonic oscillator

$$H = -\frac{\hbar^2}{2m}\frac{d^2}{dx^2} + \frac{1}{2}kx^2$$

The quantity

$$E[u_0] = \frac{< u_0|H|u_0 >}{< u_0|u_0 >}$$

is given by (since $< u_0|u_0 >= 1$)

$$E[u_0] = \int_{-\infty}^{\infty} u_0^* H u_0 dx$$

$$= \sqrt{\frac{2c}{\pi}} \int_{-\infty}^{\infty} e^{-cx^2} \left(-\frac{\hbar^2}{2m}\frac{d^2}{dx^2} + \frac{1}{2}kx^2\right) e^{-cx^2} dx$$

$$= \sqrt{\frac{2c}{\pi}} \left[\left(\frac{k}{2} - \frac{2\hbar^2 c^2}{m}\right) \int_{-\infty}^{\infty} x^2 e^{-2cx^2} dx + \frac{c\hbar^2}{m} \int_{-\infty}^{\infty} e^{-2cx^2} dx\right]$$

Since

$$\frac{d^2}{dx^2}\left(e^{-cx^2}\right) = (4c^2 x^2 - 2c)e^{-cx^2}$$

from tables,

$$\int_0^{\infty} x^{2n} e^{-ax} dx = \frac{1 \cdot 3 \cdot 5 \cdots (2n-1)}{2^{n+1}a^n}\sqrt{\frac{\pi}{a}}$$

Therefore

$$E[u_0] = \sqrt{\frac{2c}{\pi}} \left[\left(\frac{k}{2} - \frac{2\hbar^2 c^2}{m}\right)\frac{1}{4c}\sqrt{\frac{\pi}{2c}} + \frac{c\hbar^2}{m}\sqrt{\frac{\pi}{2c}}\right]$$

$$= \frac{k}{8c} + \frac{c\hbar^2}{2m}$$

The minimum value of $E[u_0]$ is found by setting

$$\frac{dE}{dc} = 0 \quad \rightarrow \quad c^2 = \frac{km}{4\hbar^2} \quad \rightarrow \quad c = \frac{\pi}{h}\sqrt{km} = \frac{\alpha}{2}.$$

The variational method shows that

$$u_0 = \left(\frac{2c}{\pi}\right)^{1/4} e^{-cx^2} = \left(\frac{\alpha}{\pi}\right)^{1/4} e^{-\frac{\alpha}{2}x^2}$$

This is the exact ground state eigenfunction for the oscillator because of the form chosen for u_0. The energy associated with u_0:

$$E[u_0] = \frac{k}{8\pi}\frac{h}{\sqrt{km}} + \frac{\pi}{h}\sqrt{km}\frac{\hbar^2}{2m}$$

$$= \frac{1}{2\pi}\sqrt{\frac{k}{m}}\left(\frac{h}{2}\right) = \frac{h}{2}\left(\frac{1}{2\pi}\sqrt{\frac{k}{m}}\right) = \frac{h\nu_0}{2}.$$

Example - 14.2 : The exact solution for an attractive delta function well, $V(x) = -\alpha\delta(x)$, is the following:

$$\psi(x) = \frac{\sqrt{m\alpha}}{\hbar}e^{-m\alpha|x|/\hbar^2} \quad ; \quad E_{exact} = -\frac{m\alpha^2}{2\hbar^2}$$

(There is only one bound state.) Considering the trial wave function $\psi(x) = Ae^{-bx^2}$ calculate the ground state energy of the delta function potential. Here b is the variational parameter.

Solution : The normalization constant A can be determined from

$$|A|^2 \int_{-\infty}^{\infty} e^{-2bx^2}\,dx = |A|^2\sqrt{\frac{\pi}{2b}} = 1 \quad \rightarrow \quad A = \left(\frac{2b}{\pi}\right)^{1/4}$$

$$<H> = <T> + <V>$$

$$<T> = -\frac{\hbar^2}{2m}|A|^2\int_{-\infty}^{\infty}e^{-bx^2}\frac{d^2}{dx^2}\left(e^{-bx^2}\right)dx = \frac{\hbar^2 b}{2m}$$

$$<V> = -\alpha|A|^2\int_{-\infty}^{\infty}e^{-2bx^2}\delta(x)dx = -\alpha\sqrt{\frac{2b}{\pi}}$$

Therefore

$$<H> = \frac{\hbar^2 b}{2m} - \alpha\sqrt{\frac{2b}{\pi}} \quad , \quad \frac{\partial <H>}{\partial b} = 0 \quad \rightarrow \quad b = \frac{2m^2\alpha^2}{\pi\hbar^4}$$

So $\quad <H>_{min} = -\dfrac{m\alpha^2}{\pi\hbar^2} > E_{exact}$ (as expected).

Example - 14.3 : Using the variational method, find an approximate energy and wave function for the $2s$ state of the H–atom.

Solution : Since the $2s$ state wave function has spherical symmetry and vanishes as $r \to \infty$, we may suppose that it contains a factor $e^{-br/a}$, where b is an adjustable parameter and a is the radius of the first Bohr orbit. We need another parameter in order to be able to apply the variational method to the first excited level, and this can be introduced by multiplying $e^{-br/a}$ by a factor $(1 + \gamma r/a)$. We therefore take as a reasonable trial function for the $2s$ state wave function the form

$$\psi_{2s}(r) = A\left(1 + \gamma\frac{r}{a}\right)e^{-b\frac{r}{a}}$$

From

$$E_{2s} = \frac{<\psi_{2s}|H|\psi_{2s}>}{<\psi_{2s}|\psi_{2s}>} \quad , \quad \int |\psi_{2s}|^2 d\tau = 1$$

we obtain

$$\gamma = \frac{1}{3}(1+b) \quad , \quad A = \left[\frac{3b^5}{\pi a^3(a^2 - ab + b^2)}\right]^{1/2}$$

and

$$E_{2s} = \frac{e^2}{a}\left[-\frac{b}{2} + \frac{7b^2}{6} - \frac{b^2}{2(b^2 - b + 1)}\right] \quad ; \quad \frac{\partial E_{2s}}{\partial b} = 0 \quad \rightarrow \quad b = \frac{1}{2}$$

and thus

$$E_{2s} = -\frac{e^2}{8a} \quad , \quad \psi_{2s}(r) = \frac{1}{\sqrt{8\pi a^3}}\left(1 - \frac{r}{2a}\right)e^{-\frac{r}{2a}}$$

Example - 14.4 : Apply the variation method to the particle in a box problem in one–dimension. Let $V = 0$ for $-1 \le x \le +1$ and $V = \infty$ otherwise. Then use $f_1 = 1 - x^2$ and $f_2 = 1 - x^4$ to construct the trial function $u = c_1 f_1 + c_2 f_2$. (a) Calculate the energy with this function and compare with the exact solution. (b) Determine the best values of c_1 and c_2.

Solution : (a) The secular determinant is in the form:

$$\begin{vmatrix} H_{11} - ES_{11} & H_{12} - ES_{12} \\ H_{21} - ES_{21} & H_{22} - ES_{22} \end{vmatrix} = 0$$

The matrix elements are defined as

$$H_{ij} = <f_i|H|f_j> = \int_{-1}^{+1} f_i H f_j dx \; ; \; S_{ij} = <f_i|f_j> = \int_{-1}^{+1} f_i f_j dx$$

where

$$H = -\frac{\hbar^2}{2m}\frac{d^2}{dx^2}$$

Using the given functions, one gets

$$H_{11} = \frac{4\hbar^2}{3m} \quad , \quad H_{12} = H_{21} = \frac{8\hbar^2}{5m} \quad , \quad H_{22} = \frac{16\hbar^2}{7m}$$

$$S_{11} = \frac{16}{15} \; , \; S_{12} = S_{21} = \frac{128}{105} \; , \; S_{22} = \frac{64}{45}$$

Letting $E' = Em/\hbar^2$ secular determinant gives

$$E'^2 - 14E' + 15.75 = 0 \quad \rightarrow \quad E' = 1.23 \; , \; 12.77$$

Eigenvalues:

$$E_1 = 1.23\frac{\hbar^2}{m} \quad , \quad E_3 = 12.77\frac{\hbar^2}{m}$$

Only even functions of x are involved in this problem, since both f_1 and f_2 are even functions.

Exact eigenvalues:

$$E_n = \frac{n^2 h^2}{8ma^2} = \frac{n^2\hbar^2\pi^2}{2ma^2} = \left(\frac{n^2\pi^2}{2a^2}\right)\frac{\hbar^2}{m}$$

$$E_{n=1} = \frac{\pi^2}{8}\frac{\hbar^2}{m} \cong 1.23\frac{\hbar^2}{m} \; ; \quad E_{n=3} = \frac{9\pi^2}{8}\frac{\hbar^2}{m} \cong 11.10\frac{\hbar^2}{m}$$

(b) From the secular equation

$$(H_{11} - ES_{11})c_{1n} + (H_{12} - ES_{12})c_{2n} = 0$$

$$\left(\frac{4}{3} - E'\frac{16}{15}\right)c_{1n} + \left(\frac{8}{5} - E'\frac{128}{105}\right)c_{2n} = 0$$

For $E_1' = 1.23$:

$$[1.3333 - 1.0666(1.23)]c_{11} + [1.6 - 1.2191(1.23)]c_{21} = 0$$

$$c_{21} = -0.2118c_{11}$$

$$u_1 = c_{11}f_1 + c_{21}f_2 = c_{11}[f_1 - 0.2118f_2]$$

$$< u_1|u_1 >= 1 \quad \rightarrow \quad c_{11} = 1.276$$

For $E_3' = 12.77$:

$$[1.3333 - 1.0666(12.77)]c_{13} + [1.6 - 1.2191(12.77)]c_{23} = 0$$

$$c_{23} = -0.8798c_{13}$$

$$u_3 = c_{13}f_1 + c_{23}f_2 = c_{13}[f_1 - 0.8798f_2]$$

$$< u_3|u_3 >= 1 \quad \rightarrow \quad c_{13} = 6.6625$$

14.4 Problems

Problem - 14.1 : Calculate the ground state energy of a hydrogen atom using the trial wave function $\psi = e^{-r/a}$, where a is an adjustable parameter.

Problem - 14.2 : Calculate the ground state energy of helium atom.

Problem - 14.3 : Consider a simple harmonic oscillator having the Hamiltonian

$$H_0 = \frac{p^2}{2\mu} + \frac{1}{2}kx^2$$

which is perturbed by the quartic potential ax^4 where $ax^4 \ll H_0$. Assume that the ground state wave function can be approximated by

$$\psi_0 = c_0\psi_0^{(0)} + c_2\psi_2^{(0)}$$

where $\psi_0^{(0)}$ and $\psi_2^{(0)}$ are the lowest even eigenfunctions for the unperturbed oscillator. Use normalization and the variation method to determine the best values of the constants c_0 and c_2, and calculate the approximate ground state energy for the perturbed oscillator.

Problem - 14.4 : Using the variational method, find an approximate energy and wave function for the 2s state of the hydrogen atom.

Problem - 14.5 : To describe the ground state of the H-atom we choose the trial wave functions

$$\psi_1(r, \alpha) = e^{-\alpha r} \ , \quad \psi_2(r, \alpha) = e^{-\alpha r^2} \ , \quad \psi_3(r, \alpha) = (1 + \alpha r)^{-2}$$

where α is a parameter to be determined from a minimization of the ground state energy. (a) Calculate the corresponding energies for these functions. (b) Determine the values of α that correspond to a minimum for $E_i(\alpha)$, $i = 1, 2, 3$. (c) Determine the minimum ground state energy for the three trial functions. (d) Comment on the accuracy of the energies determined in (c). (d) Comment on the behavior of the wave functions for various r values.

Chapter 15

The WKB Approximation

15.1 Turning points

The WKB approximation, named after Wentzel, Kramers, and Brillouin, is also called the classical approximation, since it deals with situations in which $\hbar$ is small compared to the action. Although this method is usually applied to one–dimensional problems, it can also be applied to three–dimensional problems, if the potential is spherically symmetric and a radial differential equation can be separated.

Here we will discuss the method briefly for one–dimensional case. Consider the Schrödinger equation for a particle in one–dimension

$$\frac{d^2\psi}{dx^2} + \frac{2m}{\hbar^2}[E - V(x)]\psi = 0 \tag{15.1}$$

This equation can be written in the form

$$\frac{d^2\psi}{dx^2} + \frac{p^2}{\hbar^2}\psi = 0 \tag{15.2}$$

where p is the classical momentum at the point x:

$$p = \sqrt{2m[E - V(x)]} \tag{15.3}$$

If the energy is high enough so that the wavelength $\lambda = h/p$ is very short in the classical region, and if the potential energy function changes smoothly, then the wave function can be approximated by

$$\psi(x) = \phi(x)\exp\left[\mp\frac{i}{\hbar}\int^x p(x)dx\right] \tag{15.4}$$

where $\phi(x)$ is a slowly varying function. This is the basis for the WKB method.

By the substitution of Eq.(15.4) into Eq.(15.2), the differential equation for the function $\phi(x)$ is obtained:

$$\frac{\hbar}{ip}\frac{d^2\phi}{dx^2} \mp \left(2\frac{d\phi}{dx} + \frac{1}{p}\frac{dp}{dx}\phi\right) = 0 \tag{15.5}$$

It is assumed that $\hbar/p$ is small, compared to the other dimensions of the problem, and that ϕ varies slowly. Hence, we neglect the first term in Eq.(15.5) and obtain

$$\frac{2}{\phi}\frac{d\phi}{dx} + \frac{1}{p}\frac{dp}{dx} = \frac{d}{dx}\ln(\phi^2 p) = 0 \tag{15.6}$$

which yields

$$\phi = \frac{K}{\sqrt{p}} \quad (K = \text{a constant}) \tag{15.7}$$

The approximate wave function is therefore

$$\psi_{WKB} = \frac{K}{\sqrt{p}} \exp\left(\mp \frac{i}{\hbar} \int^x p(x)dx \right) \tag{15.8}$$

The classical approximation is expected to hold in regions where the fractional change in p in one wavelength is small, that is, where

$$\left| \frac{p'\lambda}{p} \right| = \left| \frac{hp'}{p^2} \right| \ll 1 \tag{15.9}$$

The WKB approximation is valid under similar conditions. This condition, Eq.(15.9), will, in general, be fulfilled for problems where the mass is large, the energy high, and the potential smooth. However, it is clear that the WKB solutions cannot be valid near a classical turning point, where the momentum is zero (see Figure 15.1).

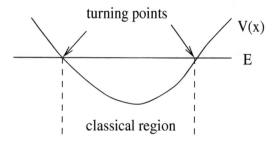

Fig. 15.1. Relative positions of potential $V(x)$, and total energy E of a particle under the influence of a potential.

Classical region : $E \geq V(x)$
Turning points : $E = V(x)$, $p(x) = 0$

Those points x, at which $E = V(x)$, are where the kinetic energy changes sign. The way of handling solutions near turning points is a litle too technical. The basic idea is as follows: Let us consider a turning point as shown in Figure 15.2.

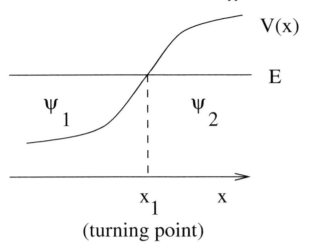

(turning point)

Fig. 15.2. Solutions in the region of turning point.

We have a solution to the left of the turning point, where $E > V(x)$, of the form

$$\psi_1(x) = A \exp\left(-\frac{i}{\hbar}\int_x^{x_1} p(x)dx\right) \tag{15.10}$$

and a solution to the right of the turning point, where $E < V(x)$, of the form

$$\psi_2(x) = A \exp\left(-\frac{1}{\hbar}\int_{x_1}^x p(x)dx\right) \tag{15.11}$$

and what we need is a formula that interpolates between them. In the vicinity of the turning point one can approximate $p/\hbar$ by a straight line over a small interval, and solve the Schrödinger equation exactly. Since it is a second order equation, there are two adjustable constants, one of which is fixed by fitting the solution to ψ_1 and the other by fitting to ψ_2.

15.2 The connection formulas

At the turning points, where $E = V(x)$, the kinetic energy changes sign. Suppose that x_1 is a turning point with the allowed region $x > x_1$ (see Figure 15.3).

In the region of turning point we can approximate $V(x)$ by the tangent to $V(x)$ at $x = x_1$. Thus we have near $x = x_1$

$$V(x) \cong E - \frac{\hbar^2}{2m}c^2(x - x_1) \tag{15.12}$$

This leads to

$$k^2(x) = \frac{2m}{\hbar^2}(E - V) = c^2(x - x_1) \quad , \quad x > x_1 \tag{15.13}$$

$$\kappa^2(x) = \frac{2m}{\hbar^2}(V - E) = -c^2(x - x_1) \quad , \quad x < x_1 \tag{15.14}$$

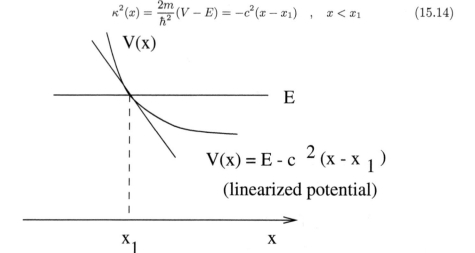

$$V(x) = E - c^2(x - x_1)$$

(linearized potential)

Fig. 15.3. Linearized potential in the region of turning point.

Substituting these linear approximations for k^2 and κ^2 into the Schrödinger equation we find

$$\frac{d^2\psi}{dx^2} + c^2(x - x_1)\psi = 0 \quad , \quad x > x_1 \tag{15.15}$$

with solution

$$\psi_a^\pm(x) = A_\pm y^{1/3} J_{\pm\frac{1}{3}}(y) \tag{15.16}$$

where $J(y)$ is an ordinary Bessel function for allowed region,

$$y = \int_{x_1}^x k(x')dx' = \frac{2}{3}c(x - x_1)^{3/2} \quad , \quad x > x_1 \tag{15.17}$$

$$\frac{d^2\psi}{dx^2} - c^2(x - x_1)\psi = 0 \quad , \quad x < x_1 \tag{15.18}$$

with solution

$$\psi_f^\pm(x) = B_\pm z^{1/3} I_{\pm\frac{1}{3}}(z) \tag{15.19}$$

where $I(z)$ is a modified Bessel function for forbidden region,

$$z = \int_x^{x_1} \kappa(x')dx' = \frac{2}{3}c(x - x_1)^{3/2} = \frac{2}{3}c|x - x_1|^{3/2} \quad , \quad x < x_1 \tag{15.20}$$

Considering the asymptotic forms of Bessel functions we obtain the connection formulas at x_1 (after some algebra):

$$\kappa^{-1/2}e^{-z} \quad \rightarrow \quad 2k^{-1/2}\cos(y - \frac{\pi}{4}) \tag{15.21}$$

$$k^{-1/2}\cos(y - \frac{\pi}{4}) \quad \rightarrow \quad \frac{1}{2}\kappa^{-1/2}e^{-z} \tag{15.22}$$

$$k^{-1/2}\cos(y - \frac{\pi}{4} + \phi) \quad \rightarrow \quad \kappa^{-1/2}\sin\phi e^z \quad , \quad \phi \neq 0 \tag{15.23}$$

These three connection formulas are applicable to all problems. In using them it is important to remember two things: y and z are so defined as to increase as we move away from the turning points, and the formulas may only be used to connect the solution on the left of the arrow to the solution on the right of the arrow and never in the reverse direction.

15.3 The WKB approximation to a potential well

Consider a potential well as shown in Figure 15.4 rising to infinity for both very large x and $-x$.

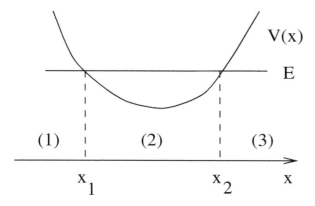

Fig. 15.4. A potential well in 1D.

For the energy E shown there are two turning points at x_1 and x_2. Regions 1 and 3 are classically forbidden and region 2 is classically allowed. Thus we want exponentially damped solutions in regions 1 and 3.

The possible solutions in these three regions may be

$$\psi_0 = \frac{A}{\sqrt{\kappa(x)}} e^{-\int_x^{x_1} \kappa(x')dx'} \quad , \quad x < x_1 \tag{15.24}$$

$$\psi_0 = \frac{2A}{\sqrt{k(x)}} \cos\left[\int_x^{x_1} k(x')dx' - \frac{\pi}{4}\right] \quad , \quad x_1 < x < x_2 \tag{15.25}$$

$$\psi_0 = \frac{B}{\sqrt{\kappa(x)}} e^{-\int_{x_2}^x \kappa(x')dx'} \quad , \quad x > x_2 \tag{15.26}$$

where

$$\kappa(x) = \frac{1}{\hbar}\sqrt{2m(V(x) - E)} \quad , \quad k(x) = \frac{1}{\hbar}\sqrt{2m(E - V(x))}$$

and hence by applying the connection formula at x_2

$$\frac{1}{\sqrt{\kappa}} e^{-z} \quad \longrightarrow \quad \frac{2}{\sqrt{k}} \cos\left(y - \frac{\pi}{4}\right)$$

we get that for $x_1 < x < x_2$

$$\psi_0 = \frac{B}{\sqrt{k(x)}} \cos \left[\int_x^{x_2} k(x')dx' - \frac{\pi}{4} \right] \tag{15.27}$$

$$= \frac{B}{\sqrt{k(x)}} \cos \left[\int_x^{x_1} k(x')dx' - \frac{\pi}{4} + \int_{x_1}^{x_2} k(x')dx' \right]$$

$$= \frac{B}{\sqrt{k(x)}} \cos \left[-\int_{x_1}^{x} k(x')dx' + \frac{\pi}{4} + \int_{x_1}^{x_2} k(x')dx' - \frac{\pi}{2} \right]$$

$$= \frac{B}{\sqrt{k(x)}} \cos \left[\int_{x_1}^{x} k(x')dx' - \frac{\pi}{4} - \alpha \right] \quad , \quad x_1 < x < x_2$$

where

$$\alpha = \int_{x_1}^{x_2} k(x')dx' - \frac{\pi}{2}$$

This solution must coincide with the solution in Eq.(15.25) of region 2 $(x_1 < x < x_2)$; for this region we require that $\alpha = n\pi$; $n = 0,1,2,\cdots$ and $B = (-1)^n A$. Hence we get the energy quantization

$$\int_{x_1}^{x_2} k(x')dx' = (n + \frac{1}{2})\pi \quad , \quad n = 0,1,2,\cdots \tag{15.28}$$

But

$$k(x) = \frac{1}{\hbar}\sqrt{2m(E-V)} = \frac{2\pi}{h}p$$

so that $\alpha = n\pi$ reads

$$2\int_{x_1}^{x_2} pdx = (n + \frac{1}{2})h \tag{15.29}$$

Eq.(15.29) can be written as $\oint pdx = (n+1/2)h$, the symbol $\oint$ denotes the integral taken over a complete cycle of the classical motion. This is the Bohr–Sommerfeld quantization condition, except that n is replaced by $n + 1/2$. Since the classical approximation is reliable only when n is large, this modification is not of great significance.

15.4   The WKB approximation to a potential barrier

Consider a potential barrier as shown in Figure 15.5.

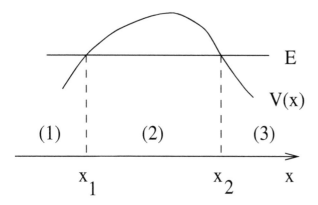

Fig. 15.5. A potential barrier in 1D.

Regions 1 and 3 are classically allowed, and region 2 is classically forbidden. Consider a particle with energy E incident from the left. In region 1 we have the WKB solution

$$\psi_1(x) = k^{-1/2} \exp\left[-i\left(\int_x^{x_1} k(x')dx' - \frac{\pi}{4}\right)\right] \tag{15.30}$$

$$+Rk^{-1/2}\exp\left[i\left(\int_x^{x_1} k(x')dx' - \frac{\pi}{4}\right)\right]$$

or

$$\psi_1(x) = k^{-1/2}(1+R)\cos\left[\int_x^{x_1} k(x')dx' - \frac{\pi}{4}\right] \tag{15.31}$$

$$-k^{-1/2}i(1-R)\sin\left[\int_x^{x_1} k(x')dx' - \frac{\pi}{4}\right]$$

Using the connection formulas, Eqs.(15.22) and (15.23), with $\phi = -\frac{\pi}{2}$, ψ_1 connects to the solution in region 2 given by

$$\psi_2(x) = \kappa^{-1/2}\left[\frac{1}{2}(1+R)e^{-\int_{x_1}^x \kappa(x')dx'} + i(1-R)e^{\int_{x_1}^x \kappa(x')dx'}\right] \tag{15.32}$$

we have in region 3

$$\psi_3(x) = Tk^{-1/2}\exp\left[i\left(\int_{x_2}^x k(x')dx' - \frac{\pi}{4}\right)\right] \tag{15.33}$$

or

$$\psi_3(x) = Tk^{-1/2} \cos\left[\int_{x_2}^{x} k(x')dx' - \frac{\pi}{4}\right] \tag{15.34}$$

$$+ Tk^{-1/2}i\sin\left[\int_{x_2}^{x} k(x')dx' - \frac{\pi}{4}\right]$$

Using the connection formulas, Eqs.(15.22) and (15.23), with $\phi = -\frac{\pi}{2}$, the WKB solution in region 2 becomes

$$\psi_2(x) = T\kappa^{-1/2}\left[\frac{1}{2}e^{-\int_x^{x_2}\kappa(x')dx'} - ie^{\int_x^{x_2}\kappa(x')dx'}\right] \tag{15.35}$$

To compare Eq.(15.35) with Eq.(15.32) we write

$$-\int_x^{x_2}\kappa(x')dx' = -\int_{x_1}^{x_2}\kappa(x')dx' + \int_{x_1}^{x}\kappa(x')dx' \tag{15.36}$$

Defining

$$S = e^{-\int_{x_1}^{x_2}\kappa(x')dx'} \tag{15.37}$$

Eq.(15.35) can be written as

$$\psi_2(x) = T\kappa^{-1/2}\left[\frac{1}{2}Se^{\int_x^{x_2}\kappa(x')dx'} - iS^{-1}e^{-\int_{x_1}^{x}\kappa(x')dx'}\right] \tag{15.38}$$

Since this expression must coinside with Eq.(15.32) we find that

$$\frac{1}{2}(1+R) = -iTS^{-1} \quad, \quad i(1-R) = \frac{1}{2}ST \tag{15.39}$$

Solving for T and R we find

$$T = \frac{iS}{1 + \frac{S^2}{4}} \cong iS \quad, \quad R = \frac{1 - \frac{S^2}{4}}{1 + \frac{S^2}{4}} \cong 1 - \frac{S^2}{2} \tag{15.40}$$

Here $S \ll 1$, WKB approximation requires this condition. The quantities T and R are the transmission and reflection amplitudes, respectively. The corresponding probabilities, transmission probability, $|T|^2$, and reflection probability, $|R|^2$, can be expressed as follows:

$$|T|^2 = S^2 \quad, \quad |R|^2 = |1 - \frac{S^2}{2}|^2 \cong 1 - S^2 \tag{15.41}$$

These probability values satisfy the general relation,

$$|T|^2 + |R|^2 = 1 \tag{15.42}$$

15.5 Worked examples

Example - 15.1 : Calculate the energy values of the half–harmonic oscillator.

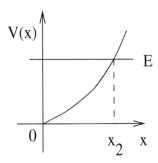

Solution : In this case $\psi(0) = 0$, where

$$\psi(x) = \frac{2A}{\sqrt{k(x)}} \cos\left[\int_x^{x_2} k(x')dx' - \frac{\pi}{4}\right]$$

$$k(x) = \frac{1}{\hbar}\sqrt{2m(E - V(x))}$$

The boundary condition $\psi(0) = 0$ requires the relation

$$\int_0^{x_2} k(x')dx' - \frac{\pi}{4} = (n + \frac{1}{2})\pi \quad ; \quad n = 0, 1, 2, \cdots$$

$$\int_0^{x_2} \sqrt{2m(E - \frac{1}{2}mw^2x^2)}dx = \left[(n + \frac{1}{2})\pi + \frac{\pi}{4}\right]\hbar = (n + \frac{3}{4})\pi\hbar$$

$$\frac{1}{2}mw^2x^2 = E$$

$$\int_0^{x_2} \sqrt{2m\frac{1}{2}mw^2(x_2^2 - x^2)}dx = (n + \frac{3}{4})\pi\hbar$$

$$mw\int_0^{x_2} \sqrt{x_2^2 - x^2}dx = mw\frac{\pi}{4}x_2^2 = \frac{\pi E}{2w} = (n + \frac{3}{4})\pi\hbar$$

Therefore,

$$E = 2(n + \frac{3}{4})\hbar w = (2n + \frac{3}{2})\hbar w \quad ; \quad n = 0, 1, 2, \cdots$$

$$E = \left(\frac{3}{2}, \frac{7}{2}, \frac{11}{2}, \cdots\right)\hbar w$$

In this case the WKB approximation gives the exact odd energies of the full har–monic oscillator.

Example - 15.2 : Analyze the cold emission of electrons from a metal.

Solution : In the absence of an external electric field, the electrons are bound by a potential as shown in the figure (a).

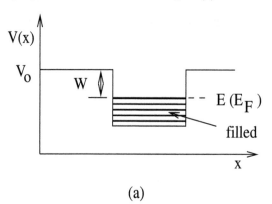

(a)

W is the work function, which is the energy required to remove an electron from the highest occupied state (Fermi level). When an external electric field $\mathcal{E}$ is applied to the metal, the potential at the surface takes the form as shown in figure (b).

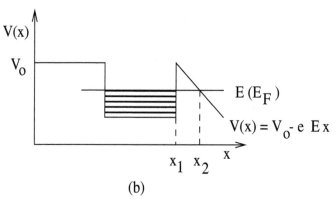

(b)

In this case the potential barrier has a finite width, and electrons are able to escape. The variation of cold emission with work function and applied field is obtained from transmission probability,

$$T^2 = S^2 = e^{-2\int_{x_1}^{x_2} \kappa(x)dx} \quad , \quad \kappa(x) = \frac{1}{\hbar}\sqrt{2m(V(x) - E_F)}$$

We set $x_1 = 0$, and find x_2 as follows:

$$V_0 - e\mathcal{E}x_2 = V_0 - W \quad \rightarrow \quad x_2 = \frac{W}{e\mathcal{E}}$$

Also,

$$V - E_F = V_0 - e\mathcal{E}x - E_F = W - e\mathcal{E}x$$

This is the condition for transmission at the Fermi level. The transmission probability therefore is

$$T^2 = e^{-\frac{2}{\hbar}\int_{x_1}^{x_2}\sqrt{2m(V(x)-E_F)}dx}$$

$$= \exp\left\{ -\frac{2}{\hbar} \int_0^{\frac{W}{e\mathcal{E}}} \sqrt{2m(W - e\mathcal{E}x)}dx \right\} = \exp\left\{ -\frac{4}{3} \frac{\sqrt{2m}}{\hbar} \frac{W^{3/2}}{e\mathcal{E}} \right\}$$

This equation is referred to as the Fowler–Nordheim equation.

Example - 15.3 : Use the WKB approximation to find the bound state energy for the potential $V(x) = -\frac{\hbar^2 a^2}{m} \operatorname{sech}^2(ax)$, where a is a positive constant.

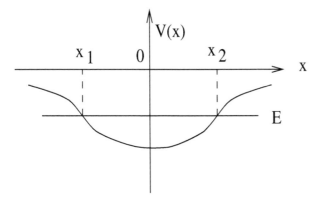

Solution : From the quantization condition

$$\int_{x_1}^{x_2} p(x)dx = (n + \frac{1}{2})\pi\hbar \ ; \ \ n = 0, 1, 2, \cdots$$

$$(n + \frac{1}{2})\pi\hbar = 2\int_0^{x_2} \sqrt{2m(E + \frac{\hbar^2 a^2}{m} \operatorname{sech}^2(ax))}dx$$

$$= 2\sqrt{2}\hbar a \int_0^{x_2} \sqrt{\operatorname{sech}^2(ax) + \frac{mE}{\hbar^2 a^2}}dx$$

$E = V(x_2)$ defines x_2 , $E = -\frac{\hbar^2 a^2}{m} \operatorname{sech}^2(ax_2)$, let $b = -\frac{mE}{\hbar^2 a^2}$, $z = \operatorname{sech}^2(ax)$.
Then (after some transformations and simplifications)

$$(n + \frac{1}{2})\pi = \sqrt{2} \int_b^1 \frac{1}{z} \sqrt{\frac{z - b}{1 - z}}dz = \sqrt{2}\pi(1 - \sqrt{b})$$

Therefore, one obtains $\sqrt{b} = 1 - \frac{1}{\sqrt{2}}(n + \frac{1}{2})$. Since $b > 0$, $(n + \frac{1}{2}) < \sqrt{2} \rightarrow$ $n < -\frac{1}{2} + \sqrt{2} \cong 0.914$. So the only possible n is 0 (there is only one bound state). Therefore, $n = 0$ and $\sqrt{b} = 1 - \frac{1}{\sqrt{2}}(0 + \frac{1}{2}) = 1 - \frac{1}{2\sqrt{2}} \Rightarrow b \cong 0.42$.

$$E = -\frac{\hbar^2 a^2}{m}b = -0.42\frac{\hbar^2 a^2}{m} \ ; \ \ E_{exact} = -0.5\frac{\hbar^2 a^2}{m}$$

15.6   Problems

Problem - 15.1 : Obtain the energy values of a harmonic oscillator by the WKB method.

Problem - 15.2 : Considering the WKB approximation determine the eigenvalues of stationary motion of a particle moving in the gravitational field $V(z) = mgz$ above the surface of the earth at $z = 0$.

Problem - 15.3 : Find the condition under which the wave function of a particle moving in a one–dimensional potential can be written as

$$\psi(x) \approx \frac{A}{\sqrt{p}} e^{\frac{i}{\hbar} \int p \, dx} + \frac{B}{\sqrt{p}} e^{-\frac{i}{\hbar} \int p \, dx} \quad (1)$$

Problem - 15.4 : Derive the conditions for the validity of the WKB approximation for the one–dimensional time–independent Schrödinger equation, and show that the approximation must fail in the immediate neighborhood of a classical turning point.

Problem - 15.5 : A particle of mass m moves with zero angular momentum in a spherically symmetric attractive potential V(r). Write down the differential equation of radial motion, defining your radial wave function carefully and specifying the boundary conditions on it for bound states. What is the WKB eigenvalue condition for s–states in such a potential?

Chapter 16

Time–Dependent Perturbation Theory

16.1 Time–dependent Schrödinger equation

Systems described by a time–independent potential energy function, $V(\mathbf{r}, t) = V(\mathbf{r})$, are the subject of **quantum statics**. Such systems may be studied by the time–dependent Schrödinger equation

$$H\Psi = i\hbar \frac{\partial \Psi}{\partial t} \qquad (16.1)$$

here the wave function depends explicitly on time,

$$\Psi(\mathbf{r}, t) = \psi(\mathbf{r}) e^{-\frac{i}{\hbar} E t} \qquad (16.2)$$

where $\psi(\mathbf{r})$ satisfies the time–independent Schrödinger equation

$$H\psi = E\psi \qquad (16.3)$$

The absolute square of $\Psi(\mathbf{r}, t)$, $|\Psi|^2$, is time independent; all probabilities and expectation values are constant in time.

If a transition takes place between two energy levels, one can describe such a system by introducing a time–dependent potential, which is the subject of **quantum dynamics**.

If the time–dependent portion of the Hamiltonian is small compared to the time–independent part, it can be treated as a perturbation. Such problems are studied by the time–dependent perturbation theory. A typical example is the emission and absorption of radiation by an atom.

Consider an atomic system with two energy levels. Let us suppose that there are just two states of the unperturbed system, ψ_a and ψ_b. They are eigenstates of the unperturbed Hamiltonian H_0:

$$H_0 \psi_a = E_a \psi_a \quad , \quad H_0 \psi_b = E_b \psi_b \qquad (16.4)$$

and they are orthonormal

$$< \psi_a | \psi_b > = \delta_{ab} \qquad (16.5)$$

Any state can be expressed as a linear combination of them,

$$\Psi(0) = c_a \psi_a + c_b \psi_b \qquad (16.6)$$

In the absence of any perturbation, each component evolves with its characteristic exponential factor:

$$\Psi(t) = c_a \psi_a e^{-\frac{i}{\hbar} E_a t} + c_b \psi_b e^{-\frac{i}{\hbar} E_b t} \qquad (16.7)$$

We say that $|c_a|^2$ is the *probability that the particle is in state ψ_a with energy E_a.* Normalization of Ψ requires that

$$|c_a|^2 + |c_b|^2 = 1 \qquad (16.8)$$

At any time t, $t > 0$, let us represent the time–dependent perturbation part of the Hamiltonian as $H'(t)$. Since ψ_a and ψ_b form a complete set, the wave function $\Psi(t)$ can still be expressed as a linear combination of them. In this case the combination coefficients c_a and c_b are also time dependent,

$$\Psi(t) = c_a(t) \psi_a e^{-\frac{i}{\hbar} E_a t} + c_b(t) \psi_b e^{-\frac{i}{\hbar} E_b t} \qquad (16.9)$$

The problem is now to determine these coefficients. If initially, that is at $t = 0$, the system is in state ψ_a, then $c_a = 1$ and $c_b = 0$. At some later time t_1 if we find that $c_a(t_1) = 0$ and $c_b(t_1) = 1$, this means that the system underwent a transition from ψ_a to ψ_b.

We solve for $c_a(t)$ and $c_b(t)$ by demanding that $\Psi(t)$ satisfy the time–dependent Schrödinger equation,

$$H\Psi = i\hbar \frac{\partial \Psi}{\partial t} \quad , \quad H = H_0 + H'(t) \qquad (16.10)$$

From Eq.(16.9) and (16.10), we find

$$c_a(t)(H_0 \psi_a) e^{-\frac{i}{\hbar} E_a t} + c_b(t)(H_0 \psi_b) e^{-\frac{i}{\hbar} E_b t} + c_a(t)(H' \psi_a) e^{-\frac{i}{\hbar} E_a t} \qquad (16.11)$$

$$+ c_b(t)(H' \psi_b) e^{-\frac{i}{\hbar} E_b t} = i\hbar \left[\dot{c}_a(t) \psi_a e^{-\frac{i}{\hbar} E_a t} + \dot{c}_b(t) \psi_b e^{-\frac{i}{\hbar} E_b t} \right]$$

$$i\hbar \left[+ c_a(t) \psi_a (-\frac{i}{\hbar} E_a) e^{-\frac{i}{\hbar} E_a t} + c_b(t) \psi_b (-\frac{i}{\hbar} E_b) e^{-\frac{i}{\hbar} E_b t} \right]$$

Considering Eq.(16.4), the first two terms on the left of Eq.(16.11) cancel the last two terms on the right, and hence

$$c_a(t)(H' \psi_a) e^{-\frac{i}{\hbar} E_a t} + c_b(t)(H' \psi_b) e^{-\frac{i}{\hbar} E_b t} \qquad (16.12)$$

$$= i\hbar \left[\dot{c}_a(t) \psi_a e^{-\frac{i}{\hbar} E_a t} + \dot{c}_b(t) \psi_b e^{-\frac{i}{\hbar} E_b t} \right]$$

Taking the inner product of Eq.(16.12) with ψ_a, and considering the orthogonality of ψ_a and ψ_b, Eq.(16.5), we obtain

$$c_a(t) < \psi_a | H' | \psi_a > e^{-\frac{i}{\hbar} E_a t} + c_b(t) < \psi_a | H' | \psi_b > e^{-\frac{i}{\hbar} E_b t} \qquad (16.13)$$

$$= i\hbar \dot{c}_a(t) e^{-\frac{i}{\hbar} E_a t}$$

Using the notation $H'_{ij} = < \psi_i | H' | \psi_j >$ we may rewrite Eq.(16.13) as

$$\dot{c}_a(t) = -\frac{i}{\hbar} \left[c_a(t) H'_{aa} + c_b(t) H'_{ab} e^{-\frac{i}{\hbar}(E_b - E_a)t} \right] \qquad (16.14)$$

Similarly, the inner product of Eq.(16.12) with ψ_b picks out $\dot{c}_b(t)$;

$$c_a(t) < \psi_b|H'|\psi_a > e^{-\frac{i}{\hbar}E_a t} + c_b(t) < \psi_b|H'|\psi_b > e^{-\frac{i}{\hbar}E_b t} \tag{16.15}$$

$$= i\hbar \dot{c}_b(t) e^{-\frac{i}{\hbar}E_b t}$$

and hence

$$\dot{c}_b(t) = -\frac{i}{\hbar} \left[c_b(t) H'_{bb} + c_a(t) H'_{ba} e^{\frac{i}{\hbar}(E_b - E_a)t} \right] \tag{16.16}$$

Eqs.(16.14) and (16.16) determine $c_a(t)$ and $c_b(t)$; taken together, they are completely equivalent to the time–dependent Schrödinger equation, for a two–level system.

16.2   Time–dependent perturbation approximations

To be able to solve the equations, Eqs.(16.14) and (16.16), we have to do some approximations. Usually, the diagonal matrix elements of H' vanish, although there are some cases in which the diagonal terms are not zero. Let us assume that $H'_{aa} = H'_{bb} = 0$. In this case the equations simplify:

$$\dot{c}_a(t) = -\frac{i}{\hbar} H'_{ab} e^{-i\omega_0 t} c_b(t) \tag{16.17}$$

$$\dot{c}_b(t) = -\frac{i}{\hbar} H'_{ba} e^{i\omega_0 t} c_a(t) \tag{16.18}$$

where

$$\omega_0 = \frac{E_b - E_a}{\hbar} \tag{16.19}$$

Here we assume that $E_b > E_a$, so $\omega_0 > 0$. If H'_{ij} is small, we can solve Eqs.(16.17) and (16.18) by a process of successive approximations. Suppose that initially, at $t = 0$, the system (electron, say, in the two level atom) is in the lower state,

$$c_a(0) = 1 \quad , \quad c_b(0) = 0 \tag{16.20}$$

If there were no perturbation at all, they would stay this way forever. This condition may be considered as **zeroth–order** approximation,

$$c_a^{(0)}(t) = 1 \quad , \quad c_b^{(0)}(t) = 0 \tag{16.21}$$

To calculate the first–order approximation, we insert these values on the right side of Eqs.(16.17) and (16.18).

$$\frac{dc_a(t)}{dt} = 0 \quad \rightarrow \quad c_a^{(1)}(t) = 1 \tag{16.22}$$

$$\frac{dc_b(t)}{dt} = -\frac{i}{\hbar} H'_{ba} e^{i\omega_0 t} \quad \rightarrow \quad c_b^{(1)}(t) = -\frac{i}{\hbar} \int_0^t H'_{ba}(t') e^{i\omega_0 t'} dt' \tag{16.23}$$

Now we insert these expressions on the right of Eqs.(16.17) and (16.18) to obtain the second–order approximation.

$$\frac{dc_a(t)}{dt} = -\frac{i}{\hbar} H'_{ab} e^{-i\omega_0 t} (-\frac{i}{\hbar}) \int_0^t H'_{ba}(t') e^{i\omega_0 t'} dt' \quad \rightarrow \tag{16.24}$$

$$c_a^{(2)}(t) = 1 - \frac{1}{\hbar^2} \int_0^t H'_{ab}(t') e^{-i\omega_0 t'} dt' \left[\int_0^{t'} H'_{ba}(t'') e^{i\omega_0 t''} dt'' \right] dt'$$

while $c_b(t)$ is unchanged, $c_b^{(2)}(t) = c_b^{(1)}(t)$.

In principle, we could continue this iteration indefinitely, always inserting the nth–order approximation into the right side of Eqs.(16.17) and (16.18), and solving for the $(n+1)^{th}$–order. Notice that $c_a(t)$ is modified in every *even* order, and $c_b(t)$ is every *odd* order. These coefficients must, of course, obey the condition, at every iteration,

$$|c_a^{(n)}(t)|^2 + |c_b^{(n)}(t)|^2 = 1 \tag{16.25}$$

16.3   Sinusoidal perturbations

Let us suppose that the perturbation has sinusoidal, that is harmonic, time dependence of the form

$$H'(\mathbf{r}, t) = V(\mathbf{r}) \cos(\omega t) \tag{16.26}$$

so that

$$H'_{ab} = V_{ab} \cos(\omega t) \tag{16.27}$$

where

$$V_{ab} = <\psi_a|V|\psi_b> \tag{16.28}$$

We assume that the diagonal matrix elements vanish. In most practical applications first–order approximation is enough as perturbation correction. To first–order we have (from Eq.(16.23))

$$c_b(t) \cong -\frac{i}{\hbar} V_{ab} \int_0^t \cos(\omega t') e^{i\omega_0 t'} dt' \tag{16.29}$$

$$= -\frac{iV_{ba}}{2\hbar} \int_0^t \left[e^{i(\omega_0+\omega)t'} + e^{i(\omega_0-\omega)t'} \right] dt'$$

$$= -\frac{V_{ba}}{2\hbar} \left[\frac{e^{i(\omega_0+\omega)t} - 1}{\omega_0 + \omega} + \frac{e^{i(\omega_0-\omega)t} - 1}{\omega_0 - \omega} \right]$$

Here ω is the driving frequency, and ω_0 is the transition frequency or natural frequency. If we consider that ω's are very close to ω_0, we assume

$$\omega_0 + \omega \gg |\omega_0 - \omega| \tag{16.30}$$

with this condition the first term in Eq.(16.29) may be omitted, then we have

$$c_b(t) \cong -\frac{V_{ba}}{2\hbar} \frac{e^{i(\omega_0-\omega)t/2}}{\omega_0 - \omega} \left[e^{i(\omega_0-\omega)t/2} - e^{-i(\omega_0-\omega)t/2} \right] \tag{16.31}$$

$$-i\frac{V_{ba}}{\hbar}\frac{\sin[(\omega_0-\omega)t/2]}{\omega_0-\omega}e^{i(\omega_0-\omega)t/2}$$

The transition probability, the probability that a particle which started out in the state ψ_a (at $t=0$) will be found, at time t, in the state ψ_b, is

$$P_{a\to b}(t)=|c_b(t)|^2\cong\frac{|V_{ba}^2|}{\hbar^2}\frac{\sin^2[(\omega_0-\omega)t/2]}{(\omega_0-\omega)^2}\tag{16.32}$$

The most remarkable feature of this result is that the transition probability oscillates sinusoidally with respect to time t (see Figure 16.1).

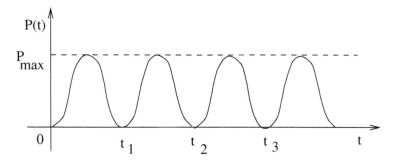

Fig. 16.1. Transition probability as a function of time, for a sinusoidal perturbation. $P_{max}=|V_{ba}|^2/\hbar^2(\omega_0-\omega)^2$; $t_1=2\pi/(\omega_0-\omega)$, $t_2=4\pi/(\omega_0-\omega)$, $t_3=6\pi/(\omega_0-\omega)$, $\cdots$.

The probability of a transition is greatest when the driving frequency is close to the natural frequency ω_0 (see Figure 16.2).

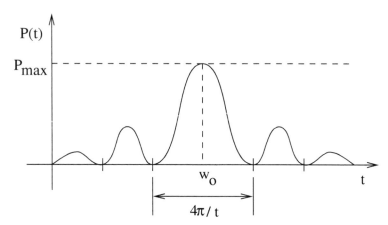

Fig. 16.2. Transition probability as a function of driving frequency. $P_{max}=|V_{ba}|^2t^2/4\hbar^2$.

The central peak, located at ω_0, gets higher and narrower as time goes on. In practice this approximation works for relatively small t.

16.4 Emission and absorption of radiation

Radiation is another name for electromagnetic wave. An electromagnetic wave consists of transverse and mutually perpendicular oscillating electric and magnetic fields (see Figure 16.3).

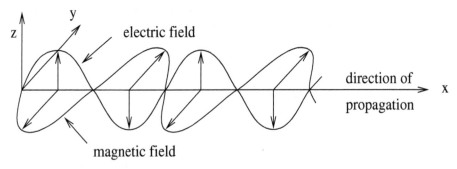

Fig. 16.3. An electromagnetic wave.

When an atom is put in an electromagnetic field, in other words, when an atom interacts with an electromagnetic wave, the atom responds primarily to the electric field component of the field. If the wavelength of the field is larger than the size of the atom, the spatial variation (distortion, let us say) in the field can be ignored, the atom, then, is exposed to a sinusoidally oscillating electric field

$$\mathbf{E} = E_0 \cos(\omega t)\hat{k} \tag{16.33}$$

Let us assume that the radiation is monochromatic (single valued wavelength) and polarized along the z–direction, that is the electric field oscillations take place along z–axis. The perturbation Hamiltonian, which describes the interaction of an atom with radiation, actually the interaction of an atomic electron (bound electron in the atom) with the radiation, may be expressed as (from general relation, $U = -\vec{\mu} \cdot \vec{E}$)

$$H' = eE_0z \cos(\omega t) \tag{16.34}$$

The matrix element of H' becomes

$$H'_{ba} = \wp E_0 \cos(\omega t) \tag{16.35}$$

where the electric dipole moment $\wp$ is defined as

$$\wp = e < \psi_b|z|\psi_a > \tag{16.36}$$

Typically, ψ is an even or odd function of z; in either case $z|\psi|^2$ is odd, and integrates to zero. This shows that the assumption that the diagonal matrix elements of H' vanish is physically meaningful. Therefore, the interaction of radiation with atoms is governed by the simple relation

$$V_{ba} = \wp E_0 \tag{16.37}$$

Absorption, Stimulated Emission, and Spontaneous Emission:

In atomic and molecular systems there are basicaly three types of transitions, which are called absorption, stimulated emission, and spontaneous emission. These transitions are shown schematically in Figure 16.4.

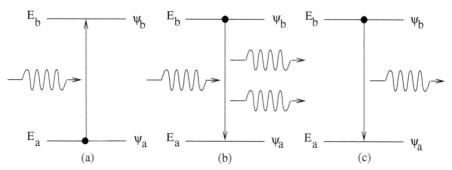

Fig. 16.4. Possible transitions in an atom. (a) Absorption, (b) stimulated emission, (c) spontaneous emission.

If an atom starts out in the lower state ψ_a, and we shine a polarized monochromatic beam of light (radiation) on it, the probability of a transition to the upper state ψ_b is given by Eq.(16.32), which (in view of Eq.(16.37)) takes the form

$$P_{a \to b}(t) = \left(\frac{\wp E_0}{\hbar} \right)^2 \frac{\sin^2[(\omega_0 - \omega)t/2]}{(\omega_0 - \omega)^2} \tag{16.38}$$

In this process, the atom absorbs energy $E_b - E_a = \hbar\omega_0$ from the electromagnetic field. We say that it has **absorbed a photon**.

We could go back and do the whole derivation for a system that starts off in the upper state, that is $c_a(0) = 0$, $c_b(0) = 1$. In this case we get the probability of a transition down to the lower level:

$$P_{b \to a}(t) = |c_a(t)|^2 = \left(\frac{\wp E_0}{\hbar} \right)^2 \frac{\sin^2[(-\omega_0 + \omega)t/2]}{(-\omega_0 + \omega)^2} \tag{16.39}$$

This expression is exactly the same as given in Eq.(16.38). If the particle is in the upper state, and we shine light on it, it can make a transition to the lower state; the probability of such a transition is exactly the same as for a transition upward from the lower state. This process is called **stimulated emission**.

In the case of stimulated emission the electromagnetic field gains energy $\hbar\omega_0$ from the atom; we say that one photon went in and two photons came out, the original one that caused the transition plus another one from the transition itself. This raises the possibility of amplification. This is the principle behind the laser (**L**ight **A**mplification by **S**timulated **E**mission of **R**adiation). It is essential (for laser action) to get a majority of the atoms into the upper state (a so–called population inversion), because absorption (which costs one photon) competes with stimulated emission (which produces one).

There is a third mechanism (in addition to absorption and stimulated emission) by which radiation interacts with matter; it is called **spontaneous emission**. Here an atom in the upper state makes a transition downward, with the release of a photon but without any applied radiation to initiate the process. This is the mechanism that accounts for the normal decay of an atomic excited state.

16.5 Incoherent perturbations

The energy density in a radiation field is given by

$$u = \frac{1}{2}\varepsilon_0 E_0^2 \tag{16.40}$$

So the transition probability, Eqs.(16.38) and (16.39), is proportional to the energy density of the field:

$$P_{b \to a}(t) = \frac{2u}{\varepsilon_0 \hbar^2} \wp^2 \frac{\sin^2[(\omega - \omega_0)t/2]}{(\omega - \omega_0)^2} \tag{16.41}$$

This is for a monochromatic perturbation, consisting of a single frequency ω. In many applications the system is exposed to electromagnetic waves at a whole range of frequencies; in this case $u \to \rho(\omega)d\omega$ is the energy density in the frequency range $d\omega$, and the resultant transition probability is calculated by integration,

$$P_{b \to a}(t) = \frac{2}{\varepsilon_0 \hbar^2} \wp^2 \int_0^\infty \rho(\omega) \left\{ \frac{\sin^2[(\omega - \omega_0)t/2]}{(\omega - \omega_0)^2} \right\} d\omega \tag{16.42}$$

The term in curly brackets is sharply peaked about ω_0, whereas $\rho(\omega)$ is relatively broad; in that case we may as well replace $\rho(\omega)$ by $\rho(\omega_0)$ and take it outside the integral,

$$P_{b \to a}(t) = \frac{2\wp^2}{\varepsilon_0 \hbar^2} \rho(\omega_0) \int_0^\infty \frac{\sin^2[(\omega - \omega_0)t/2]}{(\omega - \omega_0)^2} d\omega \tag{16.43}$$

$$= \frac{2\wp^2}{\varepsilon_0 \hbar^2} \rho(\omega_0) \frac{\pi t}{2} = \frac{\pi \wp^2}{\varepsilon_0 \hbar^2} \rho(\omega_0) t$$

This time the transition probability is proportional to t, not oscillatory as in the case of monochromatic radiation. The **transition rate**, that is the transition per unit time, $R = \frac{dP}{dt}$, is now a constant,

$$R_{b \to a} = \frac{P_{b \to a}(t)}{dt} = \frac{2\wp^2}{\varepsilon_0 \hbar^2} \rho(\omega_0) \tag{16.44}$$

So far, we have assumed that the perturbing wave is coming in along the x–direction, and polarized in the z–direction. If radiation is coming from all directions, and with all possible polarizations, the energy in the field, $\rho(\omega)$, is shared equally among these different modes. In this case we must take the average of $|\hat{n} \cdot \vec{\wp}|^2$ instead of $\wp^2$, where $\vec{\wp}$ is the general form of Eq.(16.36),

$$\vec{\wp} = e < \psi_b | bfr | \psi_a > \tag{16.45}$$

The average must be taken over both polarization ($\hat{n}$) and over all incident directions. This averaging can be carried out as follows:

Averaging over polarization:

For propagation in the z–direction, the two possible polarizations are $\hat{i}$ and $\hat{j}$, so the polarization average is

$$(\hat{n} \cdot \vec{\wp})_p^2 = \frac{1}{2}[(\hat{i} \cdot \vec{\wp})^2 + (\hat{j} \cdot \vec{\wp})^2] = \frac{1}{2}(\wp_x^2 + \wp_y^2) = \frac{1}{2}\wp^2 \sin^2\theta \qquad (16.46)$$

where θ is the angle between $\vec{\wp}$ and the direction of propagation.

Averaging over propagation direction:

Let us set polar axis along $\vec{\wp}$ and integrate over all propagation directions to get the polarization–propagation average,

$$(\hat{n} \cdot \vec{\wp})_{pp}^2 = \frac{1}{4\pi} \int (\frac{1}{2}\wp^2 \sin^2\theta) \sin\theta d\theta d\phi \qquad (16.47)$$

$$= \frac{\wp^2}{4} \int_0^\pi \sin^3\theta d\theta = \frac{\wp^2}{3}$$

So the transition rate for stimulated emission from state b to state a, under the influence of incoherent, unpolarized light incident from all directions, is

$$R_{b\to a} = \frac{\pi}{3\varepsilon_0\hbar^2}|\vec{\wp}|^2 \rho(\omega_0) \qquad (16.48)$$

This is a special case of Fermi's **golden rule** for time–dependent harmonic perturbation in first order.

Spontaneous emission:

Let us consider a container, which contains N atoms, N_a of them in the lower state (ψ_a), and N_b of them in the upper state (ψ_b), $N = N_a + N_b$; N is very large, so N_a and N_b are also very large. Let us denote the **spontaneous emission rate** by A, so that the number of particles leaving the upper state per unit time by this process is $N_b A$. The transition rate for stimulated emission, that is the **stimulated emission rate**, is proportional to the energy density of the radiation field (external electromagnetic field). Let us represent this rate by $B_{ba}\rho(\omega_0)$. The number of particles leaving the upper state per unit time by this process is $N_b B_{ba}\rho(\omega_0)$. The **absorption rate** is also proportional to $\rho(\omega_0)$. Let us represent this by $B_{ab}\rho(\omega_0)$. The number of particles joining the upper level per unit time is $N_a B_{ab}\rho(\omega_0)$. Therefore, the particle rate in each state may be expressed as

$$-\frac{dN_a}{dt} = \frac{dN_b}{dt} = N_a B_{ab}\rho(\omega_0) - N_b A - N_b B_{ba}\rho(\omega_0) \qquad (16.49)$$

If the system, atoms in the container and external radiation field, is in thermal equilibrium, the number of particles in each level (or **level population**) is *almost* constant. In that case $\frac{dN_a}{dt} = \frac{dN_b}{dt} = 0$, and it follows that

$$\rho(\omega_0) = \frac{A}{\frac{N_a}{N_b}B_{ab} - B_{ba}} \qquad (16.50)$$

On the other hand, the number of particles with energy E, in thermal equilibrium at temperature T, is proportional to the Boltzmann factor, e^{-E/k_BT}, so

$$\frac{N_a}{N_b} = frac e^{-E_a/k_BT} e^{-E_b/k_BT} = e^{\hbar\omega_0/k_BT} \tag{16.51}$$

and hence

$$\rho(\omega_0) = \frac{A}{B_{ab}e^{\hbar\omega_0/k_BT} - B_{ba}} \tag{16.52}$$

According to Planck's blackbody radiation law, the energy density of thermal radiation is given as

$$\rho(\omega) = \frac{\hbar}{\pi^2 c^3} \frac{\omega^3}{e^{\hbar\omega/k_BT} - 1} \tag{16.53}$$

Comparing the two equations, Eqs.(16.52) and (16.53), we conclude that

$$B_{ab} = B_{ba} \tag{16.54}$$

and

$$A = \frac{\hbar\omega^3}{\pi^2 c^3} B_{ba} \tag{16.55}$$

Eq.(16.54) shows that the transition rate for stimulated emission is the same as for absorption. The quantity $B_{ba}\rho(\omega_0)$, the stimulated emission rate, is actually identical to the expression given in Eq.(16.48)

$$B_{ba}\rho(\omega_0) = R_{b\to a} = \frac{\pi}{3\varepsilon_0\hbar^2}|\vec{\wp}|^2\rho(\omega_0) \tag{16.56}$$

and it follows that the spontaneous emission rate is

$$A = \frac{\omega^3\hbar}{\pi^2 c^3} B_{ba} = \frac{\omega^3|\vec{\wp}|^2}{3\pi\varepsilon_0\hbar c^3} \tag{16.57}$$

The product of A and $\hbar\omega$ defines the **radiation power**,

$$P_{ba} = A\hbar\omega = \frac{\omega^4|\vec{\wp}|^2}{3\pi\varepsilon_0 c^3} \tag{16.58}$$

Let us now consider a container full of atoms, with $N_b(t)$ of them in the excited state (upper level). As a result of spontaneous emission, this number will decrease as time goes on. This is possible if the system is not in thermal equilibrium. In a time interval dt we will lose a fraction Adt of them,

$$dN_b(t) = -N_b(t)dt \tag{16.59}$$

Solving for $N_b(t)$, we find

$$N_b(t) = N_b(0)e^{-At} \tag{16.60}$$

It is clear that the number remaining in the excited state decreases exponentially, with a time constant

$$\tau = \frac{1}{A} \tag{16.61}$$

We call this the **lifetime** of the state; it is the time it takes for $N_b(t)$ to reach $1/e$ of its initial value.

There is another quantity, which is known as the **half–life** $(t_{1/2})$ of an excited state. Half–life is the time it would take for half the atoms in a large sample to make a transition. The relation between $t_{1/2}$ and τ can be found from Eq.(16.59),

$$N_b(t) = N_b(0)e^{-t/\tau} \tag{16.62}$$

After $t_{1/2}$, $N_b(t) = \frac{1}{2}N_b(0)$, so $\frac{1}{2} = e^{-t/\tau}$, or $2 = e^{t/\tau}$, so $\frac{t}{\tau} = \ln 2$, or

$$t_{1/2} = \tau \ln 2 \tag{16.63}$$

The spontaneous emission formula, Eq.(16.57), gives the transition rate for $\psi_b \rightarrow \psi_a$ regardless of any other allowed states (transition between two states only, transition for a two level system). Actually an excited atom has many different decay modes, that is, ψ_b can decay to a large number of different lower energy states, $\psi_{a_1}, \psi_{a_1}, \cdots$ In that case the transition rates add

$$A = \sum_i A_i \tag{16.64}$$

and the net lifetime is

$$\tau = \frac{1}{\sum_i A_i} \tag{16.65}$$

An important parameter in radiation analysis is the **oscillator strength**, defined as the dimensionless form

$$f_{ba} = \frac{2m\omega}{3\hbar e^2}|\wp|^2 \tag{16.66}$$

The spontaneous emission rate, Eq.(16.57), and the radiation power, Eq.(16.58), can be expressed in terms of oscillator strength,

$$A = \frac{\omega^3|\wp|^2}{3\pi\varepsilon_0\hbar c^3} = \frac{\omega^2 e^2}{2\pi\varepsilon_0 c^3 m}f_{ba} \tag{16.67}$$

$$P_{ba} = \frac{\omega^4|\wp|^2}{3\pi\varepsilon_0 c^3} = \frac{\hbar\omega^3 e^2}{3\pi\varepsilon_0 c^3 m}f_{ba} \tag{16.68}$$

Oscillator strengths obey the sum rule,

$$\sum_{n'} f_{nn'} = 1 \tag{16.69}$$

16.6 Selection rules

The calculation of spontaneous emission rates has been reduced to a matter of evaluating matrix elements of the form $< \psi_b|\mathbf{r}|\psi_a >$. These integrals are very often zero; they are usually called dipole integrals. If the integral is nonzero then the transition is said to be **allowed transition**; if it is zero then the transition is said to be **forbidden transition**. Let us try to find the conditions (or rules) for the allowed transitions in hydrogenic atoms; these rules are usually called **selection rules** for electric dipole transitions. We may specify the states with the usual quantum numbers n, l, and m; $\psi_{nlm}(\mathbf{r}) \equiv |nlm>$. The matrix elements (dipole integrals) are $< n'l'm'|\mathbf{r}|nlm >$. The angular momentum commutation relations and the hermiticity of the angular momentum operators give a set of constraints on these matrix elements.

There are three different electric dipole transition integrals, one for each coordinate component: $< n'l'm'|x|nlm >$, $< n'l'm'|y|nlm >$, and $< n'l'm'|z|nlm >$. For hydrogenic wave functions these integrals are nonzero only if the following conditions are satisfied:

$$\Delta l = l' - l = \mp 1 \tag{16.70}$$

these are the electric dipole transition selection rules in hydrogenic atoms for l, and

$$\Delta m = m' - m = 0, \mp 1 \tag{16.71}$$

these are the electric dipole transition selection rules in hydrogenic atoms for m. The physical meaning of these rules may be interpreted as follows: The photon carries spin 1, so the rules for addition of angular momentum would allow $l' = l + 1$ or $l' = l - 1$. On the other hand conservation of angular momentum requires that the atom give up whatever the photon takes away.

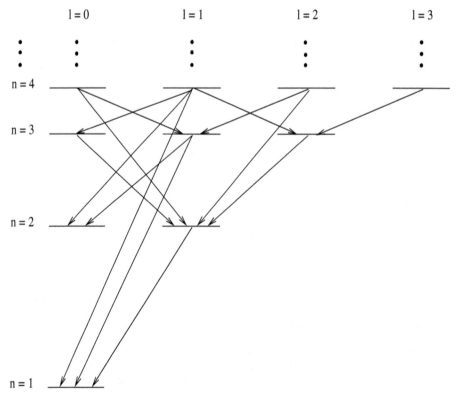

Fig. 16.5. Allowed transitions (electric dipole type) in hydrogenic atoms.

Evidently not all transitions to lower–energy states can proceed by spontaneous emission; some are forbidden by the selection rules. Some of these transitions are shown in Figure 16.5. Note that there is no transition from $2s$ state. It can not decay, because there is no lower–energy state with $l = 1$. It is called a **metastable state**, and its lifetime is much longer than that of the $2p$ states. Metastable states also decay either by collisions or by what are called forbidden transitions, or by multiphoton emission.

16.7 Worked examples

Example - 16.1 : Suppose a delta function type perturbation of the form $H' = U\delta(t - t_0)$. Assume that $U_{aa} = U_{bb} = 0$, and let $U_{ab} = \alpha$. If $c_a(-\infty) = 1$ and $c_b(-\infty) = 0$, find $c_a(t)$ and $c_b(t)$, and the transition probability $P_{a \to b}$.

Solution : From the equations

$$\dot{c}_a = -\frac{i}{\hbar} H'_{ab} e^{-iw_0 t} c_b \quad \text{and} \quad \dot{c}_b = -\frac{i}{\hbar} H'_{ba} e^{iw_0 t} c_a$$

$$\frac{dc_a}{dt} = -\frac{i}{\hbar} \alpha \delta(t - t_0) e^{-iw_0 t}$$

$$c_a(t) = c_a(-\infty) - \frac{i}{\hbar} \alpha \int_{-\infty}^{t} \delta(t' - t_0) e^{-iw_0 t'} c_b(t') dt'$$

$$c_a(t) = 1 - \frac{i}{\hbar} \alpha e^{-iw_0 t} c_b(t_0) \theta(t - t_0)$$

$$\frac{dc_b}{dt} = -\frac{i}{\hbar} \alpha^* \delta(t - t_0) e^{iw_0 t}$$

$$c_b(t) = c_b(-\infty) - \frac{i}{\hbar} \alpha^* \int_{-\infty}^{t} \delta(t' - t_0) e^{-iw_0 t'} c_a(t') dt'$$

$$c_b(t) = -\frac{i}{\hbar} \alpha^* e^{iw_0 t} c_a(t_0) \theta(t - t_0)$$

Taking $\theta(0) = 1/2$, then

$$c_a(t_0) = 1 - \frac{i\alpha}{2\hbar} e^{-iw_0 t_0} c_b(t_0) \quad , \quad c_b(t_0) = -\frac{i\alpha^*}{2\hbar} e^{iw_0 t_0} c_a(t_0)$$

so

$$c_a(t_0) = 1 - \frac{|\alpha|^2}{4\hbar^2} c_a(t_0) \quad \text{or} \quad c_a(t_0) = \frac{1}{1 + \frac{|\alpha|^2}{4\hbar^2}}$$

$$c_b(t_0) = -\frac{i\alpha^*}{2\hbar} \frac{e^{iw_0 t_0}}{1 + \frac{|\alpha|^2}{4\hbar^2}}$$

Therefore,

$$c_a(t) = 1 - \left(\frac{\frac{|\alpha|^2}{2\hbar^2}}{1 + \frac{|\alpha|^2}{4\hbar^2}} \right) \theta(t - t_0)$$

$$c_b(t) = -\frac{i\alpha^*}{\hbar} e^{iw_0 t_0} \frac{1}{1 + \frac{|\alpha|^2}{4\hbar^2}} \theta(t - t_0)$$

For $t < t_0$, $c_a(t) = 1$, $c_b(t) = 0$, so $|c_a|^2 + |c_b|^2 = 1$
For $t > t_0$,

$$c_a(t) = 1 - \frac{\frac{|\alpha|^2}{2\hbar^2}}{1 + \frac{|\alpha|^2}{4\hbar^2}} \quad , \quad c_b(t) = -\frac{i\alpha^*}{\hbar} \frac{e^{iw_0 t_0}}{1 + \frac{|\alpha|^2}{4\hbar^2}}$$

again $|c_a|^2 + |c_b|^2 = 1$. Transition probability is

$$|c_b|^2 = \frac{|\alpha|^2/\hbar^2}{(1 + |\alpha|^2/4\hbar^2)^2}$$

Example - 16.2 : Calculate the lifetime for each of the four $n = 2$ states of hydrogen.

Solution : All matrix elements of z are zero except

$$< 100|z|210 >= \frac{2^8}{\sqrt{2}3^5} a$$

For x and y components the nonzero matrix elements are

$$< 100|x|21 \mp 1 >= \pm\frac{2^7}{3^5} a \quad , \quad < 100|y|21 \mp 1 >= -i\frac{2^7}{3^5} a.$$

So $\quad < 100|\mathbf{r}|200 >= 0 \quad , \quad < 100|\mathbf{r}|210 >= \frac{2^7\sqrt{2}}{3^5} a\mathbf{k} \quad ,$

$$< 100|\mathbf{r}|21 \mp 1 >= \frac{2^7}{3^5} a(\pm\mathbf{i} - i\mathbf{j})$$

Therefore, $|\mathcal{P}|^2 = 0$ for $(200 \;\rightarrow\; 100)$ and

$$|\mathcal{P}|^2 = (qa)^2 \frac{2^{15}}{3^{10}} \quad \text{for} \quad (210 \;\rightarrow\; 100 \quad \text{and} \quad 21 \mp 1 \;\rightarrow\; 100)$$

$$w = \frac{E_2 - E_1}{\hbar} = \frac{1}{\hbar}\left(\frac{E_1}{4} - E_1\right) = -\frac{3E_1}{4\hbar}$$

for the three $l = 1$ states.

$$A = -\frac{3^3 E_1^3 (ea)^2 2^{15}}{2^6 \hbar^3} \frac{1}{e^{10}} \frac{1}{3\pi\varepsilon_0 \hbar c^3} = \frac{2^{10}}{3^8} \left(\frac{E_1}{mc^2}\right)^2 \frac{c}{a}$$

$$= \frac{2^{10}}{3^8} \left(\frac{13.6}{0.511 \times 10^6}\right)^2 \frac{3 \times 10^8}{0.529 \times 10^{-10}} \cong 6.27 \times 10^8 \; s^{-1}$$

$\tau = 1/A \cong 1.60 \times 10^{-9}$ s for the three $l = 1$ states (all have the same lifetime), $\tau = \infty$ for the $l = 0$ state.

Example - 16.3 : A mercury lamp emits radiation of wavelength 254 nm, with a fractional wavelength spread of 10^{-5}. If the output flux is 1 kW/m^2, estimate the ratio of stimulated to spontaneous emission processes in the lamp.

Solution : The ratio of stimulated to spontaneous emission processes is $r = (B/A)I$, where B is coefficient of stimulated emission, A is coefficient of spontaneous emission, and I is the energy density per unit range of angular frequency of the radiation at the w value of the lamp.

$$\frac{B}{A}I = \frac{\pi^2 c^3}{\hbar w^3}I$$

The output flux is

$$W = (\text{energy density}) \cdot c = I\Delta wc$$

where Δw is the spread in angular frequency of the radiation. If $\Delta\lambda$ is the spread in wavelength

$$\frac{\Delta\lambda}{\lambda} = \frac{\Delta w}{w} \quad , \quad r = \frac{\lambda^4 W}{16\pi^2 c^2 \hbar}\frac{\lambda}{\Delta\lambda}$$

Inserting the values $\lambda = 2.54 \times 10^{-7}$ m, $\Delta\lambda/\lambda = 10^{-5}$, and $W = 10^3$ W/m^2 we obtain $r = 2.8 \times 10^{-4}$.

Example - 16.4 : Analyze the spontaneous emission of an oscillating charge in one–dimension both quantum mechanically and classical mechanically.

Solution : Suppose a charge q is attached to a spring and constrained to oscillate along the x–axis. Say it starts out in the state $|n>$ and decays by spontaneous emission to state $|n'>$. The electric dipole integral is

$$\mathcal{P} = q < n|x|n' > \mathbf{i} = q \left[i\sqrt{\frac{\hbar}{2m\bar{w}}}(\sqrt{n'}\delta_{n,n'-1} - \sqrt{n}\delta_{n',n-1}) \right] \mathbf{i}$$

$$= -iq\sqrt{\frac{n\hbar}{2m\bar{w}}}\delta_{n',n-1}\mathbf{i}$$

Transition occurs only to states one step lower, and the frequency of the emitted photon is

$$w = \frac{E_n - E_{n'}}{\hbar} = \frac{1}{\hbar}[(n+\frac{1}{2})\hbar\bar{w} - (n'+\frac{1}{2})\hbar\bar{w})] = (n-n')\bar{w} = \bar{w}$$

The system radiates at the classical oscillator frequency. The transition rate is

$$A = \frac{w^3|\mathcal{P}|^2}{3\pi\varepsilon_0\hbar c^3} = \frac{w^3}{3\pi\varepsilon_0\hbar c^3}\frac{q^2 n\hbar}{2mw} = \frac{nq^2 w^2}{6\pi\varepsilon_0 mc^3}$$

The lifetime of the n th state is

$$\tau_n = \frac{1}{A} = \frac{6\pi\varepsilon_0 mc^3}{nq^2 w^2}$$

Each radiated photon carries an energy $\hbar w$, so the power radiated is

$$P = A\hbar w = \frac{q^2 w^2}{6\pi\varepsilon_0 mc^3}(n\hbar w)$$

or, since the energy of an oscillator in the n th state is $E = (n+1/2)\hbar w$,

$$P = \frac{q^2 w^2}{6\pi\varepsilon_0 mc^3}(E - \frac{1}{2}\hbar w)$$

This is the average power radiated by a quantum oscillator with initial energy E.

According to classical electrodynamics, the power radiated by an accelerated charge q is given by (the Larmor formula),

$$P = \frac{q^2 a^2}{6\pi\varepsilon_0 c^3}$$

For a harmonic oscillator with amplitude x_0, $x(t) = x_0 \cos(wt)$, the acceleration is $a = -x_0 w^2 \cos(wt)$. Averaging over a full cycle, we obtain

$$P = \frac{q^2 x_0^2 w^4}{12\pi\varepsilon_0 c^3}$$

Energy of the oscillator in this case is

$$E = \frac{1}{2} mw^2 x_0^2 \ , \quad \text{so} \quad x_0^2 = \frac{2E}{mw^2},$$

and hence

$$P = \frac{q^2 w^2}{6\pi\varepsilon_0 mc^3} E$$

This is the average power radiated by a classical oscillator with energy E. In the classical limit ($\hbar \to 0$) the classical and quantum formulas agree; however, the quantum formula protects the ground state: If $E = \hbar w/2$ the oscillator does not radiate.

16.8   Problems

Problem - 16.1 : As a mechanism for downward transitions, spontaneous emission competes with thermally stimulated emission. Show that at room temperature ($T = 300 \ K$) thermal stimulation dominates for frequencies well below $5 \times 10^{12} \ Hz$, whereas spontaneous emission dominates for frequencies well above $5 \times 10^{12} \ Hz$. Which mechanism dominates for visible light?

Problem - 16.2 : An electron in the state $|300 >$ of hydrogen atom decays by a sequence of electric dipole transitions to the state $|100 >$. a) What decay routes are open to it? b) If there are many number of atoms in a container in this state, what fraction of them would decay via each route? c) What is the lifetime of this state?

Problem - 16.3 : A time–varying Hamiltonian $H^{(1)}(t')$ brings about transitions of a system from a state k at $t' = 0$ to a state j at $t' = t$ with probability $p_{k \to j}(t)$. Use first–order time–dependent perturbation theory to show that, if $p_{j \to k}(t)$ is the probability that the same Hamiltonian brings about the transition $j \to k$ in the same time interval, then

$$p_{j \to k}(t) = p_{k \to j}(t).$$

Chapter 17

The Adiabatic Approximation

17.1 The Adiabatic processes

Consider a perfect pendulum oscillating in a vertical plane, which is supported in a box (see Figure 17.1).

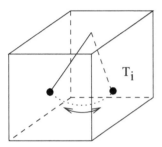

Fig. 17.1. An oscillating pendulum in a box.

If we shake the box arbitrarily, the pendulum will swing around in a chaotic fashion. But if we slowly and steadily move the box, the pendulum will continue to swing in the same direction with the same amplitude. This gradual change in the external conditions characterizes an **adiabatic process**. In adiabatic processes there are two characteristic times involved: T_i, the *internal time*, representing the motion of the system itself (in the pendulum–box case, the period of the pendulum's oscillations), and T_e, the *external time*, over which the parameters of the system change appreciably (in the pendulum–box case, if the box were put on a rotating platform, T_e would be the period of the platform's motion); see Figure 17.2. An adiabatic process is one for which $T_e \gg T_i$.

The basic strategy for analyzing an adiabatic process is first to solve the problem with the external parameters hold fixed, and only at the end of the calculation allow them to change with time. For example, the classical period of a pendulum of constant length L is $T = 2\pi\sqrt{L/g}$. If the length is changing gradually, the period would be $T = 2\pi\sqrt{L(t)/g}$.

Such an approximation is done very often in molecular calculations. Electronic motions are solved by assuming a fixed nuclei. In molecular physics this technique is known as the Born–Oppenheimer approximation. The motion of electrons is separated from that of nuclei.

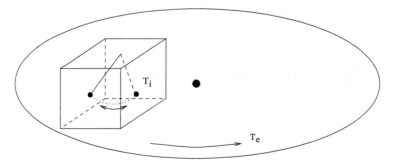

Fig. 17.2. An oscillating pendulum in a box which is on a rotating platform.

17.2   The adiabatic theorem

In quantum mechanics, the adiabatic approximation is formulated by *the adiabatic theorem*. Suppose that the Hamiltonian changes gradually from some initial form H^i to some final form H^f. The **adiabatic theorem** states that if the particle was initially in the nth eigenstate of H^i, it will be carried (under the Schrödinger equation) into the nth eigenstate of H^f, but if it does not make any transitions, the system remains in the nth state.

Proof: Suppose the time–dependent part of the Hamiltonian can be written in the form

$$H'(t) = V f(t) \tag{17.1}$$

where $f(t)$ is a function that starts out zero (at $t = 0$) and increases to 1 (at $t = T$) (see Figure 17.3).

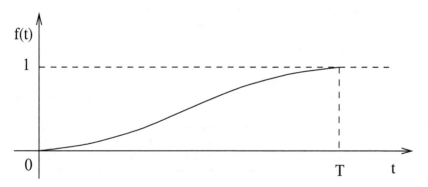

Fig. 17.3. A slowly varying function of time.

Assume that the particle starts out in the nth eigenstate of the original Hamiltonian

$$\Psi(0) = \psi_n^i \tag{17.2}$$

and evolves into some state $\Psi(t)$. The problem is to show that

$$| < \Psi(T)|\psi_m^f > |^2 = \begin{Bmatrix} 1 \ if \ m = n \\ 0 \ if \ m \neq n \end{Bmatrix} \tag{17.3}$$

Assume for the moment that V is small, so we can use first–order time–independent perturbation theory to determine ψ_m^f,

$$\psi_m^f \cong \psi_m + \sum_{k \neq m} \frac{< \psi_k|V|\psi_m >}{(E_m - E_k)} \psi_k \tag{17.4}$$

Meanwhile, we use first–order time–dependent perturbation theory to determine $\Psi(T)$,

$$\Psi(t) = \sum_l c_l(t)\psi_l e^{-\frac{i}{\hbar}E_l t} \tag{17.5}$$

where

$$c_n \cong 1 - \frac{i}{\hbar}V_{nn} \int_0^t f(t')dt' \tag{17.6}$$

and

$$c_l \cong -\frac{i}{\hbar}V_{ln} \int_0^t f(t')e^{\frac{i}{\hbar}(E_l - E_n)t'} dt' \quad , \quad l \neq n \tag{17.7}$$

After evaluating the integral in Eq.(17.7) we get

$$c_l(t) \cong -\frac{V_{ln}}{E_l - E_n} \times \left\{ f(t)e^{\frac{i}{\hbar}(E_l - E_n)t} - \int_0^t \frac{df}{dt'} e^{\frac{i}{\hbar}(E_l - E_n)t'} dt' \right\} \tag{17.8}$$

Considering the adiabatic approximation, we assume that

$$\frac{df}{dt} \ll \frac{|E_l - E_n|}{\hbar} f \tag{17.9}$$

then the last term in Eq.(17.18) may be neglected, so that

$$\Psi(T) \cong \left[\left(1 - \frac{i}{\hbar}V_{nn}G\right)\psi - \sum_{l \neq n} \frac{< \psi_l|V|\psi_n >}{E_l - E_n}\psi_l \right] e^{-\frac{i}{\hbar}ET} \tag{17.10}$$

where $G = \int_0^T f(t)dt$ is a constant. Putting together Eqs.(17.4) and (17.10) and considering the orthonormality of the initial eigenfunctions, we get

$$< \Psi(T)|\psi_n^f > = \left[1 + \frac{i}{\hbar}V_{nn}G + \sum_{k \neq n} \frac{|< \psi_k|V|\psi_n >|^2}{(E_n - E_k)^2} \right] e^{\frac{i}{\hbar}ET} \tag{17.11}$$

To first order, we have

$$< \Psi(T)|\psi_n^f > = \begin{Bmatrix} (1 + \frac{i}{\hbar}V_{nn}Ge^{\frac{i}{\hbar}E_n t} \ , \ m = n \\ 0 \qquad\qquad\qquad , \ m \neq n \end{Bmatrix} \tag{17.12}$$

Hence $| < \Psi(T)|\psi_n^f > |^2 = 1$ and $| < \Psi(T)|\psi_m^f > |^2 = 0$.

17.3   Nonholonomic processes

Systems that do not return to their original state when transported even adiabatically around a closed path are said to be nonholonomic. The Foucault pendulum is an example of this sort of adiabatic transport around a closed loop on a sphere. Every cyclical engine may be considered as a nonholonomic device.

From the quantum mechanical point of view for the nonholonomic adiabatic processes an interesting question could be: How does the final state differ from the initial state, if the parameters in the Hamiltonian are carried adiabatically around some closed cycle?

If the Hamiltonian is independent of time, then a particle which starts out in the nth eigenstate $\psi_n(x)$,

$$H\psi_n(x) = E_n\psi_n(x) \tag{17.13}$$

remains in the nth eigenstate (picking up a phase factor)

$$\Psi_n(x,t) = \psi_n(x)e^{-\frac{i}{\hbar}E_n t} \tag{17.14}$$

If the Hamiltonian changes with time, then the eigenfunctions and eigenvalues themselves are time dependent:

$$H(t)\psi_n(x,t) = E_n(t)\psi_n(x,t) \tag{17.15}$$

But the adiabatic theorem tells us that when H changes very gradually, a particle which starts out in the nth eigenstate will remain in the nth eigenstate (picking up a time dependent phase factor)

$$\Psi_n(x,t) = \psi_n(x,t)e^{-\frac{i}{\hbar}\int_0^t E_n(t')dt'}e^{i\gamma_n(t)} \tag{17.16}$$

The term

$$\theta_n(t) = -\frac{1}{\hbar}\int_0^t E_n(t')dt' \tag{17.17}$$

is known as the **dynamic phase**; it generalizes the standard factor $(-E_n t/\hbar)$ to the case where E_n is a function of time. Any extra phase, $\gamma_n(t)$, is called the **geometric phase**. According to Eq.(17.16) the particle is still in the nth eigenstate, whatever the value of γ_n. However, energy is not conserved here.

If we put Eq.(17.16) into the time–dependent Schrödinger equation,

$$i\hbar\frac{\partial\Psi_n(x,t)}{\partial t} = H(t)\Psi_n(x,t) \tag{17.18}$$

one can get a simple formula for the time development of the geometric phase:

$$i\hbar\left[\frac{\partial\psi_n}{\partial t}e^{i\theta_n}e^{i\gamma_n} - \frac{i}{\hbar}E_n\psi_n e^{i\theta_n}e^{i\gamma_n} + i\frac{d\gamma_n}{dt}\psi_n e^{i\theta_n}e^{i\gamma_n}\right] \tag{17.19}$$

$$= [H\psi_n]e^{i\theta_n}e^{i\gamma_n} = E_n\psi_n e^{i\theta_n}e^{i\gamma_n}$$

so we get

$$\frac{\partial\psi_n}{\partial t} + i\psi_n\frac{\partial\gamma_n}{\partial t} = 0 \tag{17.20}$$

Taking the inner product of Eq.(17.20) with ψ_n, we obtain

$$\frac{d\gamma_n}{dt} = i < \psi_n|\frac{\partial\psi_n}{\partial t} > \qquad (17.21)$$

If the time dependence of the Hamiltonian is represented by a parameter, say $R(t)$, then the time dependence of the wave function, $\psi_n(x,t)$, may be expressed as

$$\frac{\partial\psi_n}{\partial t} = \frac{\partial\psi_n}{\partial R}\frac{\partial R}{\partial t} \qquad (17.22)$$

so that

$$\frac{d\gamma_n}{dt} = i < \psi_n|\frac{\partial\psi_n}{\partial R} > \frac{\partial R}{\partial t} \qquad (17.23)$$

and hence

$$\gamma_n(t) = i \int_0^t < \psi_n|\frac{\partial\psi_n}{\partial R} > \frac{\partial R}{\partial t'}dt' = i \int_{R_i}^{R_f} < \psi_n|\frac{\partial\psi_n}{\partial R} > dR \qquad (17.24)$$

where R_i and R_f are the initial and final values of $R(t)$. In particular, if the Hamiltonian returns to its original form after time T, so that $R_f = R_i$, then $\gamma_n(T) = 0$.

Suppose there are N parameters $R_1(t), R_2(t), \cdots, R_N(t)$ changing with time in the Hamiltonian; in that case

$$\frac{\partial\psi_n}{\partial t} = \sum_{i=1}^N \frac{\partial\psi_n}{\partial R_i}\frac{\partial R_i}{\partial t} = (\vec{\nabla}_R\psi_n) \cdot \frac{\partial\mathbf{R}}{\partial t} \qquad (17.25)$$

where $\mathbf{R} = (R_1, R_2, \cdots, R_N)$, and $\vec{\nabla}_R$ is the gradient with respect to these parameters. This time we have

$$\gamma_n(t) = i \int_{R_i}^{R_f} < \psi_n|\vec{\nabla}_R\psi_n > \cdot d\mathbf{R} \qquad (17.26)$$

and if the Hamiltonian returns to its original form after a time T, the net geometric phase change is

$$\gamma_n(t) = i \oint < \psi_n|\vec{\nabla}_R\psi_n > \cdot d\mathbf{R} \qquad (17.27)$$

This is a line integral around a close loop in parameter space, and it is not, in general, zero. $\gamma_n(T)$ is called **Berry's phase**. Eq.(17.27) was first obtained by M.V. Berry in 1984. Notice that $\gamma_n(T)$ depends only on the path taken, not on how fast that path is traversed, provided that it is slow enough to validate the adiabatic hypothesis. By contrast, the accumulated dynamic phase,

$$\theta_n(T) = -\frac{1}{\hbar} \int_0^T E_n(t')dt' \qquad (17.28)$$

depends critically on the elapsed time.

Some properties of $\gamma_n(t)$:

- $\gamma_n(t)$ is real.

- $\gamma_n(t)$ is measurable.

- Eq.(17.21) can be obtained from Eq.(17.20), but the reverse may not be possible.

- When the parameter space is three dimensional, $\mathbf{R} = (R_1, R_2, R_3)$, Berry's formula, Eq.(17.27), is similar to the expression for magnetic flux, Φ, in terms of the vector potential $\mathbf{A}$.

$$\Phi = \int_S \mathbf{b} \cdot d\mathbf{a} = \int_S (\vec{\nabla} \times \mathbf{A} \cdot d\mathbf{a} = \oint_C \mathbf{A} \cdot d\mathbf{r} \tag{17.29}$$

Thus Berry's phase can be thought of as the flux of a magnetic field.

$$\mathbf{B} = i\vec{\nabla}_R \times\ <\psi_n|\vec{\nabla}_R\psi_n> \tag{17.30}$$

then Berry's phase can be written as a surface integral,

$$\gamma_n(T) = i \int [\vec{\nabla}_R \times\ <\psi_n|\vec{\nabla}_R\psi_n> \cdot d\mathbf{a} \tag{17.31}$$

This equation is a convenient alternative expression for $\gamma_n(T)$ given in Eq.(17.27).

17.4 Experimental evidences of nonholonomic processes

In classical electrodynamics the scalar and vector potentials φ and $\mathbf{A}$, respectively, are not directly measurable; the measurable physical quantities are the electric and magnetic fields, which are expressed in terms of φ and $\mathbf{A}$ through the Maxwell's equations:

$$\mathbf{E} = -\vec{\nabla}\varphi - \frac{\partial \mathbf{A}}{\partial t} \quad , \quad \mathbf{B} = \vec{\nabla} \times \mathbf{A} \tag{17.32}$$

One can make a transformation on φ and $\mathbf{A}$ such that it has no effect at all on the fields,

$$\varphi \to \varphi' = \varphi - \frac{\partial \Lambda}{\partial t} \quad , \quad \vec{A} \to \vec{A}' = \vec{A} + \vec{\nabla}\Lambda \tag{17.33}$$

where Λ is an arbitrary function of position and time, $\Lambda = \Lambda(\mathbf{r}, t)$. Such a transformation, Eq.(17.33), is called a **gauge transformation**.

In quantum mechanics the potentials play a more significant role. For instance the Hamiltonian of a charged particle in an electromagnetic field is expressed in terms of φ and $\mathbf{A}$, not $\mathbf{E}$ and $\mathbf{B}$:

$$H = \frac{1}{2m}(-i\hbar\vec{\nabla} - q\mathbf{A})^2 + q\varphi \tag{17.34}$$

This Hamiltonian is still invariant under gauge transformations.

We may not expect any electromagnetic influences on charged particles in regions where $\mathbf{E}$ and $\mathbf{B}$ are zero. But in 1959 Aharonov and Bohm showed that the vector potential can affect the quantum behavior of a charged particle that never encounters an electromagnetic field.

Imagine a particle constrained to move in a circle of radius b. Along the axis runs a solenoid of radius a, $(a < b)$, carrying a magnetic field **B**. If the solenoid is extremely long, the field inside is uniform, and the field outside is zero (see Figure 17.4).

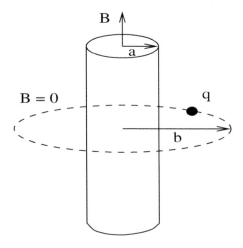

Fig. 17.4. A rotating charged particle around a solenoid.

The vector potential outside the solenoid is not zero,

$$\mathbf{A} = \frac{\Phi}{2\pi r}\hat{\phi} \quad , \quad (r > a) \tag{17.35}$$

where $\Phi = \pi a^2 B$ is the magnetic flux through the solenoid. Meanwhile, the solenoid is uncharged, so the scalar potential ϕ is zero. In this case the Hamiltonian, Eq.(17.34), becomes

$$H = \frac{1}{2m}[-\hbar^2\nabla^2 + q^2 A^2 + i2\hbar q\mathbf{A}\cdot\vec{\nabla}] \tag{17.36}$$

But the wave function depends only on the azimuthal angle ϕ, $(\theta = \pi/2, r = b)$, so $\vec{\nabla} \to \frac{\hat{\phi}}{b}\frac{d}{d\phi}$, and the Schrödinger equation becomes

$$\frac{1}{2m}\left[-\frac{\hbar^2}{b^2}\frac{d^2}{d\phi^2} + \left(\frac{q\Phi}{2\pi b}\right)^2 + i\frac{\hbar q\Phi}{\pi b^2}\frac{d}{d\phi}\right]\psi(\phi) = e\psi(\phi) \tag{17.37}$$

This is a linear differential equation with constant coefficients:

$$\frac{d^2\psi}{d\phi^2} - i2\beta\frac{d\psi}{d\phi} + \epsilon\psi = 0 \tag{17.38}$$

where

$$\beta = \frac{q\Phi}{2\pi\hbar} \quad , \quad \epsilon = \frac{2mb^2 E}{\hbar^2} - \beta^2 \tag{17.39}$$

Solutions are of the form

$$\psi = Ae^{i\lambda\phi} \tag{17.40}$$

with

$$\lambda = \beta \mp \sqrt{\beta^2 + \epsilon} = \beta \mp \frac{b}{\hbar}\sqrt{2mE} \tag{17.41}$$

Continuity of $\psi(\phi)$, at $\phi = 2\pi$, requires that λ be an integer:

$$\beta \mp \frac{b}{\hbar}\sqrt{2mE} = n \tag{17.42}$$

then one can write

$$E_n = \frac{\hbar^2}{2mb^2}\left(n - \frac{q\Phi}{2\pi\hbar}\right)^2 \quad , \quad n = 0, \mp 1, \mp 2, \cdots \tag{17.43}$$

Positive n, representing a particle traveling in the same direction as the current in the solenoid, has a lower energy (assuming q is positive) than negative n, describing a particle traveling in the opposite direction. The allowed energies depend on the field inside the solenoid, even though the field at the location of the particle is zero.

Suppose a particle is moving through a region where $\mathbf{B}$ is zero, so $\nabla \times \mathbf{A} = 0$, but $\mathbf{A}$ itself is not. The time–dependent Schrödinger equation, with potential V, may in general be expressed as

$$\left[\frac{1}{2m}(-i\hbar\vec{\nabla} - q\mathbf{A})^2 + V\right]\Psi = i\hbar\frac{\partial\Psi}{\partial t} \tag{17.44}$$

This equation can be simplified by writing

$$\Psi = e^{ig}\Psi' \tag{17.45}$$

where

$$g(\mathbf{r}) = \frac{q}{\hbar}\int^r \mathbf{A}(\mathbf{r}) \cdot d\mathbf{r}' \tag{17.46}$$

This is meaningful if $\vec{\nabla} \times \mathbf{A} = 0$, otherwise the integral in Eq.(17.46) becomes path dependent. Eq.(17.44) takes the form after substituting Eq.(17.45),

$$-\frac{\hbar^2}{2m}\nabla^2\Psi' + V\Psi' = i\hbar\frac{\partial\Psi'}{\partial t} \tag{17.47}$$

Evidently Ψ' satisfies the Schrödinger equation without $\mathbf{A}$.

In 1959 Aharonov and Bohm proposed an experiment in which a beam of electrons is split in two, and passed either side of a long solenoid, before being recombined (see Figure 17.5).

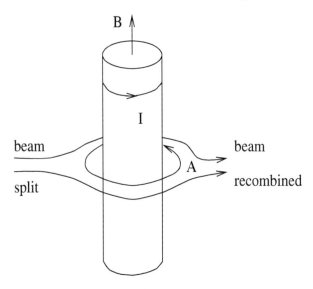

Fig. 17.5. Splitting of electron beam around a current carrying solenoid.

The beams are kept well away from the solenoid itself, so they encounter only regions where $\mathbf{B} = 0$. But $\mathbf{A} = \frac{\Phi}{2\pi r}\hat{\phi}$ is not zero, and two beams arrive with different phases:

$$g = \frac{q}{\hbar} \int \mathbf{A} \cdot d\mathbf{r} = \frac{q\Phi}{2\pi\hbar} \int \left(\frac{1}{r}\hat{\phi}\right) \cdot \left(r\hat{\phi}d\phi\right) = \pm\frac{q\Phi}{2\hbar} \qquad (17.48)$$

The plus sign applies to the electrons traveling in the same direction as $\mathbf{A}$, or in the same direction as the current in the solenoid. The beams arrive out of phase by an amount proportional to the magnetic flux their paths encircle:

$$\text{Phase difference} = \frac{q\Phi}{\hbar} \qquad (17.49)$$

This phase shift leads to measurable interference. Therefore the Aharonov–Bohm effect can be regarded as an example of geometric phase, or Berry's phase.

17.5 Worked examples

Example - 17.1 : Suppose a charged particle is confined to a box, which is centered at point $\mathbf{R}$ outside a solenoid, by a potential $V(\mathbf{r} - \mathbf{R})$. If the box is transported around the solenoid determine the geometric phase, or Berry's phase.

Solution : The Schrödinger equation is

$$\left\{\frac{1}{2m}[-i\hbar\nabla - q\mathbf{A}(\mathbf{r})]^2 + V(\mathbf{r} - \mathbf{R})\right\}\psi_n = E_n\psi_n$$

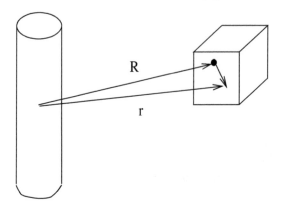

Let $\psi_n \;\; \rightarrow \;\; e^{ig}\psi'_n$, where

$$g = \frac{q}{\hbar}\int_R^r \mathbf{A}(\mathbf{r'}) \cdot d\mathbf{r'}$$

ψ'_n satisfies the same Schrödinger equation, only with $\mathbf{A} \rightarrow 0$:

$$\left[-\frac{\hbar^2}{2m}\nabla^2 + V(\mathbf{r} - \mathbf{R})\right]\psi'_n = E_n\psi'_n$$

To determine Berry's phase we must evaluate $< \psi_n|\nabla_R|\psi_n >$,

$$\nabla_R\psi_n = \nabla_R\left[e^{ig}\psi'_n(\mathbf{r} - \mathbf{R})\right]$$

$$= -i\frac{q}{\hbar}\mathbf{A}(\mathbf{R})e^{ig}\psi'_n(\mathbf{r} - \mathbf{R}) + e^{ig}\nabla_R\psi'_n(\mathbf{r} - \mathbf{R})$$

$$< \psi_n|\nabla_R\psi_n >= \int e^{-ig}\left[\psi'_n(\mathbf{r} - \mathbf{R})\right]^* e^{ig}$$

$$\times \left[-i\frac{q}{\hbar}\mathbf{A}(\mathbf{R})\psi'_n(\mathbf{r} - \mathbf{R}) + \nabla_R\psi'_n(\mathbf{r} - \mathbf{R})\right]d^3\mathbf{r}$$

using $\nabla_R = -\nabla_r$

$$< \psi_n|\nabla_R\psi_n >= -i\frac{q}{\hbar}\mathbf{A}(\mathbf{R}) - \int \left[\psi'_n(\mathbf{r} - \mathbf{R})\right]^* \nabla_r\psi'_n(\mathbf{r} - \mathbf{R})d^3\mathbf{r}$$

$$= -i\frac{q}{\hbar}\mathbf{A}(\mathbf{R})$$

Here the integral $< \psi'_n|\nabla_r\psi'_n >= 0$.

Put $in\gamma_n(T) = i\oint < \psi_n|\nabla_R\psi_n > \cdot d\mathbf{R}$

$$\gamma_n(T) = \frac{q}{\hbar}\oint \mathbf{A}(\mathbf{R}) \cdot d\mathbf{R} = \frac{q}{\hbar}\int (\nabla \times \mathbf{A}) \cdot d\mathbf{a} = \frac{q\Phi}{\hbar}$$

This also shows that the Aharonov–Bohm effect is a particular instance of geometric phase.

Example - 17.2 : a) Use

$$\gamma_n(t) = i \int_{R_i}^{R_f} <\psi_n|\frac{\partial \psi_n}{\partial R} > dR$$

to calculate the geometric phase change when the infinite square well expands adiabatically from width w_1 to w_2. b) If the expansion occurs at a constant rate $(dw/dt = v)$, what is the dynamic phase change for this process? c) If the well now contracts back to its original size, what is geometric phase (Berry's phase) for the cycle?

Solution : a)

$$\psi_n(x) = \sqrt{\frac{2}{w}} \sin\left(\frac{n\pi}{w}x\right) \quad , \quad R = w$$

$$\frac{\partial \psi_n}{\partial R} = \sqrt{2}\left(-\frac{1}{2}\frac{1}{w^{3/2}}\right)\sin\left(\frac{n\pi}{w}x\right) + \sqrt{\frac{2}{w}}\left(-\frac{n\pi x}{w^2}\right)\cos\left(\frac{n\pi}{w}x\right)$$

$$<\psi_n|\frac{\partial \psi_n}{\partial R} >= \int_0^w \psi_n \frac{\partial \psi_n}{\partial R} dx$$

$$= -\frac{1}{w^2}\int_0^w \sin^2\left(\frac{n\pi}{w}x\right)dx - \frac{2n\pi}{w^2}\int_0^w x\sin\left(\frac{n\pi}{w}x\right)\cos\left(\frac{n\pi}{w}x\right)dx$$

$$= -\frac{1}{w^2} + \frac{1}{w^2} = 0$$

Therefore, $\gamma_n(t) = 0$. If the eigenfunctions are real, the geometric phase vanishes.

b)

$$\theta_n(t) = -\frac{1}{\hbar}\int_0^t \frac{n^2\pi^2\hbar^2}{2mw^2}dt' = -\frac{n^2\pi^2\hbar}{2m}\int \frac{1}{w^2}\frac{dt'}{dw}dw$$

$$= -\frac{n^2\pi^2\hbar}{2mv}\int_{w_1}^{w_2}\frac{1}{w^2}dw = \frac{n^2\pi^2\hbar}{2mv}\left(\frac{1}{w}\right)\Big|_{w_1}^{w_2} = \frac{n^2\pi^2\hbar}{2mv}\left(\frac{1}{w_2} - \frac{1}{w_1}\right)$$

c) $\gamma_n(T) = 0$.

17.6   Problems

Problem - 17.1 : For real wave functions $(\psi_n(x,t))$, the geometric phase vanishes $(\gamma_n(t) = 0)$. Take

$$\psi_n'(x,t) = e^{i\phi_n}\psi_n(x,t)$$

where $\phi_n(\mathbf{R})$ is an arbitrary real function. Try this function to find geometric phase, then put it back to

$$\Psi_n(x,t) = \psi_n(x,t)e^{-\frac{i}{\hbar}\int_0^t E_n(t')dt'}e^{i\gamma_n(t)}.$$

Check it for a closed loop.

Problem - 17.2 : The delta function well, $V(x) = -\alpha\delta(x)$, supports a single bound state

$$\psi(x) = \frac{\sqrt{m\alpha}}{\hbar}e^{-m\alpha|x|/\hbar^2} \quad ; \quad E = -\frac{m\alpha^2}{2\hbar^2}$$

Calculate the geometric phase change when α gradually increases from α_1 to α_2. If the increase occurs at a constant rate $(d\alpha/dt = c)$, what is the dynamic phase change for this process?

Problem - 17.3 : The case of an infinite square well whose right wall expands at a constant velocity, v, can be solved exactly,

$$\Phi_n(x,t) = \sqrt{\frac{2}{w}}\sin\left(\frac{n\pi}{w}x\right)e^{i(mvx^2 - 2E_n^i at)/2\hbar w}$$

where $w(t) = a + vt$ is the width of the well and $E_n^i = n^2\pi^2\hbar^2/2ma^2$ is the n th allowed energy of the original well, width a. The general solution is

$$\Psi(x,t) = \sum_{n=1}^{\infty} c_n\Phi_n(x,t)$$

the coefficients c_n are independent of t. a) Check that $\Phi_n(x,t)$ satisfies the time-independent Schrödinger equation. b) Suppose a particle starts out, at $t = 0$, in the ground state of the initial well

$$\Psi(x,0) = \sqrt{\frac{2}{a}}\sin\left(\frac{\pi}{a}x\right)$$

Show that the coefficients can be written as

$$c_n = \frac{2}{\pi}\int_0^\pi e^{-i\alpha z^2}\sin(nz)\sin(z)dz \quad \text{where} \quad \alpha = \frac{mva}{2\pi^2\hbar}$$

c) Suppose the well expands to twice its original width $(a \to 2a)$. So the external time is given by $w(T_e) = 2a$. The internal time (T_i) is the period of the time-dependent exponential factor in the initial ground state. Determine T_e and T_i, and show that the adiabatic regime corresponds to $\alpha \ll 1$, so that $e^{-i\alpha z^2} \approx 1$ over the domain of integration. Use this to determine c_n. Construct $\Psi(x,t)$, and confirm that it is consistent with the adiabatic theorem. d) Show that the phase factor in $\Psi(x,t)$ can be written in the form

$$\theta(t) = -\frac{1}{\hbar}\int_0^t E_1(t')dt' \quad \text{where} \quad E_n = \frac{n^2\pi^2\hbar^2}{2mw^2}$$

is the n th instantaneous eigenvalue at time t.

Chapter 18

Path–Integration Method

18.1 An approximation to time–evolution for a free particle

Path–integration method is based on the Lagrangian formulation of classical mechanics and connects, in some sense, the classical mechanics and the quantum mechanics.

We know that the quantum problem is fully solved if the propagator is known. In quantum mechanics we first find the eigenvalues and eigenfunctions of Hamiltonian, H, and then express the propagator $U(t)$ in terms of these. Once we have $U(t)$, we can write

$$|\psi(t) >= U(t)|\psi(0) > \tag{18.1}$$

Let us now construct an explicit expression for $U(t)$ in terms of $|E >$ the normalized eigenket of H with eigenvalue E which obey

$$H|E >= E|E > \tag{18.2}$$

This is called the time–independent Schrödinger equation. Assume that we have solved it and found the kets $|E >$. If we expand $|\psi >$ as

$$|\psi(t) >= \sum |E >< E|\psi(t) >= \sum a_E(t)|E > \tag{18.3}$$

the equation for $a_E(t)$ follows if we act on both sides with $(i\hbar\frac{\partial}{\partial t} - H)$:

$$0 = (i\hbar\frac{\partial}{\partial t} - H)|\psi(t) >= \sum(i\hbar\dot{a}_E - Ea_E)|E > \tag{18.4}$$

or

$$i\hbar\dot{a}_E = Ea_E \tag{18.5}$$

where we have used the linear independence of the kets $|E >$. The solution to Eq.(18.5) is

$$a_E(t) = a_E(0)e^{-\frac{i}{\hbar}Et} \tag{18.6}$$

or

$$< E|\psi(t) >=< E|\psi(0) > e^{-\frac{i}{\hbar}Et} \tag{18.7}$$

so that

$$|\psi(t) >= \sum_E |E >< E|\psi(0) > e^{-\frac{i}{\hbar}Et} \tag{18.8}$$

We can now extract $U(t)$:

$$U(t) = \sum_E |E><E|e^{-\frac{i}{\hbar}Et} \tag{18.9}$$

We have been assuming that the energy spectrum is discrete and nondegenerate. If E is degenerate, one must first introduce an extra label α to specify the states. In this case

$$U(t) = \sum_\alpha \sum_E |E,\alpha><E,\alpha|e^{-\frac{i}{\hbar}Et} \tag{18.10}$$

If E is continuous, the sum must be replaced by an integral. There exists another expression for $U(t)$ besides the sum, Eq.(18.9) and (18.10), and that is

$$U(t) = e^{-\frac{i}{\hbar}Ht} \tag{18.11}$$

If this exponential series converges, this form of $U(t)$ can be very useful. It is easy to show that

$$|\psi(t)>= e^{-\frac{i}{\hbar}Ht}|\psi(0)> \tag{18.12}$$

satisfies Schrödinger's equation.

Since H (the energy operator) is Hermitian, it follows that $U(t)$ is unitary. We may therefore think of the time evolution of a ket $|\psi(t)>$ as a *rotation* in Hilbert space, thus the norm $<\psi(t)|\psi(t)>$ is invariant:

$$<\psi(t)|\psi(t)>=<\psi(0)|U^\dagger(t)U(t)|\psi(0)>=<\psi(0)|\psi(0)> \tag{18.13}$$

so that a state, once normalized, stays normalized. Whether or not H depends on time, the propagator satisfies the following conditions:

$$U(t_3,t_2)U(t_2,t_1) = U(t_3,t_1) \tag{18.14}$$

$$U^\dagger(t_2,t_1) = U^{-1}(t_2,t_1) = U(t_1,t_2) \tag{18.15}$$

$$U(t,t) = I \tag{18.16}$$

The equality given in Eq.(18.14) is the **group property** for propagators. In the path–integral approach one computes $U(t)$ directly. For a single particle in one dimension, the procedure is the following. To find $U(x,t;x',t')$:

i) Draw all paths in the $x - t$ plane connecting (x',t') and (x,t).
ii) Find the action $S[x(t)]$ for each path $x(t)$.
iii) Construct the propagator in terms of actions as

$$U(x,t;x',t') = A \sum_\alpha e^{\frac{i}{\hbar}S[x_\alpha(t)]} \tag{18.17}$$

where A is an overall normalization factor, and α runs overall paths. For any path $x(t)$ connecting (x,t) and (x',t'), the action $S[x(t)]$ is defined as

$$S[x(t)] = \int_{t'}^{t} L(x,\dot{x})dt'' \tag{18.18}$$

where L is the classical Lagrangian, given by $L = T - V$, T and V being the kinetic and potential energies of the particle, respectively. The classical path, $x_{cl}(t)$, is one on which S is a minimum. It is costumary to refer to this condition as the **principal of least action**, (see Figure 18.1).

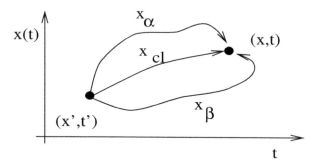

Fig. 18.1. Some of the paths that contribute to the propagator, $U(t)$.

The correct way to sum over all the paths, that is to say, *path–integration*, is actually quite complicated. We have to add the contributions

$$Z_\alpha = e^{\frac{i}{\hbar} S[x_\alpha(t)]} \tag{18.19}$$

from each path $x_\alpha(t)$. Since each path has a different action, it contributes with a different phase, and the contributions from the paths essentially cancel each other, until we come near the classical path, $x_{cl}(t)$. Since S is stationary here, the Z's add constructively and produce a large sum. As we move away from $x_{cl}(t)$, destructive interference sets in. It is obvious that $U(t)$ is dominated by the paths near $x_{cl}(t)$. Thus the classical path is important, not because it contributes a lot by itself, but because in its vicinity the paths contribute coherently.

An approximation to $U(t)$ for a free particle:

As a crude approximation we may ignore all but the classical path and its neighbors in calculating $U(t)$. Assuming that each of these paths contributes the same amount $e^{\frac{i}{\hbar} S_{cl}}$, since S is stationary, we get

$$U(t) = A' e^{\frac{i}{\hbar} S_{cl}} \tag{18.20}$$

where A' is some normalizing factor which measures the number of paths in the coherent range. Let us find $U(t)$ for a free particle in this approximation.

The classical path for a free particle is just a straight line in the $x - t$ plane:

$$x_{cl}(t'') = x' + \frac{x - x'}{t - t'}(t'' - t') \tag{18.21}$$

corresponding to motion with uniform velocity $v = (x - x')/(t - t')$. Since $L = \frac{1}{2}mv^2$ is a constant,

$$S_{cl} = \int_{t'}^{t} L dt'' = \frac{1}{2} m \frac{(x - x')^2}{(t - t')} \tag{18.22}$$

so that

$$U(x, t; x', t') = A' \exp\left[\frac{im(x - x')^2}{2\hbar(t - t')}\right] \tag{18.23}$$

To find A', we use the fact that as $t - t' \to 0$, $U \to \delta(x - x')$, namely

$$\delta(x - x') = \lim_{\Delta \to 0} \frac{1}{\sqrt{\pi \Delta^2}} e^{-(\frac{x-x'}{\Delta})^2}$$

then we get

$$A' = \left[\frac{m}{2\pi \hbar i (t - t')} \right]^{1/2} \tag{18.24}$$

so that

$$U(x, t; x', 0) \equiv U(x, t; x') = \left(\frac{m}{2\pi \hbar i t} \right)^{1/2} \exp \left[\frac{im(x - x')^2}{2\hbar t} \right] \tag{18.25}$$

This is the exact answer. Let us now obtain the exact answer considering the quantum mechanical approach. The propagator for a free particle, from Eq.(18.9), is

$$U(t) = \int_{-\infty}^{\infty} |E><E| e^{-\frac{i}{\hbar} Et} dE \tag{18.26}$$

or, in terms of linear momentum

$$U(t) = \int_{-\infty}^{\infty} |p><p| e^{-\frac{i}{\hbar} \frac{p^2}{2m} t} dp \tag{18.27}$$

The propagator $U(t)$ can be evaluated explicitly in the x basis:

$$U(x, t; x') = <x|U(t)|x'> = \int_{-\infty}^{\infty} <x|p><p|x'> e^{-\frac{i}{\hbar} \frac{p^2}{2m} t} dp \tag{18.28}$$

$$= \frac{1}{2\pi \hbar} \int_{-\infty}^{\infty} e^{\frac{i}{\hbar} p(x-x')t} e^{-\frac{i}{\hbar} \frac{p^2}{2m} t} dp$$

$$= \left(\frac{m}{2\pi \hbar i t} \right)^{1/2} \exp \left[\frac{im(x - x')^2}{2\hbar t} \right]$$

In a crude approximation we have managed to get the exact answer, Eq.(18.25), just computing the classical action. However, the assumptions $U(t) = A(t) e^{\frac{i}{\hbar} S_{cl}}$ and $U(x, 0; x') = \delta(x - x')$ may not work in general.

18.2 Path integral evaluation of the free particle propagator

Although the crude approximation yielded the exact free–particle propagator, we will now repeat the calculation without any approximation to illustrate path integration.

Consider $U(x_N, t_N; x_0, t_0)$. The problem is to perform the path integral

$$\int_{x_0}^{x_N} e^{\frac{i}{\hbar} S[x(t)]} D[x(t)] \tag{18.29}$$

where $\int_{x_0}^{x_N} D[x(t)]$ is a symbolic way of saying *integrate over all paths connecting* x_0 *to* x_N (*in the interval* t_0 *to* t_N). Now, a path $x(t)$ is fully specified by an infinity of numbers $x(t_0), \cdots, x(t_N)$, namely, the values of the function $x(t)$ at every point t in the interval t_0 to t_N. To sum over all paths we must integrate over all possible values of these infinite variables, except of course $x(t_0)$ and $x(t_N)$, which will be kept fixed at x_0 and x_N, respectively. We trade the function $x(t)$ for a discrete approximation which agrees with $x(t)$ at the $N+1$ points $t_n = t_0 + n\varepsilon$, $n = 0, \cdots, N$, where $\varepsilon = (t_N - t_0)/N$. In this approximation each path is specified by $N+1$ numbers $x(t_0), x(t_1), \cdots, x(t_N)$. The gaps in the discrete function are interpolated by straight lines (see Figure 18.2). If we take the limit $N \to \infty$ at the end we will get a result that is insensitive to these approximations.

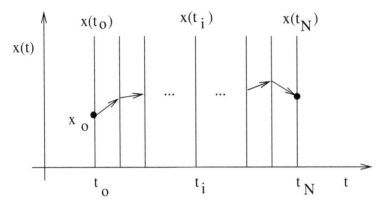

Fig. 18.2. The discrete approximation to a path $x(t)$.

Now that the paths have been discretized, we must also do the same to the action integral. We replace the continuous path definition

$$S = \int_{t_0}^{t_N} L(t)dt = \int_{t_0}^{t_N} \frac{1}{2}m\dot{x}^2 dt \tag{18.30}$$

by

$$S = \sum_{i=0}^{N-1} \frac{m}{2} \left(\frac{x_{i+1} - x_i}{\varepsilon} \right)^2 \varepsilon \tag{18.31}$$

where $x_i = x(t_i)$. We wish to calculate

$$U(x_N, t_N; x_0, t_0) = \int_{x_0}^{x_N} e^{\frac{i}{\hbar} S[x(t)]} D[x(t)] \tag{18.32}$$

$$= \lim_{N \to \infty, \varepsilon \to 0} A \int_{-\infty}^{\infty} \cdots \int_{-\infty}^{\infty} \exp\left[\frac{im}{\hbar 2} \sum_{j=0}^{N-1} \frac{(x_{j+1} - x_j)^2}{\varepsilon} \right] dx_1 \cdots dx_{N-1}$$

The factor A in the front is to be chosen at the end such that we get the correct scale for U when the limit $N \to \infty$ is taken.

Let us take the change of variables

$$y_j = \left(\frac{m}{2\hbar\varepsilon}\right)^{1/2} x_j \tag{18.33}$$

We may rewrite Eq.(18.32) as

$$\lim_{N\to\infty} A' \int_{-\infty}^{\infty} \cdots \int_{-\infty}^{\infty} \exp\left[-\sum_{j=0}^{N-1} \frac{(y_{j+1} - y_j)^2}{j}\right] dy_1 \cdots dy_{N-1} \tag{18.34}$$

where

$$A' = A\left(\frac{2\hbar\varepsilon}{m}\right)^{(N-1)/2} \tag{18.35}$$

We consider the following procedure in the evaluation of the multiple integral:

Let us begin by doing the y_1 integration. Considering just the part of the integrand that involves y_1, we get

$$\int_{-\infty}^{\infty} \exp\left(-\frac{1}{i}[(y_2 - y_1)^2 + (y_1 - y_0)^2]\right) dy_1 \tag{18.36}$$

$$= \left(\frac{i\pi}{2}\right)^{1/2} \exp -\frac{(y_2 - y_0)^2}{2i}$$

Consider next the integration over y_2. Bringing in the part of the integrand involving y_2 and combining it with the result above we compute next

$$\left(\frac{i\pi}{2}\right)^{1/2} \int_{-\infty}^{\infty} \exp -(y_3 - y_2)^2/i \exp -(y_2 - y_0)^2/2i \, dy_2 \tag{18.37}$$

$$= \left(\frac{i\pi}{2}\right)^{1/2} \exp -(2y_3^2 + y_0^2)/2i \left(\frac{2\pi i}{3}\right)^{1/2} \exp (y_0 + 2y_3)^2/6i$$

$$= \left[\frac{(i\pi)^2}{3}\right]^{1/2} \exp -(y_3 - y_0)^2/3i$$

By comparing this result to the one from the y_1 integration, we deduce the pattern: If we carry out this process $N - 1$ times so as to evaluate the integral in Eq.(18.34), it will become

$$\frac{(i\pi)^{(N-1)/2}}{N^{1/2}} \exp -(y_N - y_0)^2/Ni \tag{18.38}$$

or

$$\frac{(i\pi)^{(N-1)/2}}{N^{1/2}} \exp -\frac{m(x_N - x_0)^2}{2\hbar\varepsilon Ni} \tag{18.39}$$

Bringing in the factor $A(2\hbar\varepsilon/m)^{(N-1)/2}$ from up front, we get

$$U = A\left(\frac{2\pi\hbar\varepsilon i}{m}\right)^{N/2} \left(\frac{m}{2\pi\hbar i N\varepsilon}\right)^{1/2} \exp\left[\frac{im(x_N - x_0)^2}{2\hbar N\varepsilon}\right] \tag{18.40}$$

If we now let $N \to \infty$, $\varepsilon \to 0$, $N\varepsilon \to t_n - t_0$, we get the right answer provided

$$A = \left(\frac{2\pi\hbar\varepsilon i}{m}\right)^{-N/2} = B^{-N} \tag{18.41}$$

It is convenient to associate a factor $\frac{1}{B}$ with each of the $N-1$ integrations and the remaining $\frac{1}{B}$ with the overall process. In other words, the precise meaning of the statement *integrate over all paths* is

$$\int D[x(t)] = \lim_{\varepsilon \to 0, N \to \infty} \frac{1}{B} \int_{-\infty}^{\infty} \cdots \int_{-\infty}^{\infty} \frac{dx_1}{B} \cdots \frac{dx_{N-1}}{B} \qquad (18.42)$$

where

$$B = \left(\frac{2\pi\hbar\varepsilon i}{m} \right)^{1/2} \qquad (18.43)$$

18.3   Equivalence to the Schrödinger equation

The relation between the Schrödinger and path integration formalisms is similar to that between the Newtonian and the least action formalisms of classical mechanics, in that the former approach is local in time and deals with time evolution over infinitesimal periods while the latter is global and deals directly with propagation over finite times.

In the Schrödinger formalism, the change in the state vector $|\psi>$ over an infinitesimal time ε is

$$|\psi(\varepsilon)> - |\psi(0)> = \frac{-i\varepsilon}{\hbar} H |\psi(0)> \qquad (18.44)$$

which becomes in the x basis

$$\psi(x,\varepsilon) - \psi(x,0) = \frac{-i\varepsilon}{\hbar} \left[\frac{-\hbar^2}{2m} \frac{\partial^2}{\partial x^2} + V(x) \right] |\psi(0)> \qquad (18.45)$$

to first order in ε. To compare this result with path integral prediction to the same order in ε, we begin with

$$\psi(x,\varepsilon) = \int_{-\infty}^{\infty} U(x,\varepsilon; x')\psi(x',0)dx' \qquad (18.46)$$

The calculation of $U(\varepsilon)$ is simplified by the fact that there is no need to do any integrations over intermediate x's since there is just one slice of time ε between the start and finish. So

$$U(x,\varepsilon; x') = \left(\frac{m}{2\pi\hbar i\varepsilon} \right)^{1/2} \exp\left(\frac{i}{\hbar} \left[\frac{m(x-x')^2}{2\varepsilon} - \varepsilon V(\frac{x+x'}{2}) \right] \right) \qquad (18.47)$$

where the $(m/2\pi\hbar i\varepsilon)^{1/2}$ factor up front is just the $1/B$ factor from Eq.(18.43). So

$$\psi(x,\varepsilon) = \left(\frac{m}{2\pi\hbar i\varepsilon} \right)^{1/2} \int_{-\infty}^{\infty} \exp\left[\frac{im(x-x')^2}{2\varepsilon\hbar} \right] \qquad (18.48)$$

$$\times \exp\left[-\frac{i\varepsilon}{\hbar} V(\frac{x+x'}{2}) \right] \psi(x',0)dx'$$

The factor $\exp\left[im(x - x')^2/2\varepsilon\hbar\right]$ in the integral, in Eq.(18.48), oscillates very rapidly as $x - x'$ varies; ε is infinitesimal and $\hbar$ is so small. The substantial contribution comes from the region where the phase is stationary. In this case the only stationary point is $x = x'$, where the phase has the minimum value of zero. In terms of $\eta = x - x'$, the region of coherence is

$$\frac{m\eta^2}{2\varepsilon\hbar} \leq \pi \tag{18.49}$$

or

$$|\eta| \leq \left(\frac{2\varepsilon\hbar\pi}{m}\right)^{1/2} \tag{18.50}$$

Consider now

$$\psi(x,\varepsilon) = \left(\frac{m}{2\pi\hbar i\varepsilon}\right)^{1/2} \int_{-\infty}^{\infty} e^{\frac{im\eta^2}{2\hbar\varepsilon}} e^{-\frac{i}{\hbar}eV(x+\frac{\eta}{2})}\psi(x+\eta,0)d\eta \tag{18.51}$$

Let us now work to first order in ε and therefore to second order in η (see Eq.(18.50)). We expand

$$\psi(x+\eta,0) = \psi(x,0) + \eta\frac{\partial\psi}{\partial x} + \frac{\eta^2}{2}\frac{\partial^2\psi}{\partial x^2} + \cdots \tag{18.52}$$

$$e^{-\frac{i}{\hbar}eV(x+\frac{\eta}{2})} = 1 - \frac{i\varepsilon}{\hbar}V(x+\frac{\eta}{2}) + \cdots \tag{18.53}$$

$$= 1 - \frac{i\varepsilon}{\hbar}V(x) + \cdots$$

Since terms of order $\eta\varepsilon$ are to be neglected, Eq.(18.51) now becomes

$$\psi(x,\varepsilon) = \left(\frac{m}{2\pi\hbar i\varepsilon}\right)^{1/2} \int_{-\infty}^{\infty} e^{\frac{im\eta^2}{2\hbar\varepsilon}} \tag{18.54}$$

$$\times \left[\psi(x,0) - \frac{i\varepsilon}{\hbar}V(x)\psi(x,0) + \eta\frac{\partial\psi}{\partial x} + \frac{\eta^2}{2}\frac{\partial^2\psi}{\partial x^2}\right]d\eta$$

After evaluating the integrals, we get

$$\psi(x,\varepsilon) = \left(\frac{m}{2\pi\hbar i\varepsilon}\right)^{1/2} \tag{18.55}$$

$$\times \left[\psi(x,0) - \frac{\hbar\varepsilon}{2im}\frac{\partial^2\psi}{\partial x^2} - \frac{i\varepsilon}{\hbar}V(x)\psi(x,0)\right]\left(\frac{2\pi\hbar i\varepsilon}{m}\right)^{1/2}$$

or

$$\psi(x,\varepsilon) - \psi(x,0) = \frac{-i\varepsilon}{\hbar}\left[\frac{-\hbar^2}{2m}\frac{\partial^2}{\partial x^2} + V(x)\right]\psi(x,0) \tag{18.56}$$

which agrees with the Schrödinger prediction, Eq.(18.44).

Sometimes a different notation is used to represent propagators. So far we have used the notation $U(x,t;x',t')$ for the propagator. If we look at Eq.(18.1), $|\psi(t)> = U(t)|\psi(0)>$, we see that the propagator $U(t)$ actually describes the quantum mechanical amplitude of the ket $|\psi(0)>$ at time t. The quantum mechanical amplitude is sometimes represented by the notation $K(x,t;x',t')$, which is called as **kernel**. U and K are identical. With $K(x,t;x',t')$ we want to denote the transition amplitude for a particle that is emitted at x' at time t', and is being detected at x at time t.

$$\psi(x,t) = \int dx' K(x,t;x',t')\psi(x',t') \tag{18.57}$$

This is the fundamental dynamical equation of the theory. Although it is an integral equation, we have seen that it is completely equivalent to the Schrödinger equation. The integral form of the propagator given in Eq.(18.17) may now be expressed as

$$K(x,t;x',t') = \int_{x'(t')}^{x(t)} [dx''(t'')] \exp\left[\frac{i}{\hbar}\int_{t'}^{t} dt'' L(x''(t''), \dot{x}''(t''), t'')\right] \tag{18.58}$$

The group property of propagator given in Eq.(18.14) can be generalized as $[b \equiv (x_b, t_b), a \equiv (x_a, t_a)]$

$$K(b,a) = \int_{-\infty}^{\infty} dx_{N-1} \cdots \int_{-\infty}^{\infty} dx_1 \tag{18.59}$$

$$\times K(b, N-1)K(N-1, N-2) \cdots K(2,1)K(1,a)$$

The contents of this chapter closely follow the treatment of path integrals in Shankar's book given in the bibliography; more detailed discussions and various applications of the subject may be found in that book.

18.4   Worked examples

Example - 18.1 : Obtain the path–integral solution of a particle in a one–dimensional infinite square well of width L: $V(x) = 0$ for $0 < x < L$, $V(x) = \infty$ for $x \leq 0$, $x \geq L$.

Solution : The end point (x_f, t_f) is reached via the image point $(-x_f, t_f)$, starting at (x_i, t_i). Therefore the propagator consists of the sum of the two contributions of the classical paths:

$$K_L(x_f, t_f; x_i, t_i) = \left[\frac{m}{2\pi i\hbar(t_f - t_i)}\right]^{1/2} \left[e^{\frac{i}{\hbar}\frac{m}{2}\frac{(x_f - x_i)^2}{(t_f - t_i)}} - e^{\frac{i}{\hbar}\frac{m}{2}\frac{(-x_f - x_i)^2}{(t_f - t_i)}}\right]$$

$$= K(x_f, t_f; x_i, t_i) - K(-x_f, t_f; x_i, t_i)$$

the K's are free propagators. Since a particle bounces infinitely many times between the walls, one may also construct the propagator by forming the sum of all these infinitely many classical paths. The contribution of each classical path is the free particle propagator, multiplied by (-1) for every collision against the wall. Therefore the general form of the propagator may be expressed as

$$K_L(x_f, t_f; x_i, t_i) = \sum_{r=-\infty}^{\infty} (-1)^r K(-x_r, t_f; x_i, t_i)$$

$$= \sum_{r=-\infty}^{\infty} \int \frac{dp}{2\pi\hbar} \left[\exp\left(-\frac{i}{\hbar}\frac{p^2}{2m}(t_f - t_i) + \frac{i}{\hbar}(2rL + x_f - x_i)p \right) \right.$$

$$\left. - \exp\left(-\frac{i}{\hbar}\frac{p^2}{2m}(t_f - t_i) + \frac{i}{\hbar}(2rL - x_f - x_i)p \right) \right]$$

At every turning point we pick up a phase $2Lp/\hbar$

$$= \int \frac{dp}{2\pi\hbar} e^{-\frac{i}{\hbar}\frac{p^2}{2m}(t_f-t_i)} e^{-\frac{i}{\hbar}p x_i} 2i \left(\frac{e^{\frac{i}{\hbar}x_f p} - e^{-\frac{i}{\hbar}x_f p}}{2i} \right)$$

$$\times \left(\sum_{r=-\infty}^{\infty} \exp\left(\frac{2irLp}{\hbar} \right) \right)$$

Using

$$\frac{e^{\frac{i}{\hbar}x_f p} - e^{-\frac{i}{\hbar}x_f p}}{2i} = \sin\left(\frac{p}{\hbar}x_f\right)$$

and Poisson's formula

$$\sum_{r=-\infty}^{\infty} e^{\frac{i}{\hbar}2Lpr} = \sum_{n=-\infty}^{\infty} \delta\left(p\frac{L}{\pi\hbar} - n\right) = \frac{\pi\hbar}{L} \sum_{n=-\infty}^{\infty} \delta\left(p - n\frac{\pi\hbar}{L}\right)$$

$$K_L(x_f, t_f; x_i, t_i) = \frac{i}{L} \sum_{n=-\infty}^{\infty} e^{-\frac{i}{\hbar}E_n(t_f - t_i)} e^{-ik_n x_i} \sin(k_n x_f)$$

δ–function implies energy and momentum quantization:

$$E_n = \frac{1}{2m}\frac{\pi^2\hbar^2}{L^2}n^2 \quad , \quad k_n = \frac{\pi n}{L}$$

Since the term $n = 0$ vanishes in the last summation, we can combine positive and negative terms to get ($t_f > t_i$)

$$K_L(x_f, t_f; x_i, t_i) = \frac{i}{L} \left(\sum_{n=1}^{\infty}(\cdots) + \sum_{n=-\infty}^{-1}(\cdots) \right)$$

considering the relations $E_n = E_{-n}$ and $-k_n = k_{-n}$ we may write

$$K_L(x_f, t_f; x_i, t_i) = \frac{i}{L} \sum_{n=1}^{\infty} e^{-\frac{i}{\hbar}E_n(t_f - t_i)} e^{ik_n x_i} \left(-\sin(k_n x_f) \right)$$

$$+ \frac{i}{L} \sum_{n=-\infty}^{-1} e^{-\frac{i}{\hbar}E_n(t_f - t_i)} e^{-ik_n x_i} \sin(k_n x_f)$$

or

$$K_L(x_f, t_f; x_i, t_i) = \frac{i}{L} \sum_{n=1}^{\infty} e^{-\frac{i}{\hbar}E_n(t_f - t_i)} 2i \left(\frac{e^{-ik_n x_i} - e^{ik_n x_i}}{2i} \right) \sin(k_n x_f)$$

so the final result $(t_f > t_i)$ takes the form:

$$K_L(x_f, t_f; x_i, t_i) = \frac{2}{L} \sum_{n=1}^{\infty} e^{-\frac{i}{\hbar} E_n(t_f - t_i)} \sin(k_n x_i) \sin(k_n x_f)$$

Example - 18.2 : Obtain the path–integral solution of a particle on a circular orbit.

Solution : Let the ring be parameterized by s and its length be L. We are dealing with a problem with periodic boundary conditions. We can write the propagator of the particle moving on a closed path as

$$K_L(s_f, t_f; s_i, t_i) = \left(\frac{m}{2\pi i \hbar \tau} \right)^{1/2} e^{\frac{i}{\hbar} \frac{m}{2} \frac{(\bar{s} + nL)^2}{\tau}}$$

where $\tau = t_f - t_i$ and $\bar{s} = s_f - s_i$. Using the identity

$$\frac{1}{\sqrt{\tau}} e^{\frac{i}{\tau} x^2} = \sqrt{\frac{i}{4\pi}} \int dp\, e^{-i\frac{\tau}{4} p^2 + ipx}$$

we obtain

$$K_L(\bar{s}, \tau) = \int \frac{dp}{2\pi\hbar} e^{-i \frac{\tau p^2}{2m\hbar} + \frac{i}{\hbar} p(\bar{s} + nL)}$$

We must consider all possible orbittings, therefore the propagator takes the form

$$K_L(\bar{s}, \tau) = \int \frac{dp}{2\pi\hbar} \left(\sum_{n=-\infty}^{+\infty} e^{\frac{i}{\hbar} pLn} \right) e^{-\frac{i}{\hbar} \frac{p^2}{2m} \tau + \frac{i}{\hbar} p\bar{s}}$$

Using Poisson's summation formula

$$\sum_{n=-\infty}^{+\infty} e^{i2\pi n x} = \sum_{\mu=-\infty}^{+\infty} \delta(\mu - x)$$

and taking $x = pL/2\pi\hbar$, the sum becomes

$$\sum_{n=-\infty}^{+\infty} e^{\frac{i}{\hbar} pLn} = \sum_{\mu=-\infty}^{+\infty} \delta\left(\frac{pL}{2\pi\hbar} - \mu \right) = \frac{2\pi\hbar}{L} \sum_{\mu=-\infty}^{+\infty} \delta\left(p - \frac{2\pi\hbar}{L} \mu \right)$$

The propagator then becomes

$$K_L(\bar{s}, \tau) = \frac{1}{L} \int dp \sum_{n=-\infty}^{+\infty} \delta\left(p - \frac{2\pi\hbar}{L} n \right) e^{-\frac{i}{\hbar} \frac{p^2}{2m} \tau + \frac{i}{\hbar} p\bar{s}}$$

Performing the p–integration we obtain

$$K_L(\bar{s}, \tau) = \frac{1}{L} \sum_{n=-\infty}^{+\infty} e^{-\frac{i}{\hbar} \frac{1}{2m} \left(\frac{2\pi\hbar}{L} n \right)^2 \tau + 2\pi i n \frac{\bar{s}}{L}}$$

For a circular motion with radius R, the natural parameter is the angle ϕ. Using $\bar{s} = R(\phi_f - \phi_i) = R\phi$, $L = 2\pi R$, $\tau = t_f - t_i$, $I = mR^2$, the propagator for a particle moving on a circular orbit becomes

$$K_L(R\phi, \tau) = \frac{1}{2\pi R} \sum_{n=-\infty}^{+\infty} e^{-i\frac{\hbar}{2I}n^2\tau + in\phi}$$

or introducing $K_L(R\phi, \tau) = \frac{1}{R}K(\phi, \tau)$, the final result becomes

$$K(\phi_f, t_f; \phi_i, t_i) = \frac{1}{2\pi} \sum_{n=-\infty}^{+\infty} e^{-i\frac{\hbar}{2I}n^2\tau + in\phi}$$

The delta function above implies the quantization of the orbital angular momentum and the energy:

$$p = \frac{2\pi\hbar}{L}n \quad , \quad \frac{pL}{2\pi} = \hbar n \quad ; \quad p = mR\dot{\phi} \quad , \quad L = 2\pi R$$

$$\frac{pL}{2\pi} = mR^2\dot{\phi} = L_z = n\hbar \quad ; \quad E_n = \frac{p^2}{2m} = \frac{\hbar^2}{2mR^2}n^2 = \frac{\hbar^2}{2I}n^2$$

18.5 Problems

Problem - 18.1 : Suppose a wave function ψ has expectation values for position and momentum x_1 and p_1, and is sharply concentrated near these values, for example, in a minimum uncertainty wave packet. Let the system evolve under the Hamiltonian $H = p^2/2m + V$ from time t_1 to time t_2. Show that for $\hbar \to 0$, at time t_2 the wave function will be concentrated at x_2, p_2 where these are the values of position and momentum the classical system would have reached at t_2, starting at x_1, p_1 at t_1.

Problem - 18.2 : Prove the following equality

$$\left[\int_0^t f(\tau)d\tau \right]^N = N! \int_0^t d\tau_1 \int_0^{\tau_1} d\tau_2 \cdots \int_0^{\tau_{N-1}} d\tau_N\, f(\tau_1)f(\tau_2)\cdots f(\tau_N)$$

Chapter 19

Scattering Theory

19.1 Classical scattering theory

The phenomenon of scattering is important in the study of classical systems. But when the size of the systems become so small that we cannot directly see them, or when the forces involved are so short–ranged that we cannot directly measure them, scattering becomes an indispensable probe. In the structural analysis of matter scattering is the primary method of investigation, particularly at high energies. Scattering is also important in the study of transport phenomena such as conductivity, specific heat of bulk matter, because these properties are affected by the scattering of free electrons and phonons inside the material.

The simplest form of scattering may be thought of as a beam of particles (with well–defined momenta) from a fixed center of force. Classically, the trajectories of the particles change due to the force experienced, and the deflection of the trajectories from the initial direction defines the **scattering angle**. A scattering experiment may consist of measuring the trajectories of individual particles; the result is expressed in terms of the **differential cross section**, which is determined from the number of particles scattered into a solid angle. This statistical measurement gives information on the shape of the potential, the nature of the force, and so on.

In classical mechanics, given the initial conditions, one can theoretically calculate the trajectory exactly. However, in quantum mechanics the notion of a trajectory does not exist, and one must solve the Schrödinger equation for the scattering problem. The differential cross section is obtained from the scattering solutions.

In a scattering experiment the number of particles involved is quite large so it is impractical to measure the details of each individual trajectory. What one measures is the initial velocity of the particles (they are usually prepared to be monoenergetic) and their final velocity.

The goemetry and the parameters of a scattering configuration are shown in Figure 19.1.

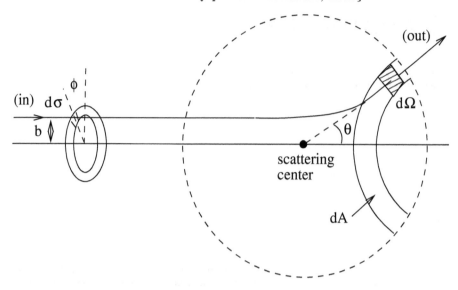

Fig. 19.1. Geometry of a scattering configuration.

b	:	impact parameter
θ	:	scattering angle
ϕ	:	azimuthal angle
$d\sigma$	:	differential (scattering) cross section, $d\sigma = b\,db\,d\phi$
$d\Omega$	:	solid angle, $d\Omega = \sin\theta\,d\theta\,d\phi$
dA	:	area of the ring perpendicular to the scattered trajectory, $dA = 2\pi r \sin\theta\,d\theta$

If the number of incident particles per unit area per unit time is N, and ΔN of them scatter by an angle (θ, ϕ) into the solid angle $\Delta\Omega$ in unit time, then the differential cross section is defined to be

$$d\sigma \rightarrow \sigma(\theta, \phi) = \lim_{\Delta\Omega \rightarrow 0} \frac{1}{N} \frac{\Delta N}{\Delta\Omega} = \frac{d\sigma(\theta, \phi)}{d\Omega} \tag{19.1}$$

The differential cross section is a useful quantity since it gives information about the shape of the interaction potential or the nature of the force, the larger the differential cross section, the stronger the force. In most scattering phenomena, the potentials are spherically symmetric, therefore physical observables cannot have an azimuthal dependence (no ϕ dependence). Hence the differential cross section for scattering into a ring around the beam axis can be defined as

$$\sigma(\theta) \rightarrow \int d\phi\, \sigma(\theta, \phi) = 2\pi\sigma(\theta) \tag{19.2}$$

One can define the total cross section as

$$\sigma_T = \sigma_{tot} = \int \sigma(\theta, \phi)\, d\Omega \tag{19.3}$$

Total cross section measures the total number of scattered particles (at all angles) per unit time when one particle is incident per unit area in unit time. Cross section has the dimension of area. The unit of cross section used in practice (in nuclear and high–energy physics) is called **barn**. 1 $barn = 10^{-24}\ cm^2$.

In scattering experiments a beam of particles is incident on a fixed target of finite thickness. The total cross section in such cases may be calculated as follows: See Figure 19.2.

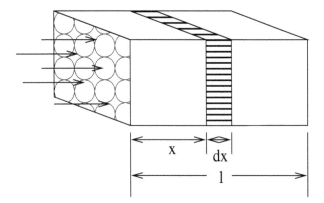

Fig. 19.2. Incident of a beam of particles on a fixed target of finite thickness.

Each target particle acts as a scattering source. Let the shaded area around each target particle represent the total cross section. If the incident particle strikes the shaded area it will scatter, otherwise it will not interact. If the target is of infinitesimal thickness dx and has a density of ρ particles per unit volume, the fraction of the transverse area that is shaded is

$$\frac{dA}{A} = \sigma_T \rho dx \tag{19.4}$$

The probability that an incident particle will scatter in the interval dx is

$$P(x)dx = \frac{dA}{A} = \sigma_T \rho dx \tag{19.5}$$

Let us define N_0 as the number of incident particles on the target, $N(x)$ as the number of incoming particles at a distance x from the edge of the target. Whenever a particle scatters, $N(x)$ decreases such that $dN_s(x) = -dN(x)$, where $dN_s(x)$ represents the number of scattered particles at the distance x (in the interval dx). We may write the relation

$$dN(x) = -N(x)P(x)dx = -N(x)\sigma_T \rho dx \tag{19.6}$$

or

$$N(x) = N_0 e^{-\sigma_T \rho x} \tag{19.7}$$

The number of particles scattered from a target of thickness l is

$$N_s(l) = N_0 - N(l) = N_0 - N_0 e^{-\sigma_T \rho l} = N_0(1 - e^{-\sigma_T \rho l}) \tag{19.8}$$

If $\sigma_T \rho l \ll 1$, then we can write

$$N_s(l) \cong N_0 \sigma_T \rho l \tag{19.9}$$

so that the total cross section in this case is given by

$$\sigma_T = \frac{N_s(l)}{N_0 \rho l} \tag{19.10}$$

This result is obtained with the assumption that the transverse dimensions of the beam are smaller than those of the target; this is the case in most experimental arrangements.

In defining the cross section we have implicitly used the concept of flux of incoming and outgoing particles. The flux **j** gives the number of particles crossing the unit area normal to their direction of motion per unit time. If **v** is the velocty of the particles and ρ their density, then

$$\mathbf{j} = \rho \mathbf{v} \tag{19.11}$$

If the amplitude (ψ) for the incoming particles is normalized to one per unit volume, then the flux coincides with the quantum mechanical probability current density and is given by

$$\mathbf{j} = -\frac{i\hbar}{2m}(\psi^* \vec{\nabla} \psi - (\vec{\nabla} \psi^*)\psi) \tag{19.12}$$

This measures the probability that a particle will cross a unit area normal to the direction of the current in unit time.

19.2   Center–of–mass and laboratory frames

Transformation from the center–of–mass frame to the laboratory frame:

When the potentials involved are central (noncentral potentials arise in the presence of spin–dependent forces), the motion can be described in a more simple way in center–of–mass (CM) frame; because the motion of the CM is not subject to any forces and therefore in the CM frame the Hamiltonian under study becomes much simpler. However, the actual observations are made in the laboratory frame (L). Thus, we must know how to transform various quantities such as the scattering angle and differential cross section back into the L frame. The total cross section σ_T is the same in both frames since it represents the total probability that an incident particle will scatter; hence it must be independent of the choice of the reference frame.

Let us consider the scattering of two particles with masses m_1 and m_2. Let m_2 be stationary in the L frame and let m_1 be moving along the z–axis with a velocity $v \ll c$ (see Figure 19.3). This is the usual case in experiments except for colliding–beam arrangements.

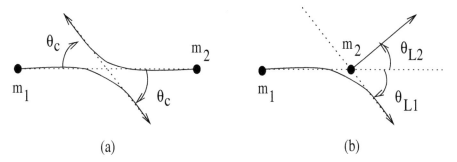

Fig. 19.3. (a) Scattering in the CM system. (b) The same scattering process as observed in the laboratory.

For such a system, the reduced mass (μ) and the velocity of the CM ($\mathbf{v}$) are, respectively,

$$\mu = \frac{m_1 m_2}{m_1 + m_2} \tag{19.13}$$

$$\mathbf{v}_{CM} = \frac{m_1}{m_1 + m_2}\mathbf{v}_1 = \frac{m_1}{m_1 + m_2}\mathbf{v}_{1L} \tag{19.14}$$

In the CM frame the particles have velocities

$$\mathbf{v}_{1C} = \mathbf{v}_{1L} - \mathbf{v}_{CM} = \frac{m_2}{m_1 + m_2}\mathbf{v}_{1L} \tag{19.15}$$

$$\mathbf{v}_{2C} = \mathbf{v}_{2L} - \mathbf{v}_{CM} = -\frac{m_1}{m_1 + m_2}\mathbf{v}_{1L} \tag{19.16}$$

In the CM frame the two particles move toward each other with equal and opposite momenta. After the scattering they must again move with equal and opposite momenta. Thus, in the CM frame the two particles scatter by equal and opposite angles.

If we denote by primes the velocities after scattering and scattering is elastic, Eq.(19.15) gives

$$\mathbf{v}'_{1C} = \mathbf{v}'_{1L} - \mathbf{v}_{CM} \tag{19.17}$$

In terms of components (along the beam direction),

$$v_{1C}\cos\theta_C = v_{1L}\cos\theta_{L1} - v_{CM} \tag{19.18}$$

or

$$v_{1L}\cos\theta_{L1} = v_{CM} + v_{1C}\cos\theta_C \tag{19.19}$$

and transverse to the beam direction

$$v_{1C}\sin\theta_C = v_{1L}\sin\theta_{L1} \tag{19.20}$$

The ratio of Eq.(19.20) and Eq.(19.18) gives

$$\tan\theta_{L1} = \frac{\sin\theta_C}{\cos\theta_C + \frac{v_{CM}}{v_{1C}}} = \frac{\sin\theta_C}{\cos\theta_C + \gamma} \tag{19.21}$$

where

$$\gamma = \frac{v_{CM}}{v_{1C}} = \frac{m_1}{m_2} \tag{19.22}$$

one can also write

$$\theta_{L2} = \frac{1}{2}(\pi - \theta_C) \qquad (19.23)$$

The relation between the differential cross section in the CM frame and the L frame can be obtained from the simple observation that the same particles that go into the solid angle $d\Omega_C$ at θ_C in the CM frame go into the solid angle $d\Omega_L$ at θ_L in the L frame. With the assumption of rotational symmetry (the differential cross section does not depend on ϕ), one can write

$$\sigma_L(\theta_L)\sin\theta_L d\theta_L = \sigma_C(\theta_C)\sin\theta_C d\theta_C \qquad (19.24)$$

or

$$\sigma_L(\theta_L) = \sigma_C(\theta_C)\left|\frac{d\cos\theta_C}{d\cos\theta_L}\right| \qquad (19.25)$$

The relation for the cross sections turns out to be

$$\sigma_L(\theta_1) = \sigma_C(\theta_C)\frac{[1 + 2\gamma\cos\theta_C + \gamma^2]^{3/2}}{|1 + \gamma\cos\theta_C|} \qquad (19.26)$$

or

$$\sigma_L(\theta_1) = \sigma_C(\theta_C)\frac{\left\{[1 - \gamma^2\sin^2\theta_{L1}]^{1/2} + \gamma\cos\theta_{L1}\right\}^2}{[1 - \gamma^2\sin^2\theta_{L1}]^{1/2}} \qquad (19.27)$$

$$\sigma_L(\theta_2) = \sigma_C(\theta_C)4\cos\theta_{L2} \qquad (19.28)$$

Using Eq.(19.28) and Eq.(19.23), one can show that the total cross section in the L frame is the same as in the CM frame; that is,

$$(\sigma_T)_L = (\sigma_T)_C \qquad (19.29)$$

For the special case of equal mass particles, i.e., $m_1 = m_2$ ($\gamma = 1$), the transformation equations reduce to

$$\theta_{L1} = \frac{\theta_C}{2} \quad , \quad \theta_{L1} + \theta_{L2} = \frac{\pi}{2} \quad , \quad \sigma_L(\theta_{1,2}) = \sigma_C(\theta_C)4\cos\theta_{L1,2} \qquad (19.30)$$

When the target particle is much more massive than the beam particle, $\gamma \to 0$ and the L frame and CM frame are equivalent.

With these kinematic relations we can now solve the scattering equation in the CM frame.

19.3 Quantum scattering theory

In the CM frame, scattering is governed by the same Hamiltonian as for the bound state of the system,

$$H = -\frac{\hbar^2}{2\mu}\nabla^2 + V(r) \qquad (19.31)$$

where μ is the reduced mass of the system. The stationary states of the Hamiltonian take the same form as for the bound states

$$\Psi(\mathbf{r}, t) = \psi(\mathbf{r})e^{-\frac{i}{\hbar}Et} \qquad (19.32)$$

In this case the energy of the system E is positive and has a continuous spectrum. Furthermore, if we consider elastic scattering, the energy E does not change, and hence we need only study the space part of the wave function that satisfies the time–independent Schrödinger equation

$$H\psi(\mathbf{r}) = E\psi(\mathbf{r}) \tag{19.33}$$

or

$$\left[\nabla^2 + k^2 - \frac{2\mu}{\hbar^2}V(r)\right]\psi(\mathbf{r}) = 0 \quad , \quad k^2 = \frac{2\mu E}{\hbar^2} \tag{19.34}$$

If the potential vanishes for infinite separation and if we assume that initially the particle was far from the scattering center, the solutions before the scattering ($t \to -\infty$) must satisfy the free–particle equation,

$$(\nabla^2 + k^2)\psi_{in}(\mathbf{r}) = 0 \quad , \quad (r \to \infty \, , \, t \to -\infty) \tag{19.35}$$

If the particle is incident along the z–axis we can write the incident wave function as a plane wave

$$\psi_{in}(\mathbf{r}) = e^{ikz} \tag{19.36}$$

In the scattering region, the wave function must depend on the form of the potential and must satisfy

$$\left[\nabla^2 + k^2 - \frac{2\mu}{\hbar^2}V(r)\right]\psi(\mathbf{r}) = 0 \tag{19.37}$$

After scattering the wave function must again have a free–particle form far from the scattering center, and must satisfy the free–particle equation,

$$(\nabla^2 + k^2)\psi_f(\mathbf{r}) = 0 \quad , \quad (r \to \infty \, , \, t \to +\infty) \tag{19.38}$$

If we assume that the scattering potential is spherically symmetric, then at very far distances from the scattering source we will have a spherically outgoing scattered wave (see Figure 19.4).

Thus, the form of the total wave function at large distances from the scattering center is given by (if the potential is centered at the origin)

$$\psi_f(\mathbf{r}) \longrightarrow \psi_{in}(\mathbf{r}) + \psi_{sc}(\mathbf{r}) = e^{ikz} + f(\theta, \phi)\frac{e^{ikr}}{r} \tag{19.39}$$

Here $f(\theta, \phi)$ measures the angular distribution of the scattered wave and is known as the **scattering amplitude**. The wave function will be normalized to one particle per unit volume, and $\psi_{sc}(\mathbf{r})$ also satisfies the free–particle equation.

The flux associated with the incident as well as the scattered wave functions for large distance may be expressed as

$$\mathbf{j}_{in} = -\frac{i\hbar}{2\mu}(\psi_{in}^*\vec{\nabla}\psi_{in} - (\vec{\nabla}\psi_{in}^*)\psi_{in}) = \frac{\hbar k}{\mu}\vec{n}_z \tag{19.40}$$

Similarly, the flux of the scattered wave is defined as

$$\mathbf{j}_{sc} = -\frac{i\hbar}{2\mu}(\psi_{sc}^*\vec{\nabla}\psi_{sc} - (\vec{\nabla}\psi_{sc}^*)\psi_{sc}) \tag{19.41}$$

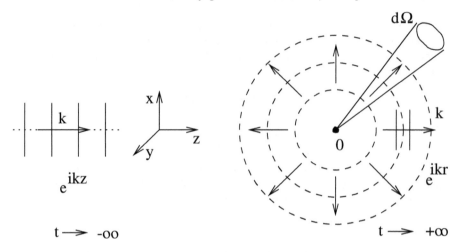

Fig. 19.4. Scattering of a plane wave from a spherically symmetric potential.

Using spherical coordinates the radial component of the flux at large distances becomes

$$(\mathbf{j}_{sc})_r = -\frac{i\hbar}{2\mu}\left[\psi_{sc}^*\frac{\partial}{\partial r}\psi_{sc} - (\frac{\partial}{\partial r}\psi_{sc}^*)\psi_{sc}\right] \tag{19.42}$$

or

$$(\mathbf{j}_{sc})_r \xrightarrow{r\to\infty} \frac{\hbar k}{\mu}\frac{|f(\theta,\phi)|^2}{r^2} + O(\frac{1}{r^3}) \tag{19.43}$$

Eq.(19.43) shows that the probability of particles scattering across an area dS of a sphere of radius r is (as $r\to\infty$)

$$\mathbf{j}_{sc}\cdot d\mathbf{S} = \frac{\hbar k}{\mu}|f(\theta,\phi)|^2\frac{dS}{r^2} = \frac{\hbar k}{\mu}|f(\theta,\phi)|^2 d\Omega \tag{19.44}$$

From Eqs.(19.40) and (19.44) we see that the differential cross section is given by

$$\sigma(\theta,\phi) = |f(\theta,\phi)|^2 \tag{19.45}$$

This is the fundamental relation between the observable differential cross section and the quantum mechanical scattering amplitude.

19.4   The method of partial waves (low–energy case)

Solving the scattering equation one can find the scattering amplitude. There are several ways of doing this; here we will discuss **the method of partial waves**. The method consists of writing the wave function as a superposition of various angular momentum components. In this method the scattering amplitude is expressed as a sum over all partial–wave contributions, and, in principle, the result is exact.

If the range of the potential is a_0 and the momentum of the particle is **p**, then only those angular momentum components would contribute that satisfy

$$\hbar[l(l+1)]^{1/2} \cong \hbar l \ll a_0 p \quad or \quad l \ll a_0 k \tag{19.46}$$

The method of partial wave analysis is therefore useful for low–energy scattering.

To obtain the scattering amplitude we do the following procedure in three steps:
i) Find the free–particle eigenfunctions expressed in spherical coordinates and obtain the spherical waves.
ii) Expand the incident particle wave function $\psi_{in}(\mathbf{r})$ as a plane wave into angular momentum eigenstates.
iii) Expand the final wave function $\psi_f(\mathbf{r})$.

Free–particle solutions in spherical coordinates:

Since the scattering phenomena mostly involve rotationally invariant potentials (spherically symmetric potential), the total wave function may be decomposed into angular momentum eigenstates.

$$\psi_{lm}(r,\theta,\phi) = N_l R_l(r) Y_{lm}(\theta,\phi) = N_l \frac{u_l(r)}{r} Y_{lm}(\theta,\phi) \tag{19.47}$$

Here $Y_{lm}(\theta,\phi)$ are the spherical harmonics, and they are related to the associated Legendre polynomials through

$$Y_{lm}(\theta,\phi) = \varepsilon \left[\frac{2l+1}{4\pi} \frac{(l-|m|)!}{(l+|m|)!} \right]^{1/2} P_{lm}(\cos\theta) e^{im\phi} \tag{19.48}$$

where $\varepsilon = (-1)^m$ for $m > 0$, 1 otherwise. Furthermore, if $\psi_{lm}(r,\theta,\phi)$ satisfies the free–particle equation, i.e.,

$$(\nabla^2 + k^2)\psi_{lm}(r,\theta,\phi) = 0 \tag{19.49}$$

then it follows that the radial functions satisfy

$$\frac{d^2 u_l}{dr^2} + \left[k^2 - \frac{l(l+1)}{r^2} \right] u_l(r) = 0 \tag{19.50}$$

with the boundary condition

$$u_l(r) \to 0 \quad as \quad r \to 0 \tag{19.51}$$

If we define the variable

$$\rho = kr = \left(\frac{2\mu E}{\hbar^2} \right)^{1/2} r \tag{19.52}$$

then Eq.(19.50) becomes

$$\frac{d^2 u_l}{d\rho^2} + \left[1 - \frac{l(l+1)}{\rho^2} \right] u_l(\rho) = 0 \tag{19.53}$$

The solutions to this equation are related to the spherical Bessel functions $j_l(\rho)$, and $n_l(\rho)$, such that

$$u_l(\rho) = \rho j_l(\rho) \quad or \quad u_l(\rho) = \rho n_l(\rho) \tag{19.54}$$

The most general solution is

$$u_l(\rho) = u_l(kr) = kr[a_1 j_l(kr) + a_2 n_l(kr)] \tag{19.55}$$

From the limiting form of these functions as $\rho \to 0$

$$j_l(\rho) \propto \rho^l \quad , \quad n_l(\rho) \propto \rho^{-l-1} \quad ; \quad \rho \to 0 \tag{19.56}$$

The boundary condition of Eq.(19.51) requires that $a_2 = 0$. The solution to the radial equation becomes

$$u_l(kr) = a_1 kr j_l(kr) \tag{19.57}$$

and hence the free–particle solution in spherical coordinates is given by

$$\psi_{lm}(r, \theta, \phi) = N_l R_l(r) Y_{lm}(\theta, \phi) = \tilde{N}_l j_l(kr) Y_{lm}(\theta, \phi) \tag{19.58}$$

The normalization constant $\tilde{N}_l$ is obtained from the orthonormality relations

$$\int d\Omega Y_{lm}^*(\theta, \phi) Y_{lm}(\theta, \phi) = 1 \tag{19.59}$$

$$\int r^2 dr j_l(kr) j_l(k'r) = \frac{\pi}{2k^2} \delta(k - k') \tag{19.60}$$

The normalization constant is

$$\tilde{N}_l = \left(\frac{2k^2}{\pi}\right)^{1/2} \tag{19.61}$$

Therefore the normalized free–particle solutions in spherical coordinates are given by

$$\psi_{lm}(r, \theta, \phi) = \left(\frac{2k^2}{\pi}\right)^{1/2} j_l(kr) Y_{lm}(\theta, \phi) \tag{19.62}$$

19.5   Expansion of a plane wave into spherical waves

The wave functions given in Eq.(19.61) form a complete basis and any wave function can be expressed in terms of the angular momentum components. We know that

$$\psi_{in} = e^{ikz} = e^{ikr \cos \theta} \tag{19.63}$$

This does not depend on the azimuthal angle ϕ, which simply reflects the fact that the wave has no angular momentum along the z–axis. Therefore, its expansion in the spherical basis would only involve $m = 0$ components. In the limit $r \to \infty$ we obtain:

$$\psi_{in} = e^{ikz} = e^{ikr \cos \theta} = \sum_{l=0}^{\infty} a_l \psi_{l,0}(r, \theta, \phi) \tag{19.64}$$

$$= \sum_{l=0}^{\infty} a_l \left(\frac{2k^2}{\pi}\right)^{1/2} j_l(kr) Y_{l,0}(\theta, \phi)$$

$$= \left(\frac{2k^2}{\pi}\right)^{1/2} \sum_{l=0}^{\infty} a_l \left(\frac{2l+1}{4\pi}\right)^{1/2} j_l(kr) P_l(\cos\theta)$$

Multiplying Eq.(19.63) by $P_l(\cos\theta)$ and integrating over all solid angles we obtain

$$\int d\Omega P_l(\cos\theta) e^{ikr\cos\theta} \tag{19.65}$$

$$= \left(\frac{2k^2}{\pi}\right)^{1/2} \sum_{l'=0}^{\infty} a_{l'} \left(\frac{2l'+1}{4\pi}\right)^{1/2} j_{l'}(kr) \int d\Omega P_l(\cos\theta) P_{l'}(\cos\theta)$$

If we use the orthogonality relation of the Legendre functions and the integral representation of the spherical Bessel functions

$$\int_0^{\pi} P_l(\cos\theta) P_{l'}(\cos\theta) \sin\theta d\theta = \frac{2}{2l+1}\delta_{ll'} \tag{19.66}$$

$$\frac{1}{2i^l} \int_0^{\pi} e^{ix\cos\theta} P_l(\cos\theta) \sin\theta d\theta = j_l(x) \tag{19.67}$$

we obtain the expansion coefficients a_l as

$$2\pi 2i^l j_l(kr) = \left(\frac{2k^2}{\pi}\right)^{1/2} a_l \left(\frac{2l+1}{4\pi}\right)^{1/2} j_l(kr) 2\pi \frac{2}{2l+1} \tag{19.68}$$

or

$$a_l = 2\pi i^l \left(\frac{2l+1}{2k^2}\right)^{1/2} \tag{19.69}$$

The incident wave in the spherical basis becomes

$$\psi_{in} = e^{ikz} = \sum_{l=0}^{\infty} 2\pi i^l \left(\frac{2l+1}{2k^2}\right)^{1/2} \left(\frac{2k^2}{\pi}\right)^{1/2} \left(\frac{2l+1}{4\pi}\right)^{1/2} j_l(kr) P_l(\cos\theta) \tag{19.70}$$

$$= \sum_{l=0}^{\infty} (2l+1) i^l j_l(kr) P_l(\cos\theta)$$

Using the asymptotic form of the spherical Bessel functions

$$j_l(kr) \overset{kr\to\infty}{\longrightarrow} \frac{1}{kr} \sin(kr - \frac{l\pi}{2}) \tag{19.71}$$

we obtain the incident wave for large distances

$$\psi_{in} \overset{r\to\infty}{\longrightarrow} \sum_{l=0}^{\infty} \frac{2l+1}{kr} i^l \sin(kr - \frac{l\pi}{2}) P_l(\cos\theta) \tag{19.72}$$

This confirms the statement that a plane wave is a superposition of an infinite number of spherical waves of various angular momenta.

19.6 Expansion of the scattering amplitude

If the potential $V(r)$ falls off faster than $1/r^2$, then for large distances the final wave function would also be a free–particle solution. This is because the centrifugal barrier in this case would dominate over the potential energy term. However, the final wave function at large distances after scattering would undergo a phase change relative to the incident wave.

If the scattering potential is attractive the particle will be accelerated, and consequently the wavelength would be shorter near the scattering source. On the other hand, if the potential is repulsive, then the particle would be decelerated and thus would have a large wavelength in the scattering region. In either case, when the particle emerges from the scattering region, its phase would be different from the case when there is no scattering. Thus, the final wave function for large distances must have the form

$$\psi_f \overset{r\to\infty}{\longrightarrow} \sum_{l=0}^{\infty} A_l \frac{2l+1}{kr} i^l \sin(kr - \frac{l\pi}{2} + \delta_l) P_l(\cos\theta) \tag{19.73}$$

where δ_l is the phase shift in the lth partial wave. It is positive if the potential is attractive, and negative if the potential is repulsive. The constant A_l is determined from the observation that

$$\psi_f \overset{r\to\infty}{\longrightarrow} e^{ikz} + f(\theta,\phi)\frac{e^{ikr}}{r} \tag{19.74}$$

Using Eq.(19.72) and Eq.(19.74) we can write

$$f(\theta,\phi)\frac{e^{ikr}}{r} = \sum_{l=0}^{\infty} A_l \frac{2l+1}{kr} i^l \sin(kr - \frac{l\pi}{2} + \delta_l) P_l(\cos\theta) \tag{19.75}$$

$$- \sum_{l=0}^{\infty} \frac{2l+1}{kr} i^l \sin(kr - \frac{l\pi}{2}) P_l(\cos\theta)$$

The left hand side of Eq.(19.75) has no e^{-ikr} term; the same must be true of the right hand side. By writing

$$\sin(kr - \frac{l\pi}{2} + \delta_l) = \frac{1}{2i}\left[e^{i(kr - \frac{l\pi}{2} + \delta_l)} - e^{-i(kr - \frac{l\pi}{2} + \delta_l)} \right] \tag{19.76}$$

one can see that the above condition can be satisfied only if $A_l = e^{i\delta_l}$. In that case the scattering wave takes the form

$$f(\theta,\phi)\frac{e^{ikr}}{r} = \frac{e^{ikr}}{r} \sum_{l=0}^{\infty} \frac{2l+1}{2ik}(e^{2i\delta_l} - 1) P_l(\cos\theta) \tag{19.77}$$

This gives an expression for the scattering amplitude in terms of the phase shifts

$$f(\theta,\phi) = \sum_{l=0}^{\infty} \frac{2l+1}{2ik}(e^{2i\delta_l} - 1) P_l(\cos\theta) \tag{19.78}$$

$$= \sum_{l=0}^{\infty} \frac{2l+1}{k} e^{i\delta_l} \sin \delta_l P_l(\cos \theta) = \sum f_l(\theta, \phi)$$

where $f_l(\theta, \phi)$ can be thought of as the scattering amplitude for the lth partial wave. The differential cross section is given by

$$\sigma(\theta, \phi) = |f(\theta, \phi)|^2 = \frac{1}{k^2} \left| \sum_{l=0}^{\infty} (2l+1) e^{i\delta_l} \sin \delta_l P_l(\cos \theta) \right|^2 \qquad (19.79)$$

The total cross section becomes

$$\sigma_T = \int d\Omega |f(\theta, \phi)|^2 \qquad (19.80)$$

$$= \sum_{l=0}^{\infty} \sum_{l'=0}^{\infty} \frac{2l+1}{k} \frac{2l'+1}{k} e^{i(\delta_l - \delta_{l'})} \sin \delta_l \sin \delta_{l'} \int d\Omega P_l(\cos \theta) P_{l'}(\cos \theta)$$

$$= \sum_{l=0}^{\infty} \sum_{l'=0}^{\infty} \frac{2l+1}{k} \frac{2l'+1}{k} e^{i(\delta_l - \delta_{l'})} \sin \delta_l \sin \delta_{l'} \frac{4\pi}{2l+1} \delta_{ll'}$$

$$= \frac{4\pi}{k^2} \sum_{l=0}^{\infty} (2l+1) \sin^2 \delta_l$$

We see that if we know the phase shift for every partial wave we know everything about the scattering.

If we calculate the forward scattering amplitude from Eq.(19.78) we obtain

$$f(\theta = 0) = \sum_{l=0}^{\infty} \frac{2l+1}{k} e^{i\delta_l} \sin \delta_l P_l(1) = \sum_{l=0}^{\infty} \frac{2l+1}{k} e^{i\delta_l} \sin \delta_l \qquad (19.81)$$

The imaginary part of Eq.(19.81) may be expressed as

$$Im[f(\theta = 0)] = \sum_{l=0}^{\infty} \frac{2l+1}{k} \sin^2 \delta_l = \frac{k}{4\pi} \sigma_T \qquad (19.82)$$

or

$$\sigma_T = \frac{4\pi}{k} Im[f(\theta = 0)] \qquad (19.83)$$

Eq.(19.83) represents an important result that relates the total cross section to the imaginary part of the forward amplitude and is known as the **optical theorem**. When scattering occurs, part of the energy carried by the incoming wave is radiated into all angles. This energy must be removed from the incident wave. Consequently the energy flowing in the forward direction is reduced and this modifies the scattering amplitude in that direction.

After the development of the formalism of the partial wave expansion let us now apply the method for some simple potentials.

19.7   Scattering from a delta potential

Let the potential be of the form (see Figure 19.5).

$$U(r) = \gamma\delta(r - a) \tag{19.84}$$

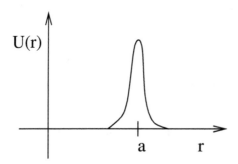

Fig. 19.5. A delta potential.

Here γ measures the strength of the potential. Let us consider extremely low energy particles so that only $l = 0$ components contribute to the scattering.

For $r < a$, the radial equation is given by

$$\frac{d^2u}{dr^2} + k^2u = 0 \tag{19.85}$$

where $k^2 = 2\mu E/\hbar^2$ and the solution is in the form

$$u(r) = A\sin kr \tag{19.86}$$

For $r > a$, the radial equation again is in the same form as given in Eq.(19.85). However, the solution for this case is given by

$$u(r) = B\sin(kr + \delta_0) \tag{19.87}$$

Here δ_0 is the phase change in the wave. The solutions must match at the boundary and this gives the condition

$$A\sin ka = B\sin(ka + \delta_0) \tag{19.88}$$

Now, the radial equation for all values of r including $r = a$ is given by

$$\frac{d^2u}{dr^2} + k^2u = \frac{2\mu}{\hbar^2}\gamma\delta(r - a)u \tag{19.89}$$

Integrating Eq.(19.89) between $a - \epsilon$ and $a + \epsilon$ we have

$$\lim_{\epsilon \to 0} \int_{a-\epsilon}^{a+\epsilon} dr \left(\frac{d^2u}{dr^2} + k^2u\right) = \lim_{\epsilon \to 0} \int_{a-\epsilon}^{a+\epsilon} dr \frac{2\mu}{\hbar^2}\gamma\delta(r - a)u \tag{19.90}$$

or

$$\lim_{\epsilon \to 0} \left(\frac{du}{dr}\Big|_{a+\epsilon} - \frac{du}{dr}\Big|_{a-\epsilon} \right) = \frac{2\mu\gamma}{\hbar^2} u(a) \tag{19.91}$$

Using Eqs.(19.86) and (19.87) we have

$$kB \cos(ka + \delta_0) - kA \cos ka = \frac{2\mu\gamma}{\hbar^2} B \sin(ka + \delta_0) \tag{19.92}$$

or

$$kB \cos(ka + \delta_0) - \frac{2\mu\gamma}{\hbar^2} B \sin(ka + \delta_0) = kA \cos ka \tag{19.93}$$

Dividing Eq.(19.93) by Eq.(19.88) we obtain

$$k \cot(ka + \delta_0) - \frac{2\mu\gamma}{\hbar^2} = k \cot ka \tag{19.94}$$

or

$$k \frac{\cot ka \cot \delta_0 - 1}{\cot ka + \cot \delta_0} = \frac{2\mu\gamma}{\hbar^2} + k \cot ka \tag{19.95}$$

which we simplify to

$$\cot \delta_0 = -\cot ka - \frac{k\hbar^2}{2\mu\gamma} \csc^2 ka \tag{19.96}$$

Since we have assumed the particles to have low energy, i.e., $ka \ll 1$, we can expand the trigonometric functions, and this gives

$$\cot \delta_0 = -\frac{1 + \frac{2\mu\gamma a}{\hbar^2}}{\frac{2\mu\gamma}{\hbar^2} ka^2} = -\frac{\frac{\hbar^2}{2\mu\gamma a} + 1}{ka} \tag{19.97}$$

For $l = 0$ (S–wave scattering), the differential cross section is isotropic. The total cross section can be written as (using Eq.(19.80))

$$\sigma_T = \frac{4\pi}{k^2} \sin^2 \delta_0 = \frac{4\pi}{k^2} \frac{1}{\cot^2 \delta_0 + 1} \tag{19.98}$$

Using Eq.(19.97) and expanding in the small quantity $ka \ll 1$, we obtain

$$\sigma_T = 4\pi a^2 \left(\frac{1}{1 + \frac{\hbar^2}{2\mu\gamma a}} \right)^2 \tag{19.99}$$

For low energy scattering the scattering cross section is independent of the energy.

19.8 Scattering from a square–well potential

Consider the spherical square–well potential in three dimensions (see Figure 19.6), given by

$$U(r) = \left\{ \begin{array}{ll} -U_0 , & r < a \\ 0 , & r > a \end{array} \right\} \tag{19.100}$$

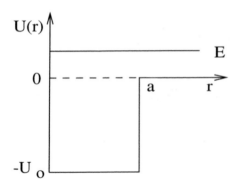

Fig. 19.6. A square–well potential.

Let us consider a shallow potential that does not allow for any bound states to exist. Consider low energy particles such that only $l = 0$ waves scatter, and $E > 0$. The radial equation for $r < a$ is given by

$$\frac{d^2u}{dr^2} + (k^2 + k_0^2)u = 0 \tag{19.101}$$

where $k^2 = 2\mu E/\hbar^2$ and $k_0^2 = 2\mu E_0/\hbar^2$. The solution to Eq.(19.101) is given by

$$u(r) = A\sin k_1 r \quad , \quad k_1^2 = k^2 + k_0^2 \tag{19.102}$$

The radial equation for $r > a$ is the free–particle equation

$$\frac{d^2u}{dr^2} + k^2u = 0 \tag{19.103}$$

The solution to Eq.(19.103) is given by

$$u(r) = B\sin(kr + \delta_0) \tag{19.104}$$

The solutions and their derivatives have to be continuous at the boundary,

$$A\sin k_1 a = B\sin(ka + \delta_0) \tag{19.105}$$

$$k_1 A\cos k_1 a = kB\cos(ka + \delta_0) \tag{19.106}$$

The ratio of Eq.(19.106) to (19.105) gives

$$k\cot(ka + \delta_0) = k_1\cot k_1 a \tag{19.107}$$

In that k_1 is finite as $k \to 0$,

$$\cot(ka + \delta_0) = \frac{k_1\cot k_1 a}{k} \tag{19.108}$$

$\cot(ka + \delta_0)$ grows large so that $\sin(ka + \delta_0)$ grows small and we may set

$$\sin(ka + \delta_0) \cong ka + \delta_0 \tag{19.109}$$

Since $ka \ll 1$, Eq.(19.109) implies that $\delta_0 \ll 1$ as well. Under these conditions Eq.(19.107) reduces to

$$k_1 \cot k_1 a \cong \frac{k}{ka + \delta_0} \tag{19.110}$$

or equivalently

$$\delta_0 = ka \left(\frac{\tan k_1 a}{k_1 a} - 1 \right) \tag{19.111}$$

Since we are considering only S–wave scattering, the cross section is isotropic and the total cross section is given by

$$\sigma_T = \frac{4\pi}{k^2} \sin^2 \delta_0 \cong \frac{4\pi}{k^2} \delta_0^2 = \frac{4\pi}{k^2} k^2 a^2 \left(\frac{\tan k_1 a}{k_1 a} - 1 \right)^2 \tag{19.112}$$

$$= 4\pi a^2 \left(\frac{\tan k_1 a}{k_1 a} - 1 \right)^2 \cong 4\pi a^2 \left(\frac{\tan k_0 a}{k_0 a} - 1 \right)^2$$

We can write the last equality because we assumed that the particles are of low energy $(ka \ll 1)$, and the potential is shallow $(k_1 a \cong k_0 a)$.

19.9   Scattering from a hard sphere

The potential in this case has the form (see Figure 19.7)

$$U(r) = \left\{ \begin{array}{ll} U_0 \,, & r < a \\ 0 \,, & r > a \end{array} \right\} \tag{19.113}$$

The low energy scattering cross section from a repulsive square–well potential can be obtained simply from Eq.(19.112).

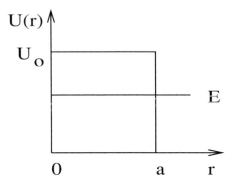

Fig. 19.7. A hard sphere potential.

The difference between an attractive and a repulsive potential amounts to re-placing k_0 by $i\kappa_0$. For a repulsive potential the low–energy scattering cross section becomes

$$\sigma_T = 4\pi a^2 \left(\frac{\tan(i\kappa_0 a)}{i\kappa_0 a} - 1 \right)^2 = 4\pi a^2 \left(\frac{\tanh(\kappa_0 a)}{\kappa_0 a} - 1 \right)^2 \tag{19.114}$$

$\kappa_0 = (2\mu U_0/\hbar^2)^{1/2} \to \infty$ since $U_0 \to \infty$. Hence the cross section becomes

$$\sigma_T = 4\pi a^2 \tag{19.115}$$

The cross section does not depend on the energy for low–energy scattering and equals four times the classical value of πa^2. The classical result in the limit of high–energy scattering may be obtained as follows:

The largest l value that contributes to scattering is given by $l_{max} = ak$. Therefore,

$$\sigma_T(k \to \infty) = \sum_{l=0}^{l_{max}} \frac{4\pi}{k^2}(2l+1)\sin^2 \delta_l \tag{19.116}$$

If we assume that the phase shifts are completely random, then we can replace $\sin^2 \delta_l$ by its average value $< \sin^2 \delta_l >= \frac{1}{2}$. Then Eq.(19.116) becomes

$$\sigma_T(k \to \infty) = \frac{4\pi}{k^2} \sum_{l=0}^{ak} \frac{1}{2}(2l+1) = \frac{4\pi}{k^2}\frac{1}{2}(ka+1)^2 \cong 2\pi a^2 \tag{19.117}$$

The discrepancy with the classical geometrical value in both cases arises because of the wave nature of particles.

19.10   Scattering of identical particles

The formalism introduced so far for scattering was valid in the low–energy case and the target particle was different from the incident particle. The formulae for scattering when the particles involved are indistinguishable need modification.

Classically we can follow the trajectory of each particle, so that dealing with a system of identical particles does not pose a special problem. In quantum mechanics, however, the notion of a trajectory does not exist. Therefore, if we are studying the scattering of identical particles, it is impossible to distinguish the target particle from the incident particle in the final state (see Figure 19.8).

The wave function of a system containing identical particles has to be symmetrized or antisymmetrized depending on whether the system consists of identical bosons or identical fermions.

The symmetry properties of the system of identical particles may be handled in the following way: The total wave function of the two–particle system is given by

$$\psi_{tot}(\mathbf{r}_1, \mathbf{r}_2) = \psi_C(\mathbf{R})\psi(\mathbf{r}) \tag{19.118}$$

where $\mathbf{R} = (\mathbf{r}_1 + \mathbf{r}_2)/2$ is the CM coordinate (with $m_1 = m_2$), and $\mathbf{r} = \mathbf{r}_1 - \mathbf{r}_2$ is the relative coordinate. Under an exchange of the two particles $\mathbf{r}_1 \leftrightarrow \mathbf{r}_2$: $\mathbf{R} \to \mathbf{R}$ and $\mathbf{r} \to -\mathbf{r}$. The CM wave function remains invariant under the exchange of particles, $\psi_C(\mathbf{R}) \to \psi_C(\mathbf{R})$. The symmetry properties of the system are completely determined by the symmetry of the wave function for the reduced system.

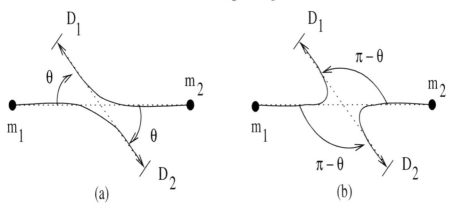

Fig. 19.8. Scattering of two particles in their CM system. (a) m_2 is detected in $D1$, (b) m_2 is detected in $D2$. If the two particles were identical the two processes (a) and (b) could not be distinguished.

For example, in the case of two identical spinless bosons, the total wave function has to be symmetric under the exchange of the particles $\psi(\mathbf{r}) \rightarrow \psi(-\mathbf{r}) = \psi(\mathbf{r})$. This means that the final state wave function must be symmetrized. $\mathbf{r} \rightarrow -\mathbf{r}$ implies $\mathbf{r} \rightarrow \mathbf{r}$, $\theta \rightarrow \pi + \theta$, $\phi \rightarrow \pi + \phi$. The scattering wave function in this case takes the form

$$\psi_{sc}(\mathbf{r}) \rightarrow f(\theta, \phi)\frac{e^{ikr}}{r} + f(\pi - \theta, \pi + \phi)\frac{e^{ikr}}{r} \tag{19.119}$$

$$= [f(\theta, \phi) + f(\pi - \theta, \pi + \phi)]\frac{e^{ikr}}{r}$$

The scattering amplitude, therefore, is symmetric and is given by

$$f_{sym}(\theta, \phi) = f(\theta, \phi) + f(\pi - \theta, \pi + \phi) \tag{19.120}$$

The differential cross section is obtained as

$$\sigma_C(\theta, \phi) = |f_{sym}(\theta, \phi)|^2 \tag{19.121}$$

$$= |f(\theta, \phi)|^2 + |f(\pi - \theta, \pi + \phi)|^2 + 2Re[f^*(\theta, \phi)f(\pi - \theta, \pi + \phi)]$$

The first two terms on the right hand side of Eq.(19.121) are what we would obtain if the two particles were indistinguishable. The cross term represents the quantum mechanical interference in any system containing identical particles.

For example, in the low–energy scattering of two identical nuclei having spin zero and atomic number Z the scattering amplitude is

$$f(\theta, \phi) = \frac{Ze^2}{4\pi\varepsilon}\frac{\mu}{2p_{CM}^2}\frac{1}{\sin^2\frac{\theta}{2}} \tag{19.122}$$

the differential cross section for this process then takes the form

$$\sigma_C(\theta, \phi) = \left(\frac{Ze^2}{4\pi\varepsilon}\right)\frac{\mu^2}{4p_{CM}^4}\left(\frac{1}{\sin^4\frac{\theta}{2}} + \frac{1}{\cos^4\frac{\theta}{2}} + \frac{2}{\sin^2\frac{\theta}{2}\cos^2\frac{\theta}{2}}\right) \tag{19.123}$$

where θ is the CM scattering angle.

As a second example, let us consider the scattering of two identical fermions, say electrons. Since an electron has spin $1/2$, such a system can be in the triplet or singlet spin state. The total wave function has to be antisymmetric under the exchange of the two particles. Let us assume that the electrons are in the triplet state. In this case the spin part is symmetric and the space part must be antisymmetric, that is,

$$\psi_{triplet}(\mathbf{r}) \rightarrow \psi_{triplet}(-\mathbf{r}) = -\psi_{triplet}(\mathbf{r}) \tag{19.124}$$

The corresponding scattering amplitude is also antisymmetric and takes the form

$$f_{triplet}(\theta, \phi) = [f(\theta, \phi) - f(\pi - \theta, \pi + \phi)] \tag{19.125}$$

Therefore, the differential cross section for the triplet state becomes

$$\sigma_{triplet}(\theta, \phi) = |f_{triplet}(\theta, \phi)|^2 \tag{19.126}$$

$$= |f(\theta, \phi)|^2 + |f(\pi - \theta, \pi + \phi)|^2 - 2Re[f(\theta, \phi)^* f(\pi - \theta, \pi + \phi)]$$

On the other hand, if the electrons are in the singlet state, then the spin part is antisymmetric and space part is symmetric. Thus

$$\psi_{singlet}(\mathbf{r}) \rightarrow \psi_{singlet}(-\mathbf{r}) = \psi_{singlet}(\mathbf{r}) \tag{19.127}$$

The scattering amplitude becomes symmetric in this case, and takes the form

$$f_{singlet}(\theta, \phi) = [f(\theta, \phi) + f(\pi - \theta, \pi + \phi)] \tag{19.128}$$

Therefore, the differential cross section is given by

$$\sigma_{singlet}(\theta, \phi) = |f_{singlet}(\theta, \phi)|^2 \tag{19.129}$$

$$= |f(\theta, \phi)|^2 + |f(\pi - \theta, \pi + \phi)|^2 + 2Re[f(\theta, \phi)^* f(\pi - \theta, \pi + \phi)]$$

In many scattering experiments unpolarized particles are used. In that case the two fermions can be in a triplet or singlet state. Consequently, one defines a spin–averaged cross section. Thus, the spin–averaged differential cross section is given by

$$\sigma_{av}(\theta, \phi) = \frac{1}{4}[3\sigma_{triplet}(\theta, \phi) + \sigma_{singlet}(\theta, \phi)] \tag{19.130}$$

$$= |f(\theta, \phi)|^2 + |f(\pi - \theta, \pi + \phi)|^2 - Re[f(\theta, \phi)^* f(\pi - \theta, \pi + \phi)]$$

19.11 Energy dependence and resonance scattering

The scattering amplitude, and hence the cross section, varies as a function of energy. From the expression of σ_T in the partial wave analysis (Eq.(19.80)) we recognize that if the peak occurs in the lth partial wave it implies that $\sin \delta_l$ tends to its maximum value at that energy, namely, $\delta_l = \frac{\pi}{2}, 3\frac{\pi}{2}, \cdots, (2n + 1)\frac{\pi}{2}$, etc. In such a case we say that the scattering amplitude has a **resonance** at that particular energy.

Let us look at the S–wave scattering from an attractive square–well potential. If we consider only low–energy scattering, i.e., $ka \ll 1$, from Eq.(19.112), the cross section is in the form

$$\sigma_T = \frac{4\pi}{k^2} \sin^2 \delta_0 = 4\pi a^2 \left(\frac{\tan k_1 a}{k_1 a} - 1 \right)^2 \qquad (19.131)$$

When $k_1 a$ is an odd multiple of $\pi/2$, $\tan k_1 a$ is infinite and the cross section becomes singular. The general behavior of cross section with respect to $k_1 a$ looks like as shown in Figure 19.9.

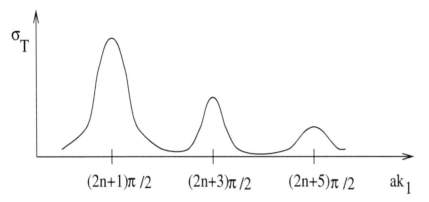

Fig. 19.9. Dependence of cross section on energy.

When $k_1 a \to (2n + 1)\pi/2$, $\sin^2 \delta_0 \to 1$ and the maximum cross section at these S–wave resonances becomes

$$\sigma_T = \frac{4\pi}{k^2} \qquad (19.132)$$

Eq.(19.131) shows resonant scattering at odd multiples of $\pi/2$. It also indicates that the attractive scattering will become transparent to the incident beam at values of $k_1 a$ which satisfy the transcendental relation, $\tan k_1 a = k_1 a$. Such resonant transparency of an attractive well is experimentally confirmed in the scattering of low–energy electrons ($E \cong 0.7$ eV) by rare gas atoms. Electrons can move through the medium with practically no scattering. This is known as the **Ramsauer–Townsend effect**. The Ramsauer–Townsend effect cannot occur for a repulsive potential since in that case it is not possible to satisfy simultaneously the conditions $ka \ll 1$ and $k_1 a = \pi$.

Let us consider the behavior of the cross section near a resonance in general. We assume that the phase shift δ_l in the lth partial wave reaches the value $\pi/2$ at the particular energy E_0. Therefore,

$$\delta_l(E_0) = \frac{\pi}{2} \ , \quad \cos \delta_l(E_0) = 0 \ , \quad \sin \delta_l(E_0) = 1 \qquad (19.133)$$

If $E \cong E_0$, then we can make a Taylor expansion around E_0 so that

$$\sin \delta_l(E) \cong \sin \delta_l(E_0) + \left[\cos \delta_l(E) \frac{d\delta_l(E)}{dE} \right]_{E=E_0} (E - E_0) = 1 \qquad (19.134)$$

and

$$\cos \delta_l(E) \cong \cos \delta_l(E_0) - \left[\sin \delta_l(E) \frac{d\delta_l(E)}{dE} \right]_{E=E_0} (E - E_0) \tag{19.135}$$

$$= -\frac{d\delta_l(E)}{dE} \Big|_{E=E_0} (E - E_0) = -\frac{2}{\Gamma}(E - E_0)$$

Here we define the rate of change of the phase shift with respect to energy near the resonance as

$$\frac{d\delta_l(E)}{dE} \Big|_{E=E_0} = \frac{2}{\Gamma} \tag{19.136}$$

with Γ a real constant. The scattering amplitude for the lth partial wave is

$$f_l(\theta, \phi) = \frac{2l+1}{k} e^{i\delta_l(E)} \sin \delta_l(E) P_l(\cos \theta) \tag{19.137}$$

The energy dependence of the scattering amplitude is given by

$$f_l(\theta, \phi) = e^{i\delta_l(E)} \sin \delta_l(E) = \frac{\sin \delta_l(E)}{\cos \delta_l(E) - i \sin \delta_l(E)} \tag{19.138}$$

If we use the relations of Eqs.(19.133) and (19.134) the amplitude near the resonance becomes

$$f_l(E) \cong \frac{1}{-\frac{2}{\Gamma}(E - E_0) - i} = -\frac{\Gamma/2}{(E - E_0) + i\Gamma/2} \tag{19.139}$$

Therefore, the energy dependence of the cross section is given by

$$|f_l(E)|^2 = \frac{\Gamma^2/4}{(E - E_0)^2 + \Gamma^2/4} \tag{19.140}$$

This result is known as the **Breit–Wigner formula** and has the typical shape of a resonance curve (see Figure 19.10).

The scattering amplitude equals unity when $E = E_0$, and Γ represents the full width of the curve at its half maximum. The total scattering cross section for the lth partial wave for the scattering of spinless particles near the resonance is given by

$$\sigma_{lT} = \frac{4\pi}{k^2}(2l+1)|f_l(E)|^2 = \frac{4\pi}{k^2}(2l+1)\frac{\Gamma^2/4}{(E - E_0)^2 + \Gamma^2/4} \tag{19.141}$$

The peaks observed in the scattering cross section as a function of the incident energy are more definite as the ratio Γ/E_0 becomes smaller. In the limit $\Gamma \to 0$, the scattering amplitude in the resonating wave becomes singular at $E = E_0$. This is the fundamental property of the discrete energies that characterize the stationary states of a bound system. However, bound states have negative energy and the energy in a scattering process is positive. Thus, these peak positions in the scattering amplitude cannot be bound states and we call such states **resonances**. The energy of the resonances in the scattering system is not sharp but has a width Γ.

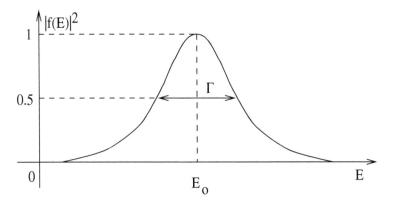

Fig. 19.10. Breit–Wigner function profile.

Using the uncertainty relation one can determine the time interval over which one can observe such a system is

$$\Delta t = \frac{\hbar/2}{\Delta E} = \frac{\hbar/2}{\Gamma/2} = \frac{\hbar}{\Gamma} = \tau \qquad (19.142)$$

We call $\tau = \hbar/\Gamma$ the lifetime of the **resonance state** or **quasi–stationary state**, and they are eigenstates of the equation of motion with complex eigenvalues. The wave function for resonance states may be expressed as

$$\psi_l(t) = \int_0^\infty f_l(E) e^{-\frac{i}{\hbar} E t} dE \qquad (19.143)$$

$$= -\int_0^\infty \frac{\Gamma/2}{(E - E_0) + i\Gamma/2} e^{-\frac{i}{\hbar} E t} dE = -\frac{i\pi\Gamma}{4} e^{-\frac{i}{\hbar}(E_0 - i\frac{\Gamma}{2})t}$$

The total wave function can be written as

$$\psi_l(\mathbf{r}, t) = \psi_l(\mathbf{r})\psi_l(t) = N_l\psi_l(\mathbf{r}) e^{-\frac{i}{\hbar}(E_0 - i\frac{\Gamma}{2})t} \qquad (19.144)$$

Remembering that $H\psi = i\hbar\frac{\partial\psi}{\partial t}$, we can express the time–independent equation that $\psi_l(\mathbf{r})$ obeys as

$$H\psi_l(\mathbf{r}) = (E_0 - i\frac{\Gamma}{2})\psi_l(\mathbf{r}) = E_l\psi_l(\mathbf{r}) \qquad (19.145)$$

The probability of finding the system in the state ψ_l takes the form

$$P_l(t) = |\psi_l(\mathbf{r}, t)|^2 = |\psi_l(\mathbf{r})|^2 e^{-\frac{\Gamma}{\hbar} t} = |\psi_l(\mathbf{r})|^2 e^{-t/\tau} \qquad (19.146)$$

The probability decreases exponentially with a time constant $\tau = \hbar/\Gamma$, where τ is the lifetime of the state. The state is not stationary since the probability of finding the system in that state depends on time.

19.12 The Lippman–Schwinger equation (high–energy case)

The partial wave analysis is not suitable for the study of high–energy scattering since the number of waves that contribute to the scattering increases. The method of Lippman and Schwinger, more commonly known as the integral solution to the scattering problem, becomes useful in this case.

The central idea in the method of Lippman and Schwinger is that rather than analyzing each angular momentum component separately, we try to obtain the scattering amplitude as a whole by solving an integral equation.

The time–independent Schrödinger equation for scattering is given by

$$(\nabla^2 + k^2)\psi(\mathbf{r}) = \frac{2\mu}{\hbar^2}U(r)\psi(\mathbf{r}) \tag{19.147}$$

where $k^2 = 2\mu E/\hbar^2$. This equation has a close similarity to the Poisson equation of electrostatics

$$\nabla^2\phi = -4\pi\rho \tag{19.148}$$

The best way to solve such equations is by Greens method: One defines a Green function $G(\mathbf{r})$ for the Laplacian through

$$\nabla^2 G(\mathbf{r}) = -4\pi\delta^3(\mathbf{r}) \tag{19.149}$$

It is known that $G(\mathbf{r}) = \frac{1}{|\mathbf{r}|}$, from which it follows that

$$\phi(\mathbf{r}) = \int d^3\mathbf{r}'G(\mathbf{r} - \mathbf{r}')\rho(\mathbf{r}') = \int d^3\mathbf{r}'\frac{\rho(\mathbf{r}')}{|\mathbf{r} - \mathbf{r}'|} \tag{19.150}$$

The solution of Poisson's equation can be given as an integral of the sources over all space.

To be able to apply this method to the problem of scattering, we have to define a Greens function for the problem. Furthermore, the solution for large spatial distances must have the form

$$\psi(\mathbf{r}) \overset{r\to\infty}{\longrightarrow} e^{i\mathbf{k}\cdot\mathbf{r}} + f(\theta,\phi)\frac{e^{ikr}}{r} \tag{19.151}$$

Let us assume that $G(\mathbf{r}, \mathbf{r}')$ is the Greens function for the scattering problem,

$$(\nabla^2 + k^2)G(\mathbf{r}, \mathbf{r}') = \delta^3(\mathbf{r}, \mathbf{r}') \tag{19.152}$$

The solution to the Schrödinger equation, Eq.(19.147), can be written as

$$\psi(\mathbf{r}) = \frac{2\mu}{\hbar^2}\int d^3r'G(\mathbf{r}, \mathbf{r}')U(\mathbf{r}')\psi(\mathbf{r}') \tag{19.153}$$

This is because

$$(\nabla^2 + k^2)\psi(\mathbf{r}) = \frac{2\mu}{\hbar^2}\int d^3r'(\nabla^2 + k^2)G(\mathbf{r}, \mathbf{r}')U(\mathbf{r}')\psi(\mathbf{r}') \tag{19.154}$$

$$= \frac{2\mu}{\hbar^2} \int d^3r' \delta^3(\mathbf{r}, \mathbf{r}') U(\mathbf{r}') \psi(\mathbf{r}') = \frac{2\mu}{\hbar^2} U(\mathbf{r}) \psi(\mathbf{r})$$

Therefore, the solution $\psi(\mathbf{r})$ can be written as an integral of the source over the whole space, except that here the source depends on the solution $\psi(\mathbf{r})$ itself. Such a solution is known as an **integral equation**.

When $U = 0$, i.e., when there is no scattering, the solution must have the form

$$\psi(\mathbf{r}) \overset{U=0}{\longrightarrow} \psi^{(0)}(\mathbf{r}) = e^{i\mathbf{k}\cdot\mathbf{r}} \tag{19.155}$$

The general solution of Eq.(19.147) may be expressed as

$$\psi(\mathbf{r}) = \psi^{(0)}(\mathbf{r}) + \frac{2\mu}{\hbar^2} \int d^3r' G(\mathbf{r}, \mathbf{r}') U(\mathbf{r}') \psi(\mathbf{r}') \tag{19.156}$$

or

$$\psi(\mathbf{r}) = e^{i\mathbf{k}\cdot\mathbf{r}} + \frac{2\mu}{\hbar^2} \int d^3r' G(\mathbf{r}, \mathbf{r}') U(\mathbf{r}') \psi(\mathbf{r}') \tag{19.157}$$

$$= e^{i\mathbf{k}\cdot\mathbf{r}} + \psi_{sc}(\mathbf{r}) \overset{r \to large}{\longrightarrow} e^{i\mathbf{k}\cdot\mathbf{r}} + f(\theta, \phi)\frac{e^{ikr}}{r}$$

From Eq.(19.157) we see that the scattered wave is given by

$$\psi_{sc}(\mathbf{r}) = \frac{2\mu}{\hbar^2} \int d^3r' G(\mathbf{r}, \mathbf{r}') U(\mathbf{r}') \psi(\mathbf{r}') \tag{19.158}$$

The integral solution to the Schrödinger equation given in Eq.(19.157) is known as the Lippman–Schwinger equation. To be able to make use of this solution we must know the form of the Greens function.

19.13 The Greens function for the scattering problem

We define the Greens function through

$$(\nabla^2 + k^2)G(\mathbf{r}, \mathbf{r}') = \delta^3(\mathbf{r}, \mathbf{r}') \tag{19.159}$$

From translational invariance $G(\mathbf{r}, \mathbf{r}')$ must have the form

$$G(\mathbf{r}, \mathbf{r}') = G(\mathbf{r} - \mathbf{r}') \tag{19.160}$$

Let us introduce the Fourier transforms

$$G(\mathbf{r} - \mathbf{r}') = \int d^3q \tilde{G}(\mathbf{q}) e^{i\mathbf{q}\cdot(\mathbf{r}-\mathbf{r}')} \tag{19.161}$$

$$\delta^3(\mathbf{r}, \mathbf{r}') = \frac{1}{(2\pi)^3} \int d^3q e^{i\mathbf{q}\cdot(\mathbf{r}-\mathbf{r}')} \tag{19.162}$$

If we substitute Eqs.(19.161) and (19.162) into Eq.(19.159) the differential equation reduces to an algebraic equation for $\tilde{G}(\mathbf{q})$,

$$(-q^2 + k^2)\tilde{G}(\mathbf{q}) = \frac{1}{(2\pi)^3} \tag{19.163}$$

or

$$\tilde{G}(\mathbf{q}) = -\frac{1}{(2\pi)^3}\frac{1}{q^2 - k^2} \tag{19.164}$$

The Greens function is obtained as

$$G(\mathbf{r} - \mathbf{r}') = \int d^3q\, \tilde{G}(\mathbf{q})e^{i\mathbf{q}\cdot(\mathbf{r}-\mathbf{r}')} \tag{19.165}$$

$$= -\frac{1}{(2\pi)^3}\int d^3q\, \frac{1}{q^2 - k^2}e^{i\mathbf{q}\cdot(\mathbf{r}-\mathbf{r}')}$$

We define the variable $\mathbf{r} - \mathbf{r}' = \mathbf{R}$, so that

$$G(\mathbf{R}) = -\frac{1}{(2\pi)^3}\int d^3q\, \frac{1}{q^2 - k^2}e^{i\mathbf{q}\cdot\mathbf{R}} \tag{19.166}$$

$$= -\frac{1}{(2\pi)^3}\int q^2\,dq\sin\theta d\theta d\phi\, \frac{1}{q^2 - k^2}e^{iqR\cos\theta}$$

$$= -\frac{1}{(2\pi)^3 iR}\int_{-\infty}^{\infty} dq\, \frac{qe^{iqR}}{q^2 - k^2}$$

The integrand has poles at $q = \mp k$. This integral can be evaluated using Cauchy's residue theorem, but to do that we must specify the contour of the integration (see Figure 19.11).

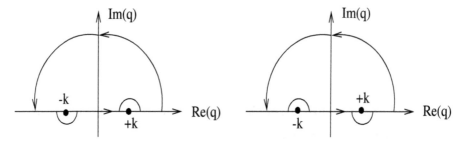

Fig. 19.11. Integration contours: (a) Contour yielding an incoming wave solution, (b) Contour for outgoing wave solution.

If we choose the contour where the pole at $k = -q$ will contribute, then the Greens function will have the form, which corresponds to an incoming wave solution,

$$G(\mathbf{r} - \mathbf{r}') = G(\mathbf{R}) = (2\pi i)\lim_{q\to -k}\left[-\frac{1}{(2\pi)^2 iR}\frac{(q+k)qe^{iqR}}{q^2 - k^2}\right] \tag{19.167}$$

$$= -\frac{1}{2\pi R}\frac{(-k)e^{-ikr}}{-2k} = -\frac{1}{4\pi R}e^{-ikR} = -\frac{1}{4\pi|\mathbf{r} - \mathbf{r}'|}e^{-ik|\mathbf{r}-\mathbf{r}'|}$$

If we choose the contour where the pole at $k = q$ will contribute, then the Greens function will have the form, which corresponds to an outgoing wave solution,

$$G(\mathbf{r} - \mathbf{r}') = G(\mathbf{R}) = (2\pi i)\lim_{q\to k}\left[-\frac{1}{(2\pi)^2 iR}\frac{(q-k)qe^{iqR}}{q^2 - k^2}\right] \tag{19.168}$$

$$= -\frac{1}{2\pi R}\frac{ke^{ikR}}{2k} = -\frac{1}{4\pi R}e^{ikR} = -\frac{1}{4\pi|\mathbf{r}-\mathbf{r}'|}e^{ik|\mathbf{r}-\mathbf{r}'|}$$

We must choose the contour of integration such that the Greens function has an outgoing form. Therefore the Greens function has the form

$$G(\mathbf{r}-\mathbf{r}') = -\frac{1}{4\pi|\mathbf{r}-\mathbf{r}'|}e^{ik|\mathbf{r}-\mathbf{r}'|} \tag{19.169}$$

The solution to the scattering problem is then given by

$$\psi(\mathbf{r}) = e^{i\mathbf{k}\cdot\mathbf{r}} + \frac{2\mu}{\hbar^2}\int d^3r' G(\mathbf{r}-\mathbf{r}')U(\mathbf{r}')\psi(\mathbf{r}') \tag{19.170}$$

$$= e^{i\mathbf{k}\cdot\mathbf{r}} - \frac{\mu}{2\pi\hbar^2}\int d^3r' \frac{e^{ik|\mathbf{r}-\mathbf{r}'|}}{|bfr-\mathbf{r}'|}U(\mathbf{r}')\psi(\mathbf{r}')$$

Eq.(19.170) is not easy to solve. It is not always possible to solve an integral equation exactly. However, one can solve it iteratively. One of the methods used often is the Born Approximation.

19.14 Born approximation

Let us take the general solution of the scattering problem and introduce the short hand notation

$$\psi(\mathbf{r}) = \psi^{(0)}(\mathbf{r}) + \frac{2\mu}{\hbar^2}\int d^3r' G(\mathbf{r}-\mathbf{r}')U(\mathbf{r}')\psi(\mathbf{r}') \tag{19.171}$$

or

$$\psi = \psi^{(0)} + \frac{2\mu}{\hbar^2}GU\psi \tag{19.172}$$

Here we have replaced $e^{i\mathbf{k}\cdot\mathbf{r}}$ by $\psi^{(0)}$. If we now substitute the lowest–order solution $(\psi^{(0)})$ to the scattering equation on the right hand side of Eq.(19.172), we obtain to first order

$$\psi^{(1)} = \psi^{(0)} + \frac{2\mu}{\hbar^2}GU\psi^{(0)} \tag{19.173}$$

or, explicitly

$$\psi^{(1)}(\mathbf{r}) = \psi^{(0)}(\mathbf{r}) + \frac{2\mu}{\hbar^2}\int d^3r' G(\mathbf{r}-\mathbf{r}')U(\mathbf{r}')\psi^{(0)}(\mathbf{r}') \tag{19.174}$$

If we substitute the first–order solution on the right hand side of Eq.(19.172) we obtain to second–order

$$\psi^{(2)} = \psi^{(0)} + \frac{2\mu}{\hbar^2}GU\psi^{(0)} + \left(\frac{2\mu}{\hbar^2}\right)^2 GUGU\psi^{(0)} \tag{19.175}$$

or, explicitly

$$\psi^{(2)}(\mathbf{r}) = \psi^{(0)}(\mathbf{r}) + \frac{2\mu}{\hbar^2}\int d^3r' G(\mathbf{r}-\mathbf{r}')U(\mathbf{r}')\psi^{(0)}(\mathbf{r}') \tag{19.176}$$

$$+ \left(\frac{2\mu}{\hbar^2}\right)^2 \int d^3r' d^3r'' G(\mathbf{r} - \mathbf{r}') U(\mathbf{r}') G(\mathbf{r} - \mathbf{r}'') U(\mathbf{r}'') \psi^{(0)}(\mathbf{r}'')$$

We can iterate to as many orders as we wish. This expansion is known as the **Born approximation**. We restrict ourselves now to the first Born approximation so that

$$\psi(\mathbf{r}) = \psi^{(0)}(\mathbf{r}) + \frac{2\mu}{\hbar^2} \int d^3r' G(\mathbf{r} - \mathbf{r}') U(\mathbf{r}') \psi^{(0)}(\mathbf{r}') \tag{19.177}$$

$$= e^{i\mathbf{k}\cdot\mathbf{r}} - \frac{\mu}{2\pi\hbar^2} \int d^3r' \frac{e^{ik|\mathbf{r}-\mathbf{r}'|}}{|\mathbf{r} - \mathbf{r}'|} U(\mathbf{r}') e^{i\mathbf{k}\cdot\mathbf{r}'}$$

The scattered wave in this approximation is given by

$$\psi_{sc}(\mathbf{r}) = -\frac{\mu}{2\pi\hbar^2} \int d^3r' \frac{e^{ik|\mathbf{r}-\mathbf{r}'|}}{|\mathbf{r} - \mathbf{r}'|} U(\mathbf{r}') e^{i\mathbf{k}\cdot\mathbf{r}'} \tag{19.178}$$

To examine the behavior of the scattered wave as $r \to \infty$, we make some assumptions and definitions: Since the range over which the potential is nonzero is finite, $r' \ll r$, and hence

$$|\mathbf{r} - \mathbf{r}'| = (r^2 + r'^2 - 2\mathbf{r} \cdot \mathbf{r}')^{1/2} = r \left(1 + \frac{r'^2}{r^2} - 2\frac{\mathbf{r} \cdot \mathbf{r}'}{r^2}\right)^{1/2} \tag{19.179}$$

$$\cong r \left(1 - \frac{\mathbf{r} \cdot \mathbf{r}'}{r^2}\right)$$

$$\frac{1}{|\mathbf{r} - \mathbf{r}'|} \cong \frac{1}{r} \tag{19.180}$$

$$e^{ik|\mathbf{r}-\mathbf{r}'|} \cong e^{ikr(1-\frac{\mathbf{r}\cdot\mathbf{r}'}{r^2})} = e^{ikr - i\mathbf{k}_f \cdot \mathbf{r}'} \tag{19.181}$$

where $\mathbf{k}_f = \frac{k\mathbf{r}}{r}$ is the momentum of the outgoing wave. The scattered wave for large distances takes the form

$$\psi_{sc}(\mathbf{r}) \xrightarrow{r \to \infty} -\frac{\mu}{2\pi\hbar^2} \int d^3r' \frac{e^{ikr - i\mathbf{k}_f \cdot \mathbf{r}'}}{r} U(\mathbf{r}') e^{i\mathbf{k}_i \cdot \mathbf{r}'} \tag{19.182}$$

$$= -\frac{\mu}{2\pi\hbar^2} \frac{e^{ikr}}{r} \int d^3r' = f(\theta, \phi) \frac{e^{ikr}}{r}$$

Here $\mathbf{k}_i$ represents the momentum of the incident wave. Thus, in the first Born approximation the scattered amplitude is given by

$$f(\theta, \phi) = f(\mathbf{k}_i, \mathbf{k}_f) = -\frac{\mu}{2\pi\hbar^2} \frac{e^{ikr}}{r} \int d^3r' e^{i(\mathbf{k}_i - \mathbf{k}_f) \cdot \mathbf{r}'} U(\mathbf{r}') \tag{19.183}$$

The scattering amplitude in the Born approximation is proportional to the Fourier transform of the potential with respect to the momentum transfer $\mathbf{q} = \mathbf{k}_i - \mathbf{k}_f$. The wave function for the scattering problem is given by, in general,

$$\psi(\mathbf{r}) = \psi_{in}(\mathbf{r}) + \psi_{sc}(\mathbf{r}) \tag{19.184}$$

$$= e^{i\mathbf{k}_i \cdot \mathbf{r}} - \frac{\mu}{2\pi\hbar^2} \int d^3r' \frac{e^{ik|\mathbf{r}-\mathbf{r}'|}}{|\mathbf{r} - \mathbf{r}'|} U(\mathbf{r}') \psi(\mathbf{r}')$$

In the first Born approximation we replace the wave function under the integral sign by the incident wave,

$$\psi_B(\mathbf{r}) = e^{i\mathbf{k}_i \cdot \mathbf{r}} - \frac{\mu}{2\pi\hbar^2} \int d^3 r' \frac{e^{ik|\mathbf{r}-\mathbf{r}'|}}{|\mathbf{r}-\mathbf{r}'|} U(\mathbf{r}') e^{i\mathbf{k}_i \cdot \mathbf{r}'} \tag{19.185}$$

This approximation is a good approximation if in the range of the potential

$$|\psi_{sc}(\mathbf{r})| \ll |e^{i\mathbf{k}_i \cdot \mathbf{r}}| = 1 \tag{19.186}$$

Since the influence of the potential is the strongest at the origin, if

$$|\psi_{sc}(0)| \ll 1 \tag{19.187}$$

then this approximation should be reliable. For a spherically symmetric potential, that is, $U(\mathbf{r}') = U(r')$, then the Born approximation is valid if [from Eq.(19.186)]

$$\left| \frac{\mu}{2\pi\hbar^2} \int d^3 r' \frac{e^{ikr'}}{r'} U(\mathbf{r}') e^{i\mathbf{k}_i \cdot \mathbf{r}'} \right| \ll 1 \tag{19.188}$$

or

$$\left| \frac{\mu}{2\pi\hbar^2} 2\pi \int r'^2 dr' \sin\theta' d\theta' \frac{e^{ikr'}}{r'} U(\mathbf{r}') e^{ikr' \cos\theta'} \right| \ll 1 \tag{19.189}$$

or

$$\frac{2\mu}{\hbar^2 k} \left| \int_0^\infty dr' e^{ikr'} U(\mathbf{r}') \sin kr' \right| \ll 1 \tag{19.190}$$

This is the condition for validity of the Born approximation. At low energies $kr' \to 0$, thus $\sin kr' \cong kr'$ and $e^{ikr'} \cong 1$; so that the condition for validity becomes

$$\frac{2\mu}{\hbar^2} \left| \int_0^\infty dr' r' U(r') \right| \ll 1 \tag{19.191}$$

If the potential has a height U_0 and range r_0, then the Born approximation is valid at low energies only if

$$\frac{\mu}{\hbar^2} |U_0| r_0^2 \ll 1 \tag{19.192}$$

On the other hand, at high energies $kr \to \infty$, the exponential in Eq.(19.190) oscillates rapidly and picks up contributions only from $r' \leq \frac{1}{k}$, so that the condition for validity at high energies becomes

$$\frac{2\mu}{\hbar^2 k} \frac{|U_0|}{2k} \ll 1 \tag{19.193}$$

or

$$\frac{\mu}{\hbar^2} |U_0| r_0^2 \ll (kr_0)^2 \tag{19.194}$$

This shows that if the Born approximation is valid at low energies it is also valid at high energies; the converse, however, is not true.

As a conclusion the Born approximation can be applied only when the effect of the scattering potential is weak and the incident particles have high momentum. If the range of the potential is a and its strength U, then the condition for the validity of the Born approximation is

$$\frac{a}{\hbar} \left[(p^2 - 2mU)^{1/2} - p \right] \ll 1 \tag{19.195}$$

Eqs.(19.192) and (19.194) are equivalent to the condition of validity given in (19.195). Let us now consider the scattering of a charged particle in a Coulomb potential. The Coulomb potential is a long range potential, its range is infinite, and as a result the conventional phase–shift analysis does not work. Even at low energies the partial wave analysis is not the right method to deal with Coulomb scattering, and the Born approximation for Coulomb scattering is not valid. For the Coulomb potential

$$U(r) = \frac{Ze^2}{4\pi\varepsilon r} \tag{19.196}$$

the condition for validity given by Eq.(19.191) is not fulfilled since

$$\int_0^\infty dr\, rU(r) = \int_0^\infty dr\, r\frac{Ze^2}{4\pi\varepsilon r} \to \infty \tag{19.197}$$

This result arises because of the long–range nature of the Coulomb potential. In practice, however, the Coulomb potential is screened, so we can get around the difficulty of Eq.(19.196) by modifying the potential to be

$$U(r) = \frac{Ze^2}{4\pi\varepsilon r}e^{-mr} \tag{19.198}$$

The parameter m defines the inverse range of the potential. Now the Born approximation would be valid if

$$\frac{2\mu}{\hbar^2}\left|\int_0^\infty dr\, rU(r)\right| = \frac{2\mu}{\hbar^2}\left|\int_0^\infty dr\, r\frac{Ze^2}{4\pi\varepsilon r}e^{-mr}\right| \tag{19.199}$$

$$= \frac{2\mu}{\hbar^2}\frac{Ze^2}{m(4\pi\varepsilon)} \ll 1$$

This condition can be met by an appropriate choice of the parameter m. Thus, we can do all scattering calculations with the modified potential given in Eq.(19.198) and at the end take the limit $m \to 0$ to obtain the result for Coulomb scattering.

The Born amplitude for this process is

$$f_B(\theta,\phi) = f_B(\mathbf{k_i}, \mathbf{k_f}) = -\frac{\mu}{2\pi\hbar^2}\int d^3r' e^{\frac{i}{\hbar}\mathbf{q}\cdot\mathbf{r'}}U(\mathbf{r'}) \tag{19.200}$$

$$= -\frac{\mu}{2\pi\hbar^2}\int d^3r' e^{\frac{i}{\hbar}\mathbf{q}\cdot\mathbf{r'}}\frac{Ze^2}{4\pi\varepsilon r'}e^{-mr}$$

Here $\mathbf{q} = \hbar(\mathbf{k_i} - \mathbf{k_f})$ so that for elastic scattering

$$q^2 = \hbar^2(\mathbf{k_i} - \mathbf{k_f})^2 = \hbar^2(k_i^2 + k_f^2 - 2\mathbf{k_i}\cdot\mathbf{k_f}) \tag{19.201}$$

$$= \hbar^2(k^2 + k^2 - 2k^2\cos\theta) = 4p^2\sin^2\frac{\theta}{2}$$

or

$$q = 2p\sin\frac{\theta}{2} \tag{19.202}$$

here θ is the CM scattering angle. The explicit form of the integral in Eq.(19.200) becomes

$$f_B(\theta,\phi) = -\frac{\mu}{2\pi\hbar^2}\frac{Ze^2}{4\pi\varepsilon}\int r'^2 dr'\sin\theta' d\theta' d\phi' e^{\frac{i}{\hbar}qr'\cos\theta'}\frac{e^{-mr'}}{r'} \tag{19.203}$$

$$= -\frac{Ze^2\mu 2\pi}{2\pi\hbar(4\pi\varepsilon)} \int_0^\infty r' dr' e^{-mr'} \frac{1}{iqr'} \left(e^{\frac{i}{\hbar}qr'} - e^{-\frac{i}{\hbar}qr'} \right)$$

$$= -\frac{Ze^2\mu}{i\hbar q(4\pi\varepsilon)} \int_0^\infty dr' \left[e^{-(m-\frac{i}{\hbar}q)r'} - e^{-(m+\frac{i}{\hbar}q)r'} \right]$$

$$= -\frac{Ze^2\mu}{i\hbar q(4\pi\varepsilon)} \left[\frac{1}{m - \frac{i}{\hbar}q} - \frac{1}{m + \frac{i}{\hbar}q} \right]$$

$$= -\frac{Ze^2\mu}{i\hbar^2 q(4\pi\varepsilon)} \frac{2iq}{m^2 + \frac{q^2}{\hbar^2}} = -\frac{2\mu}{\hbar^2} \frac{1}{m^2 + \frac{q^2}{\hbar^2}} \frac{Ze^2}{4\pi\varepsilon}$$

Now we let $m \to 0$, and thus the differential cross section is

$$\sigma_B(\theta,\phi) = |f_B(\theta,\phi)|^2 = \left(\frac{Ze^2}{4\pi\varepsilon}\right)^2 \left(\frac{2\mu}{q^2}\right)^2 = \left(\frac{Ze^2}{4\pi\varepsilon}\right)^2 \frac{\mu^2}{4p^4} \frac{1}{\sin^4\frac{\theta}{2}} \qquad (19.204)$$

Note that in this calculation we did not consider specific wave functions for the initial and final particles.

19.15 Inelastic scattering

The final wave function in a scattering process can be written as

$$\psi_f(\mathbf{r}) = \psi_{in}(\mathbf{r}) + \psi_{sc}(\mathbf{r}) \qquad (19.205)$$

$$\psi_f(\mathbf{r}) \overset{r\to\infty}{\longrightarrow} \sum_{l=0}^\infty \frac{2l+1}{2ikr} i^l \left[e^{2i\delta_l} e^{i(kr-l\frac{\pi}{2})} - e^{-i(kr-l\frac{\pi}{2})} \right] P_l(\cos\theta) \qquad (19.206)$$

$$= \sum_{l=0}^\infty \frac{2l+1}{2ikr} \left[e^{2i\delta_l} e^{ikr} + (-1)^{l+1} e^{ikr} \right] P_l(\cos\theta)$$

Here the phase shifts δ_l are all real since the potential is real. If we define

$$S_l = e^{2i\delta_l} \qquad (19.207)$$

then the wave function for large distances can be written as

$$\psi_f(\mathbf{r}) \overset{r\to\infty}{\longrightarrow} \sum_{l=0}^\infty \frac{2l+1}{2ikr} \left[S_l e^{ikr} + (-1)^{l+1} e^{-ikr} \right] P_l(\cos\theta) \qquad (19.208)$$

Since the phase shift is real

$$|S_l| = 1 \qquad (19.209)$$

The normalization of outgoing wave is the same as for the incoming wave. This physically implies that the total number of particles that come in is equal to the total number of particles going out. The radial flux at large distances is given by

$$j_r = -\frac{i\hbar}{2\mu} \left[\psi_f^* \left(\frac{\partial \psi_f}{\partial r}\right) - \left(\frac{\partial \psi_f^*}{\partial r}\right) \psi_f \right] \qquad (19.210)$$

$$j_r \xrightarrow{r \to \infty} \frac{\hbar k}{\mu} \sum_{l,l'} \frac{2l+1}{2ikr} \frac{2l'+1}{2ikr} \left[S_l^* S_{l'} + (-1)^{l+l'+1} \right] P_l(\cos\theta) P_{l'}(\cos\theta) \quad (19.211)$$

The flux of probability out of a sphere of large radius R is given by

$$\int_R \mathbf{j} \cdot d\mathbf{S} = \int j_r R^2 \sin\theta d\theta d\phi = 2\pi R^2 \int_0^\pi j_r \sin\theta d\theta \quad (19.212)$$

If we use the expression for j_r from Eq.(19.211) and the orthonormality relations for the Legendre polynomials we obtain the expression for the net flux out of a large sphere as

$$\int_R \mathbf{j} \cdot d\mathbf{S} = \frac{\pi\hbar}{\mu k} \sum_{l=0}^\infty (2l+1)(|S_l|^2 - 1) \quad (19.213)$$

If the phase shifts are real so that $|S_l|^2 = 1$, Eq.(19.209), then the net flux out of the sphere is zero. This is a statement of conservation of probability. It says that the number of particles that enter into the interaction region is the same as the number of particles that exist; this is the case for elastic scattering. However, in the case of inelastic scattering, where the internal structure of the system changes, the net flux out of a sphere need not vanish anymore. This implies that in the presence of inelastic scattering

$$|S_l|^2 < 1 \quad (19.214)$$

Thus, the net flux becomes negative since we are losing a fraction of the incoming beam to other processes. In this case we can no longer write $S_l = e^{2i\delta_l}$ with δ_l real. But if we let

$$\delta_l \to \delta_l + i\eta_l \quad (19.215)$$

with δ_l and η_l real we obtain

$$S_l = e^{-2\eta_l} e^{2i\delta_l} \quad (19.216)$$

and this expression can now describe processes where the particle flux is not conserved. It also shows that in the presence of inelastic scattering the phase shifts become complex. Let us now consider a real potential $U(r)$ and write the time–dependent Schrödinger equations

$$i\hbar \frac{\partial\psi}{\partial t} = \left(-\frac{\hbar^2}{2\mu}\nabla^2 + U \right)\psi \quad (19.217)$$

$$-i\hbar \frac{\partial\psi^*}{\partial t} = \left(-\frac{\hbar^2}{2\mu}\nabla^2 + U \right)\psi^* \quad (19.218)$$

Multiplying Eq.(19.217) from left by ψ^* and Eq.(19.218) from right by ψ and subtracting the second from the first we obtain the following

$$i\hbar \left(\psi^* \frac{\partial\psi}{\partial t} + \frac{\partial\psi^*}{\partial t}\psi \right) = \psi^* \left(-\frac{\hbar^2}{2\mu}\nabla^2 + U \right)\psi - \left(-\frac{\hbar^2}{2\mu}\nabla^2 + U \right)\psi^*\psi \quad (19.219)$$

or

$$i\hbar \frac{\partial}{\partial t}(\psi^*\psi) = -\frac{\hbar^2}{2\mu}[\psi^*\nabla^2\psi - (\nabla^2\psi^*)\psi] \quad (19.220)$$

$$= -\frac{\hbar^2}{2\mu}\vec{\nabla} \cdot [\psi^*\vec{\nabla}\psi - (\vec{\nabla}\psi^*)\psi]$$

or

$$\frac{\partial}{\partial t}(\psi^*\psi) = \frac{i\hbar}{2\mu}\vec{\nabla}\cdot[\psi^*\vec{\nabla}\psi - (\vec{\nabla}\psi^*)\psi] \qquad (19.221)$$

which we recognize as the continuity equation

$$\frac{\partial}{\partial t}P(\mathbf{r},t) = -\vec{\nabla}\cdot\mathbf{j} \qquad (19.222)$$

Integrating over the volume of a large sphere we have

$$\int_\Omega d^3r\frac{\partial}{\partial t}P(\mathbf{r},t) = \int_\Omega d^3r(-\vec{\nabla}\cdot\mathbf{j}) = -\int_S d\mathbf{S}\cdot\mathbf{j} \qquad (19.223)$$

or

$$\frac{\partial}{\partial t}\int_\Omega d^3r P(\mathbf{r},t) = -\int_S d\mathbf{S}\cdot\mathbf{j} \qquad (19.224)$$

This tells us that if the flux out of a closed surface is zero, then the particles are in a stationary state and the total probability of finding them in the enclosed volume does not change with time. That is, there are no sources or sinks of particles. This result was derived by assuming that the potential is real. In the case of a complex potential the continuity equation then becomes

$$i\hbar\frac{\partial}{\partial t}(\psi^*\psi) = -\frac{\hbar^2}{2\mu}\vec{\nabla}\cdot[\psi^*\vec{\nabla}\psi - (\vec{\nabla}\psi^*)\psi] + (U - U^*)\psi^*\psi \qquad (19.225)$$

If we write

$$U = U_R - iU_I \qquad (19.226)$$

where U_R and U_I are real, then the continuity equation becomes

$$i\hbar\frac{\partial}{\partial t}P(\mathbf{r},t) = -\frac{\hbar^2}{2\mu}\vec{\nabla}\cdot[\psi^*\vec{\nabla}\psi - (\vec{\nabla}\psi^*)\psi] - 2iU_I P(\mathbf{r},t) \qquad (19.227)$$

or

$$\frac{\partial}{\partial t}P(\mathbf{r},t) = -\vec{\nabla}\cdot\mathbf{j} - \frac{2}{\hbar}U_I P(\mathbf{r},t) \qquad (19.228)$$

and finally

$$\frac{\partial}{\partial t}P(\mathbf{r},t) + \vec{\nabla}\cdot\mathbf{j} = -\frac{2}{\hbar}U_I P(\mathbf{r},t) \qquad (19.229)$$

If the net flux out of a closed surface vanishes, then $\vec{\nabla}\cdot\mathbf{j} = 0$ and

$$\frac{\partial}{\partial t}P(\mathbf{r},t) = -\frac{2}{\hbar}U_I P(\mathbf{r},t)$$

or

$$P(\mathbf{r},t) = e^{-2U_I t/\hbar} \qquad (19.230)$$

That is, the probability of finding particles in the enclosed volume changes with time. Therefore, particles are no longer in stationary states. As it is seen from Eq.(19.230) that if $U_I > 0$ the potential acts as a sink, whereas if $U_I < 0$ the potential behaves like a source of particles.

In scattering theory, however, we assume the wave functions to be stationary states. This leads to another possibility, namely, $\frac{\partial}{\partial t}P(\mathbf{r},t) = 0$, and hence

$$\vec{\nabla}\cdot\mathbf{j} = -\frac{2}{\hbar}U_I P(\mathbf{r},t) \qquad (19.231)$$

Integrating over a large volume, we obtain

$$\int_\Omega d^3r \vec{\nabla} \cdot \mathbf{j} = -\frac{2}{\hbar} \int_\Omega d^3r U_I P(\mathbf{r}, t) \tag{19.232}$$

or

$$\int_S \mathbf{j} \cdot d\mathbf{S} = -\frac{2}{\hbar} \int_\Omega d^3r U_I |\psi|^2 \tag{19.233}$$

This means that the inelastic processes such as absorption in scattering can be described by introducing complex potentials which in turn lead to complex phase shifts and result in a nonzero flux out of a closed surface. The left hand side of Eq.(19.233) measures the flux removed from the incident beam. Hence,

$$\sigma_{abs} = \sigma_{inel} = -\frac{\mu}{\hbar k} \int_S \mathbf{j} \cdot d\mathbf{S} = \frac{\pi}{k^2} \sum_{l=0}^{\infty} (2l + 1)(1 - |S_l|^2) \tag{19.234}$$

Here we have used Eq.(19.213) for the net flux out of a closed sphere. From the definition of the scattering amplitude

$$f(\theta) = \sum_{l=0}^{\infty} \frac{2l + 1}{2ik} (e^{2i\delta_l} - 1) P_l(\cos\theta) \tag{19.235}$$

$$= \sum_{l=0}^{\infty} \frac{2l + 1}{2ik} (S_l - 1) P_l(\cos\theta)$$

We obtain the total cross section for elastic scattering as

$$\sigma_{el} = \int \sin\theta d\theta d\phi |f(\theta)|^2 \tag{19.236}$$

$$= \frac{\pi}{k^2} \sum_{l=0}^{\infty} (2l + 1) |S_l - 1|^2$$

The total cross section, which is the sum of the elastic and inelastic scattering, is given by

$$\sigma_{tot} = \sigma_{el} + \sigma_{inel} \tag{19.237}$$

$$= \frac{\pi}{k^2} \sum_{l=0}^{\infty} (2l + 1)[|S_l - 1|^2 + 1 - |S_l|^2]$$

$$= \frac{\pi}{k^2} \sum_{l=0}^{\infty} (2l + 1)[1 - Re(S_l)]$$

This reduces to the familiar expression for total elastic scattering when the phase shifts are real (Eq.(19.80)). One can see from Eqs.(19.235) and (19.237) that

$$\sigma_{tot} = \frac{4\pi}{k} Im(f(\theta = 0)) \tag{19.238}$$

This means that the optical theorem remains valid even in the presence of inelastic scattering.

From Eqs.(19.234), (19.236), and (19.237) we see that when $S_l = 1$, there is no scattering whatsover in the lth wave. When $S_l = 0$, there is complete absorption in that wave and we have

$$\sigma_{el} = \sigma_{inel} = \frac{1}{2}\sigma_{tot} = \frac{\pi}{k^2}\sum_{l=0}^{\infty}(2l+1) \tag{19.239}$$

If the absorbing potential has a range a, then in the limit of very high energies $l_{max} = ka$. Using this in Eq.(19.239) we obtain

$$\sigma_{el} = \sigma_{inel} = \pi a^2 \quad , \quad \sigma_{tot} = 2\pi a^2 \tag{19.240}$$

This is referred to as scattering from a **black disk**, and the elastic scattering is called **shadow scattering**.

We see that whenever inelastic scattering takes place it is always accompanied by elastic scattering. However, the converse is not true, since if $|S_l| = 1$, $\sigma_{inel} = 0$ but $\sigma_{el} \propto \sin^2 \delta_l$.

19.16   Worked examples

Example - 19.1 : Consider the Rutherford scattering of an incident particle of charge q_1 and kinetic energy E scatters off a heavy stationary particle of charge q_2. a) Derive the formula relating the impact parameter to the scattering angle. b) Determine the differential scattering cross–section. c) Show that the total cross–section for Rutherford scattering is infinite.

Solution :

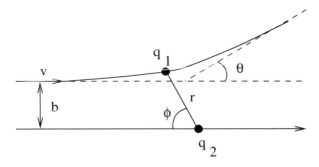

a) Energy conservation:

$$E = \frac{1}{2}m(\dot{r}^2 + r^2\dot{\phi}^2) + V(r) \quad , \quad V(r) = \frac{q_1 q_2}{4\pi\epsilon_0 r}.$$

Angular momentum conservation:

$$J = mr^2\dot\phi \;, \quad \dot\phi = \frac{J}{mr^2} \;\; ; \;\; \dot r^2 + \frac{J^2}{m^2r^2} = \frac{2}{m}(E - V)$$

let $u = \frac{1}{r}$

$$\dot r = \frac{dr}{dt} = \frac{dr}{du}\frac{du}{d\phi}\frac{d\phi}{dt} = (-\frac{1}{u^2})\frac{du}{d\phi}\frac{J}{m}u^2 = -\frac{J}{m}\frac{du}{d\phi}$$

$$(-\frac{J}{m}\frac{du}{d\phi})^2 + \frac{J^2}{m^2}u^2 = \frac{2}{m}(E - V)$$

or

$$(\frac{du}{d\phi})^2 = \frac{2m}{J^2}(E - V) - u^2 \;\; ; \;\; \frac{du}{d\phi} = \sqrt{\frac{2m}{J^2}(E - V) - u^2}$$

$$d\phi = \frac{du}{\sqrt{\frac{2m}{J^2}(E - V) - u^2}} = \frac{du}{\sqrt{I(u)}} \;\;, \quad I(u) = \frac{2m}{J^2}(E - V) - u^2$$

The particle q_1 starts out at $r = \infty$ $(u = 0)$, $\phi = 0$. The point of closest approach is r_{min} (u_{max});

$$\phi = \int_0^{u_{max}} \frac{du}{\sqrt{I(u)}}$$

q_1 swings through an angle ϕ on the way out, so $\phi + \phi + \theta = \pi$ or $\theta = \pi - 2\phi$,

$$\theta = \pi - 2\int_0^{u_{max}} \frac{du}{\sqrt{I(u)}}$$

$$I(u) = \frac{2mE}{J^2} - \frac{2m}{J^2}\frac{q_1q_2}{4\pi\epsilon_0}u - u^2 = (u_2 - u)(u - u_1),$$

where u_1 and u_2 are the two roots. Since $\frac{du}{d\phi} = \sqrt{I(u)}$, $u_{max} =$ one of the roots, setting $u_2 > u_1$, $u_{max} = u_2$

$$\theta = \pi - 2\int_0^{u_2} \frac{du}{\sqrt{(u_2 - u)(u - u_1)}} = \pi + 2\arcsin\left(\frac{-2u + u_1 + u_2}{u_2 - u_1}\right)\Big|_0^{u_2}$$

$$= \pi + 2\left(\arcsin(-1) - \arcsin\left(\frac{u_1 + u_2}{u_2 - u_1}\right)\right) = \pi + 2\left(-\frac{\pi}{2} - \arcsin\left(\frac{u_1 + u_2}{u_2 - u_1}\right)\right)$$

$$\theta = -2\arcsin\left(\frac{u_1 + u_2}{u_2 - u_1}\right)$$

$J = mvb$, $E = \frac{1}{2}mv^2$, so $J^2 = m^2b^2\frac{2E}{m} = 2mb^2E$, hence $\frac{2m}{J^2} = \frac{1}{b^2E}$, so

$$I(u) = \frac{1}{b^2} - \frac{1}{b^2}\left(\frac{1}{E}\frac{q_1q_2}{4\pi\epsilon_0}\right)u - u^2$$

Let $A = \frac{q_1q_2}{4\pi\epsilon_0 E}$, so

$$-I(u) = u^2 + \frac{A}{b^2}u - \frac{1}{b^2}$$

To get the roots, $u^2 + \frac{A}{b^2}u - \frac{1}{b^2} = 0$

$$u = \frac{1}{2}\left(-\frac{A}{b^2} \pm \sqrt{\frac{A^2}{b^4} + \frac{4}{b^2}}\right) = \frac{A}{2b^2}\left[-1 \pm \sqrt{1 + (\frac{2b}{A})^2}\right]$$

Thus,

$$u_2 = \frac{A}{2b^2}\left[-1 + \sqrt{1 + (\frac{2b}{A})^2}\right] \quad , \quad u_1 = \frac{A}{2b^2}\left[-1 - \sqrt{1 + (\frac{2b}{A})^2}\right] ;$$

$$\frac{u_1 + u_2}{u_2 - u_1} = \frac{-1}{\sqrt{1 + (\frac{2b}{A})^2}} \quad , \quad \theta = 2\arcsin\left(\frac{1}{\sqrt{1 + (\frac{2b}{A})^2}}\right)$$

or

$$\frac{1}{\sqrt{1 + (\frac{2b}{A})^2}} = \sin(\frac{\theta}{2}) \quad , \quad 1 + (\frac{2b}{A})^2 = \frac{1}{\sin^2(\frac{\theta}{2})}$$

$$(\frac{2b}{A})^2 = \frac{1 - \sin^2(\theta/2)}{\sin^2(\theta/2)} = \frac{\cos^2(\theta/2)}{\sin^2(\theta/2)} \quad ; \quad \frac{2b}{A} = \cot(\theta/2)$$

$$b = \frac{q_1 q_2}{8\pi\epsilon_0 E}\cot(\frac{\theta}{2})$$

b)

$$\sigma(\theta, \phi) = \frac{b}{\sin\theta}\left|\frac{db}{d\theta}\right| = \frac{b}{\sin\theta}\frac{q_1 q_2}{8\pi\epsilon_0 E}\frac{1}{2\sin^2(\theta/2)}$$

$$= \frac{1}{2\sin(\theta/2)\cos(\theta/2)}\left(\frac{q_1 q_2}{8\pi\epsilon_0 E}\right)^2 \frac{\cot(\theta/2)}{2\sin^2(\theta/2)}$$

$$= \left(\frac{q_1 q_2}{16\pi\epsilon_0 E \sin^2(\theta/2)}\right)^2$$

c)

$$\sigma_T = \int \sigma(\theta, \phi)\sin\theta d\theta d\phi = 2\pi\left(\frac{q_1 q_2}{16\pi\epsilon_0 E}\right)^2 \int_0^\pi \frac{\sin\theta d\theta}{\sin^4(\theta/2)} \rightarrow \infty$$

Example - 19.2 : Consider a structureless particle of mass m interacting with a fixed scattering center through the potential $V(r)$. The particle is located initially at $x = -\infty$ with the velocity v_R. a) Use the conservation of energy and angular momentum to verify the following equations for the relative velocity

$$|v_R| = [(\dot{r}^2 + r^2\dot{\phi}^2) + \frac{2V(r)}{m}]^{1/2} \quad \text{and} \quad v_R = \frac{r^2}{b}\dot{\phi}$$

b) At the point of closest approach

$$r = r_m \quad \text{and} \quad \left(\frac{r^2}{b}\right)\left[1 - \left(\frac{b}{r}\right)^2 + \frac{2V(r)}{mv_R^2}\right]^{1/2} = 0$$

Derive this equation by evaluating $\frac{dr}{d\phi}$. This equation can in principle be solved to obtain r_m. c) Show that

$$\theta(b, v_R) = \pi - 2b \int_{r_m}^\infty \frac{r^{-2}dr}{\left[1 - \frac{2V(r)}{mr^2} - \left(\frac{b}{r}\right)^2\right]^{1/2}}$$

where θ is the angle of deflection.

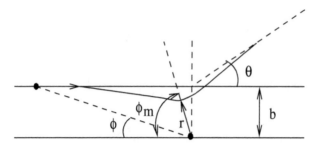

Solution : a) The classical Hamiltonian for the internal motion of the reduced two–particle system is

$$H = \frac{1}{2}m(\dot{r}^2 + r^2\dot{\phi}^2) + V(r)$$

At $r = \infty$, $V(r) = 0$ and $E_{tot} = E_{kin}$, thus

$$\frac{1}{2}mv_R^2 = \frac{1}{2}m(\dot{r}^2 + r^2\dot{\phi}^2) + V(r) \quad ; \quad |v_R| = \left[(\dot{r}^2 + r^2\dot{\phi}^2) + \frac{2}{m}V(r)\right]^{1/2}$$

The initial angular momentum: $|\mathbf{L}| = |\mathbf{r} \times \mathbf{p}| = bmv_R$; by definition

$$L = \frac{\partial \mathcal{L}}{\partial \dot{\phi}} = mr^2\dot{\phi} \quad \text{or} \quad bmv_R = mr^2\dot{\phi} \quad \Rightarrow \quad v_R = \frac{r^2}{b}\dot{\phi}.$$

b) Note that $\frac{dr}{d\phi} = \frac{dr}{dt}\frac{dt}{d\phi} = \frac{dr/dt}{d\phi/dt}$. From part (a) we have

$$\dot{r} = \frac{dr}{dt} = \pm\left[v_R^2 - \frac{b^2}{r^2}v_R^2 - \frac{2}{m}V(r)\right]^{1/2} \quad , \quad \dot{\phi} = \frac{d\phi}{dt} = \frac{b}{r^2}v_R$$

Thus

$$\frac{dr}{d\phi} = \pm\left(\frac{r^2}{b}\right)\left[1 - \left(\frac{b}{r}\right)^2 - \frac{2V(r)}{mv_R^2}\right]^{1/2} = \pm f(r).$$

At the point of closest approach $dr/d\phi = 0$.

c) Since $\theta = \pi - 2\phi_m$, we must evaluate ϕ_m. We have $dr/d\phi = -f(r)$ (from part (b)); $(-)$ sign is chosen here because particle is approaching. Then,

$$d\phi = -\frac{dr}{f(r)} \quad \text{and} \quad \phi_m = \int_0^{\phi_m} d\phi = -\int_{-\infty}^{r_m} \frac{dr}{f(r)}$$

$$\theta = \pi - 2b\int_{r_m}^{\infty}\left[1 - \left(\frac{b}{r}\right)^2 - \frac{2V(r)}{mv_R^2}\right]^{-1/2} r^{-2}dr$$

Example - 19.3 : Determine the scattering amplitude $f(\theta)$ in the first Born approximation for the potential $V(r) = -Ae^{-br^2}$, where $A, b > 0$.

Solution : For a central potential the scattering amplitude in the Born approximation is given by

$$f(\theta) = -\frac{2m}{\hbar^2} \int_0^\infty \frac{\sin(Kr)}{Kr} V(r) r^2 dr$$

$$= -\frac{2m}{\hbar^2} \int_0^\infty \frac{\sin(Kr)}{Kr} \left(-Ae^{-br^2}\right) r^2 dr$$

$$= \frac{2mA}{\hbar^2 K} \int_0^\infty \sin(Kr) e^{-br^2} r dr$$

integrating by parts: $u = \sin(Kr)$, $dv = e^{-br^2} r dr$ gives

$$f(\theta) = \frac{2mA}{\hbar^2 K} \left[-\sin(Kr) \frac{e^{-br^2}}{2b} \Big|_0^\infty + \int_0^\infty \frac{e^{-br^2}}{2b} K \cos(Kr) dr \right]$$

$$= \frac{2mA}{\hbar^2 K} \left[0 + \frac{K}{2b} \frac{\sqrt{\pi}}{2\sqrt{b}} e^{-K^2/4b} \right] = \frac{A\sqrt{\pi m}}{2\hbar^2 b^{3/2}} e^{-K^2/4b}$$

Example - 19.4 : For the potential $V(r) = \alpha\delta(r-a)$, where α and a are constants, a) calculate $f(\theta)$, $D(\theta)$, and σ), in the low energy Born approximation, b) calculate $f(\theta)$ for arbitrary energies, in the Born approximation.

Solution : a)

$$f(\theta, \phi) = -\frac{m}{2\pi\hbar^2} \int V(\mathbf{r}) d^3\mathbf{r}$$

$$= -\frac{m}{2\pi\hbar^2} \alpha 4\pi \int_0^\infty \delta(r-a) r^2 dr = -\frac{2m\alpha}{\hbar^2} a^2$$

$$D(\theta, \phi) \equiv \sigma(\theta, \phi) = |f(\theta, \phi)|^2 = \left(\frac{2m\alpha}{\hbar^2} a^2\right)^2$$

$$\sigma = \sigma_T = \int \sigma(\theta, \phi) d\Omega = 4\pi D = 4\pi \left(\frac{2m\alpha}{\hbar^2} a^2\right)^2$$

b)

$$f(\theta, \phi) = -\frac{2m}{\hbar^2} \int V(r) \frac{\sin(Kr)}{Kr} r^2 dr$$

$$= -\frac{2m\alpha}{\hbar^2 K} \int_0^\infty \delta(r-a) r \sin(Kr) dr = -\frac{2m\alpha}{\hbar^2 K} a \sin(Ka)$$

$$K = 2k \sin(\theta/2)$$

Example - 19.5 : Use

$$\psi(x) = \psi^{(0)}(x) - \frac{im}{\hbar^2 k} \int_{-\infty}^\infty e^{ik|x-x'|} V(x')\psi(x')dx'$$

to develop the Born approximation for one–dimensional scattering. Choose $\psi^{(0)}(x) = e^{ikx}$ and assume $\psi(x') \cong \psi^{(0)}(x')$. Show that the reflection coefficient takes the form

$$R \approx \left(\frac{m}{\hbar^2 k}\right)^2 \left|\int_{-\infty}^{\infty} e^{2ikx} V(x) dx\right|^2$$

Solution :

$$\psi^{(0)}(x) = e^{ikx} \quad , \quad \psi(x) = e^{ikx} \quad \text{then}$$

$$\psi(x) \cong e^{ikx} - \frac{im}{\hbar^2 k} \int_{-\infty}^{\infty} e^{ik|x-x'|} V(x') e^{ikx'} dx'$$

$$= e^{ikx} - \frac{im}{\hbar^2 k} \int_{-\infty}^{x} e^{ik(x-x')} V(x') e^{ikx'} dx'$$

$$- \frac{im}{\hbar^2 k} \int_{x}^{\infty} e^{ik(x'-x)} V(x') e^{ikx'} dx'$$

$$= e^{ikx} - \frac{im}{\hbar^2 k} e^{ikx} \int_{-\infty}^{x} V(x') dx' - \frac{im}{\hbar^2 k} e^{-ikx} \int_{x}^{\infty} e^{2ikx'} V(x') dx'$$

Assume $V(x)$ is localized, for large positive x, the third term is zero, and

$$\psi(x) = e^{ikx} \left(1 - \frac{im}{\hbar^2 k} \int_{-\infty}^{\infty} V(x') dx'\right)$$

This is the transmitted wave. For large negative x the middle term is zero, and

$$\psi(x) = e^{ikx} - \frac{im}{\hbar^2 k} e^{-ikx} \int_{-\infty}^{\infty} e^{2ikx'} V(x') dx'$$

The first term is the incident wave and the second term is the reflected wave. There-fore, the reflection coefficient is

$$R = \left(\frac{m}{\hbar^2 k}\right)^2 \left|\int_{-\infty}^{\infty} e^{2ikx} V(x) dx\right|^2$$

Example - 19.6 : Show that electron scattering with small momentum transfer permits the determination of the total charge and mean quadratic radius of atomic nuclei only.

Solution : Let us consider the elastic scattering of an electron by an atomic nu-cleus. Let $|\psi_i>$ represent the whole system before scattering, and $|\psi_f>$ represents after the scattering.

$$V(\mathbf{r}, \mathbf{R}) = \frac{Ze^2}{|\mathbf{r} - \mathbf{R}|} e^{-\frac{1}{d}|\mathbf{r} - \mathbf{R}|}$$

is the screened Coulomb potential between electron and nucleus. The probability for the transition $|\psi_i> \longrightarrow |\psi_f>$ gives some information about the structure (charge distribution) of the nucleus.

$$\text{Transition probability} \quad \propto \quad |<\psi_f|V|\psi_i>|^2$$

$$|\psi_i> = |\mathbf{k_0}> |i> = e^{\frac{i}{\hbar}\mathbf{p_0}\cdot\mathbf{r}}\phi_i(\mathbf{R}) \quad , \quad |\psi_f> = |\mathbf{k}>|f> = e^{\frac{i}{\hbar}\mathbf{p}\cdot\mathbf{r}}\phi_f(\mathbf{R})$$

$$<\psi_f|V|\psi_i> = Ze^2 \int \phi_f^*(\mathbf{R})\phi_i(\mathbf{R})\frac{e^{-\frac{1}{d}|\mathbf{r}-\mathbf{R}|}}{|\mathbf{r}-\mathbf{R}|}e^{i(\mathbf{k_0}-\mathbf{k})\cdot\mathbf{r}}dv_e dv_n$$

Let $\mathbf{s} = \mathbf{k_0} - \mathbf{k}$, therefore $\hbar\mathbf{s} = \mathbf{p_0} - \mathbf{p}$ is the momentum transfer from the electron to the nucleus during the scattering process. $\hbar^2 s^2 = p_0^2 + p^2 - 2p_0 p\cos\theta$. If $p \sim p_0$, then $\hbar^2 s^2 = 2p_0 p(1 - \cos\theta) = 4p_0 p\sin^2(\theta/2)$.
The integral over dv_e:

$$J_e = \int \frac{e^{-\frac{1}{d}|\mathbf{r}-\mathbf{R}|}}{|\mathbf{r}-\mathbf{R}|}e^{i(\mathbf{k_0}-\mathbf{k})\cdot\mathbf{r}}dv_e = e^{i\mathbf{s}\cdot\mathbf{R}}\frac{4\pi}{s^2 + \frac{1}{d^2}}$$

If $s^2 d^2 \gg 1$, then $1/d^2$ term can be neglected. This means that the momentum transfer must not become too small. Hence

$$J_e(\mathbf{R}) \sim e^{i\mathbf{s}\cdot\mathbf{R}}\frac{4\pi}{s^2}$$

If $\phi_f \sim \phi_i \sim \phi$; then $Z\phi^*\phi = \rho(\mathbf{R})$ with $\int \rho(\mathbf{R})dv_n = Z$.

$$<\psi_f|V|\psi_i> = Ze^2\frac{4\pi}{s^2} \int \phi_f^*(\mathbf{R})e^{i\mathbf{s}\cdot\mathbf{R}}\phi_i(\mathbf{R})dv_n$$

$$= \frac{4\pi e^2}{s^2} \int \rho(\mathbf{R})e^{i\mathbf{s}\cdot\mathbf{R}}dv_n = \frac{4\pi e^2}{s^2}F(\mathbf{s})$$

The quantity $F(\mathbf{s})$ is called a form factor. It is a Fourier transformed charge distribution and reflects the deviation of the nuclear charge distribution from point structure. If $\rho(\mathbf{R}) = \delta^3(\mathbf{R})$, then $F = 1$.

$$F(\mathbf{s}) = \int \rho(\mathbf{R})e^{i\mathbf{s}\cdot\mathbf{R}}dv_n$$

$dv_n = R^2\sin\theta dr d\theta d\phi$, $\mathbf{s}\cdot\mathbf{R} = sR\cos\theta$.

$$F(s) = 2\pi \int_0^\infty \int_0^\pi \rho(R)e^{isR\cos\theta}R^2\sin\theta dRd\theta$$

$$= 2\pi \int_0^\infty \rho(R)\frac{1}{isR}\left(e^{isR} - e^{-isR}\right)R^2 dR = \frac{4\pi}{s}\int_0^\infty \rho(R)\sin(sR)RdR$$

Assuming small momentum transfer s, $sR \ll 1$; $\sin(sR) \sim sR - (sR)^3/6$. Thus

$$F(s) = 4\pi \int_0^\infty \rho(R)R^2 dR - \frac{2\pi}{3}s^2 \int_0^\infty \rho(R)R^4 dR = Z - \frac{2\pi}{3}s^2 <R^2>$$

The first term is just the total charge Z of the nucleus, while the second one contains the mean quadratic radius.

To measure more details of the charge distribution $\rho(R)$, the momentum transfer has to be increased. This can be done by increasing the energy of the electron (or momentum).

$$\hbar^2 s^2 \sim 4p_0 p\sin^2(\theta/2) \quad (p_0 \sim p)$$

The best scattering angle possible $\theta = 180^0$ (back scattering at high energy)

$$\sin(sR) \sim sR - \frac{(sR)^3}{3!} + \frac{(sR)^5}{5!}$$

$$F(s) = 4\pi < R^2 > -\frac{2\pi}{3}s^2 < R^2 > +\frac{4\pi}{5!}s^4 < R^4 >$$

The different factors in front of the powers s^{2n} of the form factor reflect the higher moments of the charge distribution.

19.17 Problems

Problem - 19.1 : Show that for the atomic scattering of electrons by a potential given by

$$V(r) = -\frac{Ze^2}{r} + Ze^2 \int \frac{\rho(\mathbf{r}')}{|\mathbf{r} - \mathbf{r}'|} d\mathbf{r}',$$

where $\mathbf{r}$ is the position of the scattered electron and $\rho(\mathbf{r}')$ is the particle density of the atomic electrons at the position $\mathbf{r}'$, the scattering amplitude is in the form

$$f(\theta) = \frac{Ze^2}{4E} \sin^2(\theta/2)(1 - F(\mathbf{k})),$$

where E is the electron energy, $F(\mathbf{k})$ is the Fourier transform of $\rho(\mathbf{r})$, θ is the scattering angle.

Problem - 19.2 : Starting from the Fermi golden rule

$$P_{i \to j} = \frac{2\pi}{\hbar} | < i|V|f > |^2 \rho(E_f)$$

establish an expression for the differential scattering cross–section in the Born approximation.

Problem - 19.3 : Find the Green's function for the one–dimensional Schrödinger equation, and use it to construct the integral form.

Problem - 19.4 : Use the one–dimensional Born approximation to compute the transmission coefficient $(T = 1 - R)$ for scattering from a delta function $V(x) = -\alpha\delta(x)$ and from a finite square well $V(x) = -V_0$ for $-a < x < a$ and $V(x) = 0$ for $|x| > a$. Compare the results with the exact values.

Problem - 19.5 : Derive the quantum mechanical expression for the s–wave cross section for scattering from a hard sphere of radius R.

Bibliography and References

There are many textbooks about quantum mechanics. We have listed some selected books in this field, which are mainly about the nonrelativistic quantum mechanics. The list given here (in chronological order) is not meant to be exhaustive, and no book is criticized by its omission.

1. L. Pauling and E.B. Wilson, Jr., *Introduction to Quantum Mechanics*, McGraw–Hill, 1935.

2. D. Bohm, *Quantum Theory*, Prentice–Hall, 1951.

3. J.L. Powell and B. Crasemann, *Quantum Mechanics*, Addison–Wesley, 1965.

4. R.P. Feynman, R.B. Leighton, and M. Sands, *Quantum Mechanics, The Feynman Lectures on Physics, Vol. 3*, Addison–Wesley, 1965.

5. K. Gottfried, *Quantum Mechanics*, Benjamin, 1966.

6. S. Borowitz, *Fundamentals of Quantum Mechanics*, Benjamin, 1967.

7. D.S. Saxon, *Elementary Quantum Mechanics*, Holden–Day, 1968.

8. L.I. Schiff, *Quantum Mechanics*, McGraw–Hill, 1968.

9. R.H. Dicke and J.P. Wittke, *Introduction to Quantum Mechanics*, Addison–Wesley, 1969.

10. G. Baym, *Lectures on Quantum Mechanics*, Benjamin, 1969.

11. E. Merzbacher, *Quantum Mechanics*, Wiley, 1970.

12. A. Messiah, *Quantum Mechanics*, North–Holland, Vols. 1–2, 1972, 1970.

13. E.H. Wichmann, *Quantum Physics, Berkeley Physics Course, Vol. 4*, McGraw–Hill, 1971.

14. R. McWeeny, *Quantum Mechanics: Principles and Formalism*, Pergamon, 1972.

15. S. Gasiorowicz, *Quantum Physics*, Wiley, 1974.

16. P.A.M. Dirac, *The Principles of Quantum Mechanics*, Oxford University Press, 1974.

17. L.D. Landau and E.M. Lifshitz, *Quantum Mechanics*, Pergamon, 1977.

18. C. Cohen–Tannoudji, B. Diu and F. Laloe, *Quantum Mechanics*, Wiley, 1977.

19. R. Shankar, *Principles of Quantum Mechanics*, Plenum, 1980.

20. J.J. Sakurai, *Modern Quantum Mechanics*, Benjamin, 1985.

21. A.Z. Capri, *Nonrelativistic Quantum Mechanics*, Benjamin, 1985.

22. A. Das and A.C. Melissinos, *Quantum Mechanics: A Modern Introduction*, Gordon and Breach, 1986.

23. W. Greiner, *Quantum Mechanics: An Introduction*, Springer–Verlag, 1989.

24. A.P. French and E.F. Taylor, *An Introduction to Quantum Physics*, Chapman and Hall, 1990.

25. M.A. Morrison, *Understanding Quantum Physics*, Prentice Hall, 1990.

26. L.E. Ballentine, *Quantum Mechanics*, Prentice Hall, 1990.

27. D. Park, *Introduction to the Quantum Theory*, Third Ed., McGraw–Hill, 1992.

28. F. Schwabl, *Quantum Mechanics*, Springer 1995.

29. D.J. Griffiths, *Introduction to Quantum Mechanics*, Prentice Hall, 1995.

30. A. Goswami, *Quantum Mechanics*, Second Ed., Wm. C. Brown, 1997.

31. R.W. Robinett, *Quantum Mechanics*, Oxford University Press, 1997.

32. R.L. Liboff, *Introductory Quantum Mechanics*, Third Ed., Addison–Wesley, 1998.

There are also many books about the related subjects of quantum mechanics. Here we have listed some selected books in this field. The list given here (in chronological order) is not meant to be exhaustive, and again no book is criticized by its omission.

1. A.R. Edmonds, *Angular Momentum in Quantum Mechanics*, Princeton University Press, 1957.

2. M.E. Rose, *Elementary Theory of Angular Momentum*, Wiley, 1957.

3. J.C. Slater, *Quantum Theory of Atomic Structure*, McGraw-Hill, 1960.

4. R.M. Eisberg, *Fundamentals of Modern Physics*, Wiley, 1961.

5. J.C. Slater, *Quantum Theory of Molecular Structure*, McGraw-Hill, 1963.

6. F.L.Pilar, *Elementary Quantum Chemistry*, McGraw-Hill, 1968.

7. H.A. Bethe and R.W. Jackiw, *Intermediate Quantum Mechanics*, Benjamin, 1968.

8. R. McWeeny and B.T. Sutcliffe, *Methods of Molecular Quantum Mechanics*, Academic, 1969.

9. M. Mizushima, *Quantum Mechanics of Atomic Spectra and Atomic Structure*, Benjamin, 1970.

10. R. McWeeny, *Quantum Mechanics: Methods and Basic Applications*, Pergamon, 1973.

11. M.A. Morrison, T.L. Estle, and N.F. Lane, *Quantum States of Atoms, Molecules, and Solids*, Prentice Hall, 1976.

12. H.A. Bethe and E.E. Salpeter, *Quantum Mechanics of One- and Two–Electron Atoms*, Plenum, 1977.

13. L.S. Schulman, *Techniques and Applications of Path Integration*, Wiley, 1981.

14. R. Loudon, *The Quantum Theory of Light*, Oxford University Press, 1983.

15. W. Heitler, *The Quantum Theory of Radiation*, Dover, 1984.

16. M. Weissbluth, *Photon–Atom Interactions*, Academic, 1989.

17. Ş. Erkoç and T. Uzer, *Atomic and Molecular Physics*, World Scientific, 1996.

18. P.W. Atkins and R.S. Friedman, *Molecular Quantum Mechanics*, Third Ed., Oxford University Press, 1997.

There are several books containing solved problems in quantum mechanics. Here we have listed some selected books in this field. The list given here (in chronological order) is not meant to be exhaustive, and again no book is criticized by its omission.

1. S. Flugge, *Practical Quantum Mechanics*, Springer–Verlag, 1971.

2. D. ter Haar, Ed. *Problems in Quantum Mechanics*, Pion, 1975.

3. F. Constantinescu and E. Magyari, *Problems in Quantum Mechanics*, Pergamon, 1976.

4. C.S. Johnson, Jr. and L.G. Pedersen, *Problems and Solutions in Quantum Chemistry and Physics*, Addison–Wesley, 1976.

5. I.I. Gol'dman and V.D. Krivchenkov, *Problems in Quantum Mechanics*, Dover, 1993.

6. G.L. Squires, *Problems in Quantum Mechanics*, Cambridge University Press, 1995.

7. Y.K. Lim, Ed. *Problems and Solutins on Quantum Mechanics*, World Scientific, 1999.

Appendix

Fundamental physical constants:

From E.R. Cohen and B.N. Taylor, Rev. Mod. Phys. **59**, 1121(1987).

Quantity	Symbol	Value	Units
speed of light in vacuum	c	299792458.0	m/s
permeability of vacuum	μ_0	$4\pi \times 10^{-7}$	N/A^2
Planck constant	h	6.6260755	$10^{-34}\ Js$
elementary charge	e	1.60217733	$10^{-19}\ C$
electron mass	m_e	9.1093897	$10^{-31}\ kg$
electron magnetic moment	μ_e	928.47701	$10^{-26}\ JT^{-1}$
proton mass	m_p	1.6726231	$10^{-27}\ kg$
proton magnetic moment	μ_p	1.41060761	$10^{-26}\ JT^{-1}$
neutron mass	m_n	1.6749286	$10^{-27}\ kg$
neutron magnetic moment	μ_n	0.96623707	$10^{-26}\ JT^{-1}$
muon mass	m_μ	1.8835327	$10^{-28}\ kg$
muon magnetic moment	μ_μ	4.4904514	$10^{-26}\ JT^{-1}$
Avogadro constant	N_A	6.0221367	$10^{23}\ mol^{-1}$
molar gas constant	R	8.314510	$J\ mol^{-1}\ K^{-1}$

Some derived physical constants:

Planck constant : $\hbar = h/2\pi = 1.05457266 \times 10^{-34}\ Js$

permittivity of vacuum : $\epsilon_0 = 1/\mu_0 c^2 = 8.854187817 \times 10^{-12}\ F/m$

magnetic flux quantum : $\Phi_0 = h/2e = 2.06783461 \times 10^{-15}\ Wb$

fine structure constant : $\alpha = \mu_0 c e^2/2h = 7.29735308 \times 10^{-3}$

inverse structure constant : $\alpha^{-1} = 137.0359895$

Rydberg constant : $R_\infty = m_e c \alpha^2/2h = 10973731.534\ m^{-1}$

Boltzman constant : $k = R/N_A = 1.380658 \times 10^{-23}\ JK^{-1}$

Stefan–Boltzman constant : $\sigma = (\pi^2/60)k^4/\hbar^3 c^2$

$$= 5.67051 \times 10^{-8}\ Wm^{-2}K^{-4}$$

Bohr magneton : $\mu_B = e\hbar/2m_e = 9.2740154 \times 10^{-24}\ JT^{-1}$

nuclear magneton : $\mu_N = e\hbar/2m_p = 5.0507866 \times 10^{-27}\ JT^{-1}$

Bohr radius : $a_0 = \alpha/4\pi R_\infty = 0.529177249 \times 10^{-10}\ m$

electron magnetic moment anomaly : $a_e = \mu_e/\mu_B - 1$

$$= 1.159652193 \times 10^{-3}$$

electron g factor : $g_e = 2(1 + a_e) = 2.002319304386$

Energy conversion factors:

From E.R. Cohen and B.N. Taylor, Rev. Mod. Phys. **59**, 1121(1987).

	J	m^{-1}
J	1	5.0341125×10^{24}
m^{-1}	$1.9864475 \times 10^{-25}$	1
Hz	$6.6260755 \times 10^{-34}$	$3.335640952 \times 10^{-9}$
K	1.380658×10^{-23}	69.50387
eV	$1.60217733 \times 10^{-19}$	806554.10
$Hartree$	$4.3597482 \times 10^{-18}$	21947463.067
	Hz	K
J	$1.50918897 \times 10^{33}$	7.242924×10^{22}
m^{-1}	299792458.0	0.01438769
Hz	1	4.799216×10^{-11}
K	2.083674×10^{10}	1
eV	$2.41798836 \times 10^{14}$	11604.45
$Hartree$	6.5796839×10^{15}	3.157733×10^{5}
	eV	$Hartree$
J	6.2415064×10^{18}	2.2937104×10^{17}
m^{-1}	$1.23984244 \times 10^{-6}$	$4.5563352672 \times 10^{-8}$
Hz	$4.1356692 \times 10^{-15}$	$1.5198298508 \times 10^{-16}$
K	8.617385×10^{-5}	3.166829×10^{-6}
eV	1	0.036749309
$Hartree$	27.2113961	1

Atomic Units:

Quantum mechanical equations are simplified if atomic units (*a.u.*) are used. In *a.u.*

Electron mass, $m_e = 1$ *a.u.* $= 9.109530 \times 10^{-31}$ *kg*

Electron charge, $e = 1$ *a.u.* $= 1.602190 \times 10^{-19}$ *C*

Planck's constant, $\hbar = \frac{h}{2\pi} = 1$ *a.u.* $= 1.054590 \times 10^{-34}$ *Js*

Bohr radius, $a_0 = \frac{\hbar^2}{m_e e^2} = 1$ *a.u.* $= 0.529177 \times 10^{-10}$ *m*

Fine structure constant, $\alpha = \frac{1}{137.036}$ (dimensionless)

Speed of light, $c = \frac{1}{\alpha} = 137.036$ *a.u.* $= 2.99792 \times 10^8$ *m/s*

$\varepsilon_0 \mu_0 = \frac{1}{c^2}$, $\quad \frac{1}{4\pi\varepsilon_0} = 1$ *a.u.* $= 8.98755 \times 10^9$ *m/F*

Atomic velocity unit, $v_0 = \alpha c = 1$ *a.u.* $= 2.18769 \times 10^6$ *m/s*

Atomic time unit, $t_0 = \frac{a_0}{v_0} = 1$ *a.u.* $= 2.41889 \times 10^{-17}$ *s*

Atomic frequency unit, $\nu_0 = \frac{1}{2\pi t_0} = \frac{1}{2\pi}$ *a.u.* $= 6.57968 \times 10^{15}$ *s*$^{-1}$

Atomic energy unit, $E_0 = \frac{e^2}{4\pi\varepsilon_0 a_0} = m_e v_0^2 = 1$ *a.u.* $= 27.2116$ *eV*

Radius of Bohr orbit of a one–electron atom, $r_n = \frac{n^2 a_0}{Z} = \frac{n^2}{Z}$ *a.u.*

Bound state energy of a one–electron atom, $E_n = -\frac{m_e Z^2 e^4}{(4\pi\varepsilon_0)^2 2n^2 \hbar^2} = -\frac{Z^2}{2n^2}$ *a.u.*

Atomic unit of electric field, $\mathcal{E}_0 = \frac{e}{a_0^2} = 5.142 \times 10^{11}$ *V/m*

Atomic unit of magnetic field, $B_0 = 2.35 \times 10^5$ *Tesla*

Index

Milton Keynes UK
Ingram Content Group UK Ltd.
UKHW021847071024
449327UK00021B/1548